APPLIED PSYCHOMETRY

Applied Psychometry

Arun Kumar Singh, PhD.
*Formerly, Professor & Head of the
Department of Psychology
Patna University, Patna
&
Formerly, University Professor &
Director Institute of Psychological
Research & Service
Patna University, Patna*

**MOTILAL BANARSIDASS
INTERNATIONAL
DELHI**

First Edition: Delhi, 2022

© MOTILAL BANARSIDASS INTERNATIONAL
All Rights Reserved

No part of this book may be reproduced in any form or by any electronic or mechanical means including information storage and retrieval systems without permission in writing from the publishers, excepts by a reviewer who may quote brief passages in a review.

ISBN : 978-93-92510-72-4 (Paper)
ISBN : 978-93-92510-73-1 (Cloth)

Also available at
MOTILAL BANARSIDASS INTERNATIONAL
41 U.A. Bungalow Road, (Back Lane)Jawahar Nagar, Delhi - 110 007
4261 (basement) Lane #3, Ansari Road, Darya Ganj, New Delhi - 110 002
203 Royapettah High Road, Mylapore, Chennai - 600 004
12/1A, 2nd Floor, Bankim Chatterjee Street, Kolkata - 700 073
Stockist : Motilal Books, Ashok Rajpath, Near Kali Mandir, Patna - 800 004

Printed & Bound by
MOTILAL BANARSIDASS INTERNATIONAL

To
My wife
Kumud Rani

Preface

Applied Psychometry is basically a textbook for measurement course in psychology and education. The book concentrates around psychological testing and assessment in a very human, creative and collaboraative fashion. The book has been written and intrdouced into three parts. Part I is entitled as **Fundamental concepts of classroom assessment and evaluation** and it contains nine chapters. Through these nine chapters, the reader will get a clear, comprehensive and scientific knowledge about evaluation, assessment and tools of assessment especially about psychological tests. Part II is entitled as **Major domains of Psychometric assessment** which contains chapters 10 through 19. Therefore, this part contains ten chapters and a lucid attempt has been made to cover fields like achievement testing, attitude testing, intelligence testing, personality testing as well as the measurement of interest, values and attitudes and measurement of creativity. Besides this, neuropsychological assessment has also been discussed in comprehensive way. The most significant aspect of this part is that it also explains the application of psychological testing in different fields. Such scientific explanation will provide in-depth knowledge about the practical aspects of psychological testing in different fields. Part III is entitled as **Basic Statstical Concepts** and it contains chapters 20 through 27. The author has made a genuine attempt to explain the various statstical tools often used in assessment and measurement. It will help those students who are planning to start research work in the area of measurement and psychological testing.

Thanks to the members of academic community whose suggestions encouraged me to undertake this enterprise and conplete it with the belief that is will be not only be liked by them but they will also be able to expand their scientific knowledge in this field.

My special thanks are to the members of Editorial Board of Motilal Banaridas International, Delhi who carefully examined and presented the materials in error-free manner. However, the author takes the responsibility for any possible errors that may have somewhat found their way in the book. My humble request to the readers would be that they should point out such errors to me along with their healthy suggestions.

BagBhup Singh Lane
Patna 800007

ARUN KUMAR SINGH

Contents

Part–I Fundamental Concepts of Classroom Assessment and Evaluation

Chapters	*Page No.*
1. **Fundamentals of Evaluation and Assessment**	19-35

    Learning objectives
    Key Terms
Meaning and Nature of Evaluation
Types and Purpose of Evaluation
Phases of Evaluation
Functions of Evaluation
Principles of Assessment
Types of Assessment Procedure
Summary and Review
Review Questions

2. **Basic Principles of Measurement**   36-50

    Learning objectives
    Key Terms
Meaning of Measurement
Dimensions of Measurement
Levels of Scales of Measurement
Measurement and Decision
Steps in Measurement Process
Important Current Issues in Measurement
Summary and Review
Review Questions

3. **Principles of Psychological and Educational Testing**   51-53

    *Learning objectives*34
    *Key Terms*34
Psychological Tests: Meaning, Similarities and Differences
Historical Background of Psychological and Educational Testing
Major development in Psychological and Educational Testing in India
Types of Tests

Major Characteristics of a Good psychological test
Differences among psychological tests, psychological measurement and surveys
Assumptions of Psychological Tests
Locating Information about Psychological Tests
Major Assumptions underlying Psychological Testing
Uses or applications of Psychological testing
General Steps of Test Construction
Future of Psychological Testing
Ethical and Social Considerations in Testing
Computerized Testing
Computerized Adaptive Testing
Criticism of Testing
Summary and Review
Review Questions

4. **Item writing and Item analysis** — 94-118
   Learning objectives
   Key Terms
   Meaning and Nature of Item
   Types of Test items
   General Guidelines for Writing Good Essay Items
   General Guidelines for Writing Objective Items
   Meaning and Purpose of Item Analysis
   Standard Proecedure for Conducting Item analysis
   Item Analysis for Criterion-reference Test
   Item Analysis for Essay Tests
   Some Important Considerating in Item Analysis
   Item Characteristics Curves
   Item Response theory
   Item analysis in Qualitative tests
   When does the revision of the existing test become due?
   Concept of Cross-validation and Co-validation
   Summary and Review
   Review of Question

5. **Reliability and Validity of the test** — 119-164
   Learning objectives
   Key Terms
   The Concept of Reliability
   Sources of inconsistencies or Error variance

Two Modes to express reliability
Reliability Estimates
Test-retest Reliability Method
Parallel forms or Alternative forms reliability method
Internal Consistency Method
Kuder-Richardon (KR) formulas
Coefficient Alpha
Inter rater Reliability
Reliability of a difference Scores
Index of reliability
Which type of reliability is more appropriate?
Conditions affecting reliability co-efficients
The Concept of Validity
Categories of Validity
Content Validity
Criterion-related Validity
Construct Validity
Sources of Evidences of Validity
Interpreting Reliability and Validity Coefficient
Relation between reliability and validity
Factors affecting validity
Summary and Review
Review Questions

6. **Norms and Transformation of Scores**   165-178
   Learning Objectives
   Key Terms
   Meaning and Nature of raw score
   Nature of Norms
   Types of Norms
   Transformation of Raw Scores
   National Norms Vs. Local Norms
   Major Cautions in Interpreting test scores
   Summary and Review
   Review Questions

7. **Taxonomy of Educational Objectives**   179-191
   Learning Objectives
   Key Terms
   Dimensions of Educational Objectives

Texonomy of Educational Objectives
Cognitive domain : Bloom's Taxonomy of Cognitive Educational Objective
Affective domain : Krathwohl's Taxonomy of AffectiveEducational Objectives
Psychomotor domain : Simpson's Taxonomy of Psychomotor Educational Objectives
Summary and Review
Review Questions

8. **Classroom Instructional goals and objectives: Educational Decision making** 192-201
   Learning Objectives
   Key Terms
   Nature and Purpose of classroom instructional goods and objectives
   Instructional objectives as learning outcomes
   Criteria for selecting appropriate instructional objectives
   Some important ways of stating instructional objectives
   Classrrom instructional decisions
   Types of Assessment instruments
   Summary and Review
   Review Questions

9. **Understanding domains of Curriculum** 202-227
   Learning Objectives
   Key Terms
   Meaning of Curriculum
   Types of Curriculum
   Characteristics of a good curriculum
   Objectives of Curriculum
   Major Issues involved in Curriculum development
   Basic Principles of Curriculum Construction
   Curriclum and Syllabus
   Nature of Curriculum
   Foundations of Curriculum
   Functions of Curriculum
   Models of Curriculum development
   Tyler : Behavioural model
   Saylor, Alexander and Lewis : Administrative model
   Taba's instructional strategies model
   Weinstein and Fautini : Humanistic model
   Eisner: Systematic-aesthetic model

Summary and Review
Review Questions

Part II : Major Domains of Assessment

10. Planning Classroom test and Assessment   229-237
   Learning Objectives 212
   Key Terms 212
   Steps in Classroom Testing and Assessments
   Purpose of Classroom Testing and Assessment
   Preparing Specifications for Tests and Assessments
   Selecting appropriate items and assessment tasks
   Some important considrations in preparing relevant test items and assessment tasks
   How to improve learning and instruction?
   Tests used for making decisions in the classroom
   Summary and Review
   Review Questions

11. Achievement tests   238-253
   Learning Objectives
   Key Terms
   Meaning of achievement test
   Major features of Standardized achievement tests
   Standardized achievement tests Vs. Teacher-made classroom tests
   Major functions or uses of achievement testing
   Types of Standardized test
   Norm-referenced achievement test Vs. Criterion Reference Achievement test
   Steps involved in construction and standardization of achievement test
   Dark side of achievement testing
   Authentic assessment of achievement
   Summary and Review
   Review Questions

12. Aptitude tests   254-268
   Learning Objective
   Key Terms
   Meaning and Nature of aptitude test
   Aptitude test Vs. Achievement test
   Types of aptitude test
   Uses of aptitude test
   Summary and Review
   Review Questions

13. Theories and Measurement of Intelligence — 269-310
Learniong Objectives
Key Terms
Definition of Intelligence
Theories of Intelligence
Sensory Keeness theory of Galton and Cattell
Binet's theory
Spearman's Two factor theory
Thurstone's theory of mental abilities
Guilford's Structure-of-inttelect model
Jensen's theory
Wechsler's theory
Cattell-Horn G_r-G_c theory
Carroll Three-Stratum theory
Cattell Three-stratum theory
Cattell-Horn-Carroll (CHC) model of intelligence
Sternberg's Triarchic theory
Gardner's theory
Das-Naglieri Pass Model
Measurement of Intelligence
Culture Fair Intelligence Test (CFIT)
Individual test Vs. Group test of Intelligence
Advantages and Disadvantages of Group testing
Important Indian Intelligence tests in brief historical purspectives
Summary and Review
Review Questions

14. Personality Testing — 311-357
Learning Objectives
Key terms
Personality assessment
Traits, States and Types
Nature of Projectives test
Classification of Projective tests
Association techniques
Pictorial Techniques
Verbal Techniques
Expressive Techniques
Criticisms of Projective techniques

Structured Personality Tests
Logical Content Strategy
Criterion group Strategy
Factor analytic strategy
Theoretical strategy
Behavioural Assessment Methods
Summary and Review
Review Questions

15. Measurement of Interests, Values and Attitudes — 359-375
Learning Objectives
Key Terms
Measurement of interest
Value Scales
Measurement of attitudes
Summary and Review
Review Question

16. Measurement of Creativity — 376-383
Learning Objectives
Key Terms
Meaning of Creativity
Different approaches to Creativity
Measurement of Creativity
Ways of enhancing creativity
Summary and Review
Review Questions

17. Neuropsychological Assessment — 384-394
Learning Objectives
Key Terms
Nature and Goals of Neuropsychological Assessment
Tests of Neuropsychological Assessment
Characteristics of Some useful Neuropsychological Batteries
Halstead-Reitan Neuropsychological Test Batterry HRNB
Luria-Nebraska Neuropsychological Test Batterry LNMB
Some Indian Neuropsychological Tests
Standard Psychological Testing Vs. Neuropsychological Testing
Summary and Review
Review Questions

18. Grading and Reporting 395-406
Learning Objectives
Key Terms
Meaning of Grading and Reporting
Functions of Grading and Reporting System
Types of Grading and Reporting System
Multiple Grading and Reporting System
How to assign letter Grades?
Summary and Review
Review Questions

19. Application of Psychological Testing in Different Fields 407-437
Learning Objectives
Key Terms
Application of Psychological Testing in Clinical setting
Application of Psychological Testing in Counselling and guidance
Application of Psychological Testing in Educational setting
Application of Psychological Testing in Industry, Organizational Bussiness
Application of Psychological Testing in Sports
Application of Psychological Testing in Military Services
Summary and Review
Review Questions

Part III : Basic Statistical Concepts

20. Basic Ideas in Statistics 438-458
Learning Objectives
Key Terms
Meaning of Statistics
Brief History of Statistics
Popularion and Sample
Parameter and Statistics
Variable and Constant
Levels of Measurement Scale
Parametric Statistical tests Vs. Non-parametric Statistcial tests
Types of Statistical Methods
Degree of Freedom
Null Hypothesis, Alternative Hypotehsis and ResearchHypothesis
Level of Significance
Type I Error and Type II error
One-tailed test Vs. Two-tailed test

Some dirty words about statistics
Summary and Review
Review Questions

21. **Measures of Central Tendency** 459-471
    Learning Objectives
    Key Terms
    Meaning of Central Tendency
    Arithmetic Mean
    Median
    Mode
    Which of these measures of Central tendency is more appropriate?
    Summary and Review
    Review Questions

22. **Measures of Variability** 472-491
    Learning Objectives
    Key Terms
    Meaning of Variability
    Measures of Variability
    Range
    Semi-Interquartile range or Quartile deviation
    Average deviation or Mean deviation
    Standard deviation
    Variance
    Comparing Measures of Variability
    Moment about Mean
    Summary and Review
    Review Questions

23. **Normal Curve** 492-512
    Learning Objectives
    Key Terms
    Meaning of Normal Curve
    Historical Background of Normal Curve
    Equations of Normal Curve
    Properties of Normal Curve
    Normal Curve as a Statistical Model
    Applications of Normal Curve
    Measuring divergence from Normal distribution
    Why do frequency distribution deviate from Normal distribution?

Summary and Review
Review Questions

24. Percentiles and Percentile Rank — 513-520
Learning Objectives
Key Terms
Meaning of Percentile rank
Computation of Percentile
Computation of Percentile rank (PR)
Uses and limitations of Percentile rank
Concept of deciles and Quartiles
Summary and Review
Review Questions

25. Correlation and Regression — 521-545
Learning Objectives
Key Terms
Meaning and Types of Correlation
Pearson Product-Moment Correlation
Correlation and Causation
Rank difference Correlation
Partial Correlation
Multiple Correlation
Meaning of Regression and Regression Equation
Obtaining *a* and *b* Coefficient
Limitations of Regression Analysis
Comparison between Regression and Correlation
Summary and Review
Review Questions

26. t test, z test and F test — 546-559
Learning Objectives
Key Terms
z test, t test and F test: Significance of Mean differences
Significance of the difference between two sample means using z-test
z test vs. t test
Robustness of the t test
F test: Analysis of Variance
Summary and Review
Review Questions

27. Some Important Non-parametric Statistics 560-591

Learning Objectives
Key Terms
Meaning and Nature of Chi square tset
Major characteristics of chi square test
Assumptions of chi square test
Major uses of chi square U test
Chi square as a test of Equal Probability Hypothesis
Chi square as a test of independence
Chi square test for Goodness-of-fit
Major sources of Errors in Chi-square test
Advantages and disadvantages of chi square test
Nominal measures of Association based upon X^2
Features of Phi-coefficient
Limitations of Phi-coefficient
Coefficient of Contingency
Mann-Whitney test
Median test
Kendall's Coefficient of Concordance (W)
Summary and Review
Review Questions

References — 592-623
Major Events in Educational Assessment — 624-629
Glossary — 630-643
Objective Questions — 644-680

Brief Contents

Part–I Fundamental Concepts of Classroom Assessment and Evaluation

Chapters	*Page No.*
1. Fundamentals of Evaluation and Assessment	19
2. Basic Principles of Measurement	36
3. Principles of Psychological and Educational Testing	51
4. Item writing and Item analysis	94
5. Reliability and Validity of the test	119
6. Norms and Transformation of Scores	165
7. Taxonomy of Educational goals and objectives	179
8. Classroom Instructional goals and objectives : Educational Decision marking	192
9. Understanding domains of Curriculum	202

Part–II Major Domains of Assessment

10. Planning Classroom test and Assessment	228
11. Achievement tests	238
12. Aptitude tests	254
13. Theories and Measurement of Intelligence	269
14. Personality Testing	311
15. Measurement of Interests, Values and Attitudes	358
16. Measurement of Creativity	376
17. Neuropsychological Assessment	384
18. Coding and Reporting	395
19. Applications of Psychological Testing in different Fields	407

Part–III Basic Statistical Concepts

20. Basic Ideas in Statistics	438
21. Measures of Central tendency	459
22. Measures of Variability	472
23. Normal Curve	492
24. Percentiles and Percentile Rank	513
25. Correlation and Regression	521
26. t test, z test and F test	546
27. Some Important Non-parametric statistics	560
References	592
Major Events in Educational Assessment	624
Glossary	630
Objective Questions	644

Part–I Fundamental Concepts of Classroom Assessment and Evaluation

CHAPTER – 1

Fundamentals of Evaluation and Assessment

Learning objectives :
- Meaning and Nature of Evaluation
- Types and Purpose of Evaluation
- Phases of Evaluation
- Functions of Evaluation
- Difference between Assessment, Measurement, Evaluation and Testing
- Principles of Assessment

Key Terms :

Assessment, Evaluation, Measurement, Testing, Formative Evaluation, Summative Evaluation, Diagnostic Evaluation, Internal Evaluation, External Evaluation, Placement Evaluation, Norms-Referenced Assessment, Criterion-Referenced Assessment.

Evaluation Plays a very Important role in the process of teaching – learning. Teachers are required not only to measure performance of the students but they are also required to evaluate them. Only after evaluation, they arrive at some meaningful decision. Not only this, assessment is also very vital. It is an integrated process through which the nature and extent of students' learning and development is determined. Any comprehensive assessment requires a variety of procedures. In this chapter, a lucid attempt has been made to present all the fundamentals associated with evaluation and assessment.

Meaning and Nature of Evaluation

Evaluation is a wider term frequently used in teaching-learning process. Different experts have somewhat different notions about what evaluation is and they differ in the level of abstraction and show specific concerns of the person who formulated them.

In the most general way, evaluation means determining the value or worth of something. More specifically, evaluation is the process through which a value judgement or decision is made over a level of performance or achievement. In fact, many value judgements over a level of performance presuppose a set of objectives. Evaluation helps in determining the extent to which those objectives are achieved. From the educational point of view, evaluation may be understood as a systematic process of determining the extent to which educational objectives have been achieved. Therefore, it implies an assessment of educational process and its outcomes in the light of its objectives. Taking a broader view then, it can be said that evaluation includes both quantitative and qualitative description of the performance of students and the value judgement of those performances.

With some minor variations, most of the definitions of evaluation basically represent one of the two viewpoints as exemplified by the following two definitions :

1. *Evaluation is the systematic process of collecting and analyzing data for determining whether or to what degree objectives are being achieved.*
2. *Evaluation is the systematic process of collecting and analyzing data in order to take some decision.*

If we pay attention to these two classes of definitions, it becomes clear that the systematic process of data collection (that is, measurement) and the analysis of the collected data is common to both definitions. In this way, measurement is one important component of evaluation. However, the important difference between these two definitions is the issue of whether decisions or judgements are the basic component of evaluation. Those who support the definition of first category agree that the results of evaluation may be used for decision-making whereas the supporters of definition of second category consider decision-making to be one part of evaluation. The second definition is preferred to the first definition because of some obvious reasons. The second definition is more inclusive than the first one because it does not preclude the process described in the first definition. Another reason is that the definition implies that evaluation can be done for strictly descriptive purposes. However, this claim is said to be naïve at best.

Recently, Beeby (1979) has provided a still more comprehensive definition of evaluation. According to him, "Evaluation is the systematic collection and interpretation of evidence leading as a part of process to a judgement of value with a view to action." Analyzing this definition, we get a few major points as under :

(i) In evaluation the investigator collects information in a systematic way with proper planning. Such information can be gathered through the administration of various standardized tests or by means of questionnaires, interview, observational techniques and so on.

(ii) Such collected information or evidences are interpreted for locating the presence or absence of quality in them.

(iii) In evaluation there occurs judgement of value or worth. In educational evaluation the evaluator makes the judgement of value of an educational programme, curriculum or educational institution so that a decision may be taken regarding their goal. In fact, the purpose of evaluation is to determine the current status of object of evaluation and to compare the status with some standard or criteria and to select an alternative in order to take a decision. This is clearly reflected in 'with a view to action' terminology of the

Beeby' definitions. Evaluation, thus is decision-oriented. Educational evaluation, particularly is intended to lead to some basic policies and practices in the field of education by bringing important significant changes.

Types and Purposes of Evaluation

As we know, evaluation is basically a process of examining a programme to decide what is working satisfactorily, what is not working satisfactorily and why? It acts as a blueprint for judgment and improvement because it provides way for valuing the learning and training programmes.

There are different types of evaluation in psychology and education. In general, the evaluation can be classified on the basis of two major criteria :

1. How is the evaluation process applied?
2. What is being evaluated?

Based on the first criteria, there are five common types of evaluation : *formative evaluation, summative evaluation, diagnostic evaluation, placement evaluation, internal evaluation and external evaluation.* A brief discussion follows :

1. Formative evaluation

Formative evaluation is defined as an on-going process, which permits for feedback to be implemented during a cycle of programme. Such evaluation not only provides timely feedback about the various educational programmes but also concentrates on examining changing and improving various educational programme. It allows to make some adjustments for better achieving the goals. The basic purpose of formative evaluation is to provide students and teachers with feedback of students' progress towards the mastery of learning. Such evaluation strengthens or improves the object being evaluated. This type of evaluation focuses on the process and is basically done on fly. It permits the evaluator to monitor how well the instructional goals as well as the objectives are being met. Its major purpose is to catch deficiency in the on-going programme so that proper interventions may take place, which allows the learners to master the required skills and abilities. Examples of formative evaluation are homework exercises, discussion responses, pretests, etc.

2. Summative evaluation

Summative evaluation is defined as evaluation that occurs at the end of a programme cycle and it tends to provide on overall description of the on-going programmes. Such evaluation tends to find out which student, to what extent, has successfully learnt the intended learning outcomes. Summative evaluation tends to answer some important questions like these :

- What is the overall impact of the programme?
- To what extent the objectives of the programme have been met?
- Do you need to modify the existing design of the programme?
- What types of resources are needed to remove the weaknesses of the programme?

There are several versions of summative evaluations. Goal-based evaluation, outcome evaluation and impact evaluation are some of the common versions. Goal-based evaluation permits the evaluator to determine whether the intended goals of a programme were achieved. Outcome evaluation tries to investigate whether the programme has impact on some specifically defined target outcomes. For example, whether students have improved their mathematical skills after participation in say, six weeks training programme. Impact evaluation evaluates the overall or net effects of both intended and unintended impact upon the programme. Some examples of summative evaluation are final examinations, journals when completed for a course, laboratory write-ups, etc.

Scriven (1967) had first suggested an important distinction between formative evaluation and summative evaluation. According to him, the former intended to foster development and improvement within the on-going programme whereas the latter intended to assess whether the outcomes of the programme or person being evaluated met the desired goal. Likewise, Sacttler (1990) pointed out that the formative evaluation is used to refine goals and the various strategies used for achieving goals whereas summative evaluation is undertaken to verify the validity of a theory or determine the effect of an educational practice.

3. Diagnostic evaluation

Diagnostic evaluation is defined as the evaluation whose basic aim is to identify those students where learning or classroom behaviour is being adversely affected by factors not directly related to various instructional practices. In other words, diagnostic evaluation is concerned with locating the reasons for securing learning difficulties that cannot be resolved by standard corrective measures. Such diagnostic evaluation aims to find out the important causes of learning problems and plan to take some corrective or remedial actions.

4. Placement evaluation

In placement evaluation students or learners are placed according to the previous achievement or personal characteristics at the appropriate point in the instructional sequence with a more suitable teacher. In still simple words, it can be said that in placement evaluation the entry behaviour or capability of the learners are assessed for finding out whether they possess knowledge and skills needed to the begin the course of intended instruction.

5. Internal evaluation and external evaluation

Internal evaluation is done by teachers for the purpose of assessing students' cognitive, affective and psychomotor abilities. The basic purpose of such evaluation is to analyze and identify the factors that are responsible for students' development. Such evaluation may be continuous or periodic. External evaluation is one that is done by somebody other than teachers who compares the programme or course with some standards and then, come to some conclusion. Such evaluations are very important in maintaining good standards of the course and help in promoting students based on their performance.

Based on the second criterion, that is, what is being evaluated, there are several types of evaluation such as *student evaluation, curriculum evaluation, school evaluation, evaluation of projects and programmes and evaluation of personnel*. A brief discussion follows :

1. Student evaluation

In student evaluation teachers evaluate the achievement, intelligence, aptitude, personality, attitudes and interests of the students. For such evaluation, both standardized and teacher made tests are administered. Teachers use such performance data of students not only to evaluate students' progress but also to evaluate their own teaching instruction. If only a small number of students seen to experience difficulty with the teaching instruction of the teachers, it is concluded that the teachers' instructional strategy needs was successful. On the other hand, if many students experience difficulty with the teaching instruction of the teachers, it is concluded that the teachers' instructional strategy needs modification for improvement.

2. Curriculum evaluation

Curriculum evaluation is one that involves the evaluation of any instructional programme and related factors such as instructional strategy, textbooks, individual materials as well as physical and organizational arrangements. Such evaluation may involve evaluation of total package of curriculum or evaluation of one small aspect of the total curriculum. Curriculum evaluation involves both internal as well as external criteria. Internally, evaluation is basically concerned with whether the new process achieves what it intends to achieve. Externally, evaluation is concerned with whether the process does whatever it does better than some other process. Besides student achievements, other factors generally recommended for inclusion in curriculum evaluation are the attitudes of teachers and students toward this. How a teacher feels about changes in curriculum may affect the ultimate effectiveness. How do students react toward the change in curriculum? For example, students in the short run may achieve something under a system they don't like but it is highly doubtful that they would cooperate for long with an approach, which they don't like.

Curriculum evaluation has one drawback. It is very difficult to compare the effectiveness of one curriculum with another. Even when the two curriculums deal with the same subject area, they have different objectives and it is very difficult to find a measure or test, which is equally fair to both programmes.

3. School evaluation

School evaluation is one that involves the evaluation of total educational programmes of the school and requires collection of data from each such programme. The major purpose of school evaluation is to ascertain the extent to which school objectives are being met and identify the areas of strengths and weaknesses of each programme. Such information helps in shaping the future activities of the school. For example, based on school evaluation, decision might be made to eliminate the programme of mid-day meal in schools or the decision might be done to improve its quality and extend to high schools, too. One important component of school evaluation is the school testing programme. If school testing programme is carried comprehensively and genuinely, the more valuable would be the resulting data. A good school testing programme must include the measurement of achievement, aptitude, personality and interest of students. It is also essential that tests selected for a school evaluation must match the objectives of the school and not only that, they must be appropriate for the concerned students.

4. Evaluation of projects and programmes

Evaluation of some special projects and programmes include all those efforts that are, strictly speaking, not part of the regular school programme, but they are innovative in nature and this continuation is dependent upon their success. Such projects or programmes may be school specific or nationwide in scope. Whether the existing project or programme should continue, it is best decided by the evaluation. One problem often encountered here is that by its very nature, it is likely to be concerned with such goals or objectives for which there are no corresponding standardized instruments. Therefore, such instruments must be developed and standardized.

5. Evaluation of personnel

Evaluation of personnel includes evaluation of those persons responsible directly or indirectly for educational outcomes. Such persons are not only teachers but are also other persons like counsellors, administrators, etc. This area of evaluation is very slow to progress. This is not because we don't know how to evaluate rather we don't know what to evaluate. In other words, we don't know what our objectives are. Therefore, it is impossible to evaluate anything in that valid criteria against which actual performance may be compared. Although debatable, the teachers are, to a great extent, accountable for the achievement of students.

Thus there are different types of evaluations. Depending upon the purpose of evaluations, the evaluators resort to some selected modes of evaluation.

Phases of Evaluation

Contrary to the common thinking, evaluation is a continuous process. Evaluation is planned prior to its execution of any effort and should be involved throughout the course, that is, at the beginning, in the middle and at the last. There may be a series of some temporary ends in a continuous cycle. Any evaluation involves three important phases: Planning phase, Process phase, and Product phase
A discussion of these phases is as under :

1. Planning phase

This is the first phase of evaluation and in fact, it takes place prior to actual implementation and contains decisions about the course of actions likely to be taken in future. This phase consists of seven sub-phases like analysis of the situation, specification of objectives, specification of prerequisites, selection and development of appropriate measuring tools, delineation of strategies, selection of a design and preparation of time table. These seven sub-phases may be discussed as under :

(a) Analysis of the situations :

This is the first step where the authority responsible for evaluation, analyses the situation as it presently exists so that correct parameters of efforts may be determined. This step incorporates activities like collection of background information as well as assessment of present constraints. Here the evaluator may sometimes be required to prepare a review of literature, which indicates what approaches have or have not been found to be successful. Constraints and resources also need to be carefully assessed for making reasonable decisions. An evaluator does not plan something which he cannot afford. For example, a teacher of Mathematics cannot plan to use computers if the school cannot afford them. Likewise, a cricket coach cannot plan Monday morning practice if the school is closed on that day. And so forth.

(b) Specification of objectives :

When the parameters of the situation have been decided, some realistic goals and objectives are formulated. In fact, goals are general statements of the purpose and therefore, are not directly measurable. Each goal, therefore, is translated into some specific objectives, which are specific statements about what is to be done and achieved and how well. These objectives, therefore, are stated in terms of quantifiable or measurable units or outcomes.

Objectives are of two general type : they may be *process-oriented* (called process objectives) and *product-oriented* (called product objectives). Process objectives state outcomes desired or preferred during the execution of effort. In simple words, they relate to the development and execution. Product objectives describe outcomes desired as a result of effort. In fact, process objectives deal with such strategies and activities that aim at the achievement of product objectives. For example, a product objective may be that 'students' attendance in the class will be enhanced by 25 percent.' A corresponding process objective will be that 'every guardian of student will be informed about the number of days his son or daughter attended the class in the previous month'. Here the assumption is that such information given to guardian will add an additional force in attending the classes properly.

(c) Specification of prerequisites :

In most cases of educational evaluation specification of a given set of objectives is directly based upon the assumption that students have already acquired some skills and knowledge. These assumed behaviours are technically called as *prerequisites* or *entry behaviours*. For example, if teachers are teaching students how to divide numbers and if those students don't know how to add, subtract and multiply, it is very likely that they will never learn how to divide. If it is found that the some students don't possess one or more specified prerequisites capabilities, then objectives must be revised. In order to know whether or not students possess the necessary prerequisites, some tests of entry behaviour are conducted in the beginning of school year or at the beginning of any new unit of instruction.

(d) Selection and development of appropriate measuring tools :

At this sub-phase, the evaluator needs administration of one or more tests/tools so that data may be collected regarding the achievement of objectives. Sometimes, for assessing achievement of objectives, no instrument or tools are needed because secondary data are already available. However, for assessing achievement of most of the objectives, teachers need some tests, which may be selected out of many available or if not available, they may be developed and standardized. Usually, teachers prefer a post-test of the desired behaviour for assessing achievement of product objectives. However, it is also a good idea to administer a pre-test of the same material prior to the start of instruction.

(e) Delineation of strategies :

At this sub-stage, the evaluator delineates some strategies for promoting the achievement of objectives. Strategies are defined as some general approaches for promoting achievement of one or more objectives. There may be several types of strategies such as instructional strategies, programme strategies, curriculum strategies, etc. Whatever may be the general class of strategies, each strategy contains a number of specific activities. Execution of the strategies is planned to ensure availability of the necessary resources. Several instructional strategies such as sequencing, review, feedback and practice are available. These strategies can be generalized across various subject areas and objectives. Other instructional strategies such as role playing, jumping as well as the use of a given medium are considered appropriate for certain objectives only.

(f) Selection of a design :

The use of a research design is not always appropriate or feasible in evaluation studies. However, when called for, the very best one such as experimental design should be selected. Experimental design refers to a design whose purpose is to assist in making some decision concerning the effectiveness of an approach (product) and usually involves comparisons between or among groups using different approaches. A good experimental design assures initial group equivalency as well as control for unwanted factors. The one obvious way to insure initial group equality is to constitute one larger group selected randomly, if possible, and randomly divide them in two halves, and each half receiving a different treatment (or curriculum). Let us take an example. Suppose in evaluation of a new reading curriculum, the purpose is to determine whether students using the new curriculum would be achieving significantly more reading skills than students using regular or some other curriculum. This can be validly determined by using an experimental design in which the teacher may begin with two or more groups who are essentially equal on some relevant variables (such as here abilities) and treat them the same, controlling for all factors, which might influence one group differently than the other *except* that each group is getting a different treatment (curriculum). Subsequently, if the groups perform differently after some time, this difference will be said to be due to differential effectiveness of treatments. In many evaluation studies, randomization is not feasible because they are conducted in real-world setting such as school. When existing groups are to be compared, a quasi-experimental design may be selected. Such designs are not as good as true experimental designs but they do a reasonable job of controlling extraneous factors.

(g) Preparation of a time schedule :

For all types of evaluation, preparation of a realistic time schedule is important. A realistic time schedule includes a listing of major activities of the proposed evaluation effort and corresponding expected initiation and completion time for each activity. One of the advantages of such time schedule is that it helps the evaluator in assessing the feasibility of various alternative plans. Moreover, it also helps the evaluator to remain on the schedule. One of the important precautions to be taken by evaluator in propounding time schedule is that he should not make mistake of 'cutting it too thin' by providing a minimum amount of time for each activity. In fact, evaluator should allow sufficient time so that if an unforeseen minor delay occurs, he can still meet his final deadline.

2. Process phase :

In this process phase, the evaluator makes decisions based upon events, which occur during the actual implementation of the programme, the planned instruction or project. This phase is often called as *formative evaluation* and is basically concerned with analyzing the differences between intended and actual outcomes in terms of procedures and contents. The first step in the process phase is to administer pre-tests and in case of students evaluation, tests of entry behaviour are administered. Based on the results of pre-tests, decisions are made by the evaluator concerning the appropriateness of already specified objectives in planning phase. The planned strategies and activities are done in predetermined order. The basic aims of the process phase are to determine whether the effort is done as specified, to determine the extent of achievement of the process objectives and to identify ways in which improvements can be easily made.

3. Product phase :

In product phase the evaluator makes decision either at the end of the programme or at the end of one cycle of a programme or project. Such decisions are based upon pre-tests of achievement, attitude, behaviour and the like as well as on other cumulative types of data. Here decisions are made regarding the overall effectiveness of a programme or instruction or project. Here evaluator tends to determine whether and/or to what extent intended product objectives were achieved. Educators are of view that the results of product phase are used in three ways. *First*, they provide feedback and direction to all who are involved in evaluation such as students, teachers, programme directors, etc. *Second*, they also provide feedback to outside decision-makers such as parents, members of school committee and funding resources. *Third*, there are persons such as guidance and counselling personnel, school administrators educational researches who can utilize the results depending upon the type of evaluation. Psychologists and educators are of view that the results of the product phase must be interpreted very cautiously. The product phase is also known as *summative evaluation*.

Functions of Evaluation

In psychology and education, evaluation has many purposes. Some of the important ones are as under :

- To formulate objectives, goals, etc of a programme or project or policy.
- To identify the worth a value inherent in a programme or policy.
- To determine the various criteria for assessing success of the programme or students.
- To analyse and interpret the collected data and other related information.
- To identify the impact of a programme or policy on some variables like achievement, personality attitude, etc.
- To identify the influence of some unplanned or undesirable impact upon the programme or policy.
- To recommend for some modification in the existing on-going programme.
- To assess the educational value or utility of a programme or policy.
- To set up a committee, if desired, for reviewing the obtained results of a programme.

Difference between Assessment, Measurement, Evaluation, *Testing and Surveys*

In psychology and education these terms such as *assessment, measurement, evaluation, surveys* and *testing* are often get confused because they are used interchangeably. Therefore, it is essential to look into the depth of their meaning for knowing the distinction among them. *Assessment* is a general term that includes all the procedures used to gain information about an individual. It can be defined as appraising or estimating the magnitude of attributes in the person (Gregory, 2004). It includes traditional paper and pencil tests, observations, rating of performance or project as well as formation of worth or value judgment regarding learning process. Thus assessment includes both quantitative descriptions (measurement) and qualitative descriptions (non-measurement) of students. Thus it has both objective component and subjective component (Matarazzo, 1990). Assessment tends to answer the questions: *How will does the student perform*? Figure 1.1 displays the comprehensive nature of assessment and the role of various measurement and non-measurement techniques in the process of assessment.

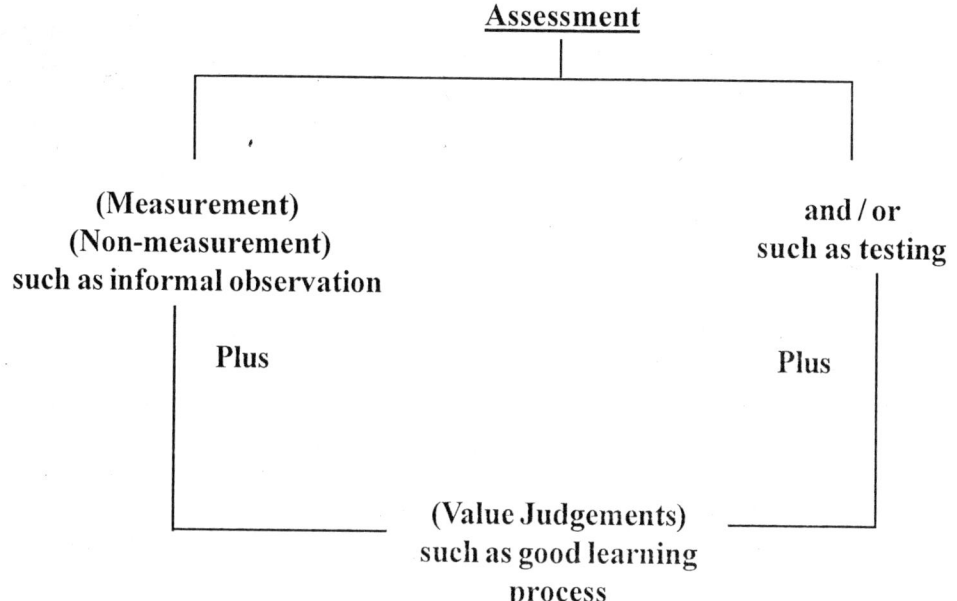

Figure 1.1 : The Comprehensive Nature of Assessment

Measurement is the process of assigning numbers to the objects, events, etc. according to some rules in such as way as to represent quantities of attributes. In other words, it is the quantitative descriptions of the attributes of the students such as Mohan has got a score 40 in an arithmetical ability test. It does not include qualitative descriptions as well as nor does it include judgement concerning the worth or value of the obtained results (Linn & Miller, 2011). In this way measurement is included within assessment. Measurement tends to answer the question : *How much* ?

Psychological test is a particular tool of assessment that consists of a set of questions administered usually during a fixed period of time in a uniform manner. Sometimes, we use the term testing and assessment together although tests are a special type of tool of assessment. For example, a clinical psychologist may conduct a psychological assessment of his patient and as a part of this assessment, he may administer a psychological test such as Rorschach test or MMPI. However, not all assessment techniques are tests. In a strict sense, any assessment technique is called a test only when its procedure for administration, scoring and interpretation are standardized and there are clear evidences for reliability and validity of the tests. Since tests are a form of assessment, they also answer the questions, *"How well does the student perform either in comparison to a domain of a performance task or in comparison with others"*.

Evaluation is a process through which a value (or worth) judgment or decision is made over a level of performance or achievement. Thus evaluation involves a process of appraisal of an object or attribute with reference to some standard. The standard may be social, scientific or cultural. The standard may also be true or arbitrary. Suppose a typist's speed of typing is seventy words per minute and after this measurement, he is evaluated as 'Grade A' typist. Here his typing speed is being evaluated against some criterion already established (say, a standard is typing fifty words per minute). Since his performance, exceeds this standard or meets this standard, he is classified as Grade 'A' typist. This is an example of evaluation where as describing his typing behaviour in terms of fifty words per minute is an example of measurement or quantitative assessment. In a nutshell, evaluation taking a broader view, includes both quantitative and qualitative description of the performance of the person.

Survey means gathering information about a large number of people by examining a few of them. Here the inspection or examination is carried out with some specific aims in mind. The research employs a number of specific kinds of assessment tools such as questionnaires, inventories or interviews for gathering information about attitudes, opinions or preferences of the persons.

Principles of Assessment

Assessment is a wider process for knowing about the nature and extent of students' learning and development. There are certain principles of assessment which, if followed, would make the process of assessment most effective one. These general principles are as under :

(i) In assessment process what is to be assessed should be on priority :

One most important general principle is that what is to be assessed should be clearly specified and defined. Such specification of the characteristics must precede the selection or development of assessment procedures. For example, when the teacher is assessing student learning, he must specify intended learning goal before selecting the assessment procedures for use. Likewise, if a teacher is going to assess students knowledge in the field of standardisation of psychological and educational test, he must emphasize upon the greater specificity for assessment process to be effective. For example, he may specify that their knowledge in the field of reliability and validity would be assessed. For such assessment, then he may prepare some multiple-choice questions, short-answer or essay questions.

(ii) The selected assessment procedures must be relevant to the characteristics or performance to be measured :

Teacher frequently selects assessment procedures on the basis of their objectively, validity, accuracy or convenience. These criteria are no doubt important, they are secondary to the main criteria, such as : Is the assessment procedure most effective method for measuring the intended learning goal? For example, if the teacher is assessing students' achievement, the assessment procedure must represent a very close match between the intended learning goals and the types of assessment tests used. If the intended goal is to assess the organisational ability of the students, a multiple-choice test would be less appropriate than student writing in different conditions such as writing projects, term papers and classroom essay tests.

(iii) Assessment, if it is comprehensive, requires a lot of procedures :

A comprehensive assessment required a variety of procedures. Different types of assessment procedures are available. Multiple-choice test, short answer test, essay test, observational techniques and self-report measures etc. are some of the popular assessment procedures. When the purpose is to assess knowledge, understanding and application outcomes, multiple-choice tests and short answer tests of achievement are considered most appropriate. But when the purpose is to assess the organisational ability and express ideas on a certain theme, essay tests and written projects are considered much more relevant. Observational techniques are considered useful for assessing performance skills and self-report measures are considered useful for assessing interest, aptitudes and personality. If the teacher wants to have a complete and comprehensive picture of the students' achievement, different assessment procedures must be used. No single assessment procedure would provide comprehensive picture of the students' achievement.

(iv) A satisfactory use of assessment procedures requires the proper awareness about their limitations, too :

Any assessment procedure or test has some limitations and a knowledge about those limitations makes it possible to use them more carefully and effectively. One limitation is that no test or assessment procedure asks all the questions or contain all the problems for assessing knowledge, etc. rather only a sample of questions or problems is included. As such, sampling error occurs in such educational and psychological assessment. For example, even a best achievement test may not perfectly and adequately sample a particular content area. A mathematical test may not cover all types of problems in the field of algebra, geometry, trigometry, etc. However, such sampling error can be checked by a careful application of established measurement procedures. Another source of limitation is caused by various chance factors that influence the results of assessment. Among these chance factors, primary ones are subjective scaring on essay tests, inconsistent responding on self-report measures and guessing on objective tests. Due to these various errors, no score on educational or psychological assessment should be treated as the correct and accurate measurement of the attribute in question. However, when the assessment procedures or tests are carefully used, these errors of measurement are controlled to a greater extent. Misinterpretation of assessment or test results is still another source of limitation. Sometimes the test users interpret results as more precise than they actually are or take the results as indicative of even those attributes, which are not assessed by the concerned test. Such misinterpretation can be easily avoided by paying a careful attention to what the test actually measures.

(v) Assessment is one means to an end (or goal) and not an end in itself :

Assessment should always be treated by users as a process of obtaining information, which should be the base of educational decision. Therefore, assessment should be considered as means to arrive at a goal or end and never a goal or end in itself.

Types of Assessment Procedure

The assessment process especially classroom assessment process in teaching include variety of procedures and such assessment procedures can be classified in different ways depending from the used frame of reference. The most popular types of classroom assessment procedures are as under:

1. On the basis of *nature of assessment*, there are two types of assessment: measures of maximum performance and measures of typical performance (Cronbach, 1990). In the category of measures of maximum performance are placed those procedures that tend to determine developed abilities or achievements of the students. Here the results of the assessment procedures indicate what the students can do when they are motivated to put their best effort. Aptitude test and achievement test are the examples of this category of measure. Another category of assessment procedure includes those measures that basically reflect a person's typical behaviour. Those measures are concerned with what the students will do rather than what they can do. Measures aimed to assess attitudes, interests, adjustment and other personality traits are placed in this category of measures of assessment. Observational techniques and peer appraisal are also included in this category.

2. On the basis of *form of assessment*, there are two types of assessment procedure: *fixed choice test* and *complex performance assessment*. Fixed choice test in form of multiple choice test items and other variations of selected response test items such as true-false, matching, etc. are very common and frequently used in standardized tests for efficient measurement of knowledge and skills. Such measurement procedures are called efficient because here students can answer to a large number of questions in brief period of time and they can be objectively scored. Their reliability is also usually high. However, fixed-choice tests are criticised because such tests tend to overemphasize factual knowledge and very low level skills at the cost of higher order problem solving conceptual and organized skills (Linn & Miller, 2011). As a consequence, a different and new approach to measurement and classroom assessment, which depend upon the extended tasks and analysis of complex performances of the students, have gradually taken shape. Written essays, projects, oral presentations and laboratory experiments are the examples of complex performance task. However, performance assessments are more time consuming to administer and score than fixed choice tests. These are also subjective in nature and require a high degree of expertise and training for its scoring.

   If we compare fixed-choice tests and complex performance assessments, it would be clear that these two procedures of assessments represent two ends of a continuum. Short answers tests fall in between these two ends.

3. Based on use or functional role in classroom instruction, there are four types of assessment: *formative assessment, summative assessments, diagnostic assessment and*

placement assessment. A discussion of these four types of measurement and assessment has already been done in this chapter. Therefore, at this point, only a very brief description and reference will follow. Airasin and Madaus (1972) have classified the assessment of student performance as formative assessment, summative assessment, diagnostic assessment and placement assessment. The correct sequence of these four assessments to be used in the classroom are : placement assessment, formative assessment, diagnostic assessment and summative assessment. Although this classification has been done about several years back, still it is considered very relevant.

Formative assessment is used to monitor the progress of learning. It basically aims at providing continuous feedback to both students and teachers regarding success and failures of learning or instruction. Formative assessment is usually done through teacher-made tests and various observational techniques. Since formative assessment is directed towards improving learning and instruction in the classroom, their results are typically not used for assigning grades to students.

Summative assessment, as explained earlier, is done at the end of the learning course or programme and aims to assess achievement of the students in the course. On the basis of the results of the summative assessment, course grades are assigned or the teacher is able to certify student mastery of the intended leaving course. Important techniques used in this type of assessment are achievement tests, ratings on various types of performance, product scales and various teacher-made survey tests.

Diagnostic assessment is one, which is done for the purpose of identifying the recurring learning difficulties of students not corrected by standard practices of formative assessment. Such assessment tends to determine intellectual, physical, emotional and environmental courses of persistent learning difficulties. Such assessment is comprehensive in nature and involves the use of specially prepared diagnostic tests as well as various observational techniques.

Placement assessment tends to determine the performance of students at the entry level. Knowing about their performance and skills already developed, they are placed in the appropriate learning course. Thus the placement assessment basically aims at answering the question : Does a student possess the knowledge and skills needed to begin a certain course ? Important techniques used for placement assessment are readiness tests, aptitude tests, pre-tests on course objectives, self-report inventories as well as various observational techniques.

4. Based on methods of interpreting results, assessment procedures are of two types : *criterion-referenced assessment* and *norm-referenced assessment*. A criterion-referenced assessment is one, which is designed to assess the performance that is interpretable in terms of clearly defined and delimited domain of learning tests. For example, assessment results may be reported in terms of some categories of performance such as

'Poor', 'Satisfactory', 'Excellent'. The categorisation of a student as meeting say 'Excellent' does not depend upon a comparison of that students' performance to other students rather it depends upon the criterion or cut-off score established by 'Excellent' performance. Such assessment process is called *standards-based assessments*. A test intended to provide criterion-referenced assessment is called *criterion-referencetest*. Norm-referenced assessment is one which is designed at providing a measure of performance interpretable in terms of an individual's relative standing in some defined and known group. In other words, in norm-referenced assessment interpretation depends upon a comparison of a student's performance with the performance of other students whose performance constitute norms. The norms may be national or local. A test intended to provide norm-referenced assessment is called *norm-reference test*. Let us take an example to illustrate norms-reference assessment. Suppose a teacher after making assessment of performance in the classroom, states that the concerned student is better than 75 percent of his class members. Here his performance is being interpreted in terms of his relative standing or position in his class. This is norm-reference assessment.

Norm-referenced assessment and criterion reference assessments may be best viewed as the two ends of a continuum rather than a clear-cut dichotomy. Figure 1.2 shows this continuum.

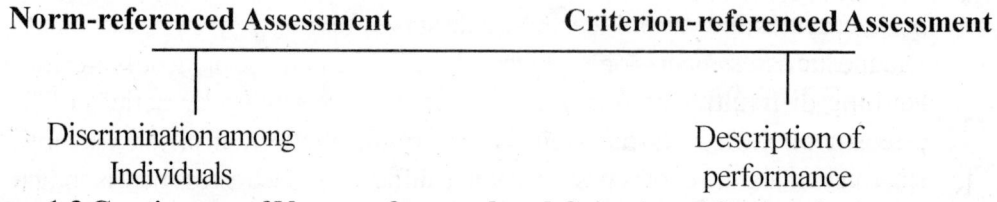

Figure:1.2 Continuum of Norm-referenced and Criterion referenced measurement

It is clear from Figure 1.2 that the norm-referenced assessment emphasizes upon discrimination among individuals whereas criterion-referenced assessment emphasizes upon description of performance. Some examples of instruments commonly used in norm-referenced assessment are standardized aptitude text, achievement test, interest inventories, adjustment inventories, teacher-made survey tests etc. whereas the common instruments used in criterion-referenced assessment are teacher-made tests, observational techniques, etc.

Thus there are different types of assessment procedures. A teacher or any researcher depending upon his purpose, resorts to any or a combination of these procedures for better and effective assessment results.

Summary and Review

- In most general sense, evaluation means determining the value or worth of something. More specifically, evaluation is a systematic process of correcting and analyzing data in order to take some decisions about the value or worth of something. It involves a comparison with some standard for determining the worth or value of something.
- The various types of evaluation can be better understood on the basis of two criteria : how is the evaluation process applied? and what is being evaluated ? Based upon the first criterion, there are different types of evaluation such as formative evaluation, summative evaluation, diagnostic evaluation, and placement evaluation. Based upon the second criteria, there are student evaluation, curriculum evaluation, school evaluation, evaluation of project and programme and evaluation of personnel.
- There are three phases of evaluation : planning phase, process phase and product phase.
- There are several functions of evaluation. Of these various functions, important ones are to formulate objectives, goal, etc. of a programme or project, to analyze and interpret the collected data and other related programme as well as to recommend for some modification in the existing on-going programme.
- There are some general principles of assessment, which must be kept in view for making any assessment procedures more effective and accurate.
- There are various types of assessment procedures. The important ones are formative assessment, summative assessment, diagnostic assessment, placement assessment, criterion-referenced assessment, norms-referenced assessment, etc.

Review Questions

1. Discuss the nature of measurement. Discuss the major principles of measurement.
2. Make distinction between measurement, evaluation, assessment and testing.
3. Define evaluation. Discuss the major types and purposes of evaluation.
4. Discuss the important phases of evaluation.
5. Point out the major functions of evaluation.
6. Citing examples make distinction between formative evaluation and summative evaluation.

Chapter – 2

Basic Principles of Measurement

Learning objectives :
- Meaning of measurement
- Dimensions of measurement
- Levels or scales of measurement
- Measurement and decisions
- Steps in measurement process
- Problems relating to the process of measurement
- Important current issues in measurement

Key terms :

Measurement, Physical measurement, Psychological measurement, Nominal scale, Ordinal scale, Interval scale, Ratio scale, Absolute Zero, Arbitrary Zero, Normative comparison, Invasion of privacy.

Individuals have to make many types of decisions. Some decisions are very simple and require no serious thoughts. For example, we all decide easily what to take up in breakfast, when to go to meet the friends, etc. However, there are some decisions that require careful thought. For example, which college one should attend? whether one should opt for Mathematics or Economics? Should one accept a job in X company?, etc. Some other decisions are done by the persons acting for the society. For example, a teacher has to decide whether he should recommend a special educational training for all those students who are not doing well in the class. A counsellor must decide what action should be taken with those students who are having difficulties in personal adjustment. A crucial element in decision making is that more and better information is likely to yield a better decision. All such decision making efforts imply effective process of measurement, which, by nature, is a quantitative approach.

Meaning of Measurement

The term *measurement* is very popular and common not only in psychology and education but also in other disciplines. Measurement refers to assigning numbers to attributes or characteristics of persons, objects, events according to rules. Stevens (1946) defined measurement as, "assignment of numerals to objects or events according to rules". Tyler (1963) similarly defined measurement as, "assignment of numerals, according to rules". But these ways of defining measurement was not considered much appreciating by many experts because it ignores a crucial aspect of measurement, namely, its connection with quantity or magnitude as well as because of the fact that it includes rule-governed assignments of numbers that don't represent quantities or magnitude. Consequently, it was defined in a more elaborate way. For example, Nunnally (1970) defined measurement as, "rules for assigning numbers to objects in such a way as to represent quantities of attributes". Thus the process of obtaining a numerical description of the degree to which a person possesses a particular characteristic or attribute is called measurement. Measurement answers the question, "*How much*"?

Analyzing the elaborate ways of defining measurement, the following three major properties or features of measurement are clearly identified:

1. In measurement the person or the researcher assigns numerical according to some rules. Rules are procedures for transforming qualities of attributes into numbers (Rosenthal et al., 2000). Rules may be explicit or implicit. In measuring physical characteristics such as height or weight, rules are explicit and standardized. Consequently, everyone understands its procedure. For example, when one says that Mohan is of 5 feet in height and Sohan is of 6 feet height, everyone knows what does it mean and clearly understands that Sohan is taller than Mohan. But suppose one is to measure the trait of dominance, or intelligence of a person, here rules are vague and implicit. These rules are not so explicit and standardized as in case of measuring the physical characteristics. In psychology and education rules for assigning numerals to the attributes are mostly vague and implicit ones.

2. In measurement the researcher is always concerned with the features or attributes of the object or the person. No person would try to measure a 'table' rather its length, height or hardness (that is, attributes). Likewise, teachers can measure achievement of the pupil in different curricula as well as the intelligence of the pupils. All these illustrate the measurement of attributes and not the individual nor the object. While measuring attributes sometimes the researcher may face the difficulty especially when the attribute being measured is poorly defined or whose existence is in controversy. For example, when the researcher is asked to measure 'well-being' of a person, he may face difficulty because the nature of the attribute is not unitary. The attribute 'well-being' incorporates many dimensions within itself and therefore, one has to measure each of those dimensions. Likewise, if the researcher is asked to measure 'telepathy', he will have to face difficulty because its existence is controversial in nature.

3. In measurement the numerals used tend to represent quantities of the attributes. Thus measurement involves the process of quantification, which tends to answer the basic question–how much the attribute is present in the person. For example, when the

researcher, after measuring intelligence of the person, states that the person has the IQ of 130, it obviously, expresses his level or extent of intelligence the person possesses.

Dimensions of Measurement

Broadly, then are two dimensions of measurement : psychological measurement and physical measurement. By psychological measurement is meant measurement of psychological attributes such as intelligence, interest, adjustment, aptitude, attitude, etc., of the person whereas by physical measurement is meant the measurement of physical qualities or attributes of objects, things, etc. For example, measurement of height, weight, size, volume, etc. is the example of physical measurement. A field of study concerned with the theory and technique of psychological measurement is called as *psychometrics*. One part of the field is concerned with the objective measurement of abilities, skills traits, knowledge and educational achievement. The other part of the field is concerned with statistical research bearing on theory of measurement (that is, item, response theory, intraclass correlation, etc.).

To understand the distinction between psychological measurement and physical measurement, some points are worth mentioning :

1. In physical measurement there is a true zero point whereas in psychological measurement, there is an arbitrary zero point. A true zero point is one that reflects the true absence of an attribute. For example, when it is said that the object has zero height, it means that the object has no height at all. By an arbitrary zero point means that the assigned zero point is not true. For example, a student of class X may get zero marks in arithmetic or mathematics, but this does not mean that his arithmetical or mathematical ability is really zero. Likewise, a person securing zero score on an intelligence test does not mean that he has zero intellectual capacity. All this happens because these zero points are not real rather arbitrary.

2. In psychological measurement the unit of measurement is not fixed and therefore, it varies from one situation to other or from one researcher to other. On the other hand, in physical measurement the opposite is true because here unit of measurement is fixed and stable. Let us take an example : Suppose a person's weight is measured today at Patna and it comes to 65 kg. This weight is then measured next morning at Delhi, his weight again will be the same. This happens because kilogram has the same fixed properties at whatever place it is used. On the other hand, suppose the researchers are interested in measuring intelligence of students. Some researchers may prefer to measure intelligence on the basis of errors done in answering the questions during specified period of time whereas some other researchers may prefer to a measure intelligence on the basis of some manipulative tasks done by the person during a fixed period of time.

3. There is a greater degree of accuracy and predictability in case of physical measurement whereas in case of psychological measurement, the degree of such accuracy and predictability are considerably low. For example, when it is said that length of the table is

six feet, it means it is actually of six feet and twice that of three feet. In this way in such measurement there is a perfect accuracy and predictability. All this happens because in physical measurement there is a true zero point. But in case of psychological measurement, the picture is different. Suppose a student has obtained a score of 40 on the arithmetic test of 100 items, each item having one score for the correct answer. On the basis of the score of 40, no strong conclusion can be drown because it possesses little accuracy. This score of 40 is not 40 points above the zero point because the zero point itself is arbitrary. Moreover, another student securing a score of 80 on the same arithmetic test can't be said to possess the arithmetical ability two times more than the first student because there is no true zero point.

4. In physical measurement the researcher can measure the entire objects or persons but in psychological measurement, the researcher keeps himself satisfied with the measurement of some limited cases. For example, if the researcher is to measure the weight of 50 students of class X, he can measure the weight of all students one by one. But suppose he want to measure intelligence of the class X students of Bihar, it is not possible to assess the intelligence of all students of class X of Bihar. He will have to satisfy himself by assessing intelligence of some selected representative students of class X.

5. Psychological measurement is indirect whereas physical measurement is direct. Where one is to measure the length of the door, one can directly measure it with the help of measuring tape and come to a conclusion. This is physical measurement that involves directedness. But if one is to measure extraversion trait of the personality, no such direct measuring instrument is available. This can be measured indirectly on the basis of responses to a set of questions.

6. Psychological measurement is more difficult than physical measurement. This is so because in psychological measurement, it is difficult to define standard unit. For example, what is the standard unit of the measurement of extraversion? Or of caste prejudice? Not only that, when a unit has been defined or well-specified, there is problem of finding out how many 'units' are contained in that which is measured. In physical measurement these difficulties are not reported because it is easy to defined a standard unit such as 1 feet means 12 inches and one kilogram means 1000 grams, etc.

7. Psychological measurement is rarely characterized by equal intervals whereas physical measurement possesses the good quality of equal intervals. For example, difference between inch 3 and inch 5 on a ruler represents the same quantity as the difference between inch 13 and inch 15. But the difference between IQ of 50 and 60 does not mean the same thing as the difference between IQs of 90 and 100 although there is difference of 10 points in both cases.

Thus physical measurement and psychological measurement differ from each other.

Levels or Scales of Measurement

Although it is just possible to conceive of a large number of scales, the common practice is to differentiate among some selected scales or levels of measurement such as *nominal scale, ordinal scale, interval scale and ratio scale* of measurement. Before these scales of measurement are considered, it is essential to understand the general principles of scales of measurement. One common property of all measurement scales is *classification* where by the individuals or objects are simply placed into different categories or class. Besides this, there are three major properties of scales of measurement that make one scale different from the other. These three properties are : magnitude, equal intervals and absolute zero.

Magnitude refers to the property of being 'more' or 'less' or 'equal' (McCall, 1994). When for example, on a scale of weight, one can say that Shyam is heavier than Ram, then this scale has the property of magnitude. Likewise, on an IQ scale if one says that Sita has higher level of IQ than Neeta, it becomes the example of magnitude.

Equal intervals are another important property of measurement scales. By equal intervals is meant that the difference between two points at any place in the scale has the same meaning as difference between two other points that tend to differ by the same number of units. For example, difference between 50 kg. and 56 k on a weight scale has the same meaning as the difference between 70 kg. And 76 kg. This illustrates equal intervals. This is example from physical measurement. In case of psychological measurement, such equal intervals are rarely achieved. For example, difference in IQ 50 and IQ 70 does not have the some meaning as difference in IQ 110 and IQ 130 although there is difference of 20 points at both stages. Researchers have shown that when any scale has the property of equal intervals, the relationship between the measured units and some outcomes can be easily described by the linear equation $Y = a + bx$. This equation clearly exhibits that an increase in equal units on a scale shows equal increases in some meaningful correlates of units.

Absolute zero is another property. By absolute zero is meant a point where nothing of property being measured exist. Therefore, it is a true or real zero point. In physical measurement absolute zero is found. For example, when the doctor finds that the heart rate is zero because the patient has died, it is absolute zero. In measuring psychological characteristics such as leadership, dominance, extraversion, introversion, intelligence, etc. it is very hard to find out a point of absolute zero because a person with zero amount of these traits cannot be located.

A discussion of four levels of scales of measurement follows :

1. Nominal Scale

In fact, nominal scales are no real scales. Their purpose is simply to name objects, persons, etc. Classification of human beings into Hindu, Muslim, Sikh, etc. are examples of nominal scale. Likewise, classification of clinical groups into schizophrenia, manic-depressive disorder, conversion hysteria, mentally retarded children are examples of nominal scale. Classification of persons into males and females are also examples of nominal scale. One important feature of nominal scale is that members of any two groups are never equivalent but all members of any one group are wholly equivalent. This equivalence relationship is reflexive (a = a for all values of a), symmetrical (if a = b, then b = a) and transitive (if a = b, and b = c, then a = c). Here admissible statistical operations are counting or frequency, percentage, proportion, coefficient of contingency, etc.

2. Ordinal Scale

Measurement on ordinal scale not only makes distinction between one person from another person but also tells about whether a person possesses more or less of one trait being observed than other persons in the group. In other words, it possesses the property of magnitude but not of equal intervals or of an absolute zero. Therefore, ordinal scale allows the researcher to rank the persons or objects but does not say anything about the differences between the ranks. Ordinal measurement indicates which person or object is smaller or larger, heavier or lighter, harder or softer, etc. In this way, the categories in ordinal scales are not only homogeneous and mutually exclusive but they stand in some kind of relationship of inequalities (higher or lower). The best example of ordinal measurement is our socio-economic status where every member of upper socio-economic status is higher in social status than every member of middle socio-economic status and every member of middle socio-economic status is higher in social status to every member of lower socio-economic status. Students' ranking in their examination such as 1st rank, 2nd rank, 3rd rank, etc. also reflect ordinal measurement. In ordinal measurement the relationship of the 'greater than' is usually irreflexive (it is not true for any a that a > a), transitive (if a > b and b > c, then a > c) and asymmetrical (if a > b, then b > a).

The limitation of ordinal measurement is that ordinal measures don't convey that the distance between the different rank points is equal because ordinal measurements are not equal-interval measurements nor do they have absolute zero point. Let-us take on example : Suppose on academic achievement test, students A, B, C and D got the scores like 70, 69, 60, 58 and are ranked 1, 2, 3 and 4 respectively. Here the difference between A and B is of 1 score, between B and C of 9 score and between C and D of 2 score. But these differences are equally spaced in term of ranks and from ranks, it is difficult to estimate the underlying differences in actual scores.

The permissible statistical operations in the measurement of ordinal scales are median, rank–difference correlations, percentiles plus all those, which are permissible for nominal measurement.

3. Interval Scale

Interval scale has all the properties of nominal scale and ordinal scale and in addition, the property of equal units or intervals. It means that equal differences in scores display equal differences in whatever is being measured. Some examples of interval scale are the temperature scales such as Fahrenheit and Celsius thermometers, time as shown in calendar, scores on intelligence tests and aptitude tests. Temperature scale such as Fahrenheit scale clearly possess the property of magnitude because 40° F is warmer than 36°F and 60°F is also warmer than 50°F. Apart from this, the difference between 40°F and 36°F is similar to 4 degree differences at any point on the scale. However, the measurement on Fahrenheit scale does not have absolute zero. A zero on Fahrenheit scale does not have any meaning because water freezes at 32° F. Similar is the story of Celsius scale. Although zero on Celsius scale represents freezing point, but it is not an absolute or true zero because there is still many things on thermometer below zero. Since these temperature scales don't have an absolute zero, statement in terms of ratios cannot be made. Therefore, a temperature of 60° F cannot be said to be twice as hot as 30° F. Likewise, a person scoring 30 on intelligence test is definitely lower than a person having a score of 40 on the same intelligence test. But a person having a score of 100 cannot be said to be twice as intelligent as the person having a score of 50.

Since interval scale is the first truly quantitative scale, the common parametric statistics such as mean, standard deviation, product-moment correlations, t test, F test can be applied. The only statistic that cannot be applied in interval measurement is the coefficient of variation because this statistic is a sort of ratio of standard deviation to the arithmetic mean.

4. Ratio scale :

This is the *highest* level of measurement and it possesses all three important properties of the scale namely, magnitude, equal intervals and an absolute zero (see Table 2.1). In ratio scale, the ratio of any two scale points is independent of the unit of measurement and therefore, it can be meaningfully equated. This is a scale upon which the common physical measurement of length, time and weight are made. Since a scale has on absolute zero point, statement of ratios are meaningfully made. Thus table of 6 feet in length can be said to be twice as long as a 3 feet. A 10 pound object is twice as heavy as 5 pound object and so on. Kelvin scale and speed of the travel are also examples of ratio scales. Since Kelvin scale is based on absolute zero point, it can be said that 20° K is twice as cold as 40° K. So far as the speed of travel is concerned, 60 miles per hour is speed, which is exactly double 30 miles per hour speed.

In behavioural sciences, of the four measurements, we frequently encounter measurement on interval scale where the experts can easily assume equal intervals and an arbitrary zero point. All the basic properties of the measurement scales have been presented in Table 2.1.

Table 2.1 Important properties of scales of measurement

Types of Scales	Classification	Magnitude	Equal Intervals	Absolute zero
Nominal	✓	×	×	×
Ordinal	✓	✓	×	×
Interval	✓	✓	✓	×
Ratio	✓	✓	✓	✓

Measurement and Decision

In fact, educational and psychological assessment has helped a lot in making decisions about people and groups. Some sections of experts like teachers, counsellors, psychologists are continuously involved in making decision about people or they help people in making decisions about themselves. The various measurement procedures help these experts by providing a lot of information that make those decisions very accurate and appropriate.

There are several types of decisions, which are made by the experts with the help of various measurement procedures. Some of these are as under :

- Some decisions are *placement* or *classification* decisions. A school teacher or principal may decide about placing a student into higher grade on the basis of his or her brilliant academic performance. Likewise, in army personnel the authority needs to decide what type of training the selected persons be given. For making placement decisions, the concerned expert needs information in making prediction about how much the person will learn from the assigned work or to what extent the person will be successful if placed in certain programme. A doctor makes a classification decision when he diagnoses that the person is suffering from pain in spinal cord and muscles of the whole body.
- Some decisions are *instructional* in nature where the authorities need to decide about the others keeping in view their skills and competency. For example, teachers and school psychologists may decide about what type of reading materials will be more helpful for a particular student or for the group of students keeping in view their skills and academic interest.
- Some decisions are *curricular* in nature. In college or school the authority may decide to bring changes in the syllabi for betterment of the future of the students. Likewise, the authority may decide to introduce computer-assisted instruction (CAI) for teaching the basic principles of arithmetical manipulation.
- Some decisions are selection-based usually done by the employers who need to take decision about more effective employees from the pool of applicants. Sometimes, they collect information from some controlled testing situations for improving the accuracy of hiring decisions. Selection decisions are also done by college authorities who need to decide which applicant should be selected for admission into a given academic course.
- Some other decisions may appropriately be called as *personal* decision, which are decisions that a person makes at the crossroad of life. Should one take a course in medical or engineering course? Should one do the job simultaneously carrying an academic course? Should one purchase the residential flat in this area? In such cases, if the person has accurate information about their own abilities, interests and skills, he is likely to take effective decision. Usually, counsellors use some standardized tests for helping the persons in making such decisions.

Steps in Measurement Process

In measurement of any field including psychology and education, there are three basic steps as under :

1. Identifying the attribute to be measured :

In measurement the researcher measures the attribute or quality of the person or thing and he never measures the thing or a person. For example, we measure intelligence of the person, the weight of a person, length of the clothes, etc. Among the various attributes, there are some attributes, which are explicit and clear whereas there are some which are vague and difficult to observe and are more abstract. Such abstract concepts are technically called as construct. For example, extraversion and intelligence are constructs.

When the researchers try to identify and define attribute, two types of situations do arise. First, he must decide which attributes are relevant and important to measure because if he chooses wrong attributes to assess, he may arrive at a wrong judgement. For example, for selecting truck drivers attributes like ability to solve arithmetical problems and verbal reasoning will be less relevant attributes but attributes like depth reception, freedom from colour blindness, eye-hand coordination will be the more relevant ones. Second, the investigators or the researchers must establish a clear meaning of the attribute or quality to be measured. The psychologists or educators usually face, while attempting to measure attributes, the problem of how to arrive at clear, precise and generally accepted definitions of these attributes. So far as the physical attributes like weight, length, etc., are concerned, their meaning is well-defined and no problem arises in their measurement. But psychological and educational constructs or attributes are difficult to define. Let us take an example. Suppose the researcher wants to measure intelligence and therefore, he needs to define it accurately. Will he define intelligence in terms of ability to carry out abstract thinking? Or will he define intelligence in terms of ability to do manipulative tasks in a given time? Or will he define it as the ability to adjust in new situation? Or will he define it in terms of ability to recall past associations? In this way, there are several points regarding which there exists differences among expert to arrive at a definition of intelligence sufficiently precise to allow measurement. Therefore, the wiser first step that the researcher should take is to identify and define the attribute precisely.

2. Deciding operations to display the defined attribute :

The next step in measurement procedure is to locate or find out a set of operations that may isolate and display the attribute of the interest. An attribute which is defined by its operations or procedures (or how it will be observed or measured) is said to have operational definitions. Let us take an example : Suppose the psychological or educators want to measure the extraversion. Now for measuring extraversion, they will identify the operations or procedures through which the trait of extraversion may be measured. Thus they may identify the person as extraverted who is sociable, active, talkative, person-oriented, optimistic, fun-loving and affectionate. Taking these observations or operations in view, several tests of extraversion have been developed. Sometimes there is found enough ambiguity in the definitions of construct or trait as well as there is enough variety in the tests or tools devised for measuring the construct. This creates confusion on the part of the researcher or the investigator who is involved in the task of measurement.

3. Quantification of attribute or quality :

This is the final step in the measurement process where the researchers quantify the outcomes of the accepted set of operations. Here the common meaning of measurement, that is measurement refers to a process of assigning numerals to the attribute according to some rules is justified. The numerals assigned indicate how much of the attribute is present in the object or person. In quantification the researcher uses a set of rules for assigning numbers that ultimately allows him to answer the question. *How much? Or How many?* These rules are explicit in case of physical measurement such as measurement of length of table or bench. We clearly ask here how many feet? Or how many meters? These feet or meters have also the trait of equality because 4 feet means double the length of 2 feet and 2 meters mean exactly double the one meter. But in case of measurement of psychological and educational attributes such as intelligence, aptitude, academic achievement, etc. the researcher has no such explicit and straightforward rule nor is there proof of equality. For these reasons, quantification of psychological and educational attribute becomes shaky.

Quantification with the help of numerals secures two major advantages. *First*, quantification makes the communication precise and understandable. For example, when it is said that Mohan's weight is of 80 kg, it provides a better understandable theme about Mohan than when he is simply described in non-quantitative ways such as he is a healthy person. Another advantage of quantification is that by adding, subtracting, multiplying and dividing the numericals, the researcher is able to arrive at some broader levels of meaning.

Problems Relating to the Process of Measurement

In psychological and educational measurement, three are there basic problems, relating to each of the three steps of the measurement process. These problems are being discussed as under :

1. First problem is related to the first step of measurement where the researcher has to select and define the attributes of the interest clearly and in mutually agreeable terms. For example, if one is to define introversion, to what extent definition should include each of the following attributes :
    - Maintains silence
    - Remains unassertive
    - Is timid
    - Is unenergetic
    - Is unadventurous
2. Second problem is related to the second step of measurement process where the

researcher has to devise procedures or operations to elicit relevant attributes. For some attributes, the operations or procedures are clear cut and explicit through which one can easily conclude that the person is displaying certain attribute or the object possesses the given attribute. Such operations are well-observed under uniform and standardized conditions of tests. For example, there are standardized tests of aptitude and intelligence where the persons are required to go through various questions with proper understanding about the quantitative relationship being asked. Based upon such operation or observations, one can easily identify the relevant attributes. However, there are attributes for which relevant operations or observations are not well-defined. For example, if one is to devise operations for a policeman duty, the standard operations are difficult to be delineated.

3. Third problem is related to the third step of the process of measurement. Since the psychological and educational measures lack equality of units, arithmetical analysis like addition, subtraction, multiplication, division are always suspected. Not only this, in measurement of attribute consistency from one occasion to other or from one appraiser to another is also not satisfactory.

Despite these problems, psychological and educational devices are yielding worthwhile information for taking effective decision. Generally, information provided by these devices or tests are more useful and accurate than from other sources.

Important Current Issues in Measurement

Psychological and educational measurements are not perfect. That is the reason why often questions arise regarding the reliability and validity of the results. In this section, attempt is being made to concentrate upon some important current issues in measurement.

1. Invasion of Privacy :

To some extent all observations, measurements and testing are invasion of privacy to the persons who prefer to be tested. The probability of invading privacy is very high whenever the testing or measurement is designed for the benefit of someone other than the persons being tested. (Thorndike & Thorndike–Christ, 2015, Dahlstrom, 1969; Sax, 1974) For example, a teacher who administers a test to his students solely to gather information for his professional purposes and advancement, may well be encroaching upon the rights of his students to privacy unless he gets informed consent from his students and parents. The various elements included in informed content are a full description of the purpose, procedure, risk (both physical and psychological), cost and potential values of the student. Besides these, the procedures used to ensure anonymity or confidentiality should also be described. During testing some students resent any question of a personal nature. Likewise, some parents strongly object to discuss issues that they prefer to remain in the privacy of the home. Ethical considerations require that teachers or assessor be sensitive to these matters, perhaps by allowing alternative activities and insuring that the elements required for obtaining informed consent be followed scrupulously. A person's privacy is invaded where such information is used inappropriately. Psychologists and educators are ethically and legally bound to maintain confidentiality and they don't have to reveal any information about the person than is necessary to serve the purpose for which testing was done.

2. Right and Responsibilities of testers and test-takers :

Both the person administering the test and test-takers have some obligations in testing and measurement situations. For example, test takers has the right to know about the details of procedures he or she will encounter and how they will be physically used but they also have the responsibility to ask questions about the aspects of testing they don't understand and participate in a genuine and responsible way in testing. In this connection, the most influential guidelines for educational and psychological testing, first published by American Psychological Association (APA) in 1954, have been revised four times and the latest guidelines were published in 1999 as a joint project of American Educational Research Association (AERA), APA, and National Council on Measurement in Education (NCME) under the title '*Standards for Educational and Psychological Testing*' briefly known as '*Standards*'. The new guidelines have clearly provided the current standards for the practice of test construction, administration and interpretation of the test results to protect the various rights of the test – takers.

3. Testing Minority and Disadvantaged Individuals :

For tribal, ethnic, linguistic minorities as well as for the disadvantaged people, the appropriateness of tests are often questioned. Most of the studies agree that students from tribal, linguistic minorities and disadvantaged homes attain lower intelligence, achievement and aptitude scores than do students from the middle and the upper socio-economic levels. Test contents are so designed that the latter type of students are more benefitted. The issue is that the test contents should not be biased rather should present such blend that both types of students be benefitted.

The motivation of the tribal, ethnic and linguistic the minorities as well as those of the disadvantaged children to do well on such tests is also an issue. Many such students are not highly motivated by promise of benefits from the school as are children of affluent classes. These students tend to put forth their best efforts when some immediate gains are visualized and they are more concerned with pleasing their pears than their teachers and parents. Students from affluent class strives to do best on tests because they are convinced that doing well now would have important and favourable impact on them in future. Perhaps, even more serious issue is raised when tests are used as a basis for deciding what a student can learn to do. An inference from score of disadvantaged student obtained at one time concerning what he can do next by a certain time in future is, in fact, more questionable inference in comparison to one that merely states that what a student can do by this time. Many intervening factors can throw off the production and several biasing factors may come up that can distort the production.

4. Normative Comparisons :

The issue of normative comparison is also an important concern for the measurement professionals. This issue comes up in test interpretation. The common practice is to compare the performance of one person with the norms representing the performance of a national or local sample. The question is to what extent such comparison is justified in taking a decision about the person. Most of the experts now realise that instead of such normative comparison, criterion-reference and mastery test as well as performance assessments can yield a better information for taking a decision about the person.

5. Creation of anxiety and interface with learning :

Another issue towards which measurement professionals have shown concern is the impact of anxiety upon performance and learning. Some important questions often raised here are : Does test anxiety lower or raise the performance of the students? Are some groups of students more vulnerable to the impact of test anxiety? If test anxiety really affects the performance systematically, what can the teacher or assessor do to minimize it? Researchers have shown that some students are badly affected by test anxiety whereas some other students perceive test anxiety as helpful one. Firke (1967) has reported in his study that less than fifteen percent gave negative responses to a wide variety of tests when their purpose was to correct research data. However, a larger percentage felt uneasy about taking tests if on the basis of the performance of the test, a crucial decision affecting their lives was involved. Thus in this study tests were perceived as favourable or unfavourable depending upon how the results were to be interpreted. In another study conducted by Feldhusen (1964) some college students were asked to respond to a questionnaire on the effect of quizzes on their achievement and attitudes. Results revealed that eighty percent students reported a favourable attitude towards quizzes because such quizzes helped them in learning whereas twenty percent reported an unfavourable attitude because they found such quizzes no more helpful. Kirkland (1971) has summarized the impact of test anxiety upon students' performance in the following way :

- Lower or mild degree of anxiety usually facilitate learning whereas high degree of anxiety hinders the performance in most cases.
- If the students are already familiar with the test, it lowers the anxiety.
- Less capable students incur more test anxiety than more capable students.
- Highly anxious students perform better than his anxious ones on tests measuring rote learning or recall. However, they perform less well on tests requiring critical thinking and flexibility.
- Although there appears to be relationship between sex and anxiety among primary school going children, high school girls reported to experience more anxiety that do their boys counterpart.
- Test anxiety increases as grade level of the students go above.

An effective teacher tends to create sufficient interest and motivation among their students to do their best without also creating the undesirable stress among them.

6. Categorization of students :

Another issue that has been a point of concern for measurement professionals is that if educational and psychological tests are misused, they do place some students permanently by rigidly categorizing them and allowing teachers to use these classifications relentlessly. For example, if teachers know that a students' IQ score is low, some teachers may decide that the child is unteachable and may assign some menial tasks for keeping them occupied. However, if tests are considered to be samples of behaviour that don't measure fixed or unchanging traits, teachers will not be likely to categorize students. Moreover, not all teachers misuse test results. The evidence suggests that when teachers are given knowledge about their pupils, they tend to assess the achievement of their pupils on individualized basis.

7. Miscellaneous factors :

Researchers have shown that measurement professionals are also concerned with some other issues like nutritional status of the pupils and their ability to make concentration because these factors also have been found to affect their tests results. Likewise, gender relationship between the test-takers and test administrators as well as the effect of some coaching on test performance also have been found to affect performance of the students on the test.

Summary and Review

1. Measurement is defined as rules for assigning numbers to objects in such a way as to represent quantities of attributes. Measurement answers the question : *How much?*
2. There are two important dimensions of measurement-physical measurement and psychological measurement. In physical measurement the measurement professionals are interested in measuring the attributes of objects or persons such as length of table, weight of a person, etc. Psychological measurement implies measurement of psychological attributes such intelligence, extraversion, introversion, etc. The important point of distinction between physical measurement and psychological measurement is that in physical measurement there is a true or absolute zero point whereas in psychological measurement there is an arbitrary zero point.
3. There are four levels or scales of measurement : nominal scale, ordinal scale, interval scale and ratio scale. The attribute of classification is found in all the four scales; attribute of magnitude is found in ordinal, interval and ratio scales, the attribute of equal intervals are found by in interval and ratio scale and the attribute of absolute zero is found only in ratio scale.
4. There are several types of decisions, which are made by the professionals with the help of various measurement procedures. Classification decision, curricular decision,

instructional decisions, selection-based decision, and personal decisions are common ones.
5. There are three basic steps in any measurement process : identifying the attribute to be measured, deciding operations to display the defined attribute and quantification of attribute or quality. There are also some problems related to each of these steps.
6. Measurement professionals have raised some important issues such as invasion of privacy, rights and responsibilities of test – takers as well as test administrators testing minorities and disadvantaged individual, normative comparisons, creation of anxiety and interference with learning, categorization of students, and other miscellaneous factors.

Review Questions

1. Give the meaning of measurement. Also discuss the different dimensions of measurement.
2. Citing examples discuss the levels or scales of measurement.
3. Discuss the major steps in the process of measurement.
4. Explain the major problems associated with the process of measurement.
5. What are the important current issues in the measurement?

Chapter – 3

Principles of Psychological and Educational Testing

Learning objectives :
- Psychological Tests: Meaning, Similarities and Differences
- Historical introduction to psychological and educational testing
- Major characteristics of a good psychological test
- Difference between Psychological tests, Psychological measurement and Surveys
- Assumptions of Psychological tests
- Locating Information about Psychological Tests
- Uses or Applications of testing
- General Steps of Test construction
- Future of Psychological testing
- Ethical and social considerations in Testing
- Computerized Testing
- Criticism of Testing

Key Terms:

Mental test, Speed Test, Power test, Standardized test, Criterion-referencing test and norms-referencing test, Reliability, Validity Norms, Invasion to privacy, Confidentiality, Deception, Informed consent, Labelling, Divided loyalties, Dehumanization, Computerized Adaptive Testing.

Psychological and educational tests are important tools. Any tool can be used to help or harm the people. In fact, test is a measuring device used to quantify behaviour or aid in the understanding and prediction of variety of behaviour. Testing has been growing day by day and it is contributing effectively in several important areas of life. But this growth has also been accompanied by some unrealistic expectation and limitations. Tests don't measure the full understanding of behaviour. This is because of the fact that test measures only a limited sample of behaviour and therefore, errors are always involved therein. Thus although tests are not perfect measures of behaviour, they do significantly contribute to the understanding and prediction of behaviour. Psychological testing is concerned with all the possible uses, applications and the major principles underlying concepts of psychological and educational test. The major uses of these tests is to locate and evaluate the individual differences in ability and personality and at the same time assume that the differences shown in the test scores reflect the actual differences among the persons. The principles of psychological testing takes into account all the basic concepts and fundamental ideas underlying all psychological and educational tests.

Psychological Tests: Meaning, Similarities and Differences

A test is simply defined as a measuring device or procedure. When the term test is prefaced with a modifier, it becomes device or procedure designed to measure a variable related to that modifier. For example, when the test is prefaced by the word psychological it becomes psychological test, which may be defined as a device or procedure designed to assess variable related to psychology such as intelligence, aptitude, personality, interests, values, etc.

Many definitions of psychological tests (or educational tests) have been offered. For example, Anastasi and Urbina (1997) have defined psychological test as, "essentially an objective and standardised measure of sample of behaviour." Sax (1974) has defined test as, " a task or series of tasks used to obtain systematic observations presumed to be the representation of educational or psychological traits or attributes." Recently, Kaplan and Saccuzzo (2001) have defined a psychological test or educational test as," a set of items designed to measure characteristics of human beings that pertain to behaviour." Considering the basic spirits of these definitions, it can be said that psychological or educational test is essentially a standardised procedure for sampling behaviour, which is often described with some scores or categories.

Psychological tests have some obvious features as under :
- Psychological test is a standardised procedure. Any test is considered to be standardised if the procedures for administering it are uniform from one situation to other or from one test administrator to other.
- Even when the test is targeted to a well-defined behaviour domain, neither the test taker nor the test administrator has sufficient time for a comprehensive testing. So, test is limited to a sample of behaviour. For example, for assessing intelligence, a test may include 100 items, each item representing a segment of behaviour related to intelligence. But reality is that there may be numerous segment of behaviours related to intelligence. But it is not possible to cover all those. Therefore, a test is limited to only a sample of behaviour.
- A psychological or educational test also permits derivative of some scores or categories. Every test furnishes one or more scores or provides evidence for the facts that the test taker belongs to one category and not to another. Thus test sums up the performance of the test takers in terms of some scores. For example, on a vocabulary test consisting of 100 words, if the test taker is able to define 80 words out of 100, he is put in category difference from that who is able to define only 20 words. If one score is provided for each correct definition, the former will get a score of 80 whereas the latter will get a score of only 20.

Besides these fundamental features, some psychological and educational tests possess norms, which are the average performance of a large and representative group of subjects. Norms help in interpreting the obtained test scores by providing a comparison with scores obtained by the representation group on the same test. Besides this, test also predicts some non-test behaviour or additional behaviours not directly targeted. For example, a child scoring very low on intelligence test, does not simply classify him as belonging to a category of low intelligence but also predicts many reformatory behaviours due to that child.

One important question arises that what all psychological tests do. What constitutes the points of similarities among various psychological tests etc. All psychological tests do at least have two common things :

- All psychological tests require that the person must perform some such behaviours, which are observable and measurable. For example, when a person is responding on multiple choice items, it is expected that he must read the item and each alternative and decide about the correct answer.
- Personal attribute, trait or characteristics that are considered important in understanding human behaviour are measured by the behaviour that the person performs. For example, the answers given by the person on intelligence test might reflect his verbal ability or quantitative reasoning.

If there exists similarities among the psychological tests on the above two points, then there must also be some point of *differences* among the psychological tests. Yes, there are some points of differences as under :

- Psychological tests may differ in terms of behaviour they require the person to perform. For example, an intelligence test may require the test takers to define words, explain what is missing from the pictures, arrange blocks to duplicate a geometrical design, etc. whereas a projective test such as TAT may require the test taker to write a story on the basis of vague pictures.
- Psychological tests may vary in terms of their psychometric qualities. Some psychological tests possess high degree of reliability and validity whereas some other psychological tests have poor degree of reliability and validity.
- Psychological tests also differ in terms of how they are administered and formatted. A psychological test may be administered individually or in group setting. Likewise, psychological test may be designed in multiple-choice format or in True-False format or in open-ended questions.
- Psychological tests also differ in terms of attribute they tend to measure. Some tests measure verbal intelligence and some tests measure non-verbal intelligence. Likewise, some tests measure personality and some other tests measure motivation, anxiety, etc.
- Psychological tests also differ in terms of their content. Two psychological tests measuring the same attribute or trait may require the individual to perform different types of behaviour

or to answer different questions. For example, questions on two intelligence tests may differ because one test constructor may define intelligence in terms of ability to carry out abstract reasoning and another test constructor may define it as the ability to profit by past experience. In fact, difference in the content of the psychological tests may also be due to different theoretical orientation of the test.
- Psychological tests differ in terms of how they are scored and interpreted. Some tests are hand-scored and some tests are computer scored. There are also tests, which are scored by the tests takers themselves. The results of some tests are easily interpreted with the help of table provided where as the results of some tests may be interpreted only by trained professionals.

Thus despite some similarities among the psychological tests, they tend to differ from each other on various points.

Historical Introduction to Psychological and Educational Testing

The history of psychological and educational testing is very much fascinating and its roots are lost in antiquity. If we heartily trace its history, it can be better presented by dividing the entire period till date into the following three areas:
1. Early Antecedents
2. Crucial Scientific development in between 1800 A.D to 1900 A.D
3. Development between 1900 A.D to 2000 A.D

Major developments during these periods and their impact upon psychological testing is briefly discussed here.

1. Early Antecedents

The origins of rudimentary form of psychological testing dates back approximately 4000 years to 2200 B.C.E. When the Chinese emperor Yushun had introduced a system of civil service examination and examined every third year to determine their fitness for office (DuBois, 1970; Bowman, 1989). However, modern Chinese scholar find little archaeological evidence to support this claim. During the Han Dynasty (206 B.C.E. – 200 C.E.), the test batteries were in common use. Tests were well developed by the Ming Dynasty (1368-1644 C.E.) during which the examination become more formal and national multistate testing programme evolved. Those who did well on testing at local level, went to provincial capitals for extensive essay examinations waiting ahead. After this second round of testing, those with highest test scores went to nation's capital for final round. Those who passed this final round of testing were finally inducted into office. Thus upon passing each level of examination, people received more titles and more power in civil service examination (Bowman, 1989). In fact, civil service examinations began in China during Chan Dynasty in 1115 B.C.E. and ended in 1905 when a reformatory measure abolished the system. At that time, the western worlds such as France, Britain and America were very much influenced by the Chinese system. This encouraged the English. East India Company in 1832 to apply the Chinese system as a method of selecting employees. Since this testing worked well, the British government adopted a similar system of testing civil service in 1855. After good results of such testing, French, German and American government followed the suit. They developed and conducted competitive examinations for selecting employees for government jobs. Among the ancient Greeks, testing was very frequent in educational process. In fact, tests were used at that time to determine the level of intellectual and physical skills. In 400 B.C.E. Plato had suggested that people should work at jobs consistent with their abilities and endowments. In 175 B.C.E. Galen conducted experiment to demonstrate that it is the brain and not heart that is the seat of intelligence.

2. Crucial scientific development in between 1800 A.D to 1900 A.D

Psychological testing, during the 19th century, flourished a lot due to a strong awakening of interest in humane treatment of mentally retarded and insane persons. In simple words, psychological testing owes much to early psychiatry. The establishment of many institutions for caring mentally retarded persons in Europe and United States of America necessicated an objective system of classifying such patients. As a consequence Esquirol, a French physician, for the first time pointed out that there are different degrees of mental retardation and for developing some system of classification of the different degrees and types of mental retardation, he developed some procedures and pinpointed to the fact that the language of the persons is the best indicator of his intellectual level. In fact, he pointed out the first classification of mental retardation. He recognized three levels of mental retardation: (a) those using short phrases, (b) those using only monosyllables, and (c) those without speech but cries. Another important French physician name Seguin, student of Esquirol, also contributed a lot. He rejected the idea that mental retardation is not curable. In 1837 he established a school where mentally retarded children were being educated and trained. For this purpose, he developed performance or nonverbal test of intelligence called *Seguin Form Board Test* in which the person is required to insert blocks of different shapes into the corresponding recesses as quickly as possible. Seguin has also studied with Itard who is well-known for study in which he attempted to train the Wild Boy of Averson, a child who had lived in the forest for the first 11 or 12 years. In 1885 the German physician Hubert Von Grashey developed what is called antecedent of memory drum for testing brain-injured patients. Shortly after this, the German psychiatrist Conrad Riegner developed test battery for identifying brain damage. But these early tests were not standardized and therefore, they were relegated to oblivions.

In a nutshell, early psychiatric attempts contributed to the development of psychological testing by emphasizing that these tests can help and reveal the nature and extent of systems in mentally ill patients as well as in brain damage patients.

If we pay attention to the details of development of that time, it will be obvious that psychological testing developed from at least two lines of inquiry: One based upon the work of Darwin, Galton and Cattell on the measurement of individual differences and the other (and probably more influential) based upon the important contributions of German psychophysicists like Herbart,

Weber, Fechner and Wundt and their psychophysical measurements. This second set of individuals and their researches have led to the development of experimental psychology and standardized testing.

Although human beings have realized very long ago that there exists individual differences among persons, a scientific approach towards studying and measuring such differences started with the publication of Charles Darwin's very important and influential book entitled, *The origin of species*' in 1859. Darwin was of view that some members of a species possess characteristics that are more adaptive or successful in a given environment than others. He further pointed out that those with best adaptive characteristics tend to survive at the expense of those who were less fit and the survivors pass their major characteristics on to the next generation. This was called by Darwin as *the survival of the fittest*. Through this process, life has evolved to its present form or level.

Later on Sir Francis Galton, a relative of Darwin, began to apply Darwin's theory to the study of individual differences among human beings. Galton extended the work of Darwin and set out to demonstrate that some people do have characteristics that made them more fit than others. His theory was developed and published in his famous book entitled *Hereditary Genius* in 1869. He concentrated on displaying through a series of experimental studies that the individual difference exists in human sensory functioning and motor functioning such as reaction time, visual acquity and physical strength. Through these experimental studies, Galton, in fact, initiated a research for gaining insight into the individual differences. Galton's another book entitled '*Inquiries into Human Faculty and its development*' published in 1883 contained a series of essays that emphasized individual differences in mental abilities. According to Boring (1950) *Inquiries* was the beginning of the mental test movement and scientific psychology of individual differences. A further contribution of Galton was in the development of statistical methods for the analysis of data obtained from individual differences. This phase of work was further carried by one of the brilliant students of Galton, that is, Karl Pearson.

The US psychologist J.M. Cattell extended the work of Galton. In fact, Cattell received enthusiastic support from Galton for studying individual differences. Cattell opened his own laboratory and developed a series of tests that were considered mainly extension and additions to Galton's test battery. In fact, Cattell's doctoral dissertation was based on Galton's work on individual differences in the field of reaction time. He also coined the term *mental test* in 1890 in his famous paper entitled *Mental tests and Measurements* (Kaplan & Saccuzzo, 2010) In this paper he described ten mental tests for use with the public for assessing their intellectual level. There tests, to be administered individually, includes measures of speed of movement, sensitivity to pain, muscular strength, weight discrimination, reaction time, memory and the like.

The second major foundation of testing (more popularly known as German stream) can be found in the field of experimental psychology where attempts were made to know about consciousness through the scientific methods. Before psychology emerged as a science, mathematical models of mind were developed and J.E. Herbart frequently used these models as the basis of educational theories, which had a great impact upon 19th century educational practice. After Herbert, E.H. Weber through several studies, was able to show that there existed a minimum necessary level to activate a sensory system. This minimum level was called as a *psychological threshold*. Thus building upon Herbart's work, Weber tried to prove the existence of a psychological threshold, claiming that a minimum of stimulus was necessary to activate a sensory system. Following Weber, G.J. Fechner proposed the law that the strength of a sensation produced by stimulus grows as the logarithm of the stimulus intensity.

Wilhelm Wundt who is credited with founding the first psychological laboratory in 1879 at the University of Leipzig (now Karl Marx University), Germany, followed the tradition of Weber and Fechner (Hearst, 1979). From his work also emerged the idea that, like experiment, testing also requires vigorous experimental control. Such control comes from administering tests under standardised conditions.

In an article published in France in 1895, Alfred Binet and Victor Henry did a pivotal review of German and American work on individual differences. They criticized most of these tests attempting to assess individual differences as being too largely sensory and as concentrating heavily on simple, specialized abilities. They expressed the idea that intelligence could be better measured with the help of higher psychological processes rather than by means of elementary sensory processes such as reaction time. They were of view that the intelligence could be better measured by covering such functions as attention, imagination, memory, comprehension, suggestibility, asthetic appreciation and the like.

3. Crucial Scientific Development between 1900 AD to 2000 AD

The above trends were clearly reflected when Alfred Binet and T. Simon constructed and published the first modern intelligence test in 1905 popularly called as *Binet-Simon Scale*. This scale consisted of 30 items of increasing difficulty values and was designed to identify mentally retarded children. In 1908, Binet and Simon published a revision of the 1905 scale. In this revision very simple and unsatisfactory items were dropped and new items were added towards the higher end of the scale. The 1908 revision had 58 problems or items, almost double the number from the 1905 scale. All problems or items of the test were grouped into age levels on the basis of performance of about 300 normal children of 3 to 13 years. One of the most important innovation of the 1908 scale was the introduction of the concept of mental level, which was soon translated as *mental age*. In 3-years level were placed all those problems or items of the test passed by 80 to 90 percept of normal 3-years olds; in 4-year level, all items similarly passed by normal 4-years olds and this practice was continued till 13 years of age. A child's score on the entire test could then be easily expressed as a mental level corresponding to the age of normal children whose performance he or she equated. In a very simplified form, a psychologist can easily think of mental age as a measurement of child's performance on the test relative to the children of that particular age group. For example, if a child of 6 years obtains the test performance, which is equal to that of 8-years, then his mental age would be of 8 years although his chronological age was that of 6 years. This concept of mental age was one of the most important contribution of the revised 1908-Binet-Simon Scale.

The third revision of Binet-Simon Scale appeared in 1911, the unfortunate year of Binet's untimely death. In this scale, some minor changes and relocation of specific items of the test were instituted. Each age level had now five problems or test items. The scale was also extended to adult level. Very shortly, William Stern in 1912 suggested that an intelligence quotient (IQ) derived from mental age divided by chronological age would provide a better measure of relative functioning of a child compared to his age mates.

Even prior to 1908 revision, Binet-Simon Scale had obtained wide popularity throughout the world. Therefore, many translations and adaptations had appeared in different countries, including USA. In USA, the first translation was done by H.H. Goddard who was then a research psychologist at Vineland Training School. He had published a translation of the 1905 scale in 1908 and the 1908 scale in 1911. Those Goddard revisions were very much instrumental in getting acceptance of intelligence test by medical experts. Others US test developers subsequently published many versions of the scale. However, it was in 1916 that under the direction of L.M. Terman at Stanford University the most significant revision was done and it was called as *Stanford-Binet Intelligence Scale* or *Stanford-Binet* or *Simply the Binet*. The 1916 Stanford-Binet was extensively revised and changed in the 1930s. Two forms of the scale Form L and Form M were produced. In 1960 a third edition of Form L and Form M was published. This revision included the best items from the 1937 revision. New norms were published in 1972. However, there was no change in the test. In fact, the first most significant and real revision of the scale during last 50 years appeared in 1986. This was the fourth edition (Thorndike, Hagen and Sattler, 1986a, 1986b). It also included performance test the first time. A fifth edition of Stanford-Binet intelligence Scale appeared in 2003 (Roid, 2003) sue fifth edition (SBS) yields scores for five factors, each based one on verbal test and one non-verbal test. Four additional scores can also be obtained: a *verbal IQ*, a *non-verbal IQ*, a *full scale IQ* as well as an *abbreviated scale IQ*.

A mere two years after 1937 revision of Stanford-Binet Scale, David Wechsler (1939) published another important intelligence scale called *Wechsler-Bellevue Intelligence Scale*, which contained several interesting innovations in intelligence testing. In fact, Wechsler challenged the supremacy of Stanford-Binet intelligence scale as a measure of intelligence. He had strong objection to the single score offered by the 1937 Binet Scale because he insisted that intelligence was concerned with so many different and varied abilities. Wechsler also capitalized on the inappropriateness of Stanford-Binet scale as a measure of adult intelligence.

Group Testing :

The Binet Scale as well as its later revisions were all individual tests because they can be administered to one person at a time. Moreover, they also need a highly trained test administrator or examiner. However, at the time of World War I, there arose a demand for quick and efficient way of evaluating the emotional and intellectual functioning of thousands of military recruits in United States of America. To meet this demand, need for group testing was realized. World War I and need for group tests had, then, added momentum to the psychological testing movement. At the request of Army, Robert Yerkes who was then the president of the American Psychological Association, headed a Committee of distinguished psychologists. This Committee soon developed two group tests of human abilities: *Army Apha Test* and *Army Beta Test*. The Army Alpha Test was meant for measuring the intelligence of literate adults and the Army Bet Test, being a nonlanguage test, was meant for measuring the intelligence of illiterate adults. Bash tests were suitable for administration to large groups. Soon after World War I was over, these Army tests were released for civilian uses. These tests also passed through many subsequent revisions and served as models for several group intelligence tests.

Aptitude Tests

Intelligence tests measured only the general intellectual level. At that time there was emerging a realization to measure some specific abilities. As a consequence, aptitude tests were developed. At that time the technique of factor analysis popularized by Charles Spearman (1904, 1927) and further refined and developed by Kelley (1928) and Thurstone (1938, 1947) paved the way for measurement of specific ability (or aptitude) in a better way. One of the chief practical outcome of factor analysis was the development of *multiple aptitude batteries*, which generally yielded in place of a total score or IQ, a separate score for several specific abilities like numerical aptitude, verbal comprehension, arithmetic reasoning, perceptual speed, spatial visualization and the like Two important multiple aptitude batteries of that time were very much influential in promoting psychological testing. They were: *Differential Aptitude Test* (DAT) Battery and *General Aptitude Test Battery* (GATB). DAT was originally published by the Psychological Corporation in 1947 as a guidance battery for using with students at secondary level and were subsequently revised in 1963, 1972 and 1990. DAT has will subtests like verbal reasoning, numerical reasoning, abstract reasoning, spatial relations, mechanical reasoning perceptual speed and accuracy, spelling and language usage. GATB was developed by the Bureau of Employment security of US Department of labour in the early1940s. In most recent forum, it measured 9 different specific abilities such as general mental ability, verbal aptitude, numerical ability, spatial aptitude, form perception, clinical perception, motor coordination, manual dexterity, finger dexterity. These nine factors are assessed through 12 tests. The factor of general mental ability is measured with the help of scores on three tests, namely, vocabulary, arithmetic reasoning and three dimensional space.

In late 1980s and early 1990s a new tendency to integrate two separate approaches to mental measurement, one represented by traditional intelligence test and other by multiple aptitude batteries was visualized (Anastasi, 1994). As a consequence, more recently developed intelligence test such as the Differential Ability tests and fourth and fifth editions of Stanford-Binet test tended to combine comprehensive coverage of multiple aptitudes with multi-level scoring for some specific testing purposes.

Achievement tests

Following World War I the development of some standard achievement tests was a very important development in the field of psychological and educational testing. Prior to standardized achievement tests, essay type tests were popular for assessing scholastic achievement. The standardized achievement tests were equipped with multiple choice questions that were standardized on a large sample to produce norms. Such tests became very popular due to the relative ease of administration, and scoring as well as due to objectivity. By 1920 it became an established fact that essay tests of assessing scholastic achievement were not only subjective and time consuming but also yield less reliable results than standardized achievement tests. In 1923 the development of standardized achievement tests culminated in publication of Stanford Achievement Test by Kelly, Ruch and Terman, the three early leaders in the test development. Among standardized achievement tests, both group standardized achievement tests and individually administered achievement tests were developed. The Stanford Achievement Test series and the California achievement tests were the two most popular group standardized achievement tests whereas Peabody Individual Achievement Test, one of the first such test, was from the individual achievement test side. In the decade of 1930s, test-scoring machines were introduced and the new objective tests easily adapted such machines for scoring. This also enhanced their objectivity. Achievement tests were not only used for educational purposes rather they were also used for selection of applicants for government jobs as well as industrial jobs.

Personality tests

Personality testing is another area of psychological testing and it is concerned with non-intellectual or affective aspect of behaviour. Although the rudimentary assessment methods of personality through free association technique was done by Francis Gulton before the turn of 20th century, it was Emil Kraepelin who made a through early attempt of personality testing through free association test in a scientific way. He used this technique with mentally ill patients who were required to respond to each selected stimulus word with the first word that comes to their mind. Later, Sommer also suggested that the technique of free association could also be used to differentiate between various forms of mental disorder.

The earliest scientific approach to assessment of personality was through the questionnaire or self-report inventory, which is still today a very popular measure of personality. The first such questionnaire was the *Personal Data Sheet* (PDS) developed by Woodworth during World War I. This test consisted of 116 questions (originally 200 questions) to be answered by the respondents by underlying *Yes* or *No*. The purpose of the test was to identify seriously mentally disturbed men who would be disqualified for military service. The Woodworth Personal Data sheet served as one of the important models for development of later personality questionnaire especially measuring emotional adjustment and it was followed by the creation of varieties of structured personality tests. After World War I, Woodworth developed new version of PDS and named it as *Woodworth*

Psychoneurotic Inventory. This inventory was designed for use with the civilians, was the first self-report tests and first widely used personality inventory.

One such important structured personality test was *Minnesota Multiphetic Personality Inventory* (MMPI) developed by Hathaway and Makinely (1940). Subsequently, MMPI has been revised and framed in two separate versions: MMPI-2 (Butche, Dahlstrom, Graham, Tellegan & Kaemmer, 1989) and the MMPI-Adolescent or MMPI-A (Butcher, Williams, Graham, Archer, Tellegen, Ben-Porath, & Kaemmer, 1992). MMPI has some clinical scales and also introduced the use of validity scales for determining fake-good, fake-bad and other random response pattern. Recently, Ben-Portath and Tellegen (2008) have published a modification of MMPI-2 called MMPI-2-RF (Restructured Form) which consists of 338 True-False items. Likewise, MMPI-A has been modified by Archer, Handel, Ben-Porath and Tellegen (2016) and called as MMPI-A-RF. This consists of 241 True-False items and is meant for 14 to 18 years of adolescents.

Using the technique of factor analysis in the early 1940, Comilford made one of the first serious attempts to develop a structured personality test. Further, Cattell (1950) developed the Sixteen Personality Factor Questionnaire (16PF) to measure sixteen source traits of personality. The inventory made a significant contribution to the development of psychological testing. Still another and very latest attempt to factorize and reduce the various lists of factors of personality has been undertaken by Costa and McCrae (1992) who developed a questionnaire called the NEO-Personality Inventory Revised (NEO-PI-R) for measuring Big Five personality factors such as neuroticism, extraversion, openness to experience, agreeableness and consciousness (OCEAN- an easy way to remember all the five dimensions).

Another approach to the assessment of personality was undertaken through use of *situational tests* or also called as *performance test*. In this test, the person is required to perform or do a task whose purpose is often disguised. One such test meant for assessing cheating, lying, stealing, cooperativeness etc. by school children was developed by Hartshrone, May and their colleagues (1928, 1929, 1930). Another important situation test meant for adult was developed during World War II in the assessment program of the office of the Strategic Service (OSS, 1948). Such test attempted to assess subtle social and emotional behaviour of the person.

Still another very significant approach to assess personality was the development of projective tests. In projective tests the test takers are given some unstructured tasks to which they respond. The assumption remains here that the test takers will project his characteristic mode of response into such a task. The various types of projective tests were developed. Among these, Words association test, Rorschach test, Thematic Apperception test and Sentence Completion Test were very popular.

Carl G. Jung as early as in 1904 used words association as method of uncovering the emotional complexes hidden in personality. He presented a standard list of 100 words to his patients. While Jung's test became popular in clinics in Europe, a list developed by Grace Kent and Aaron Rosanoff in 1910 became popular in United States. Subsequently, David Rappaport and his colleagues in 1946 prepared a list of 60 words based on psychoanalytic theory of personality.

The most highly developed and the most widely used projective test was the Rorschach Test, which consisted of 10 inkblots and was published in 1921 by Hermann Rorschach, a Swiss psychiatrist. All the ten inkbots represented a completely ambiguous stimuli to which the test taker has to respond. Unfortunately, Rorschach died only a year after his monograph had been published, that is, in 1922 and now it was up to others to complete his leftover work.

Another test that added to the momentum of the development of projective test was *Thematic Apperception Test* (TAT) developed by H. Murray and C. Morgan in 1935. TAT was developed to study normal personality. The test consists of a series of pictures that depicted some persons engaged in a vague interaction. The test-taker is asked to write a story describing past, present and future of what is depicted in the picture. Murray (1938) had assumed that the underlying various needs such as need for achievement, need for affiliation, need for aggression, etc. of the persons would be revealed by the contents of the stories.

The sentence completion tests were also developed as a projective measure during this period. In this type of test, the test takers are presented with several incomplete sentences with a request to complete them from their own side. The first serious attempt to assess personality using this technique begun in the 1930s and one of the earliest tests of this type appeared in 1940 by Amanda Rohde and Gertrude Hildreth. Later on during World War II, several attempts were made to develop sentence completion tests. Rotter Incomplete Sentence Blank is another addition in this series (Rotter & Rafferty, 1950; Rotter, Lah & Rafferty, 1992). A relatively recent addition to the family of sentence completion test is the Incomplete Sentence Task developed by Lanyon and Lanyon 1980).

At that time an entirely new approach to projective testing was undertaken by Goodenough (1926). Her *Draw-A-Man test* published in 1926 and later updated by Harris (called Goodenough-Harris version of the test) not only measures intellectual capacity but also interest and personality traits of the children by analyzing their drawings. Machover (1949) published the *Draw-A-Person* test and Buck (1948) published the House-Tree-Person test (HTP test). In HTP test the test-taker is asked to draw a house, a tree and a person and after analysis of their drawings, attempt is made to know about their personality traits.

Interest and Value Tests

Interest tests (also called interest inventories) provide information about an individual's general pattern of likes and dislikes. The root of the interest inventory goes to the Thorndike's study of developmental trends in the interest of 100 college students in 1912. However, the scientific and the most popular attempt to develop interest inventory was made by E.K. Strong who developed what was called *Strong Vocational Interest Blank* (SVIB) in 1927. Subsequently, SVIB went into several revisions and the latest edition has been completed in 1994 and renamed as simply *Strong Interest Inventory* (SII). The SII has 377 items presented in three different formats. Most of the test items require the test taker to respond to by selecting L (like), I (indifferent) or D (dislike).

A serious competitor to SII was the interest inventory developed by F. Kuder. The earliest was named as the *Kuder Preference Record – Vocational* that used forced-choice triad items and provided score for 10 broad interest areas. Later this was revised and named as the Kuder General Interest Inventory (KGIS) and it was downward extension of *Kuder Preference Record-Vocational*. A still later version came in 1991 and it was called the Kuder Occupational Interest Survey, Form DD (KOIS) (Kuder & Zytowski, 1991).

Towards this direction, another significant and the most popular work was done by Holland (1985, 1997) who developed what is called Self-Directed Search (SDS) which was a self-scoring instrument used to classify oneself according to six personality types or occupational themes (realistic, artistic, investigative, conventional, social and enterprising) that he identified in his research with E.K. Strong. This test has been recently revised in 1994 and it consisted of two bookets – *Assessment booket* and *Occupations finder* and can be completed in about 40 to 50 minutes.

Jackson (1977) developed still another popular inventory called Jackson Vocational Interest Survey (JVIS), which contains 34 basic interest scales covering 26 work roles and 8 work styles. JVIS can be administered to both sexes. A high score on any of 34 scales indicates a strong interest in the things people do in a particular field of work.

Major Development in Psychological and Educational Testing in India

Historically, work in the area of psychological and educational testing had started more or less about the same time as it had started in America and England. These early works were mainly devoted to the development of individual intelligence tests. Here, it will be worth to mention the name C.H. Rice who made the first attempt to standardize the Binet test in Urdu and Punjabi during 1930s. At about the same time, Mahalanobis attempted to standardise Binet's test of intelligence in Bengali. In fact, in India, upto 1921, work relating to construction and development of psychological tests were done by Christian missionaries, which were engaged in educational work. Among such research workers, E.L. King at Narsingpur, and D.S. Herrick at Bangalore were popular ones. Besides C.H. Rice, work on Binet test was also undertaken by Miss Gordon at Madras, Spence at Jabalpur and west at Dacca. J.M. Sen and G. Dasgupta of David Hare Training College at Calcutta were first to devise group tests based on Otis. Papers relating to this were presented at Indian Science Congress in 1925 and 1926. Some Indian psychologists started developing Indian norms of some foreign tests like Raven's Progressive Matrices, WAIS, Alexander's Passalong, Cube construction, Koh's Block Design, Good enough test and Minnesota Paper Forms Board. However, it was only since 1950 that published evidence points to the construction of Indian tests (Mitra, 1972).

In India three important institutions have the traditions of developing various psychological tests in the field of intelligence, aptitude, personality, attitudes interests and values. They are: National

Council of Educational Research and Training (NCERT) Indian Statistical Institute (ISI), Kolkata and Defence Institute of Psychological Research (DIPR), Delhi.

So far as the NCERT is concerned, it publishes handbooks on mental measurement that contains discipline about different tests. The National library of Educational and Psychological Tests (NLEPT) at NCERT has documented Indian psychological and educational tests. According to its 2016 release, NLEPT has documented about 800 Indian psychological and educational tests in the various fields such as intelligence, personality attitudes, interest, values creativity, achievement, guidance and counselling, etc. NCERT first handbook on mental measurement (1991) contained 36 tests. Of these 36 tests, 34 tests were intelligence tests and the remaining two were aptitude tests. Important intelligence tests developed by Indian psychologists are *Group Test of Intelligence* by Prayag Mehta, *Group Test of General Mental ability* by S. Jalota, *Group Test of Intelligence* by Bureau of Psychology, Allahabad, *Adaptation of Wechsler Adult Performance IntelligenceScale* by Prabha Ramalingaswamy, *Group Test of Intelligence* by Pramila Ahuja, *Bihar Test of Intelligence* by S.M. Mohsin, Draw-a-Man Test by Pramila Pathak, *Test of General Mental ability* (Hindi) by M.C. Joshi, *Bhatia Performance Test of Intelligence* by C.M. Bhatia, *Gujarati Adaptation of Wechsler Intelligence Scale for Children* by M.C. Bhatt, *Social Intelligence Scale* by M.K. Chadha and Usha Ganesen and the like. NCERT's second handbook (1998) contains different psychological tests on personality and the third (2001) contains various tests on values, attitudes and interests. The 2016 version of NLEPT has documented such 240 tests. Among important personality tests are *Differential Personality Inventory* by A.K. Singh (Author), *Indian adaptation of Mardsley Personality Inventory* by S. Jalota and S.D. Kapoor, *Indian adaptation of Cattell's 16 P.F. Questionnaire* by S.D. Kapoor, *Psychological Hardiness Scale* by A.K. Singh (author), *Psychological well-being Scale* by D.S. Sisodia and Pooja Chaudhary, *Hindi adaptation of Bell's Adjustment Inventory* by S.M. Mohsin and S. Hussain and many more. Likewise among various interest tests, *Vocational Interest Record* by S.P. Kulshreshtha, *Multiphasic Interest Inventory* by S.K. Bawa and *Non-language Preference Record* by S. Chatterjee are important ones. Similarly among value tests, the *comprehensive value scale* by K.G. Agrawal, and *Value orientation Scale* by M.S. Chauhan and S. Aurora are important. Among attitude scales, the *Attitude Scale towards Religion* by R.K. Ojha and *Attitude Scale towards Family Planning* by M.A. Hakim and Y. Singh and *Religiosity Scale* by I. Bhushan are comparatively more popular.

ISI, Kolkata has a psychological research unit where psychologists do research on the application of statistics in test development as well as in explaining other related phenomena: Psychologists of this unit developed 11 tests, which relate to the assessment of achievement for neo-literates, achievement for primary school children, non-language intelligence, managerial skills, entrepreneur task motivation, prolonged motivation, computer programming attitude, reading and writing motivation, achievement in Bengali, Mathematics and environmental studies and aptitude for screening professional outderess.

DIPR that became a full-fledged institute of Defence Research and Development Organisation (DRDO) in 1982 originally it came into existence in 1943 at Dehradun, has successfully developed various psychological tests meant for personnel selection, placement, trade allocation, intelligence, organisational behaviour and the like. It has also developed aptitude tests for assessing abilities to work in inhospitable terrains. More than 66 psychological tests in difference areas lie to the credit of DIPR. This institute has also successfully used projective tests in selection of officers in the armed forces.

Types of Tests

There are several types of tests. They have been classified on the basis several criteria. Here a brief review of those classifications is being presented.

1. On the basis of *administrative conditions*, there are two categories of tests– individual test and group test. Individual test is test, which is administered to one person at a time. Administering individual tests generally require that the test administrator have much training and experience. Such tests are often administered by school psychologists and counsellors for knowing motivations of young children as well as for observing how the child responds. Kohs Block design test is a good example of individual test. Group tests are those tests which can be administered to many persons at a time. In a typical classroom situation, group tests are adequate for assessing cognitive skills to know about pupil achievement, strengths and weaknesses. Since individual tests are time-consuming and require the services of trained and experienced test administrators or examiners, group tests, in general, are preferred.

2. On the basis of *scoring criterion*, tests may be classified as *objective tests* and subjective tests. When scoring criteria are well-defined, unambiguous and well-agreed upon so that the subjective judgements among the scorers are minimized, the test is said to be objective one. Such tests are structured tests. Multiple-choice, true-false, and matching items are often included in objective tests because they can be objectively scored according to a predetermined key. Stanford-Binet Intelligence Scale, MMPI, the NEO Personality Inventory are the examples of objective test. Subjective tests are those tests where the test takers provide short answer in a few words or sometimes provide an extended answer. They cann't be scored objectively and scores differ among themselves because they provide greater freedom in scoring. Some items that may appear to be objective may prove to be highly subjective. For example, there is an item like this on an arithmetic test:

Divide 16 by 8 or $16 \div 8 = ?$

A student solves this like :

$$\begin{array}{r} 2 \\ 8\overline{)16} \\ \underline{15} \\ 0 \end{array}$$

(65)

How many points will be given to this answer? If some scorers give it half credit and some full credit or no credit, the item is said to be *subjectively scored*. Projective tests such as Rorschach test, TAT, sentence completion tests belong to subjective test where the role of test taker is less clear as compared to objective test.

3. On the basis of the criterion of response limitations, tests are classified as *power test* and *speed test*. A power test is defined as test, which has generous time limits so that most test takers are able to attempt every item. Items of power test vary in difficulty and most of the items are difficulty in nature. Such tests tend to demonstrate how such knowledge or information a person has. The annual examination taken by school and college authorities are examples of power tests. Speed tests are those tests, which have strict time limits and the items are so easy that only a few test takers are expected to make errors. Speed tests emphasize upon the fact that how rapidly the test taker can respond within a restricted time limit to various easy items. Clerical speed and accuracy tests are examples of speed tests.

   Whether a test is power test or speed test is partly determined by the nature of tests takers for whom it was designed. For example, an arithmetic test meant for class three students will be a power test for such students but will be an example of speed test for M.Sc. students.

4. On the basis of criterion of *contents or items*, there are four types of tests: *verbal test*, *non-verbal test*, *performance test* and *non-language test*. A verbal test is one which emphasizes upon reading, writing or oral form of communication. Most of the psychological tests are verbal in nature. Such tests are usually administered in group. MMPI, Mohsin General Intelligence Test and Singh Differential Personality inventory are the examples of verbal test. A Non-verbal test is one that deemphasizes but don't altogether eliminate the role of language by using pictures, diagrams, figures or other symbolic materials. Here language is usually used in instructions but items deemphasizes the role of language. Raven's Progressive Matrices is one example of non-verbal test. Non-verbal tests are commonly used with young children and illiterates for assessing non-verbal aspects of behaviour such as spatial perception. Performance tests are those tests where the test takers are required to manipulate some concrete objects in order to show their performance. The items of the performance test have some concrete objects and instructions are usually given orally or through gesture or pantomime. Kohs Block design Test, Pass Along Test and Cube Construction Test of intelligence are the good examples of performance test. The primary difference between nonverbal test and the performance test is that the former have items in form of pictures, diagram, etc. which cannot be physically manipulated by the test takers whereas the latter contains some concrete objects such as blocks, puzzles and cubes made of wood, plastics etc. that are manipulated for solving the given problems. The similarity between non-verbal test and performance test

is that both can be easily administered on young children and illiterates with almost equal accuracy.

Non-language tests, as their name implies, are completely free from any form of language. They don't use language either in instruction or in item. Instructions are usually given through gestures and pantomime and the test takers respond by pointing at or by manipulating pictures, blocks and puzzles. Such tests are said to be most suitable for persons or children who have difficulty in communicating with the help of any form of language.

5. On the basis of the criterion of standardisation, tests have been classified as either *teacher-made tests* or *standardized tests*. Teacher-made tests are those tests which are constructed by teachers in a less formal manner for single use within the classroom. In fact these are non-standardized tests having no standardization sample. Here objectives are very specific to meet the needs of the students in a given class. The contents of such test do come from any area of curriculum and items of the tests may be modified as desired. Rules for administration and scoring are determined by the teacher. They are kept uniforms within the class. No norms are provided for such tests but they may be developed by the teacher for his own class. Evaluation of such tests is determined by the teacher. The development such tests usually includes the planning of the test item writing and item analysis (Anastasi & Urbina, 1997).

Standardized tests are those tests which are constructed by psychometricians and test specialists. They are standardized in the sense that they have been administered and scored under standard and uniform testing so that the results from different situations may be compared. Here objectives are not specific rather general. Here items are fixed and not modifyable. Rules for administration and scoring are fixed in the manual and must be followed in verbatim. Norms are also provided for making a meaningful comparison. Most of the psychological tests are standardized tests.

6. Based on the criterion of *referencing* to some external status, there are two types of tests: *criterion-referenced tests* and *norm-referenced test*. A criterion-referenced test is one in which the performance of the test taker is compared with some criterion established by expert or teacher (such as 20 words correctly spelled within two days or a minimum of 50 words typing per minute, etc.) The norm-referenced test is one in which performance of the test taker is compared with some specified reference group (such as other members of the class or group). Norms-referenced tests can demonstrate that the test taker has more or less knowledge, ability or interest in comparison to other members of the reference group. Such tests should *maximise* the differences among the individual. Maximizing individual differences mean that the different scores among individual reflect actual differences in ability and differences among scores are as large as possible. In contrast to norm-referenced test, criterion-referenced test compare the individual not in

relation to the others but in relation to the level of the performance he is expected to arrive at. For example, a student can be required to attain minimum level of mastery (such as 55 percent marks or 60 percent in Master degree or 75 percent for getting double promotion in class V) before he is allowed to proceed ahead, that is, to appear for national eligibility test or to be promoted in next class. A student's failure to reach the criterion standards may suggest diagnostic testing for discovering reasons for the lack of achievement. The criterion-referenced tests are usually helpful in the following three situations:

(a) In those areas that demand competence and proficiency, criterion-reference tests are more appropriate. For example, if a doctor is going to do surgery, he must be competent enough to meet the minimum standards. Likewise, to provide license for driving four wheelers, it is essential that the concerned person must meet the minimum standards.

(b) In those areas that are cumulative and becomes progressively complex, the students (or anybody) must reach some minimum level of proficiency on the preceding task before being promoted to the next higher level. For example, if a student of class three needs double promotion, he must be able to achieve 75 percent marks in the annual examination.

(c) In diagnostic work also criterion-referenced tests are more appropriate. Using such test, the teacher is able to know that a particular student has not been able to meet the specified criterion of academic achievement and therefore, a diagnostic testing is suggested to discover reasons for the lack of achievement.

Norm-referenced tests, on the other hand, have proved to be useful in the following circumstances:

(a) When tests are used to predict some degree of success, they should be norm-referenced tests. Since individuals are expected to differ on whatever criterion is used to predict success, norms-referenced tests are appropriate because such tests are capable of making distinctions among students.

(b) If the authority is interested in accepting the only highest-performing individuals, norms-referenced tests are more appropriate.

(c) When the situation is such that the subject-matter is not cumulative and the student is not acquired to meet a certain criterion such as a certain degree of proficiency or competence, norm-referenced tests are more appropriate.

7. On the basis of criterion of dimension measured, psychological tests are of many types such as aptitude tests, achievement tests, intelligence tests, personality tests, neurophysiological tests interest tests, creativity tests, etc. These are called dimension because they are wider and broader than single attribute or trait. A discussion of all these tests will be done in separate chapter later on.

8. Psychological tests may also be classified as tests of maximal performance, behaviour observation tests and self-report tests. Tests of maximal performance are those tests,

which require the test takers to perform the assigned task in as well-defined manner as he can maximally do. Test takers also try to do their best because their success is determined by their scores on the test. Examples are intelligence test aptitude test, driving tests, achievement tests, etc. In behaviour observation tests, the behaviours of the people are observed in a typical situation. Sometimes the person does not know that their behaviours are being observed and there are no single tasks defined for the person. For example, when the job performance of the employees in any organization is observed for performance appraisal, it constitutes the example of behaviour observation test. Self-report tests are very common and are defined as those tests where the test takers report about their feelings, beliefs, opinions, etc. Many personality inventories are examples of self-report tests.

Thus it is clear that psychological and educational tests have been classified from different angles and it is essential to have a global look on them for a better understanding.

Major characteristics of a good psychological Test

Any psychological or educational test must have some defined characteristics or attributes if it is to be qualified as a good and standardized test. The following are such important attributes:

1. Reliability

A test must be reliable one. By reliability is meant consistency of test scores. This consistency can be obtained by two administration of the same test on the same sample with a reasonable time gap. This is called as temporal consistency or external consistency. Besides, the consistency may be obtained by the single administration of the test. In such situation the test is usually divided into two equal halves and subsequently, the two halves are correlated. This is called as *internal consistency*. Reliability includes both types of consistencies. Higher the coefficient of reliability of the test scores, better is the dependability of such test.

2. Validity

Validity is another important characteristic of the test score. Validity, in simple words, refers to the truthfulness of the test. When a test correctly measures what it intends to assess, it is said to be valid one. If a test is not valid or poorly valid, the researcher has little option to proceed ahead with such test.

3. Norms

Norms are another important characteristics of a psychological or educational test. They are defined as the average performance of a typical representative sample or group. Norms provide meaning to the interpretation of scores. In absence of norms, the obtained score on the test can't be meaningful interpreted. Norms may be based on age grade percentile and standard score.

4. Practicality

A good psychological or educational test must also have the attribute of practicality. By practicality here means that the test should not be lengthy one, it should not be too difficult or too easy for the test takers, and its scoring system should not be complex and highly specialized one.

5. Objectivity

A good psychological or educational test must possess the attribute of objectivity. By objectivity is meant interpersonal agreement among the psychometricians or experts regarding the meaning of test items and their scoring. Items are said to be objective when they are interpreted by all the test takers in an uniform way and there remains no vagueness in the meaning of items. By objectivity in scoring is meant that test is unimorfy scored by different scorers at different times.

Difference among psychological tests, psychological measurements and surveys

The terms psychological test, psychologicalmeasurement and surveys are often get confused by the average student. So, it is essential to draw a line of distinction. Psychological test and psychological measurements are not synonymous although there exists overlapping in their meaning. By definition, measurement is nothing but simply assignment of numerals according to some rules. The end product of measurement is some numerical value describing the behaviour measured. In psychological tests the test takers are required to answer some questions or perform the given task for measuring personal attributes or abilities. It is important to remember here is that the test takers' answers to questions or their performance on the task are initially not expressed in terms of among physical units. However, scores are derived according to some predermined methods of scoring. In some cases the end product of a psychological test is not a derived score rather simply a verbal description of the behaviour of the person. For example, the popular personality test Myers-Briggs Type Indicator (MBTI) don't produce overall scores rather produces a profile although it has some rules for scoring and summarising information. Any psychological test can be considered psychological measurement when the sample of behaviour assessed can be expressed in a numerical score. Most psychological tests, however, produce a numerical score after assessing the sample of behaviour.

Therefore, the conclusion is that most psychological tests are measurements no doubt, but not all psychological tests strictly meet the definition of measurement.

Psychological test and surveys are also confused most of the time. The similarity between these two is that both are used to collect information about the respondents. However, the two differs. First, whereas surveys focus on group outcomes, the psychological tests focus on the individual outcomes. Since psychological tests provide information about the individual differences, they help in taking important decision about the individual or person. For example, when a psychologist comes to know on the basis of the intelligence test that the child is mentally retarded, he is able to take important decision regarding placing such child in special classes. Surveys tend to provide information about the group as a whole, and therefore, help the researcher in making decision about the group. For example, if the producer of a television serial may come to know on the basis of survey that his serial is not liked and praised by most of the viewers, he will try to stop its telecast or will bring some necessary modifications in stories, etc. Second, the overall results on the psychological tests are often reported in terms of some derived score or standard scores but the overall results of the surveys are often reported at the item or question level by calculating percentage of the respondents who selected the particular alternative answer. However, sometimes surveys tend to provide focus on individual outcomes and are constructed using scales. In such instances, surveys come nearer to the psychological test.

Assumptions of Psychological Tests

When psychological tests are used, many assumptions are made. Following are some of such important assumptions :

1. Individuals comprehend items of the test in a similar way (Wiggins, 1973). It means that the test administrator assumes that the test takers are drawing the same meaning to all the items of the test in a similar way.
2. Psychological tests measure what they intend to measure or predict what they intend to predict. This assumption is more popularly known as validity of the test. If the test intends to measure intelligence, it is assumed that the test must be measuring intelligence. If a test intends to predict about the achievement of an employee, it is assumed that it will be predicting such achievement.
3. All individuals behaviours more or less remain stable over time. This assumption is related to the reliability particularly the test-retest reliability of the test. If we administer a test today and after some days say after a fortnight, again administering the same test on the same sample, we assume that the individuals will receive more or less the same score at both administration. This is more so when we are measuring some stable traits or attributes. However, when we are measuring some variable attribute such as mood, we don't expect high test-retest reliability.
4. Persons report about their thoughts and feelings honestly (Wiggins, 1973). When people report about their feelings and thoughts, we assume that they will report their feelings and thoughts honestly. But sometimes people choose not to do so. Sometimes they think that they should respond about their feelings and thoughts in the manner test administer likes them to respond, which automatically means that they are inclined to lie. For example, if the test administrator asks the test takers whether they have ever disobey their parents, they may tell that they have not ever, when they have disobeyed many times. Why? It is because most individuals don't disobey and therefore, the testtakers think that they should say 'No' to this question and test administrator also expects to respond in this way.

5. Test scores represent not only the true ability of the testtakers but also some errors attributable to the testtakers, the test administrator, the test itself and the environment. For example, if the testtaker has obtained a score of 40 on an intelligence test, this score reflects his true ability plus several errors resulting from vague instructions, awkward working of item, errors in administration of the test, fatigue of the testtaker as well as the temperature level of the room in which the test was administered. Therefore, any obtained score on the test is assumed to reflect both true variance and error variance.

6. Testtakers will report accurately about themselves (Wiggins, 1973). When the testtakers are asked to report about their feeling, likes and dislikes, it is assumed that they will tell accurately and they possess ability to do so. For example, if the testtakers are asked to report whether they agree with the statement, "I like journey by train". It is expected that the testtakers will say agree or disagree on the basis of not only their immediate feelings but also on the basis of their past feelings and thus, whatever their response would be, that response will be assumed to be correct and accurate one.

Locating Information About Psychological Tests

There are two types of psychological and educational tests: those developed and standardized in foreign countries and those developed and standardized in India. A student must have access to these various tests by locating suitable source. In fact, prior to 1950s, the test users have very few sources for obtaining such information. Today, however, several such sources are available. Some sources provide detail information about the test including test reviews and detailed bibliographies. However, some other sources provide only general information about the test such as test's name, author name, and publisher name. Some sources include only those tests that are meant for certain groups such as children whereas other sources include a broad range of tests for different types of populations.

The most common sources of locating information about psychological tests developed in foreign country especially USA, Canada, Britain etc., are *Tests in Print* (TIP), *Mental Measurements Yearbook* (MMY), *Tests* and *Test Critiques*. *Tests in Prints* (TIP) is published in multiple volumes and each volume displays descriptive listings of published tests available for purchase. *Mental Measurements Yearbook* (MMY) is also published in multiple volumes and each volume publishes descriptive information about test and test reviews. *Tests* publishes a broad range of tests that includes test titles, author, publisher, test purpose, intended population, cost, and its availability. *Test Critiques* is also published in multiple volumes and each volume contains in-depth information about reliability, validity and procedure of test construction. Each volume contains reviews of frequently published psychological and educational tests.

Besides these popular sources of locating information about foreign-made psychological and educational tests, there are other sources such as *Personality Test and Reviews*, *Tests in Education, Testing Children, Tests and Measurements in Child Development: A Handbook* and *Personality Test and Reviews* is published in multiple volume and each volume contains a bibliography of the personality tests. *Tests in Education* publishes those educational tests that are frequently used by teachers and educational advisors in the field of education. *Testing children* publishes those tests that are used for children. The description of the test includes skills, knowledge and abilities measured by each test. *Tests and Measurements in child Development* publishes the listing of unpublished measures for use with children. Besides these, widely known test publishers are: *Educational Testing Service, Pearson, IPAT, PAR, Hogan Assessment Systems, Psytech International, PSI, Hogrefe, Wonderlic and University of Minnesota Press Division*. Some popular tests like Baron Emotional Quotient Inventory and Baylay scales of Infant Toddler were published by Pearson. Likewise, Rorschach Inkblot test was published by Hogrefe. MMPI was published by University of Minnesota Press Test Division and Cattell 16 PF Personality Factor (PF) was published by IPAT. Bender visual Motor Gestalt Test-II was also published by Pearson.

There are some unpublished psychological tests, which are also used by researchers. There are a number of print and nonprint sources for locating information about unpublished tests. Two of the most popular print sources are the *Directory of unpublished Experimental Mental Measures* and *Measures for Psychological Assessment : A Guide to 3,000 original Sources and Their Applications*. Likewise, three most popular nonprint resources for locating information about unpublished tests are *Tests in Microfiche, PsycINFO Database* and *Heath and Psychosocial Instruments Database* (HAPI).

In India the important sources of locating information about psychological and educational tests are National psychological corporation (Agra), Ankur Psychological Agency (Lucknow) Rupa Psychological Centre (Varanasi) Manasayan (New Delhi) and Psychotronics (Bangluru). All these are publishers and publish tests developed and standardized by psychologists in Indian conditions. They provide general description about the test, which includes the title of the test, author's name, a brief description about the test, their cost and availability. Of these various sources, National Psychological Corporation has a very broader source of information about different types of psychological tests and is one of the most dependable and reliable sources.

Major Assumptions Underlying psychological Testing

As we know, psychological testing refers to all the possible applications, uses as well as underlying concepts of psychological and educational tests (Kaplar & Saccuzzo, 2001). In other words, psychological testing may be understood as the process of administering, scoring and interpreting psychological tests. When we administer a psychological test for assessing something, the process called is psychological testing, which rests upon several assumptions Testing denotes everything from test administration to test interpretation and it has the following underlying major assumptions:

1. There exists psychological traits and states

One major assumption underlying psychological testing and assessment is that psychological trait and states exist. According to Guilford (1959), trait is defined as any distinguishable and *relatively enduring* way in which one person differs from the other person. The term *state* also does the same function, that is, it also distinguishes one person from the other person but it is relatively less enduring (Chaplin et al., 1988). The term psychological traits includes a wide variety of possible characteristics of human beings. Important such psychological traits are intelligence, aptitude, special abilities, personality traits, attitudes, sexual orientation, etc.

Psychological traits don't have physical existence like a circuit in brain although some psychologists have argued in favour such viewpoints (Allport, 1937; Holt, 1971). Today psychologists, in general, agree that psychological traits exist only as a *construct*. A person cann't hear, see or touch constructs and he can infer their existence from overt behaviour, which is defined as an observable action or the product of an observable action. The term *relatively enduring* included in defining trait indicates that the trait is not manifested in the behaviour hundred percent of time. Its manifestation is also determined by the situation or context. In other words, how a particular trait is manifested itself is, partly, situation dependent. For example, Shyam may be viewed as dull, unattentive and cheap by his wife but very much attractive, charming and extravagant by his business loss whom he wants to impress. Not only this, the situation or context within which the behaviour occurs also plays a significant role in assigning appropriate trait name for the observed behaviour. For example, suppose you observe the behaviour of a person who is kneeling and talking to *Hanumanjee*. If you observe such behaviour in a temple, you are most likely to describe his behaviour as *religious*. On the other hand, when you observe the same behaviour in a compartment of a train or on a railway platform, you are likely to describe his behaviour *deviant* or *paranoid*.

2. Psychological traits and states can be measured

Another assumption is that if psychological traits, construct and states exist, they can be quantified and measured. For assessing the traits or states, they need to be first defined. After defining the construct, the test developers write a pool of items that would provide insight into it. Let us take an example. Suppose the test developer has properly defined the construct intelligence and he considers knowledge of Indian history to be one component of adult intelligence, then, the item like *who was the first woman Prime Minister of India?* may appear on the test. Likewise, if he thinks that ability to adjust with new situation is indicative of adult intelligence, then the item like why are some people often afraid of travelling by air? may appear. Subsequently, such items are weighted keeping in view the technical considerations, the way construct has been defined for the purpose of the test as well as the values attached by society to the behaviours being evaluated. After developing appropriate items, the test developer also tends to determine appropriate ways to score the test and interpret the obtained test score. The test score reflects the strength of the targeted trait or construct and is frequently based upon the *cumulative scoring* in which the test takers receive a net score on the basis of cumulation of scores on the individual items.

3. Test-related behaviour predicts non-test related behaviour

Another assumption underlying psychological testing is that tests not only predicts the targeted behaviour but also tells about something that is said to be non-test related behaviour. For example, a person answering questions to 'Yes' and 'No' response options of a personality test very appropriately and quickly, not only tells something about targeted personality trait but also tells about his better power of concentration and thinking. Psychological tests have been used not only to predict test-related behaviour but to postdict, that is, to help in the understanding of many such behaviour, which have already occurred more appropriately also called as non-test related behaviour. For example, a test of criminal behaviour administered to a person who has done a crime, no doubt predicts his inclination towards crime but also, under some circumstances, throw light on the existing condition of mind at the time of crime done.

4. There exists various sources of errors in psychological testing

Another important assumption related to psychological testing and assessment is that there exists different sources of error. In psychological testing errors refer to a set of factors other than what a test attempts to measure and that influence test performance. Test scores are always influenced by such factors. Since these errors are variable, they must be taken into account in any psychological testing and assessment. These errors are called as *error variance*. There are many potential sources of error. Of these sources, three are important: test taker, test user and the test itself. Test taker himself may be a source of error variance. At the time of testing, test taker may be feeling drowsy due to last night sleeplessness or he may be having flu at the time of testing. Likewise, test users also are the source of error variance. Some test users may strictly follow the instructions stating how and under what condition a test should be administered whereas some other may be taking a liberal view towards administering the test. Besides these two sources, the test itself may be a source of variance. Some tests, may have a strong psychometric properties like reliability and validity whereas some other tests may be having a poor degree of reliability and validity.

5. Psychological tests have their strengths and weaknesses

Competent test users know in advance about the situation in which test can be appropriately administered. They are well able to anticipate about how a test was developed, what are its limitations, how the test should be interpreted and to whom and what are the correct ways of interpreting the test scores. It has also been frequently emphasized by the codes of ethics of professional that a test user must be competent enough to know about every aspect of test administration and test interpretation.

6. Psychological testing benefits society at large

Testing and assessment are boons for society because they provide very useful services to the mankind. To think about a world without tests, would be like hiring personnel on the basis of nepotism rather than on the basis of potential and documented merit. Likewise, without tests, it will not be possible to screen thousands of recruits coming for various jobs including military services. Without psychological tests, teachers could only arbitrarily place children in different types of classes. Likewise, without tests, clinical psychologists would not be able to diagnose persons with personal and emotional problems and then, they can't start proper treatment.

7. Psychological testing can be done in a fair and unbiased manner

This is one of most controversial assumption of psychological testing. There has been frequent social demands for development of fair test and fair administration and interpretation of test. In response to these demands, test developers claim to have developed fair tests and they further claim that if the test is strictly used according to the guidelines of the test manual, there is no question of doubting about its quality of fairness. But the reality is that despite these claims, several tests don't fully satisfy the criterion of test fairness.

Uses or Applications of Psychological Testing

Tests are administered with the hope that the measurements obtained from them will be helpful in making some decisions with a minimum of risk. The use of the tests and measurements to aid in better decision presumes that these measurements can predict in a much better way than would be possible without them. Important uses of tests are as under :

- **Tests are used for selection and classification decisions**

The measurements derived from tests may be used to help in deciding who will be accepted or rejected by the concerned institution. Such selection decision demands better ability to predict success and failure with minimum risks both to the individual and the institution. The ideal and best selection tests would be those that admit only those individuals who later on proved to be successful while rejecting all those individuals who would prove to be unsuccessful. Tests are also used for classification purposes where a person is assigned to one category and not to other. The classification has important effects in determining whether a person is hired for a particular job. Screening and certification are the common variants of classification. Screening refers to simple tests or procedures for identifying persons who might have some special needs or features. Certification has pass/fail qualification. When one passes a certification examination, he is conferred some privileges. For example, certification if passed, confers a right to drive a car or right to practice as a psychologist.

- **Tests are used for diagnostic and remedial purposes**

Test are also used to determine an individual's strengths and weaknesses for improving performance and thus, they serve diagnostic functions. Diagnosis is usually considered as precursor to remediation of personal distress. Tests play a significant role in such diagnostic and remedial decisions. For example, the measurements derived from personality tests are very helpful in diagnosing the external of emotional disturbances and maladjustment. Intelligence tests are helpful in diagnosing the extent of mental retardation. A proper diagnosis conveys various information like strengths, weaknesses, etiology as well as the best choices for remedial treatment. Knowing that a child's IQ is below normal range, it is diagnosed that he is mentally retarded and needs help that can provide indispensible basis for treatment.

- **Tests are used for placement decisions**

    Tests are also used for placement decisions, which are defined as the sorting of persons into different programmes appropriate to their needs or skills (Gregory, 2004). Once a person has been selected, he must be placed in a programme where he is likely to be most successful. In schools, placement decisions determine what curriculum a student will pursue, which remedial or special class he will be placed and what reading group he will be assigned. Thus placement decisions involve both the prediction (success) as well as the risk that the predictions my lead to incorrect decisions. The ideal test for placement decision is one that predicts success in one programme and rejects placement in another.

- **Tests are used for feedback or knowledge of results**

    Tests also provide feedback or acts as a potent source of self-knowledge. A feedback, the person receives from tests, can change a career path. Tests have been constructed to provide two types of feedback. *First* a person's progress may be compared with the performance of some specified reference group (that is, other members of his class). *Second* a person's progress may be compared with some specified criteria prepared by the experts or teachers. When the tests are used to compare individuals with others, they are called norm-referenced test. On the other hand, when a person's progress is compared to the criteria established by teacher or expert (such as 20 words correctly spelled out of a total of 30 words for class II students or a minimum of 40 words correctly typed per minute), it is called as criterion-referenced test.

- **Tests can be used to provide motivation**

    Tests can also be used to motivate students to learn. The very announcement that there is going to be an examination may affect the teacher either positively or negatively. Students can perceive tests as punishment or threats and can become fearful a unhappy. By using some tests for diagnostic work and mastery, teacher can help students overcome their fear of taking tests and help in preventing anxiety. One factor that affects students' motivation is the directions of test. In one famous study conducted by Yamamoto and Dizney (1965), three different sets of directions for an intelligence test were randomly given to three groups of students. It was found that the average IQ for students who were given ego-involving directions was higher than those given directions related simply to routine works (relating to how well you understand words and their meaning etc.) and to higher achievement on the test. Sometimes students intentionally score poor on the test if they believe that they might be penalized for attaining high scores by being given some more difficult assignment or task.

- **Tests are used for development of theory**

    Tests have been used to develop psychological and educational theories. By assessing the different aspects of development by using suitable tests, theories relating to growth and development have been propounded. The famous Swiss psychologist Jean Piaget developed *the theory of cognitive development* by observing children solve many unfamiliar problems. Studies such as this, led Piaget to develop a theory of growth of intelligence. Still another way in which tests contribute to theory is in developing the science of measurement more popularly called psychometry or psychometrics. Many theories of attitude personality and intelligence have been developed by studying how different groups of people respond to various types of tests.

- **Tests are used to improve programmes and curricula and also for making evaluations**

    Tests are used to evaluate programme and curricula. Programmes may be educational programme and social programme. Various educational programmes are evaluated by the use of different types of achievement tests. Curriculum evaluation is one of the most important part of educational programmes. Such evaluation depends not only on student achievement but also on the attitudes and values of the people. A particular programme may produce high levels of achievement, the school administrator may also want to know about the attitudes of students and their parents before recommending the programme. Also, he might be interested in knowing whether gains in one area of curriculum were accompanied by losses in others. All these are done by using various psychological and educational tests. One good example of social programme being much emphasized in India is the counselling programme for AIDS patients or awareness programme towards AIDS. In such programmes also tests are being used by the counsellor for evaluating their effectiveness.

    Tests are also used in formative evaluation and summative evaluation. Formative evaluation is designed to provide help to the teacher or administrator in taking effective decisions throughout the project's duration. Such evaluation provides continuous information that can be used to modify the programme to improve its effectiveness. All these works are smoothly done by various short unit tests, as well as by the measures of interest and attitude. Formative evaluation is also closely related to the feedback and diagnostic functions of testing. Summative evaluation, as we know, occurs at the end of a programme and is used to determine its overall effectiveness. Again, various types of tests are used to complete the process of summative evaluation.

- **Tests are used for in various types of behavioural researches.**

    Research is of two basic types: theoretical research and applied research. In both types of researches, tests play a very significant roles. Various variables involved in these researches are assessed by using suitable tests.

    Thus it is clear that psychological and educational tests are used for different purposes. These various uses of tests have made them widely popular.

General Steps of Test Construction

Psychological and educational tests are developed not randomly rather they pass through a series of defined steps while being constructed. In general, the following are the major steps of test construction:

- Planning of the test
- Item writing
- Preliminary administration of test
- Reliability of test
- Validity of test
- Norms of test
- Development of Test manual

A brief discussion follows:

• Planning of the test

This is the first step involved in test construction. As its name implies, here the test developer or constructor makes a planning regarding various aspects of the test. For example, he decides about the objective of test, types of contents or items to be used, types of instructions to be prepared, a probable length the test, time limit of the test, statistical methods to be used, methods of sampling etc. Thus this step fully prepares the test developer for other steps and process to be followed in this stage of planning acts like a well-prepared ground where the test is to be developed. Important questions often asked at this stage are: What is the objective of the test? Who will use this test? Who will take this test? What content the test will cover? What is the ideal format of the test? How will the test be administered? What types of responses will be required from the test takers? Who will benefit from the administration of the test? Is there any potential for harm from test administration? How will the meaning be attributed to the test scores? etc. In a nutshell, here the test developers define the testing universe, target audience and the test purpose (Miller et al., 2013).

• Item writing

This is the second step where the test developer starts writing the items of the test according to his planning. He may write objective item or essay type item. Objective items are often written in true-false, multiple choice fill-in-the blank, forced-choice and matching formats. In writing items, the item writer has to follow some rules. However, a lot depends upon the item writer's intuition, imagination, experience, practice and ingenuity. It is suggested that the item writer must have through knowledge and complete mastery of the subject-matter as well as he must have a large vocabulary. He must also have ability to convey the meaning of items in the simplest possible language. In addition to objective items, the test constructor may also write items in subjective test formats, which don't have single response that can be called as 'correct' one. Here interpretation of the response as correct is much left to the judgement of the person who scores or interprets the response of the test takers. Projective tests are examples of subjective test formats. Other examples of subjective test formats are open-ended or essay questions as well as the traditional interview questions.

After choosing the appropriate test formats, the test constructor or developer tends to specify how to administer and score the test. Whether the test will be administered orally or in writing, whether it will be administered individually to all test takers or in group, etc. Although these decisions are part of the test planning, the test developer here has to pay specific attention to these things. So far as the scoring is concerned, there are three models, which guide the test developer: *cumulative model*, *categorical model* and *ipsative model*. The cumulative model is the most common methods of scoring where the test taker is given one point or score for each correct answer and the total number of correct answers become the raw score of the test. This model yields interval-level data because it assumes that the test questions or items are comparable. The *categorical model* is one that places the test takers in a particular category or class. For example, the test taker may exhibit such response pattern that allows him to be categorized as belonging to certain psychological disorder such as phobia or paranoid group. This models yields nominal data because it places the test takers in certain category. The ipsative model of scoring is one that requires the test takers to choose among the constructs that test measures. The forced choice items tend to yield ipsative data where each statement of the item, represent different constructs. In another ipsative model, the test taker's scores on various items within the test can be compared for arriving at a profile. However, the test users should not compare the individual test scores for any decision making using ipsative model because this model assumes that individuals are unique and not comparable.

- **Preliminary administration (or Experimental tryout) of the test**

At this stage, the test developer or constructor conducts the preliminary administration of the entire written items on a sample. This preliminary administration is also called as *experimental try-out* of the test. Such preliminary administration is done for achieving so many purposes. Important ones are as under :

    (a)   To locate ambiguities, inadequacies and weakness, if any, among items
    (b)   To determine difficulty value of the item
    (c)   To determine item discrimination value of the item
    (d)   To determine a reasonable time limit for the test
    (e)   To determine appropriate length the test
    (f)   To determine the intercorrelations among items for avoiding overlapping
    (g)   To determine vagueness, if any, in the instructions of the test.

- **Reliability of the test**

At this step, the test constructor determines the reliability of the test. Reliability is defined as the consistency of test scores or statistically speaking, it is the self-correlation of the test. The consistency may be external or internal. The external consistency of the test may be determined by conducting the test twice on the sample same by a reasonable time gap such as a fortnight. This is popularly called as *test-retest reliability*. In determining internal reliability the test maybe administered on a sample and the test items may be divided on the basis of odd-even formula. The scores on odd-numbered items may be correlated against scores on all even-numbered items. The resulting correlation coefficient is the *split-half reliability* and then, using Spearman Brown prophecy formula, the whole length reliability of the test may be calculated. Such correlation becomes the *index* of the internal consistency of the test.

- **Validity of test**

After determining reliability of the test scores, the test constructor determines validity of the test. Validity means truthfulness. When a test measures what it intends to measure, the test is said to be valid one. The usual procedure of calculating validity is to correlate the test with a criterion, which measures the same trait or attribute. That is why the validity of the test is statistically defined as the correlation of the test with an outside independent criterion. There are different types of validity such as content validity, concurrent validity, predictive validity construct validity, etc.

- **Norms of the test**

After calculating validity of the test, the test developer prepares norms of the test. Norms are defined as average performance of the representative sample on the test. Norms are constructed for the purpose of comparison and adding meaning to the scores of the test. Important types of norms are age norms, grade norms, percentile norms and standard score norms.

- **Development of Test manual**

Strictly speaking, this step is not included in construction of a psychological or educational test. When the norms of the test have been developed, the test is fully constructed. However, the test constructor prepares a booklet in which he writes about all the psychometric properties of the test beginning from item writing to the development of norms of the test. This booklet is called as *manual*. Finally, keeping in view the requirement of the tests, the test constructor orders for printing of the test and booklet for further circulation among teachers, students and researchers.

Future of Psychological Testing

The future of psychological testing is promising and bright. Keeping in view the recent developments during past several decades, it can be said that psychological testing is going to build a strong and bright future in the field. Its promising future has been analyzed and discussed below.

- **Testing has a promising future**

Since the beginning of the 20th century, psychological testing has attracted the attention of both professionals and many governmental agencies. This field gained its first real status during World War I when psychological tests were used for screening the military personnel. Its field expanded during World War II when people realized the need for many types of psychological testing. Today psychological tests are used not only in research but also in schools, colleges, hospitals, industry, business, government and so forth. The new application and creative uses of the psychological tests continue to emerge in response to their various demands. It may be that some of the current tests may gradually fade away, still it is hoped that many new tests with computer base will average during the current century and it will definitely flourish the field.

- **New and improved tests will be coming up in future**

    Another prediction regarding the psychological testing is that new and improved tests will continue to cover the field. The two important intelligence tests, that is, Stanford-Binet test and Wechsler scales will be enjoying that unparallel status in future after proper revision using a representative normative samples with some special norms for particular groups as well as with the support for validity documentation research.

    So far as the projective tests are concerned, it is predicted that the use of Rorschach test will continue to diminish (Kaplan & Saccuzzo, 2001). The test is based upon the early theories of Freud and there are some doubts that the test provides clinically useful information (Hunsley & Bailey, 1999). Its psychometric properties are under attack. Even Exner's comprehensive system could fail to provide the needed support to the Rorschach Test. So far as the TAT, another projective test, is concerned, it is very difficult to say anything about its future. However, it can be said that because TAT materials have been revised, TAT may remain a respectable tool for data regarding the measurement of motives, needs and sentiment (Ritzler, Sharkey & Chudy, 1980).

    In structured personality testing, MMPI plays a significant role and with its continuous refinement and revision, MMPI-2, will continue to be one of the premier tests in the years and decades to come. Ben-Porath and Tellegen (2008) published a revision of MMPI-2 called MMPI-2-RF (Restructed Form), which consists of 338 True-False items. It is meant for 18 years and older and has completion time of 35-50 minutes. In fact, MMPI-2-RF normative sample is drawn from the MMPI-2 normative sample and it consists of 2,276 men and women between ages 18 and 80 from several regions and diverse communities in the USA. The MMPI-2-RF provides a very important alternative to MMPI-2 test. Psychologically updated, MMPI-2-RF is well-linked to the current models of psychopathology and personality. Likewise, MMPI-A has been modified by Archer, Handel Ben-Porath and Tellegen (2016) and called as MMPI-A-RF. It is meant for adolescents coming between 14 and 18 years. It consists of 241 True-False items. Its completion time is 30-45 minutes. Drawn from the MMPI-A normative group, MMPI-A norms consists of 805 boys and 805 girls of 14 to 18 years from several regions of USA. MMPI-A-RF provides a valuable alternative to MMPI-A and helps clinicians in problem identification, diagnosis and helps clinicians in problem identification, diagnosis and treatment planning for adolescents' ages 14 to 18. With such refinement and revision, it is predicted that MMPI-2-RF and MMPI-A-RF will continue to enjoy the status of premier test in future.

- **Disagreement and changes will continue in future**

    Psychologists don't agree to the fact that any one test is wholly perfect. As a consequence, there is always scope for disagreement regarding the test. Such disagreement brings new ideas, theories and also produce some classification. As a results, some changes are introduced in the test and this practice would also continue in future, making the test more useful and valuable for future.

- **Integrational approach will lead to several innovations in testings**

    These days there is emphasis upon the integration of concepts from experimental psychology, cognitive psychology computer science and psychometrics and are rapidly taking shape. Such integration has affected the future of testing. Today computer based scoring are becoming popular. It enhances its accuracy and also cuts the total time for completion. Besides, computer also offers test developers unlimited scope in developing various new technologies, which have changed both the present and future of testing.

Ethical and Social Considerations in Testing

Testing is conducted in responsible manner. For the effective, appropriate and correct uses of tests, some guideline have been prepared by American Psychological Association and other organisations (AERA/APA/NCME, 1985, 1999, 2014). We shall here concentrate upon two important considerations : ethical considerations (or issues) and social considerations (or issues). The ethical issues relating to psychological and educational testing are: *human rights, invasion of privacy, confidentiality, labelling, right of informed consent, divided loyalties and responsibility of test users and test developers*. The social issues relating to testing are: *dehumanization*, and *usefulness* of tests. A discussion of these issues follows:

Ethical issues

The field of testing is shaped by various ethical issues or considerations. Important ones are as under:

- **Human rights**

    Test users and interpreters have an ethical obligation to protect human rights. One such human right is not to be tested. A person who does not want to be a part of testing programme, should not and ethically cannot be forced to be involved in testing. Besides, the test takers have right to know the result of testing and know about their test scores, which may directly or indirectly influence their lives.

- **Invasion of privacy**

    While responding to test items, individuals have no idea about what is being revealed by their responses. However, they often feel their privacy has been invaded in some unjustified way. Although the issue of invasion of privacy is related to personality test, it logically applies to any type of test. There are two aspect of the issue of invasion of privacy (Dahlstrom, 1969). According to him, the issue of invasion of privacy is based on serious misunderstanding. Psychological tests measure only a sample of limited behaviour and therefore, they cannot invade privacy. They are measuring only behaviour sample and therefore, they don't possess mysterious powers to penetrate beyond behaviour. Thus popular fears and suspicion regarding invasion of privacy are no longer valid. Another aspect, according to Dahlstrom, is that the notion of invasion of privacy is itself vague one. In no way, it is wrong to find out something about a person. An individual's privacy is invaded when the obtained such information is used inappropriately. Moreover, psychologists are ethically and legally bound to keep the information secret and not to reveal any such information to any person than is necessary to accomplish the basic purpose of the testing.

- **Confidentiality**

The ethical code of the APA (1992) includes confidentiality, which means that the personal information obtained by psychologist from any source can be communicated to third person (other than person tested) only after person's consent. Exception to this exist in circumstances in which withholding information causes danger to the person or society or such a release is mandated by law or permitted by law for important and valid purposes. Recently, there has been growing awareness of the right of the individuals to look at or know about the findings of their own test reports. Psychologists are trying more and more to involve the test takers to be the active participant in their own assessment and they are trying to present the test results in a more understandable form. When tests are being administered by institutions such as school, college or by some employers, the test takers should be informed at the time of testing regarding how test results will be used and the purpose of the test, etc.

- **Labelling**

Labelling people with certain psychiatric disorder can be extremely damaging. When people are labelled as schizophrenic or paranoid, this may be stigmatizing perhaps for life. Not only this, such labelling may affect the person's access to help. For example, if a person is labelled as chronic schizophrenia, this may be very much stigmatizing and people will no longer be willing to help such case because nothing can be done to cure that. Labelling not only stigmatize the individual but also lower the tolerance for stress and make treatment more difficult. As a consequence, the individual should have right not to be labelled or the right to the least stigmatizing label.

- **Right of informed consent**

Test takers have right to know why they are being tested, how the test data will be used and what information, if any, will be released to whom. When all these aspects have been understood well, the test takers give their *informed consent* for being tested. If the test taker is incapable of providing an informed consent to testing, such consent be provided by his parents or by legal representative. Such consent must be in written form and not in oral form. The written form must specify the general purpose of the testing, the specific reason being undertaken in the case and the general type of instruments to be administered.

One gray area with respect to the test taker's right to the informed consent before testing involves the situations wherein the test user's complete disclosure of all facts relating to testing irrevocably contaminate the test results and data. In such situation, deception is used for creating situations that occur rarely. For example, a deception might be created to evaluate how a worker in emergency might react under situation. Sometimes deception involves the use of confederates to simulate a condition that is likely or unlikely to occur in a particular situation. The APA Ethical Principles of Psychologists and code of conduct (2002) specifies that psychologists should not use deception unless it is absolutely necessary and they should not use deception at all if it will cause emotional distress among the test takers.

- **Divided loyalties**

When the test user is trapped between two opposing forces and principles, his loyalty is said to be divided. For example, a psychologist employed in an industry may be asked by management to identify employees who have poor stress tolerance capacity and might break down under stress. He has also responsibility, on the other hand, to protect the rights and welfare of the individuals, seeking employment with the industry. Here the psychologist's loyalty is divided. Likewise, a psychologist must maintain test security properly and at the same time, must not violate the rights of the test takers to know the reason of adverse decision based on test scores.

This conflict may be solved like this: Psychologist must inform the individuals in advance how tests are to be used and describe the limits of confidentiality. On the other hand, he must provide the institution the minimum needed information and unnecessary or irrelevant personal information is kept confidential.

- **Responsibility of the test users and test developers**

This ethical issue is related to the responsibility of test users and test constructors or test developer. Because even the best test can be misused or misinterpreted, some definite stringent guidelines regarding responsible test used has been issued. Test users must be able to know the reason for using test, the important consequences for using the test as well as he must have ability to maximize the test's effectiveness and minimize the unfairness. Not only these, he must have sufficient knowledge for understanding the principles underlying the construction of the test as well as the nature of supporting research of any other test he is going to administer. He must be able to know the meaning of the psychometric qualities of the test to be used as well as the relevant literature. Jackson and Messick (1967) have rightly proposed that the test user must ask the following two questions before testing begins:
    (a)   Is the test a good and appropriate measure of attribute it intends to measure?
    (b)   Should the test be used for the given purpose?

Answer to the first question relates to the psychometric properties such as reliability and validity of the test whereas the answer to the second questions relates to ethical and social values of the test users who are to decide about the consequences of the test on the persons and their human rights.

Test developer or constructor is also responsible for providing necessary information. In particular, it is anticipated that test developers will release tests of high quality, market the test in some responsible manner and tend to restrict distribution of tests only to persons with some proper qualifications. It is also expected that they must provide a test manual and having sufficient data to permit appropriate use of test, clearly specified method of scoring and interpreting the test scores as well as some adequate information about reliability and validity of test data.

Social Issues

Besides ethical issues, some social issues also play an important role in testing. Some of the important social issues are as under :

- **Dehumanization**

One social issue in the field of testing is dehumanizing tendency especially when there is a strong ever-growing lust for computerized analysis of the test result. Many important psychological tests today have computerized scoring and interpretation, which has lessened the flexible role of human beings. Such system has minimized individual freedom and uniqueness. If this trend of dependence upon such technology of high speed computers continues, it is not far away when such machine will make important decisions about human life. Here the people now have to decide about the risk of such dehumanization against the benefits of computerization before some undesirable situation develops. When the risk is minimized and the benefits are maximized, the decision can be acceptable.

- **Usefulness of tests to society**

Another important social issue in testing is not whether any test is perfect or not but whether the test is useful to the society. The assumption underlying any test may be fundamentally incorrect and the resulting test may be far from perfect. Still that test may be useful so long as they provide useful and important information for making better predictions and understanding than can otherwise be available. However, as new concepts and new knowledge are being accumulated, the society has to compare the risk of the tests and the likely benefits of the tests. The common risks are the likely misuse of the test that may adversely affect the life of the people. The benefits are the likely enhancement in precision and fairness involved in the decision making on the basis of the test results.

Computerized Testing

Computers are now being used in psychological testing frequently. Computers are not only used in scoring and interpreting tests but they are used in test development and test administration. With the help of computers, many types of tests, particularly classroom tests are easily developed. Textbook publishers frequently offer school teachers as well as college teachers computerized test construction software programmes, which provide help in preparing course examination. In fact, these programmes have a large number of multiple-choice, true-false as well as short-answer questions that intend to assess knowledge of the students. These questions are categorized by chapter as well as by difficulty. These set of questions are technically called as *test bank*. Teachers choose from this test bank a representative sample of questions from each chapter. Finally with the help of a software, these questions are combined and printed.

Computerized test administration is also done. Test users do administer many types of assessment on computers. The assessment may be done through a more typical multiple-choice or true/false questions. One example of this is Blackboard (www.blackboard.com), which is an educational platform that allows teachers to develop classroom tests as well as administer tests to the students. Assessment is also done through online simulation where testakers are exposed to some real life situations and then, they answer questions about how they would like to respond to a situation. Still another type of assessment through computerized administration is what is called *adaptive testing*. Adaptive tests are defined as those tests which are made up of questions chosen from a large test bank for matching the skill and ability level of the students.

Besides, computerized administration, tests can be scored and interpreted through various programmes developed for computers. Today, many computer administered tests are immediately scored or the assessment data may be sent through e-mail to computerized scoring centre.

Advantages of computerized testing

Computerized testing has some obvious advantages. Some of the major ones are as under :
- Computerized tests can be very easily administered individually in a comfortable manner. Paper-and-Pencil tests are often administered in group where it becomes difficult for test users to keep a control.
- Computerized tests are efficient because computers save test users both money and time.
- Computerized tests easily eliminate human errors and therefore, facilitate standardized administration scoring and interpretation
- Computerized tests can be administered at the testtakers convenience in terms of time and place.
- Computerized tests can be easily administered in an adaptive format where all test takers start with the same set of questions (usually of moderately difficult ones). As the test progresses, the computer software chooses and presents each test taker harder or easier questions depending upon how well the test taker answered previous questions.
- Computerized tests provide more opportunities for testing people with physical and mental challenges. For example, there are some assistive technology software and equipment which allows disabilities to take online tests. Likewise, the person who has

difficulty in typing on a keyboard may use *Dragon Naturally speaking* for dictating directly to computer with words.
- Computerized tests have little scope for errors in scoring.
- Computerized tests provide better security for storing data than do ordinary paper files.
- Computerized tests permit better technologically advanced testing procedures. For examples, computers can simulate real-life situations in a better way, respond to voice-activated responses, present three-dimensional graphics and also provide on-screen calculators.
- Computerized tests can communicate using audio. By so doing, the test developers can enhance the accuracy of measurement or fidelity of the measurement. One advantage of using such as audio is that the testtakers may not attend instruction on computer screen or listen to the words of test administrator.

Disadvantages of computerized testing

There are disadvantages, which are associated with computerized testing. Some of the important ones are as under:
- Some people especially disadvantaged persons who don't have access to computers or older individuals who are less experienced with computer, are generally found to be anxious about taking a test which is administered on a computer. They are found to exhibit computer-related phobia.
- Computerized tests generally provide more general interpretation of test scores than a test taker is likely to misperceive. When a computer programme scores a test, it, in fact, selects the prearranged interpretation. If the test developer groups a range of scores for each interpretation, the computer is not able to tell the testtaker that he or she is at high or low end of the grouping. Thus computer is able to provide a general interpretation unless it is tempered by human judgement.
- Generally testtakers prefer to complete questions when they know the answers or skip questions when they find that questions is difficult and return to all difficult questions after completing all the questions to which they know the answers. This strategy works well in paper-and-pencil multiple choice tests. But this strategy does not work on most computerized tests because on many computerized tests, the testtakers must complete question presented on the screen before they go to the next. However, recently some such computerized tests have been developed that allow testtakers to skip questions, mark question for re-review and return to unanswered questions later on.

Computerized Adaptive Testing

Computerized Adaptive Testing (CAT) is getting popular day by day. In CAt all testtakers start with the same set of questions, which are generally of moderate difficulty. With the progress in the test, the computer chooses and presents each testtaker with harder or easier questions depending upon his or her previous performance, that is, answers to those beginning questions. The computer continues to present questions until it affirms the testtaker's skill level where the test ends. In fact, computer software houses the large bank of questions required by adaptive testing.

If we trace the history, the concept of adaptive testing goes back to Binet when he developed the Binet IQ test in the beginning of the 20th century. However, adaptive testing for standardized tests remained unfeasible in the most part of the 20th century until the advent of computerized testing allowed such adaptive testing to be widely used in testing situations.

In the early 21st century, some researchers presented evidence for *Computerized Adaptative Rating Scales* (CARS), one of the important applications of CAT. This CARS might help employers or managers to rate their employees more accurately than any popular type of rating scale. In CARS study the raters are presented with a pair of behavioural statements and are asked to choose one such statement, which best described the employee they observed in their work unit. Like in CAT, the computer selects the behavioural statements based on rater's previous responses. It has been found that outcome of such study clearly showed that CARS format provided more reliable and valid ratings than other rating scales such as graphic rating scale or behaviourally anchored rating scales.

Criticism of Testing

There has occurred rapid growth in the number, variety and uses of educational and psychological tests. It is also true that many educational decisions are increasingly based on such test results. As a consequence, tests have come into the scanning of public view and there are some serious allegations against the use of the test and testing in general. Some of these criticisms are as under :

- **Tests measures only a limited aspects of behaviour**

A very common criticism of tests is that they measure only a limited sample of behaviour or a sample of superficial behaviour. It has been argued that the most important traits like creativity, motivation, intelligence, character and the like are not correctly evaluated by tests on the basis of some limited questions/items relating to the concerned behavioural trait. Tests only measure less important traits, which are easier to assess. While this criticism appears to be true apparently, many tests or battery of tests in use today have resulted from various societal needs and when properly standardized and used, these tests provide a very useful information for decision-making.

- **Tests permanently categories the persons**

Sometimes psychological and educational tests are misused and as a consequence, the students are permanently categorized and later teachers uses those categories or classifications relentlessly. For example, if a student scores low on the intelligence test, he may be categorized as feeble-minded and then, teachers frequently treat him differently by assigning mental tasks that keep him simply engaged. Such categorization may force the student to think negatively about onself, causing further damage to his personality.

- **Tests tend to penalize creative and bright students**

    Psychological tests are not much sensitive to atypical and creative responses, which are usually given by bright students. Such responses are not given much credit and in this way, tests provide a discrimination against talented and creative students. By giving credit to usual and common responses provided by average students and by not paying much attention to some creative and original responses given by bright and talented students, tests discriminate against the latters.

- **Tests represent an invasion of privacy**

    Whether tests represent an invasion of privacy partly depends upon how those tests are used. When the test takers are told about how the test results would be used and subsequently, they volunteer, there is no invasion of privacy. However, when the person does not want to be tested and the psychologists persuade him for being tested, there may be an invasion of privacy. The probability of invading privacy is high when testing aims at providing benefit to someone other than the person being tested. For example, a teacher who administers test to his students in the class for preparing a research paper, thesis or dissertation, may be invading their privacy unless he obtains the informed consent of the students, their parents and school authorities. Obviously, the important elements of informed consent include a complete description of the procedures, purposes, risks, costs and potential values of the study. Risks here involves both physical and psychological trauma.

- **Tests tend to create anxiety and emotional distress, which interfere in learning**

    Another very common criticism of tests is that they create anxiety and emotional distress among students and such anxiety and distress tend to interfere with learning. The observation has been that some students are no doubt affected by tests-produced anxiety but not all students are affected negatively. Indeed, most studies have shown that students perceive the tests positively and they do better on these tests. In one earlier study conducted by Feldhusen (1964) college students were asked to respond to a test on the effect of weekly quizzes on their attitudes and achievement. Results revealed that 80% students reported that quizzes helped them to learn more and more whereas only 20% students reported that the quizzes did not provide much help. Fiske (1967), in his study, asked the respondents to indicate how favourable they feel while taking tests. It was found that less than 15 percent respondent gave negative responses to a wide variety of tests when purpose was to gather data for research. A large percentage of respondents experienced uneasy when tests required a crucial decision affecting their lives. Kirkland (1971) has arrived at some important conclusions about relationship among test, anxiety and learning as under:

(a) When students are familiar with the nature of the tests to be administered, it creates less anxiety.
(b) On rote learning highly anxious students do better than less anxious students. However, on tests requiring flexibility in thinking, they perform less well than low anxious children.
(c) The less capable students incures a higher level of test anxiety than does the more capable one.
(d) With increase in grade level of students, test anxiety tends to increase.
(e) Junior high school girls, on the average, experience more anxiety than their male counterpart.

- **Tests discriminate against poor and disadvantaged students**

Tests generally tend to favour the rich and advantaged students and discriminate against the poor and disadvantages ones. Most of the items of the tests require developed skills and capabilities, which the disadvantaged students lack. Moreover there is difference of motivation among advantaged and disadvantaged. Middle and upper-class children (advantaged ones) strive to do their best on tests because they are convinced that doing well now have a favourable impact later on their parents and teachers. Such striving is lacking on the part of the children coming from poor and disadvantaged class. Tests are said to be biased to the extent that they penalize students by rejecting them who have low test scores but would perform very satisfactorily if admitted into a course or employed in a job. Thus if both advantaged and disadvantaged would have performed equally well in a college or job but only advantaged attain a minimum acceptance test score, the test is biased against the disadvantaged applicant.

Thus tests, no doubt being popular in the field of assessment, do have their own limitations.

Summary and Review

- Psychological and Educational tests are important tools. They are defined as an objective and standardized measure of sample of behaviour.
- The roots of history of psychological and educational test are lost in antiquity. However, it has been conveniently presented by roughly dividing it into three periods: early antecedents, crucial scientific development, between 1800 A.D to 1900 A.D and developments between 1900 A.D to 2000 A.D. A separate sketch of development of psychological and educational testing in India has also been appended.
- Psychological and educational tests have been classified from many different angels. On the basis of administrative conditions, tests are of two types: individual test and group test. On the basis of scoring criterion, tests are categorized as objective test and subjective test. On the basis of criterion of response limitation, tests are classified as power test and speed test. On the basis of contents of items, there are four types of tests: verbal test, non-verbal test, performance test and non-language test. Likewise, on the basis of standardization, there are two types of tests: teacher-made tests and standardized test.

- Based on the criterion of external referencing, there are also two types of tests such as norm-referencing tests and criterion-referencing test.
- There are some major characteristics of a good psychological tests such as reliability, validity, norms, practicality and objectivity.
- There are several assumptions of psychological testing and assessment. It is essential to know about those various assumptions if one is to be fully satisfied with the different aspects of testing.
- Psychological and educational tests are used for various purposes: Tests are used for selection, remedial and diagnostic purposes, for making placement decisions, for providing feedback or knowledge of results, for improving programme and curriculum as well as for research purposes.
- In construction and development of a psychological and educational test, there are some defined steps such as planning of the test, item writing, tem analysis, reliability, validity, norms and the preparation of manual.
- The future of psychological testing is promising and bright. Many important tests are being revised and renovated and they are being used by others for various useful purpose. Despite some disagreement among psychologists, significant changes including computerized technologies are being introduced. Such changes not only have widened the scope of testing rather have made their future bright.
- There is difference between the psychological test, psychological measurement and survey.
- There are some assumptions of psychological tests. When we administer any psychological tests, we assume many such things.
- There are some ethical (or moral) and social considerations in testing and APA/AERA/NCME have issued a clear guidelines regarding this. Important ethical issues are related to human rights of being tested or not tested, invasion of privacy, confidentiality, informed content, divided loyalties and the responsibility of the test users and test developers. Important social considerations (or issues) are dehumanization as well as usefulness of tests to society.
- Although psychological and educational tests are indispensible means of assessment and measurement, they have been criticised. It is said that tests measure a limited aspects of behaviour, tests permanently categorize the persons, tests tend to penalize creative and bright students, tests represent an invasion to privacy and tests tend to create anxiety.

Review Questions

1. Define a psychological test? Also discuss the major characteristics of a good psychological test.
2. Present a historical introduction to the development of psychological and educational testing.

3. What are the major application of testing. Also discuss the important assumptions of psychological tests.
4. Outline general steps of construction of a psychological or educational test.
5. Discuss the major ethical and social considerations in psychological and educational testing.
6. Is future of psychological testing bright? Point out the major limitations of psychological test?
7. Citing relevant examples make distinction between psychological test, psychological measurement and surveys.

Chapter – 4

Item Writing and Item Analysis

Learning objectives :
- Meaning and nature of item
- Types of test items
- General guidelines for writing good essay items
- General guidelines for writing good objective items
- Standard procedures for conducting items analysis
  - Item difficulty index
  - Item discrimination Index
  - Analysis of distractors or item alternatives
  - Item-reliability index
  - Item-validity index
- Item analysis for criterion-referenced test
- Item analysis for Essay test
- Summary and Review
- Review Questions

Key Terms:

Essay item, Objective item, Selected-response item, Constructed-response item, True-False item, Multiple-choice item, Matching item, Completion item, Item-difficulty index, Item-discrimination index, Item-reliability index, Item-validity index, Distractor analysis, Item-endorsement index.

In construction of psychological and educational test, item analysis plays a significant role. The common observation has been that even the most carefully written items are susceptible to human error and when they are analyzed, they may prove to be ambiguous, too difficult, too easy or non-discriminating. In the present chapter a focus will be made on identifying these problems and some ways to resolve them. Besides this, the art of item writing would also be discussed.

Meaning and Nature of Item

In constructing a test, the test constructor writes some items for assessing the major theme for which the test is being constructed. The quality of a test is dependent upon the item. Item is a statement usually framed in question. According to Bean (1953), it is a single task that cannot be broken into small units. Thus item is a constituent part or tiniest scorable unit of a test. Nunnally (1959) had called item as the *lowest common denominator* of the test. For example, for an arithmetic test, any arithmetical problem is an example of item. Likewise, for an intelligence test, any one statement intending to assess intellectual ability is an example of item. As stated above, this single task cannot be further broken down as this is tiniest unit of the test. Therefore, an item possesses the following three basic generic features:

- Item is a statement put usually in questions form
- Item is the tiniest unit of the test
- Item can be scored

Now, the question is *What is a good item?* In the same sense that a good test is reliable and valid, a good item is that which is reliable and valid. A good test item is one that helps in discriminating between high test scorers and low test scorers. A good test item is one which is correctly answered by high scorers on the test as a whole. That item is not a good item which is not correctly answered by high scorers on the test a whole. Likewise, it can also be said that a good item is one that is wrongly answered by low scorers on the test a whole. An item that is correctly answered by the low scorers on the test as whole, cannot be said to be a good item.

Types of test items

In writing items for the test, the test constructor is faced with one of the basic questions: Which of the many different types of item formats should be employed? Various variables such as form, structure, plan, arrangement and layout of an item of the test is collectively known as *item format*. There are two major categories of item format: *Objective Item and Essay Item* (see Figure 4.1). The objective item is defined as the question or statement followed by some response options from where the testtaker selects the correct option or sometimes he is asked to construct or create his own correct answer. For example: Which is the capital of Bihar?

    (a) Patna   (b) Ranchi   (c) Gaya   (d) Buxar

Here the testtaker is required to select any one out of the four options. The content of the objective test is subjective in the sense that it reflects some judgement of what belongs to the test. It is the scoring that is objective. Essay item is one where the testtaker is required to answer the question in several sentences or in a paragraph of one or two, taking help of past associations or his memory. Also known as *produce-response item*, essay items have their potential for measuring testtakers' abilities to organize, synthesize and integrate their knowledge, as well as for using information to solve novel problems and to demonstrate original and integrative thought. An essay item is said to be useful when the test constructor wants that the testtakers should demonstrate a depth of knowledge about any single topic. For example : Make distinction between learning and maturation. Here the testtaker will answer the questions in several lines recalling materials from his memory. This illustrate essay item.

Objective item is of two types: Selection type item or select-response item or *selected-response format* and *supply-type item* or *construct-response format*. Items of selected-response format require testtakers to select a response from a set of given alternatives. There are three common types of selected-response item formats: *True-false items, multiple-choice items and matching items*.

Items prepared in constructed-response format require testtakers to supply or to create the correct answer, from his on mind. Completion item and Unstructured-short answer are two very popular types of constructed-response items.

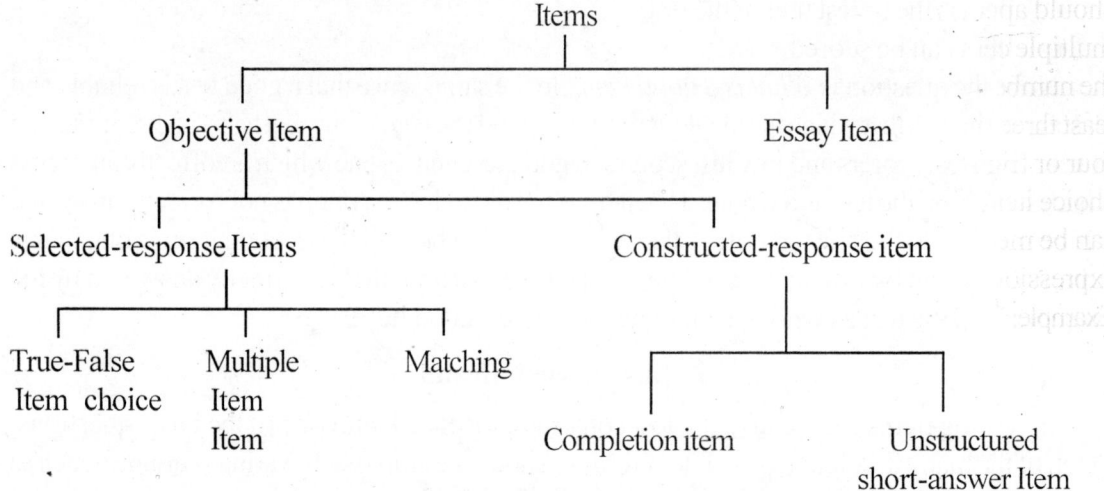

Figure 4.1 : A detailed typology of Psychological test item

A brief discussion of these various items is presented below:

1. True-False item :

Also called as *binary-choice item*, it refers to item that consists of a statement to be marked either *true* or *false*. Sometimes it may be more convenient to have binary-choice item in form of *agree* or *disagree* with a statement or answer *Yes* or *No* with a *question*. A good binary-choice item is one, which is said to contain a single idea, is not fit for debate and is not excessively long and where the correct answer must necessarily be one of two choices. One advantage of binary item is that since it does not contain distractor alternatives, it is much easier to write than multiple-choice item. It can also be answered quickly. However the disadvantage is that on binary-choice item, the testtaker has a 50% chance of getting item correct by guessing. Due to this reason, this format yields less accurate information per item about the testtaker's knowledge than do the other forms of objective items.

Example: 1. Tuberculosis (TB) is a communicable disease   *True*   False
          2. Is sun a star?   *Yes*   No

3. Nationalisation improves economy of the country.   *Agree   Disagree*

2. Multiple-choice Item :

Multiple-choice item is the most popular type of objective item. The multiple choice item consists of three parts: *stem, a correct alternative or option* and *distractors or foils*. The stem presents the problem. Stem is followed by several responses options, of which one is correct alternative and the remaining alternatives are called distractors or foils, which distracts the attention of the testtakers from the correct alternative. All distractors or foils should be plausible, that is, they should appear reasonable to the testtakers who don't really know the answers. The stem of the multiple choice item may be presented either as a question or as incomplete statement. Although the number of options used in multiple choice item differs from test to test, such item must have at least three answer choices or options. However, a typical pattern of multiple choice item is to have four or five response options so that the probability of guessing may be reduced. The multiple-choice items are commonly used to assess the achievement of any educational objectives, which can be measured by paper-and-pencil tests except those related to originality, skills in written expression and the ability to organize a response.

Example: 1. The capital of Maharashtra state is :
        (a) Ahmedabad   (b) Pune
        (c) Mumbai   (d) Solapur
    2. Type-S conditioning is associated with the name of :
        (a) Skinner   (b) Pavlov
        (c) Watson   (d) Guthrie

In general, the difficulty level of a multiple-choice item, to a large extent, depends on the cognitive processes required by the item as well as upon the closeness of the response options.

3. Matching item :

Matching item consists of two columns, *premises* on the left and *responses* on the right. The testtaker is asked to determine which response is best matched or associated with which premise. When the testtaker is child, the instruction usually directs him to draw a line from one premise to one response. On the other hand, when testtaker is adult, he may be asked to write a letter or number. For example :

		Column X		Column Y
C	1.	Psychoanalysis	A.	Skinner
B	2.	Behaviourism	B.	Watson
F	3.	Gestalt Psychology	C.	Freud
D	4.	Hormic Psychology	D.	Mc Dougall
E	5.	Humanistic Psychology	E	Maslow
			F	Wertheimer

The two columns don't have equal number of items. In column X there are five items and in column Y there are six items. If the number of items in the two columns were same, then a testtaker who does not know the answer of any one item of the premises, may deduce the correct answer by matching all other options first. In such situation, he would gain perfect score even if he does not know all answers. To minimize such possibility, more options or responses than needed are provided. Still another way to reduce the probability of guessing in such test score is to state in the direction itself that each response may be a correct answer once, more than once or not at all.

The matching item is most suitable for assessing factual information such as meaning of terms, achievement of people, symbols for chemical events, authors of books, dates of events, etc. When the test constructor needs to assess the testtaker's ability to make associations, matching items are considered more efficient than multiple-choice items. The major drawback of matching item is that it is suitable only for measuring associations and not for assessing higher levels of understanding.

4. Completion item[1] :

The completion item or also known as *fill-in item* is one where the testtaker is required to provide a word, a term, number or phrase that may complete the sentence. For example :
1. The Quartile deviation is considered a useful measure of variability
2. The first battle of Panipat was fought in the year 1526 A.D

5. Unstructured short-answer item :

Such item bears a close resemblance to the completion item. Unstructured short item and completion item, both being constructed-response item, differ only in form because when an item is written in question form, it becomes unstructured short answer item but when the same question is written in form of an incomplete statement, it constitutes a completion form of the item. For example :

Who was the first woman Prime Minister of India? (Mrs. Indira Gandhi) (Unstructured short answer item).

After rewriting this question in form of incomplete statement :

The first woman Prime Minister of India was *Mrs. Indira Gandhi* (completion form item).

6. Essay Item :

One popular misconception is that essay items require recall of information while objective items require recognition only. While it is true that essay items involve recall, objective items also emphasize upon recall. For example, identifying which of the four formulas represent standard deviation of a distribution may involve only recognition but identifying the correct solution of standard deviation of the distribution requires recall as well as the application of the formula.

General guidelines for writing good Essay items

There are some general guidelines, which must be followed while writing good essay items. Some of these guidelines are as under.

- In writing essay questions the item writer must know about mental processes he is going to assess before started writing the items.
- Item writer should use novel material or novel organization of material in writing the questions.
- Essay questions should be started with phrases such as 'Criticise', 'Compare', 'Explain', 'Differentiate', 'Explain how' etc. Such questions should not be started words such as 'when', 'what', 'who', 'list', 'which'. These words simply ask for tasks requiring only the reproduction of fact.
- The essay questions should be written in such a way that the task is clearly and unambiguously defined for the testtakers.
- The item writer should write the questions in such a way that the task to be assessed is clearly defined for the testtakers.
- A question which deals with the controversial issue should ask for and be evaluated in terms of the presentation of evidence of the position rather than in terms of simply a particular position taken.
- The test developer should try to keep the length and complexity of questions as well as the likely answers to the maturity level of the testtakers.

General Guidelines for writing Objective Items

There are some general guidelines which must be followed for writing objective items. Some of such important guidelines are as under :

- The item should have one and only one correct answer or best answer unless otherwised specified.
- The item writer should keep the reading difficulty and vocabulary level of test items at the simplest possible level.
- Item should deal with the important content of the area and should strictly avoid trivia.
- The item writer should write independent item. He should, therefore, avoid dependent item which is defined as item whose answer is dependent upon the correct answer of other item. For example

  If $X + 8 = 20$, then

  Item No. 1. $X = $ _____

  Item No. 2. $X^2 = $ _____

In this example, the correct answer to item 2 is dependent upon the correct answer to item No. 1.

- Item should not contain direct quotes because such practice encourage meaningless and

verbation memorisation.
- Item writers should avoid negative especially double negative in the item.
- Item writer should not promote irrelevant difficulty either intentionally or unintentionally.
- Item writer should avoid the use of tricky questions.
- Item writer should ensure that the problem posed by the item is clear and unambiguous.

Meaning and Purpose of item Analysis

Item analysis is one of the most important steps in the construction of psychological end educational test. Item analysis is a general term for a set of methods used to evaluate the items of the test (Kaplan & Saccuzo, 2001). As we know, items represent the test and it is considered very important for maintaining good qualities of the test. We can understand the *importance* of item analysis from the following facts :

- The primary psychometric quality of the test, that is, both the reliability and validity of the test, is dependent upon the good characteristics of the item. In fact, high reliability and validity of the test can be built into the test in advance through the item analysis.
- Since item analysis allows for substitution and revision of items, tests can be improved through item analysis.
- Item analysis is also important because it allows for shortening the test and enhance the reliability and validity of the test.

The major *purpose* of item analysis may be lined as under :
- Item analysis is conducted to identity items that are ambiguous.
- Item analysis is conducted to identify miskeyed items.
- Another purpose of item analysis is to know about too easy or too difficult items (item difficulty index).
- Still another purpose of item analysis is to know about nondiscriminating items (item discrimination index)
- Item analysis also helps in enhancing the technical quality of the test by pointing out options that are nonfunctional and that should be improved or eliminated.

Items can be analyzed qualitatively in terms of their content and form or quantitatively in terms of their various statistical properties. Qualitative analysis emphasizes upon the content validity as well as evaluation of items in terms of effective procedures of item analysis. On the other hand, quantitative analysis includes principally the determination of item difficulty and item discrimination through various statistical procedures (Anastasi & Urbina, 1997; Cohen & Swerdlik, 2005).

Standard Procedures for conducting Item Analysis

The standard procedures for concluding item analysis includes primarily the calculation of the following three indices in case of norm-referenced tests.
- Item difficulty Index
- Item discrimination Index

- Analysis of distractors or item alternatives.

A discussion follows :

Item difficulty Index :

Item difficulty index refers to the statistic showing proportion or percentage of passing or correctly answering a particular item. An item which is passed by all testtakers or an item which is failed or given wrong answer by all the testtakers can't be a good item. An item which is passed by most of the testtakers is too easy and an item which is wrongly answered by all the testtakers or most testtakers is too difficult.

An index of item difficulty is obtained by calculating the proportion or percentage of total number of testtakers who gave the right answer. A lowercase, italicized p is used for item difficulty and the subscript refers to the number of the item. For example P_1 means difficulty index of item no. 1 and P_2 means difficulty index of item two. The formula for calculating item difficulty index is as under :

$$p = \frac{R}{T}$$

where, p = Item difficulty index
R = Number of testtakers who got the item correct
T = total number of testtakers who responded to the item.

Theoretically, the value of an item-difficulty index range from 0 (zero) where no one got the correct answer to 1 (one) where everyone gave the correct answer. If 60 of the 100 testtakers gave right answer to the item number 1, then the difficulty index for this would be equal to 60/100 or 0.6. Since p indicate the percentage of people passing an item, the higher p for an item, easier is that item.[2] In case of intelligence test, aptitude test and achievement test, this statistic is referred to as item difficulty index and it is called as *item-endorsement index* in case of personality tests, adjustment and interest inventories. This is because the statistic here is not a measure of percentage of people passing but percentage of people who endorsed to 'Yes-No', 'Agree-Disagree', 'True-False' etc.

An important reason for measuring item difficulty is to choose items of suitable difficulty. An item passed by all testtakers ($p = 1.00$) or failed by all testtakers ($p = 0$) is not considered good item because it fails to provide any information about individual differences and such items need to be discarded or rewritten. Since such item does not affect the variability of test scores, they don't contribute anything to the reliability and validity of the test. If the difficulty value of an item reaches closer to 1.00 or 0, the less information about the testtakers is obtained. On the other hand, if the difficulty value of an item remains closer to .50, such item provides more differential information about the testtakers. For example, suppose an item of a test is passed by 50 persons and failed by 50 persons and there were a total of 100 persons. The difficulty value of this item is .50 and this item enables the test constructor to differentiate between each of those who passed it and each of those who failed it. Thus the test constructor has 50 × 50 = 2500 paired comparisons or bits of information about individual differences. On the other hand, if an item is passed by 60 persons and failed by persons, we have 60 × 40 = 2400 bits of differential information. Similarly, an item passed by 80 persons and failed by 20 persons provides only 80 × 20 = 1600 bits of differential information. Therefore, it is suggested that for maximum differentiation, the test constructor should choose items preferably close to p value of .50.

But here decision to have difficulty value of all items at or near .50, is not so simple as it appears rather is complicated due to the fact that the items within a test have intercorrelations. If the test is homogeneous one, such inter-item correlation may be higher. In a very extreme case where all items are perfectly intercorrelated and the difficulty values of all items were at .50 level, the same 50 persons out of 100 would pass each item. As a consequence, half of the testtakers would be getting perfect scores and other half would be getting zero. To avoid such situation created due to item intercorrelations, it is advisable to select items with moderate range of difficulty value (such as ranging from from .30 to .70 etc.) and whose, average difficulty level is .50. An index of the difficulty of average test item of a particular test may be calculated by summing the item-difficulty indices for all test items and dividing by the total number of items on the test.

The likely effect of guessing must also be considered when considering items of selected-response format such as that of true-false type, multiple-choice type, etc. With these types of item, guessing or probability of answering correctly by random guessing, becomes an important factor that affect the actual value of the difficulty level. With these items, the optimal average item difficulty is often the midpoint between 1.00 and the chance success proportion. In true-false item, the probability of guessing is 1/2 = .50. Therefore, the optimal item difficulty value would be the midpoint (or halfway) between .50 and 1.00, calculated as under :

$$.50 + 1.00 = \frac{1.5}{2} = .75$$

Likewise, in a four-option multiple choice item, the probability of guessing on any one item on the basis of chance alone is 1/4 = .25. Therefore, the optimal item difficulty will be

$$.25 + 1.00 = \frac{1.25}{2} = 0.625$$

What should be the appropriate level of difficulty value of the test, is to a greater extent, dependent upon the purpose of the test. In constructing standardized psychological tests, the maximum differentiation among testtakers at all levels is required and for such tests, items clustering around a medium difficulty value (.50) would yield maximum differentiation. But when the test has been constructed to serve some special purposes, choice of appropriate item difficulties depends upon the type of discrimination sought. For example, if a test is to be used for screening students for scholarship, items should be comparatively more difficulty than the average for that sample. Likewise, if some children with learning disabilities or slow learners are being selected for a remedial training programme, items which are easier than the average would be much appreciated.

These examples clearly show that the appropriate difficulty level of items depends upon the specific purpose of the test. It can be said that the items, in general, clustering around a medium difficulty value that is, .50 tend to yield maximum information about individual performance. The decision about item difficulty cannot be positively made ignoring the purpose of the test.

Item-Discrimination Index

Item-discrimination index measures the extent to which items are capable of measuring individual differences. If successful testtakers and unsuccessful testtakers respond equally well on the item, it has a discrimination index of 0 (zero) and is useless for measuring individual differences. On the other hand, if most successful testtakers respond correctly to an item and most unsuccessful ones incorrectly to it, the item is said to discriminate in proper direction.

Thus item-discrimination index is a measure of item discrimination, which is symbolized by a lower case italicized letter *d*. The item-discrimination index may be ordinarily determined by two methods: *by choosing a criterion* and by *constituting extreme groups.* Let us discuss these two methods in a bit detail.

Item-discrimination index can be determining by a criterion. In achievement test as well as in several aptitude tests, item discrimination is usually determined against a total score on the test itself. This total score on the test provides first approximation to the measure of ability, trait or construct under investigation. This is an example of internal criterion. Here correlation of each item with total score is calculated. If the performance pattern on a given item is similar to that on the total test, that is, high scorers get it high and low scores get it wrong, then the item is said to have good discrimination power. One major advantage of this item-total score correlation is that items may be scored right/wrong or the scores may be awarded on the basis of correctness especially when some complex item formats are used (Thorndike & Thorndike-Chirst, 2015). In item-total test score correlation, items having low correlations with total score are rejected and this provides a means of purifying the test. In this method the items with the highest average intercorrelations are retained. The correlation between total score and score on an item having right or wrong as response option and scored as 1 or 0, is the point-biserial correlation. Other statistics like biserial *r*, phi-coefficient, etc. are also calculated for calculating item-discrimination index. In determining the item-discrimination index, sometimes the criterion may be *external* rather than internal. Teachers' ratings or grade point averages are the example of external criteria. The performance on an item can be compared with performance on an external criterion. For example, if the test constructor is constructing a test to select airplane pilots, he may evaluate how well the individual items predict success on flying performance of pilot. Guttm (1950) has outlined many advantages of using external criterion in item analysis. However, external criteria are rarely used in item analysis (Linn, 1994a, 1994b).

A very popular method of determining item-discrimination index of item is by setting the *extreme groups*. In this method the estimate of item discrimination is done on the basis of comparing performance on a particular item with performance in upper and lower regions of the distribution of continuous test scores. The common practice is to have upper 27% and lower 27% of the distribution of scores provided the distribution is large and normal (Kelley, 1939). The most suitable total N for applying Kelly method is 370 of which the upper 27% is 100 and lower 27% is 100. As the distribution of test scores becomes more flat (or platykurtic), the optimal boundary lines for defining upper and lower groups gets larger and approaches 33% (Cureton, 1957). However, it has been reported that for most applications, any percentage between 25 and 33 would yield similar estimates (Allen & Yen, 1979).

The item-discrimination index (d) is basically a measure of difference between proportion of upper group (high scores) answering an item correctly and proportions of lower group (low scores) answering the item correctly. The d is usually and very simply calculated by the following formula:

$$d = \frac{N_u - N_L}{N} \quad (4.1)$$

where d = item-discrimination index
         N_u = Number of testtakers in upper group who answered correctly.
         N_L = Number of testtkers in the lower group who answered correctly
         N = Number in one group.

As the formula suggests, there can be three important possibilities as under:

(i) First possibility is that $N_U > N_L$, which means that testtakers from upper group gave more correct responses than the testtakers from the lower group. This yields a positive item-discrimination value and all those items that yield highest values (that is, over .50) are retained and those items that yield values below .20 are generally eliminated (Thorndike & Thorndike-Christ, 2015).

(ii) The second possibility may be $N_U < N_L$ which means that testtakers from the upper group gave less correct responses than the testtakers from the lower group. When the number of testtakers from the lower group answering correctly, exceeds the number of testtakers answering correctly from the upper group, the item-discrimination index is *negative* and it is an alarming situation. Such items are never *desirable* in the test because they are measuring something other than the rest of the test is measuring. Such items may be measuring nothing at all, or they may be confusing or misleading to able testtakers. The negatively discriminating items call for some action such as revision or elimination.

(iii) The third possibility may be $N_U = N_L$ which means that the number of testtakers who gave correct answers in both upper group and lower group is equal such items yielding equal or near equal correct responses in both groups, are not discriminating and they have item-discrimination vlaue of .00. Such items also serve no useful purpose.

The maximum value of positively discriminating item (d) is +1.00 and the maximum value of negatively discriminating item is –1.00. The maximum discrimination of 1.00 occurs when all of the members of the upper group got the item right and all of the members of the lower group missed it. When item-discrimination index is –1.00, it indicates that all members of the upper group failed it whereas all members of lower group passed it. This is test constructor's nightmare and such items need urgent revision or elimination.

Let us take an example to illustrate the basic facts related to item-discrimination index. In Table 4.1 the item-discrimination indices of five items of a test have been calculated and shown for illustrative purposes.

Table 4.1: Item-discrimination indices of five items of a test

Item No.	U	L	U – L	N	d
1	95	80	15	100	0.15
2	10	02	8	100	0.08
3	79	30	49	100	0.49
4	20	80	–60	100	–0.60
5	60	60	0	100	.00

In table 4.1 item-discrimination index of five hypothetical items of a test has been determined by formula 4.1. Let us see what these five items reflect:

- Item No. 1 is very easy because it is passed by 95 testtakers out of 100 in upper group and by 80 testtakers in the lower group, which also consists of 100 testtakers. The d is 0.15. This item is not worthwhile because it is least discriminating.
- Item No. 2 is very hard because only 10 in upper group (N = 100) and 2 in lower group (N = 100) and 2 in lower group (N = 100) gave the correct answer. Its d value is 0.08, which is also not a good item from the pointview of making discrimination.
- Item No. 3 is positively discriminating item, passed by 79 of upper group and 30 of lower group. This item is making proper discrimination and such items are retained in the test.
- Item No. 4 is negatively discriminating item because it is correctly answer by only 20 persons from the upper group and by 80 persons from the lower group. Its d value is – 0.60 and such items qualifies for revision or elimination.
- Item No. 5 is correctly answered by 60 of the upper group and by the same number of 60 from the upper group. It is not discriminating at all and its d is 0.00. Such items are also eliminated.

Analysis of distractors (or item alternatives):

Item difficulty indices and item discrimination indices merely tell whether an item is doing its intended job but they donot tell nothing about *why* or *why not*. If any item is found to be defective or does not work in hypothesized direction, the test developers have to look for its reason. Its cause may be very simple such as its *incorrect scoring key* but this is very rare. Problem may lie in the misunderstanding of testtakers or something wrong may be with the item itself. If the test developer finally decides that the problem lies with the item itself, he tries to look at the pattern of incorrect responses given by the testtakers and examination of this pattern often produces some clues to such problem. Although this is not much important for true-false items but it is very important for multiple choice items. The quality of each alternative in multiple choice item can be broadly evaluated with reference to the comparative performance of upper group and lower group. No formulas or statistics are needed for analysis of distractors.

A good distractor has the capacity of *distractibility*, that is, ability to distract. Therefore, a good distractor is one, which is selected by more testtakers of the lower group than in the upper group and therefore, has *negative item discrimination index*. If a distractor is not selected by anyone, it is said to have no distractibility and is ineffective. If a distractor is selected more by the testtakers of upper group than the testtakers of lower group, it has *negative distractibility*. All such distractors require looking into. They should be either replaced or their fault may be located in stem (it was ambiguous, for example) or in the instruction which might not have been clearly understood.

As we know, an examination of the alternatives selected by each group in the item analysis makes it very evident that how the distractors worked. Let us pay attention to the example of an item (item No. 10 of a test) having five response options presented in Table 4.2.

Table 4.2: Response patterns to five alternatives of item No. 10

Item No. 10	Response alternatives	Upper group N = 50	Lower group N = 50	Discrimination Index (d)
	a	0	0	.00
	b	30	45	−.30
	*c	40	20	+.40
	d	30	15	+.30
	e	12	40	−.56

*correct answer

In the hypothetical example presented in Table 4.2, there are four distractors (a, b, d and e) and c is the correct option. Distrator *a* is not effective because it is chosen by no one testtaker.

Distractor *b* is good one because it is chosen by 45 testtakers out of 50 in the lower group and only 30 out of 50 in the upper group. It has *d* of –.30. Distractor *d* is not good one because it is chosen by more testtakers of upper group than the lower group and hence, the test developer is required to know about its underlying causes in this case. This distractor has *d* value of .30. Distractor *e* is most effective one because 40 out of 50 testtakers of lower group chooses it whereas only 12 of the upper group chooses it. In this example, distractor *b* and e are effective; distractor *a* is totally ineffective and distractor *d* is also not good. Thus the test developer should pay attention to distractor *a* and *d*. Distractor a may be replaced and distractor *d* may need rewording or there may be some changes in instruction or stem, if necessary. In fact, we may not compute item-discrimination index (d) of distractors because the inspection of choices gives us the basic needed information.

Besides these two important indices, that is, item-difficulty index and item-discrimination index, test developers may also employ *item-reliability index* and *item-validity index* for analyzing and selecting items (Cohen & Swerdlik, 2005). Item-reliability index provides an indication about the *internal consistency* of the test. Higher the index of the item-reliability, greater is the test's internal consistency. Statistically, this index is equal to the product of the item-score standard deviation(s) and the correlation (r) between the item score and the total test score. The item-score standard deviation can be calculated using the index of item difficulty by the following formula.

$$s = \sqrt{P(1-P)} \quad (4.2)$$

where,
    S = Item-score standard deviation
    P = Index of item difficulty value
For example, if P of item 10 is .06, then, its item-score standard deviation will be

$$s = \sqrt{.60(1-.60)} = \sqrt{.60 \times .40} = \sqrt{.2400} = .489$$

If the item-score standard deviation is multiplied by the correlation (r) between item score and the total test score, the product is known as the *item-reliability index*, which gives the internal consistency of test (Cohen & Swerdlik, 2005). For example, if item score standard deviation is .489 and r = .75, then, its item-reliability index will be .498 × .75 = .367.

Likewise, item-validity index is a statistic that is designed to indicate the degree to which the test is measuring what it intends to measure. Higher item-validity index indicates higher criterion-related validity of the test. The item validity index can be calculated only after two statistics are know: the item-score standard deviation and the correlation between the item score and the criterion score. If these two are multiplied, the resulting product is known as the index of an item validity. When the goal of test developer is to maximize the criterion-related validity of test, the calculation of item-validity index becomes important.

Item Analysis for Criterion-reference test

In criterion-reference test (CRT) the item analysis is not so critical and important as we find in case of norm-referenced test. In CRT the concept of average difficulty has no specific meaning because here difficulty is determined by objective. On any mastery test it is expected that all items will be easy because they are based on minimum essential outcomes. However, on a nonmastery test, it is expected that items may exhibit a range of difficulties reflecting some range of intended outcomes although if on such test all the testtakers answer all items correctly, it would still be better. Here the researcher is basically concerned with the *change* in item difficulty between pretesting and posttesting. If only a few testtakers respond to item in pretesting (or prior to instruction) but many testtakers respond to item in posttesting (after the instruction), it is concluded that item is valid and the instruction was effective. Likewise, in CRT the indices of discrimination are of little interest because we are not interested in promoting variability in scores of testtakers.

In CRT the researcher is basically interested in whether or to what degree the testtakers achieve the objectives. For this purpose, the achievement rate of each of the items related to some given objectives, has to be carefully examined. If one looks at the item in isolation, it will never give any the needed information. If a test is measuring more than one objective, all the items that measure a given objective must be analyzed together. The results of such analysis are generally reported in terms of percentages called *item achievement percentages*, which can be expressed both at group as well as individual level of achievement. For item analysis the researcher's major concern will be in knowing the percentage of testtakers who responded correctly to each item as well as the pattern of percentages for a given objective.

Item Analysis for Essay Tests

Item analysis can also be done on essay tests provided the desired response for each item as well as analytical method of scoring have been clearly defined. If point values are assigned to each subpart of an essay item, each subpart can be treated as a separate item. However, the problem lies in the fact that these items, by their nature, are not *independent* because degree of success on one item may be, to a larger extent, related to the degree of success on another Table 4.3 illustrates how item analysis can be done with essay tests.

Table 4.3: Illustrates example of item analysis in Essay test.

Part C

	Number of Points				
	0	1	2	3	4
Upper testtakers	0	2	3	5	6
Lower testtakers	3	6	5	4	2

Analysis:

Upper testtakers = (0 × 0) + (1 × 2) + (2 × 3) + (3 × 5) + (4 × 6) = 47

Lower testtakers = (0 × 3) + (1 × 6) + (2 × 5) + (3 × 4) + (4 × 2) = 26

Analysis clearly shows that the item is good and performing effectively.

Suppose item No. 5 of the essay test has six parts (A, B, C, D, E & F) and each part is worth 4 points. In Table 4.3, part C has only been shown as item worth 4 points and results have been accordingly analyzed for upper scorers and lower scorers. Analysis done in Table 4.3 shows that the upper group did get much higher scores than the lower group, providing evidence for the effectiveness of the test. In the table, within parentheses, the first number refers to the number of points and the second number refers to the number of testtakers who got that points.

Some Important Consideration in Item Analysis

There are some important considerations that the test developers should have in their mind while they are doing item analysis. These considerations may be briefly discussed as under.

- **Guessing:**

Guessing is one of the important problems that must be handled properly by the test developers. This problem is more serious in case of objective type such as intelligence test, aptitude test and achievement tests. Although there have been developed many correction formulas for guessing, and some interventions at the level of test instructions, too have been proposed, none tends to yield satisfactory results. The reason is that the problem of guessing is far more complex than it first appears. In fact, there are following three criteria that any correction for guessing must meet:

1. Any correction for guessing must recognize one basic fact that the testtakers, when resort to guessing in objective test, they don't guess on a totally random basis. In fact, the testtaker's guess maybe based on some knowledge regarding the subject as well as upon his ability to rule out one or more distractors. However, this knowledge of the testtaker regarding the subject-matter may vary from item to item.
2. Some testtakers may be more lucky than others in guessing the correct choice. Any attempt for correction of guessing may underestimate or overestimate the effects of guessing for lucky or unlucky testtakers.
3. A correction of guessing may also take into consideration the omitted items. Sometimes, instead of guessing, the testtakers simply omit a response to an item. What should be done here? Should the omitted items be scored as wrong? Should omitted item be scored as if the testtakers has made a random guess? Should omitted items be excluded from item analysis?

In fact, these criteria are not fully met by any correction for guessing. Therefore, no appreciable solution to the problem of guessing has yet been proposed. However, the modern test developer tries to address the problem of guessing by including in the manual of test two things: explicit instructions to the guessing likely to be conveyed to the testtakers and specific instructions for scoring and interpreting the omitted items in the test.

So far as the guessing on personality and related tests is concerned, that is thought to be not so serious. This is because here assumption is that the testtaker indeed does select the best choice.

- **Fairness of test items :**

As we often speak of biased tests, so also we may also speak of biased test item, which is defined as an item that shows favour to one group of testtakers as compared to others even when the differences in group ability are controlled (Camilli & Shepard, 1985). Many different methods have been used to identify biased test items. Item characteristic curve (ICC) have been frequently used to identify the item. In a statistical sense, items are identified as biased if they tend to exhibit differential item functioning. The differential item functioning is exhibited by different shapes of item-characteristic curves for different groups although the two groups don't differ in total test score. (Mellenbergh, 1994). Any item that is considered fair to different groups of testtakers, should not show significant different types of item-characteristic curves for the different groups (Jensen, 1980). A statistical test of null hypothesis of no difference between item-characteristic curves of the two groups needs to be calculated. Items that show significant difference in item-characteristic curves should be either revised or dropped from the test. The major rationale of using ICC criterion as a measure of item bias is that the same proportion of persons from each group should pass any given item of the test provided all the concerned persons obtained the same total score on the test.

- **Speed tests :**

Item analysis of speed tests may be misleading. Item indices from the speed test tend to reflect the position of the item in the test rather than its difficulty or discriminative power. Items that appear towards the end in the speed test will be passed by a relatively small percentage of the total testtakers because only a few testtakers have time to reach these items. The closer an item lies to the end of the test, the more difficult it may appear. Likewise, item discrimination indices may be over-estimated for the items that have not been responded by all testtakers. Since more proficient testtakers tend to work faster, they are more likely to respond to later items in a speed test. It means that regardless of the nature of the item itself, items appearing late in the speed test are more likely to show positive item-total correlations because some proficient testtakers have reached those items.

To avoid these difficulties, the test developers should limit the analysis of each item to those persons who have reached the item. However, this is not an effective solution due to three reasons:
    (a)   The analysis of item lying towards the end of the test would be based on progressively a small number of testtakers and this will yield progressively poor reliable results.
    (b)   If the more proficient testtakers reach later items, it means that the part of the analysis

would be based on some select sample (proficient testtakers) and part of the analysis would be based upon the total testtakers.

(c) Since the more proficient testtakers tend to respond more correctly, their performance will make items appearing towards the end of the test, easier than they really are. Keeping in view these difficulties, it is suggested that if speed is not an important element of the ability being assessed by the test, the test developer ideally should develop a test to be item-analyzed with generous time limits for properly responding to all the items.

Item-characteristic Curves

A very important and valuable way to know about item difficulty and item discrimination is to graph their characteristics and this is done through what is called *item-characteristic curve* (ICC). It is a graph in which the total test score reflecting the ability of the testtakers is plotted on the horizontal (X) axis and the proportions of the testtakers who got the item correct, is plotted on vertical (Y) axis. The total test score is used as an estimate of the amount of a 'trait' possessed by the testtakers. Since the trait cannot be directly measured, the total test score is considered its best approximation. A separate ICC is graphed for each item of a test. Figure 4.2[3] presents four item-characteristic curves. Now let us discuss these four

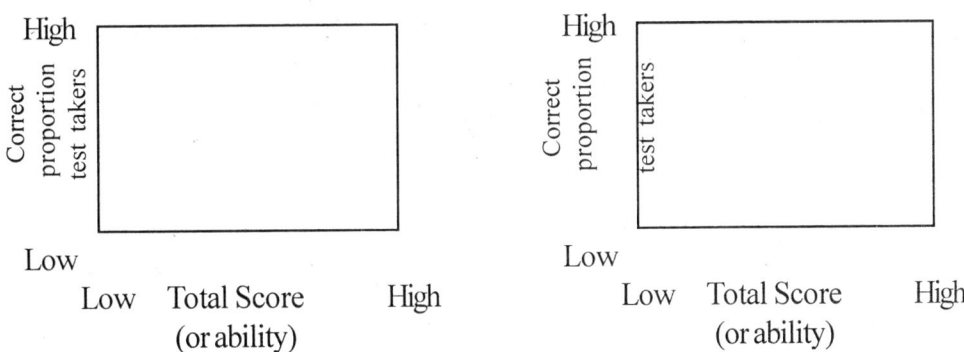

In all four ICC curves the scale of the total score or ability on X and the scale of the probability to respond on Y axis have been omitted for simplicity.

(111)

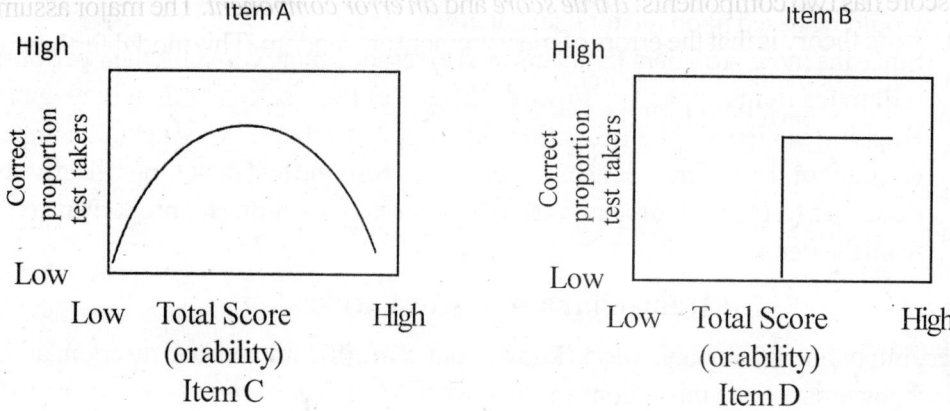

Figure 4.2: Item-characteristic curves

item-characteristic curve. What does ICC of item A tell? Obviously, it shows that the probability of giving correct response to it increases with increase in ability. So, this is definitely a good item. Moreover, it also shows that the probability of responding correctly to this item is low for low ability group and is high for high ability group. Item B is not a good item because the probability of the testtakers' answering correctly is high for low ability and low for testtakers of high ability. Item C is again not a good item because it shows that the testtakers of moderate ability have the highest probability of answering the item correctly whereas the testtakers with high ability as well as the testtakers with low ability are not likely to respond correctly to this item. Now, what's about item D? The item-characteristics curve of item D shows that this item discriminates at only one point on the scale of ability. The probability is that all testtakers at or above this point will respond correctly to this item. Conversely, the testtakers who are below that point on the scale would respond incorrectly to this item. Such item is said to have excellent discriminative ability and is considered very important and useful for the test, which is designed to select applicants on the basis of some cutoff score such as in case of various jobs in public and private section. However, such item might not be considered as a good item for a test intended to provide detailed information about the ability of the testtakers across all levels of ability. Test developer can average item-characteristic curves for creating a test characteristic curve, which provides the proportion of responses expected to be correct for each level of ability (Guion & Ironson, 1983).

Item Response Theory

Item response theory (IRT) is considered as the most important development in the field of psychological testing in the second half of the 20th century. Prior to IRT, *classical test theory* or *classical true score theory* was in vogue. The classical theory defines score as something derived from the scores of the individual's responses to various items. This theory assumes that each person has a true score that would be obtained if there were no errors in measurement. The obtained score has two components: *a true score* and *an error component*. The major assumption of this true score theory is that the errors of measurement are random. This model further focuses on the extent to which individual test items are useful in assessing individuals presumed to have various amounts of particular trait or ability. It must be made clear that the true score is *never* known. We can simply obtain a probability that the true score resides within a certain interval and a best estimate of true score can be derived. But we can never know the value of true score with full certainty. The newer approach, based on item analysis, takes into account the chances of getting particular items right or wrong. This newer approach is called item-response theory or also known by various other names such as *latent-trait theory*, the *latent-trait model* and the *Rosch Model*. Thus, item-response theory is not a single theory rather it refers to a number of models of test development that are collectively referred to by these said various names. The latent-trait model is viewed as more sophisticated model of test development than classical test theory (Mitchell, 1999) and has garnered more support from large-scale users of the tests. This model, widely used in educational set up, offers a good way to estimate the probability that a person with X ability (or total score) will be able to perform at a level of Y. In terms of personality and related psychological tests, it models the probability that a person with X amount of a particular personality trait will tend to show Y amount of that trait on a personality test intended to measure it.

Now the one question frequently asked is why the model is called as *latent-trait*. To understand this, consider the reality of the test, which is typically designed to provide an estimate of the ability or knowledge or strength of a trait possessed by the testtakers. The performance of the testtakers depends upon the variables like knowledge, ability or personality trait, which cannot be measured directly. However, their estimate of the amount of the variable is obtained through the test. Thus these latent traits are not directly measured but are reflected through the items of the test. The latent trait theory predicts that the underlying latent trait (or unobservable variable) is *undimensional* and all items of the test measure this latent trait.

There is one important difference between the latent-trait model and the classical true score test theory. For example in classical true score test theory, no assumptions are made about the frequency distribution of the test scores whereas in latent-trait model such assumptions are inherent (Allen & Yen, 1979). Using IRT, various approaches to the construction of tests have been proposed. Some of these approaches use only two dimensions, that is, item difficulty and item discrimination in test development. Some other approaches, besides these two dimensions, add a third dimension, that is, the very probability that the testtakers with very low levels of ability may get a response correct. Still there are other approaches, which only use the dimension of item difficulty.

The wide-spread use of item response theory in test developments has been questioned by some investigators. For example, difficulty with the theory arises when the tests measure multiple dimensions because in such cases the assumption of test undimensionality is violated. It is further stated that even the same item of the test may be tapping different abilities from the same testtaker in light of the varying life experiences of the testtakers.

Despite these lightening criticisms, the item-response theory or latent-trait models do play an increasingly significant role in development of new psychological tests and testing.

Item Analysis in Qualitative tests

Besides quantitative research methods, there is another class of research popularly known as qualitative research methods. In qualitative research methods, the techniques of data generation and analysis depend upon verbal rather than some mathematical or statistical techniques. *Qualitative item analysis* refers to various nonstatistical procedures designed to examine how individual test items works. Here the test developer compares individual test items to each other as well as the test as a whole. In qualitative item analysis attempt is made to explore the issues through verbal means such as *group discussion* and *interview* conducted with testtakers as well as with some relevant concern. Some of the important possible questions regarding different aspects of the test, the test developers might want to discuss has been presented in Table 4.4.

Table 4.4: Some relevant Questions for exploring the test qualitatively

- Did the test appear to measure what it intended to measure? If not, what was contrary to your expectation?
- Did you feel that any item of the test was discriminatory with respect to any group of people? If so, why?
- Did the behaviour of the test administrator affect your performance on the test? If so, how?
- How do you feel about the length of test especially with respect to number of items and the time it took in completion?
- Do you feel that the test was a fair test regarding what it sought to measure? Why or why not?
- Did you guess on any of the test item? What percentage of the items you estimate you might have guessed? Did you employ any strategy for guessing
- Did you experience difficulty in understanding the meaning of any item?
- Do you think that there was some cheating during the test? If yes, kindly describe the methods of cheating adopted?
- Do you think that your mental and physical conditions affected performance of test? If so, how?
- Did you find any part of the test entertaining, educational or otherwise rewarding?
- How did you prepare for this test? If you are asked to tell others how to prepare for this test, what you would like to tell?
- What is your overall impression of the test? What suggestions you would like to make for its further improvement?

Questions presented in Table 4.4 could be raised either orally or in writing shortly after the administration of the test. These questions may also be framed into formats like True-False or Multiple-choice depending upon the objectives of the test developer.

In qualitative item analysis the following two innovative approaches may be used :

Expert Panels :

Expert panels may provide qualitative analyses of test items. The expert makes a *sensitivity review* of the test items. The sensitivity review of test items which is conducted during the test development process is one in which items are reviewed for their fairness to all prospective testtakers and for the presence of stereotypes, offensive language or situations. Such sensitivity reviews have become a standard part of the modern process of test development (Reckase, 1996).

'Think aloud' administration of test:

In this approach the testtakers verbalize thoughts as they occur during the administration of the test. Cohen et al. (1988) proposed the use of 'think aloud' test administration as a qualitative tool for throwing light on the testtaker' thought processes during administration of the test. On a one-to-one basis with the test developer, the testtakers are requested to take the test, thinking aloud as they respond to the item. If a test is designed to measure personality, this 'think aloud' technique may yield valuable insight into the ways testtakers think, perceive, interpret and respond to the items. Likewise, in intelligence or achievement test, such technique may enable the test developer to know if some testtakers are misinterpreting a particular item and if so, why how they are misinterpreting.

On the basis of the information obtained through expert panel or testtakers themselves, the test developer may decide to modify or revise the test items. The test developers may decide to delete some items or create new items or entail rewording of item. This is called *test revision*, different from that test revision, which becomes due after a lapse of substantial period of time since the last publication of the test.

When does the Revision of Existing Test become Due?

The popular saying is that time and tide wait for no one. A person gets old and so also, tests get old, too. A time comes in the life of the most existing tests when the test will be revised in some way or its publication is likely to be discontinued. Now the question is *when is that time*? Although no hard-and-fast rule exists for the revision of the existing test, the general suggestion is that the existing test may be kept in use so long as it remains useful and it must be revised when there occurs significant changes in the domain measured by the test or when new conditions of test use and interpretation make the test somehow inappropriate (APA, 1996). However, the revision of the existing test becomes due when any of the following conditions exist and becomes obvious:

- Due to cultural changes when some words take on new meanings. In such situation certain words in the test or directions may be perceived as inappropriate and must therefore, be changed.
- The stimulus materials of the test look outdated and the testtakers fail to relate to them.
- The verbal contents including instructions, administration and the test items are perceived by the testtakers to have some outdated vocabulary, which is not readily understood by them.
- The reliability or validity of the test including the effectiveness of the test items are to be improved by the likely revision.
- Norms of the test become inadequate because the basic features of the group upon which norms had been developed, stood changed now.
- Norms of the test becomes inadequate due to age-related shifts in the trait or abilities measured and therefore, upward, downward or in both directions the age extension of norms become necessary.
- The theoretical base of the test has changed and has improved significantly. In such situation the existing test must be revised to reflect the new improvement in the theory of the test.

The steps to revise the existing test is more or less equivalent to developing a new test.

Concept of Cross-validation and Co-validation

Another important aspect in the development of all tests is *cross-validation* as well as a latest trend in test publishing called *co-validation*. The term cross-validation refers to a procedure in which the validity of the test is computed on a different sample of testtakers from that on which the items had been selected. This independent determination of test validity is known as cross-validation. Thus cross-validation refers to the revalidation of a test on a sample of testtakers other than those used for selection of items (Cohen & Swadlik, 2005). Any test validity coefficient computed on the same sample that was used in item-selection purposes may entail random sampling errors within that particular sample and therefore, will be spuriously high. In such situation, a high validity coefficient may result even though the test has little validity in predicting the criterion.

It is expected that items selected for the final version of the test, will have low item validities when administered to a second set of testtakers. This is possible due to operation of chance factor. This decrease in item validities that occurs after cross-validation is technically known as *validity shrinkage*. Such shrinkage is commonly expected and much more preferable to a situation where spuriously high item validities are reported in the manual of the test due to inappropriate use of identical sample of testtakers for the computation of test validity and cross-validation of the findings.

Co-validation that sounds much similar to cross-validation, is defined as a process of test validation conducted on two or more tests using the same sample of testtakers. This is a current trend among the test publishers who publish more than one test designed for use with the same population. When this procedure is used in the development of norms or revision of the existing norms, it is referred to as *co-norming*. Co-validation is very economical to the test publishers because both money and time are saved if the same testtakers are used in the validation studies for multiple tests. Co-validation is not only beneficial to test publishers but it is also beneficial to testtakers. When two tests to be used together are normed on the sample, the chance of sample error is greatly minimized and as a consequence, the testtakers can depend upon the test results for a better accuracy and they can become confident that the scores on the two tests are comparable.

Summary and Review

- Item is usually a statement framed in question. It is a single task that cannot be broken into smaller units.
- There are two broad types of items – objective item and essay item. Objective items are of two general types–selected-response item and constructed-response item. Multiple choice item is a good example of selected-response item and completion item is a good example of constructed-response item.
- There are some general guidelines for writing objective items as well as essay items that must be followed by the test developers.
- Item analysis refers to a set of procedures for knowing about the effectiveness status of the item. Item analysis procedures are different for norm-referenced test and criterion-reference test because the items in such tests are designed to do different things. Traditional item analysis procedure that include determination of item difficulty, item discrimination and distraction analysis are meant for norm-referenced tests.
- Item difficulty index indicates proportion or percentage of testtakers passing the item. Higher the proportion, easier the item. In achievement testing such proportion is called item-difficulty index but in personality testing, the same index is known as item-endorsement index because in personality testing this index provides not a measure of the proportions of people passing an item but a measure of percentage of testtakers who endorsed either 'Yes or No' or 'Agree-disagree' or otherwise endorsed item.
- Item discrimination index indicates to what extent the item discriminates between those who are successful (or high scorers) and those who are unsuccessful (or low scores). It is determined either by choosing a criterion or by setting up extreme groups.
- Distraction analysis is also done in item analysis especially in multiple choice format to know about how the distractors are attracting the attention of testtakers. A good distractor as one which attracts more testtakers from lower group than from the upper group. Hence, such distractor has a negative item discrimination index.
- In order to know about the internal consistency of the test, the test developers also sometimes calculate item-reliability index. For maximising criterion-related validity of test, the test developer also calculates item-validity index.

- Item analysis can also be done in criterion-referenced test where the researcher or test developer is basically interested in whether or to what extent the testtakers achieve the objects.
- Item analysis can also be done with essay items provided the desired response for each item as well as the analytical method of scoring have been clearly defined.
- In item analysis the test developers also pay attention to some other considerations like guessing, fairness of test items and whether speed in an element in test.
- Item-characteristic curve is a graphic method to know about the basic characteristics such as item difficulty and item discrimination of the test.
- Item-response theory is an important theory in the field of test development. The theory takes into account the chance of getting particular item right or wrong. The theory assumes unidimensionality of the latent trait being assessed.
- Item analysis is also done in qualitative research by framing some questions to be asked from the testtakers by the expert panel or by 'think aloud' administration of the test.
- The concept of cross-validation and co-validation are important aspects of test development. While cross-validation is an old procedure, the co-validation is the latest trend among the test publishers.

Review Questions

1. Define item. Discuss its major types.
2. Outline some general guidelines in writing objective items.
3. Define item analysis. Discuss the features of item difficulty index and item discrimination index.
4. Make distinction between item-reliability index and item-validity index.
5. Discuss the role of distractors in multiple-choice item. How are they analyzed and why are they analyzed?
6. Citing an example illustrate how is item analysis done for essay items.

■■■

Chapter – 5

Reliability and Validity of the Test

- The concept of reliability
- Sources of inconsistency or error variance
- Two modes to express reliability
- Reliability estimates
    - Test-retest reliability
    - Parallel form and Alternate form reliability
    - Internal consistency reliability
        - Split-half reliability
        - Kuder-Richardson formulas
        - Coefficient Alpha
    - Inter-scorer reliability
- Reliability of a difference score
- Index of reliability
- Which type of Reliability is more appropriate?
- Conditions affecting reliability coefficients
- The concept of validity
- Categories of validity
    - Content validity
    - Criterion-related validity
    - Construct validity
- Interpreting reliability and validity coefficient
- Relation between reliability and validity
- Factors affecting validity
- Summary and Review
- Review Questions

Key-Terms:

Internal consistency Temporal consistency, True variance, Error variance, Standard error of measurement, Test-retest reliability, Split-half reliability, Coefficient alpha, Inter-scorer reliability, Content validity, Concurrent validity, Predictive validity, Construct validity, Expectancy Table, Convergent validity, Discriminant validity, Correction for attenuation, Spearman-Brown prophecy formula.

Reliability and validity are the two most important indices of test efficiency. In day to day interaction, reliability is a synonym for *dependability*. If we are lucky, we have a reliable friend who is always ready for us in the time of need. Likewise, if we speak of a train being very reliable, we can set our watch accordingly. Likewise, in day-to-day life, we say something as valid if it is meaningful and sound. We often say it is a valid argument or valid theory because it is based on the meaningfulness of something. In the present chapter a discussion will be done regarding the reliability and validity as they apply to the process of psychological assessment.

The Concept of Reliability

In day-to-day language reliability means dependability or trustworthiness. This term means essentially the same thing with respect to test and measurement. When we ask about the reliability of the test, we never ask what the test measures but instead, ask how accurately it measures whatever it does measure. In other words, it indicates to what extent the obtained score is stable and free from error particularly variable error. Therefore, reliability refers to the accuracy or precision of the measurement procedure. The indices of reliability provide an indication of the extent to which the scores produced by a particular test or measurement are consistent and reproducible. It is in this sense Anastesi & Urbina (1997) have defined reliability as the consistency of scores obtained by the same testtakers on administration of two different occasions or with different sets of equivalent items obtained on single administration. The consistency of scores obtained from testing and retesting is called as *external consistency* or *temporal consistency* whereas consistency scores obtained from two equivalent sets of items of a singled test after a single administration is known as *internal consistency*. The correlation coefficient indicating the degree of temporal stability is known as the coefficient of stability and the correlation coefficient indicating the degree of internal consistency, is known as *coefficient of internal consistency* or *alpha coefficient*. Since in knowing about consistency, the test is correlated against its own scores, statistically reliability is defined as self-correlation of the test. The more reliable the test is, the more confidence we can show that the scores obtained from the administration of the test are essentially the same scores that could be obtained after readministration of the same test on the same group of testtakers. A unreliable test is really a useless test.

One important point regarding reliability is that it always refers to the result obtained with the test and not to the test itself. Thus, it is better to speak of *reliability of test scores* rather than of the test. Another basic point is that the estimate of reliability always refers to a particular type of consistency. Reliability is assessed mainly with statistical indices. Reliability is also a matter of degree because any test is neither wholly reliable or wholly unreliable.

Logical or Technical Meaning of Reliability

The logical or technical meaning of reliability is based upon the classical test theory or true score model. The classical test theory (or also known as theory of true and error scores) claims that a score on an ability test or personality test is presumed to reflect the true score of the testtaker as well as the error. Here error refers to the component of the observed test score that does not have to do with the ability of the testtaker. If X is used to represent observed score, T to represent a true score and E to represent error (or also called as error of measurement), then the fact that an observed score equals the true score and error score (or error of measurement) can be expressed as under :

$$X = T + E \qquad (5.1)$$

and therefore,

$$E = X - T$$

Thus any difference between an observed score and true score is the error of measurement.

Errors of measurement can be of two types–*random error* and *constant systematic error*. Random errors are sometimes positive, making the observed score too high and sometimes, they are negative, making the observed score too low. Such errors produce inconsistency in measurement and their higher proportion tends to lower the reliability of the test scores. Constant errors work in one direction always and therefore, don't lead to inconsistency. As a consequence, constant errors don't affect the reliability of the measurement. For example, if a measuring instrument such a weighing machine consistently overweighed every one who stepped on it by 2 pounds, then the relative standing of people remain unchanged and the recorded weights would consistently vary from the true weight by 2 pounds. Such systematic or constant error does not change the variability of the distribution or affect reliability.

A very important useful statistic frequently used in describing the source of the test score variability is variance (σ^2), which is defined as standard deviation squared. Variance from true differences, is true variance (σ_{tr}^2) variance from random and irrelevant sources is error variance (σ_e^2). The true variance and error variance added together constitute the total variance (σ_T^2). Thus the relationship of these variances can be expressed as

$$\sigma_T^2 = \sigma_{tr}^2 + \sigma_e^2$$

Reliability is logically or technically defined as proportion of true variance (σ_{tr}^2) in the total variance (σ_T^2). The greater the proportion of true variance, the more reliable is the test. In other words, the reliability is that part of the total variance which is the true variance. The proportion of the true variance and error variance are found by dividing the true variance and error variance respectively by the total variance. Thus, the proportion of true variance is equal to σ_{tr}^2/σ^2 and the proportion of error variance equal to σ_e^2/σ^2.

Reliability coefficient by logical definition, is then calculated by the following formula:

$$r_{tt} = \frac{\sigma_{tr}^2}{\sigma^2} \qquad (5.2)$$

where,

r_{tt} = Reliability coefficient

σ_{tr}^2 = variance of true score

σ^2 = Total variance

If true variance of the test scores is 0.67, and the total variance of the test scores 0.88, then the reliability by formula 5.2 will be equal to 0.67/0.88 = 0.76. This means that 76% of the obtained total variance is attributable to the true variance indicating a higher degree of reliability. If the ratio was 1.00, the obtained variance would be wholly attributable to true variance. If the ratio were o, there would be no relationship between the obtained scores and the corresponding true scores.

Alternatives to classical Test Model :

There are some alternative to classical test model. Important such alternatives are item-response theory (IRT), domain sampling theory and generalizability theory. A discussion of IRT has already been made in chapter 4. So we shall concentrate upon the remaining two.

During 1950s a very important alternative to classical test model called *domain sampling theory* emerged. From the beginning of 1900 to 1940s, true score model had enjoyed unparallel and unchallenged reign of acceptance. Domain sampling theory rebels against the concept of true score, which exists with the measurement of psychological construct (Tryon, 1957). In domain sampling theory attempt is made to estimate the extent to which the specific sources of variations related to the area of interest are contributing to the test score. Here the reliability of the test is taken to be an objective measure of how precisely the test score measures the domain from which the test draws a sample. Domain here refers to a very large collection of items. Some researchers prefer the term universe or population to domain for describing the same concept (Nunnally & Bernstein, 1994). When a psychological test is constructed, each item is a sample of the ability or trait or behaviour to be measured. A long test has many such sample whereas the short tests have very few such samples. The measurement emphasized in the domain sampling theory is the error induced by using a sample of items rather than the whole domain. Reliability in this model is estimated by obtaining the correlation of the observed test score with the true score. In this theory the items in the domain are considered to have the same means and variance as those in the test that sample from the domain. Measures of internal consistency are most compatible with domain sampling theory.

Generalizability theory proposed by Cronbach et al. (1972) is another theory which is the extension of true score theory. In this theory the concept of universe score, in fact, replaces the true score (Shavelson et al., 1989). This theory assumes that the testtakers' test scores vary from one testing to another due to variables involved in the testing situation. Instead of taking all types of variabilities in the testtaker's score as error, theory suggests the test developers to find out the details of the particular test situation or universe leading to the specific test score. This universe is further described in terms of its *facts*, which may include variables like the purpose of test administration, amount of training of the test developer as well as the number of items in the test. According to this theory, this universe score is practically analogous to the true score in true score theory. The theory further states that if there are the exact same conditions of all facets in the universe, there will be obtained exactly the same test score. Generalizability theory, in a nutshell, proposes for separating sources of systematic error from random error for eliminating systematic error. In fact, if the researcher is able to record the amount of random error in each measurement, the average error would be zero and with passage of time, random error would not interfere with obtaining an accurate measurement. However, the systematic error does influence the accuracy of measurement. Therefore, the goal of generalizability theory is to eliminate the systematic error.

Cronbach et al., (1972) have suggested that the test may be developed with the help of generalizability study, which should be followed by a *decision study*. A generalizability study shows to what extent the scores of a test are generalizable if the test is administered in different situations. In other words, such study examines how much of impact different facets of universe such as the time of the day in which the test was administered, group administration as opposed to individual administration of the test, etc. have on the test. The impact of particular facet on the test score is represented by coefficient of *generalizability*, which is similar to the reliability coefficient of true score theory. After generalizability study, decision study is done by the test developers. The decision study enables the test developers to examine the usefulness of test scores. As we know, the test scores are used to guide a variety of decisions such as placing a child in next grade, promoting an employee to the next cadre, diagnosing a mental patient under certain category for treatment, etc. In this way, the decision study is designed in helping test developers know how the test scores should be used and how much dependable those scores are the basis of taking some decisions.

Although generalizability theory has not been able to replace the true score theory, it is at least able to provide a clear indication that the reliability of the test does not reside in the test itself rather a test's reliability is the result of the various situations under which the test is developed, administered and interpreted.

Sources of Inconsistency or Error variance

There are many sources of inconsistency or error variables in testing. Some of the important ones are: test construction, test administration, test scoring and interpretation and some miscellaneous sources of error. A discussion of these sources of errors is presented below:

1. Test construction :

One popular source of variance during the construction of test is *item sampling* or *content sampling*. Such sampling reflects variations among items within a test as well as variables among items between tests. Let us take an example. Suppose two tests are available to assess the personality of the testtakers. There may be differences in the way the items in the two tests attempting to measure the same thing have been worded as well as in the exact content sampled. The testtaker may score high on one test as opposed to other test though both tests intend to measure the same thing. Thus content (or item) sampled on test and the way content is sampled produces variations in the score of the testtakers and become one important source of variance. In a nutshell, item selection is very crucial to the accuracy of measurement. In any well-designed and standardized test, the measurement error from item sampling would be minimal. However, a psychological test is always a sample and never the totality of an individual behaviour, ability or trait. As a consequence, item selection is always an important source of measurement error in psychological testing.

2. Test administration :

Another potential source of error can occur from the administration of a test. There can be three such important sources: *environmental variables*, *testtaker variable* and *the examiner-related variables*. Some of the physical or environmental factors that contribute to the error variance and therefore, to the unreliability are poor amount of ventilation and noise, extremely high or low temperature, seating conditions, lighting, seating arrangements, a pencil with a dull or broken points, poor writing surfaces on which responses are entered. Among the testtaker variable affecting the unreliability are physical discomfort, lack of sleep, unwanted effects of some drug or medication as well as some other emotional problems. Some other testtaker-related sources of error variance are frequent changes in mood, illness, causal life experiences, etc. All these sources make the testtakers prone to committing errors in answering the test.

There are some factors which are related to the examiners or test givers. The physical appearance of the examiner, the presence or absence of examiner and the way the examiner asks questions in oral examination are likely to influence the performance of the testtakers at the test. Sometimes by emphasizing upon the key words while putting questions, the examiner provides some unnecessary cues to the testtakers. Examiners head nodding, eye-movements or some other nonverbal gestures also provide some cues to the probability of the correctness of a response. All these become a potential source of error variance.

3. Test scoring and interpretation :

Test scoring and scoring systems are also important source of potential error variance. In recent days although there has been placed much reliance upon computer scoring, not all tests can be scored with the help of computer. There are many personality tests where the testtakers are required to supply some words, sentences, pictures as responses to stimuli. These responses are later to be evaluated quantitatively or qualitatively by the examiner. Likewise, there are tests of intelligence and creativity where the testtakers responses are to be evaluated subjectively at some point by the examiners. All such test scoring contributes to the error variance and reduces reliability. In creativity test the testtakers might be asked to create as many things as he can in response to the given stimuli. These creations have to be evaluated and interpreted by the examiners by awarding credit to some of these creations. Again the probability of the subjectivity is introduced thus contributing to the error variance. These problems in scoring agreement can, to some extent, be addressed through training to the scorers or examiners so that they may be able to show some consistency in the scoring.

4. Miscellaneous sources :

Some assessment situations themselves tend to produce varieties of systematic and nonsystematic error. Let us consider a situation in which the examiner is interested in assessing degree of agreement between husband and wife regarding the quality and quantity of physical abuses in their relationship. This is a situation in which the physical abuses between the partners occur in private situation, and only these two persons know what really goes on between them. In such situation the potential source of nonsystematic error includes failure to notice abusive behaviour, forgetting, wrong reporting (underreporting or overreporting) etc. In general, females tend to underreport physical abuse due to shame, fear, etc. However, they may overreport the physical abuse if they are expecting some help from others. Males tend to underreport abuse due to embarrassment and overreport if they are trying to justify the report. Many studies have confirmed this (Riggs et al., 1989, Strauss, 1979). All these tend to contribute to the error variance of the measuring instrument.

The systematic error arises when, unknown to the test developer, a test consistently measures something other than behaviour or trait for which it was intended. For example, if a scale for assessing introversion inadvertently also taps or measures anxiety consistently, it would be example of systematic error and contributes to the error variance. Since presence of systematic error goes initially undetected, systematic measurement errors tend to constitute a very significant problem in the psychological testing.

Two Modes to Express Reliability

There are two ways to express reliability of the test scores: standard error of measurement (SEM) and reliability coefficient. The standard error of measurement, also called as the *standard error of score* is well suited to the interpretation of individual scores. The standard error measurement allows us to estimate the extent to which a test provides inaccurate readings. In other words, it provides an estimate of the amount of error inherent in an observed score. The larger the value of standard error of measurement, the less certain we are about the precision or accuracy with which an attribute is assessed or measured. Conversely, a smaller value of standard error of measurement tells us that an individual score is probably closed to the value of the measured attribute. In general, the relationship between SEM and the reliability of a test is inverse, that is, higher the reliability of a test, the lower is the SEM.

Since SEM is a tool used to assess or infer the extent to which an observed score deviates from a true score, for many testing purposes, it is more useful than the reliability coefficient (Anastasi & Urbina, 1997). The standard error of measurement is statistically defined as the standard deviation of a theoretically normal distribution of test scores obtained by one testtaker on equivalent tests. Thus standard error of measurement is an index of the degree to which one testtaker's scores vary over tests presumed to be parallel. SEM can be indirectly computed from the reliability coefficient of the test. If the standard deviation of the distribution of test scores is known and the reliability of the test is also known, then an estimate of the standard error of a particular score can be determined with the help of the following formula:

$$SEM = \sigma \sqrt{1 - r_{tt}} \qquad (5.3)$$

where,

SEM = standard error of measurement
σ = standard deviation of test scores of the testtakers
r_{tt} = Reliability coefficient of test scores

Let us take an example to illustrate. Suppose Mohan has a score of 65 on a test and the mean of the group of the testtakers is 57 and the standard deviation 10. The reliability coefficient of the test is .84. Plugging these values in the formula, we get

$$SEM = 10 \sqrt{1 - .84} = 4.00$$

The standard error of measurement is used to create confidence interval so that the probability of range of score in which true score may lie can be estimated. Common confidence intervals in testing are the 68% interval, the 95% interval and the 99% interval. These intervals are created with the help of sigma or z score. The 95% interval is associated with a z score of 1.96. Therefore, the upper bound and the lower bound of the 95% interval would be calculated in this case as under:

Upper bound : 65 + 1.96 (4) = 72.84
Lower bound : 65 – 1.96 (4) = 57.16

Although we don't know the true score of Mohan who has obtained a score of 65, we can be 95% confident that his true score falls between 72.84 and 57.16. If the standard deviation of a test is held constant, the smaller the SEM, the more reliable is the test. As the reliability of test *increases*, SEM *decreases*. For example, when the reliability coefficient equals .88, σ = 15, the standard error of measurement equals to :

$$SEM = 15 \sqrt{1 - .88} = 5.196$$

On the other hand, when reliability coefficient equals .60, σ = 15, SEM equals to

$$SEM = 15 \sqrt{1 - .60} = 9.48$$

If the researcher is using standard error of measurement for estimating the range of true score, he makes an assumption. The assumption is that if the person is to take a large number of equivalent tests, scores on these tests would tend to be normally distributed with the person's true score as the mean.

Another way to *express* reliability to the test is the *reliability coefficient*, which is a correlation coefficient providing a statistical index of the extent to which two measurements tend to agree or to place each testtaker in the same relative position, that is, low with low and high with high. The more nearly the testtakers maintain the same set of score the second time as the first, higher the correlation and more reliable the test.

Although reliability coefficient is the correlation coefficient (product-moment correlation), it, unlike correlation, is always positive whereas the correlation may be positive or negative. The reliability coefficient of a test ranges from 0 to +1.00 whereas simple correlation varies from –1.00 to +1.00. Reliability coefficient becomes zero when measurement involves nothing but random or variable error and its value becomes +1.00 when there is no such error in the measurement.

The reliability coefficient and the standard error of measurement are very intimately related. As we know, as any obtained score is equal to true score plus error score so also variance in observed scores (that is, total variance σ) is equal to variance of true score and variance of error score . The reliability of the test is equal to the proportion of variance in the observed scores that is attributable to true variance or true differences between people (cf. formula 5.2). We can also show that the standard error of measurement is a function of the standard deviation of the observed scores and the reliability coefficient (cf. formula 5.3).

Researchers are of view that if one wants to compare the reliability of different tests, the reliability coefficient is a better measure. However, if one is interested in interpreting the *individual scores*, the standard error of measurement is more appropriate.

Reliability Estimates

There are several methods of estimating reliability of the test important ones are as under:
- Test-retest reliability method
- Parallel forms or Alternative-forms reliability method
- Internal consistency method
  - Split-half reliability
  - Kuder-Richardson formulas
  - Coefficient Alpha
- Inter-scorer reliability

A discussion of these methods follows :

• Test-retest reliability method

In test-retest reliability the researcher estimates the reliability by correlating pairs of scores from the same testtakers on two different administration of the same test. Thus a test administered today and then administered after say, a fortnight later on the same individuals and yields a high correlation between these two sets of scores, it becomes an index of test-retest reliability. The reliability coefficient thus obtained is also known as the *temporal stability coefficient*. The test-retest method of reliability is most appropriate when the test measures characteristic or trait that don't vary with timer or don't change overtime. It means that the tests that measure some constantly changing characteristic are not appropriate for test-retest estimate of reliability. A personality trait is considered relatively stable over time. Therefore, a test intending to assess such trait can be evaluated with test-retest reliability estimate. A test-retest reliability equation may be put as under:

Test score/Retest score = 1   (5.4)

or

(Test score) − (Retest score) = 0

Equation 5.4 clearly indicates that if the ratio is 1.00, it shows 100 percent reliability or if the difference between test score and retest score is zero, it also indicates 100 percent reliability. The greater the difference between the test score and retest score, power is the reliability.

One of the important factors in test-rest estimate of reliability is the determination of the appropriate time interval between the two administrations of the test. If the two administrations of the test are very close in time, there is a risk of *practice effect* and *carryover effect* and therefore, the correlation between these two sets of scores may be spuriously high. If the time between the two administrations increases, many other factors such as new learning, maturational factors may intervene and the resulting correlation may be low. Thus the passage of time can be a source of error variance in the test-retest reliability. When there elapses more than six months between the two administrations of the same test, the estimate of test-retest reliability is often referred to as the *coefficient of stability*. (Cohen & Swerdlik, 2005)

The advantage of this method is that it permits the test to be compared with itself and thus, it avoids all those problems that might arise with the use of another test. The major disadvantage is that when the time period between two administrations is too short, various factors like practice, experience, fatigue, motivation, etc. may intervene and confound the obtained measure of reliability.

- **Parallel forms or Alternative-forms reliability method**

Sometimes the test developer develops two forms of the same test. These two forms are independently constructed to meet the same specifications. These two forms usually have similar number of items and same range and level of difficulty in items. More popularly, such forms are known as parallel forms[4] or alternate form or equivalent form of the test. For estimating alternate-form reliability, the both forms of test are administered on the same sample and subsequently, the two sets of score are correlated with each other. Such correlation coefficient become the estimate of alternate form reliability. Sometimes both forms of the test are administered the same day on the sample (immediate) and sometimes these two forms of the test are administered on the same sample with some time gap such as a fortnight gap or so (delayed). When the two forms of the test

---

Although parallel forms of the test and alternate forms of the test are used interchangeably, recently some authors have tried to draw a line of distinction. Cohen and Swerdlik (2005) have pointed out that parallel forms of the test are said to come into operation when mean and the variances of observed scores for each form are equal. Alternate forms are simply different versions of a test so constructed as to be paralleled. They don't meet the basic requirements of the legitimate designation 'parallel'. Alternate forms are typically constructed to be equivalent with respect for variable like content and level of difficulty

are administered the same day, time sampling variances are reduced to zero. However, item sampling differences remain active. In other words, although the two forms are equally difficult on the average, some testtakers may find one form a bit harder (or easier) than the other form. On the other hand, when two forms of the test are administered on the same sample with a reasonable time gap, both item sampling differences and time sampling differencing become sources of variance. Such reliability coefficient becomes a measure of both temporal stability as well as consistency of response to different item samples or test forms. Thus this reliability coefficient tends to combine two types of reliabilities.

Since developing alternate forms of tests can be time-consuming as well as expensive, alternate from of reliability is comparatively less popular.

- **Internal consistency method :**

One can estimate the reliability of the test without administering it twice on the sample. One can estimate reliability of the test after single administration of the test. This is called *internal consistency estimate of reliability* or an estimate of *inter-item consistency*. There are different methods of obtaining internal consistency estimates of reliability. Of these methods, we shall here discuss only three: *Split-half reliability, Kuder-Richardson formulas* and *coefficient alpha*.

Split-half method of estimating reliability is very popular. In split-half reliability the test is administered on the sample. Subsequently the test is divided into two equal halves, scored independently and correlated with each other. On the basis of this half test correlation, the correlation for whole length test is estimated. The computation of the coefficient of split-half reliability generally entails the following four steps:

- Administer the test on the designated sample
- Divide the test into two equivalent halves
- Compute a Pearson r between the scores on the two halves of the test
- The obtained half-test correlation is corrected for full length or whole length test using Spearman-Brown prophecy formula.

The biggest challenge before the estimate of split-half reliability is how to divide the test into two equal halves. There are more than one way to split a test but there are some ways through which one should *never* split the test. Dividing the test into two halves in the middle is not recommended because this technique way unnecessarily lower or raise the reliability coefficient. Moreover, in such division, differences in item difficulty, different amount of test anxiety as well as the different amount of fatigue for the first half as opposed to the second half of the test may contribute to lowering the value of reliability coefficient. Another way to split the test is to randomly assign all items into two halves. This method should also be avoided. However, the most desired way of dividing the test is to put all odd-numbered item (1, 3, 5 etc.) into one half and all even-numbered items (2, 4, 6, 8 etc.) into other half of the test. This method yields an estimate of split-half reliability popularly known as *odd-even reliability*. In general, the major purpose of splitting the test is to develop what is called '*mini-parallel forms*' with each half nearly equal or as nearly equal as possible in format, style and other aspects.

When the test has been splitted into two equal halves, the scores on the two halves are correlated. On the basis of this half-test reliability, the reliability of the whole test is estimated using Spearman-Brown prophecy formula, which is as under:

$$r_{tt} = \frac{2r_{hh}}{1+r_{hh}} \quad (5.5)$$

where,
r_{tt} = reliability of the whole test
r_{hh} = correlation of two halves.

For example, if halves test correlation is 0.65, the correlation for whole length test by formula 5.5 would be:

$$r_{tt} = \frac{2 \times .65}{1+.65} = \frac{1.30}{1.65} = 0.788$$

Estimate of reliability coefficient based on whole test tends to be higher than those based on half of a test because usually (but not always) reliability increases with increase in the length of the test.

The use of Spearman-Brown prophecy formula is not always recommended. The use of this formula would be inappropriate for measuring the reliability of heterogeneous test and speed tests.

A popular and alternate method of finding split-half reliability was developed by Rulon (1932) called Rulon's formula, which requires only the variance of the *differences* between each testtaker's scores on the two half tests and the variance of the total score. By substituting these two values in the formula 5.6, one can estimate the reliability of whole test directly.

$$r_{tt} = \frac{\sigma_d^2}{\sigma_t^2} \quad (5.6)$$

Here it is interesting as well as appreciating to note the relationship of Rurlon formula to the definition of error. The difference between a testtaker's scores on the two half-tests represents error variance. The variance of those differences when divided by the variance of the total scores, provides an estimate of proportion of error variance in the scores. When this error variance is subtracted from 1.00, it provides the proportion of true variance for specified test use, which becomes equal to the reliability coefficient (see formula 5.2)

Kuder-Richardson (K-R) formullas:

Kunder and Richardson (1937) developed some procedures to calculate inter-item consistency. Like split-halve methods, here inter-item consistency or internal consistency of the test is found from single administration of the single test. K-R formula does not require two half-scores rather it requires examination of performance on each item. There are several K-R formulas of which K-R 20 and K-R 21 are the most popular ones. They are so called because they are twentieth and twenty-first in the series of derivations. K-R 20 is relevant in those situations in which items are scored as 0 or 1 (such as right or wrong) or according to some other all-or-none system. For the use of KR-20, items should not much vary in their indices of difficulty. The formula for K-R 20 is as under:

$$KR_{20} = \left(\frac{n}{n-1}\right)\left(\frac{\sigma_t^2 - \sum pq}{\sigma_t^2}\right) \quad (5.7)$$

where,

Kr_{20} = Reliability coefficient by K-R 20
n = number of items in the test
= variance of total scores on the test
p = Proportion of correct answer to each item
q = Proportion of incorrect answer to each item, hence, it is equal to 1-P.

Mathematically, it has been proved that K-R20 formula provide the same estimate of reliability that one would get if one takes the mean of split-half reliability estimates obtained by dividing the test in all possible ways.

Another formula for estimating reliability coefficient by Kuder and Richardson is $K-R_{21}$, which requires that all items should be of equal difficulty values or that the average difficulty level is 50%. In fact, $K-R_{21}$ is the simplification of $K-R_{20}$ and the information needed for $K-R_{21}$ is the mean of total test score, SD of the total score and the number of items in the test. The $K-R_{21}$ is expressed as under

$$KR_{21} = \left(\frac{n}{n-1}\right)\left(\frac{\sigma_t^2 - n\overline{p}\overline{q}}{\sigma_t^2}\right) \quad (5.8)$$

It is obvious that $\sum pq$ of $K-R_{20}$ has been substituted for $n\overline{p}\overline{q}$ in $K-R_{21}$. is the average of correct proportion to each item and is the average of incorrect proportions to each item in the test. Rest of symbols of formula 5.8 are defined as they had been defined in formula 5.7. Statistically, $\overline{p} = \frac{M_t}{n}$

and $q = \left(\frac{n - Mt}{n}\right)$ where M_t is mean of total score and n is the number of items.

One of the limitations of KR-21 is that in practice the assumption of the formula requiring that the difficulty indices of all items should be equal or that the average difficulty is 50% is rarely met. Therefore, $K-R_{21}$ under estimates the split-half reliability (Kaplan & Saccuzzo, 2001)

Coefficient Alpha:

Coefficient Alpha as one method of estimating internal consistency of the test was developed by Cronbach (1951) and later on elaborated by others (Novick & Lewis, 1967; Kaiser & Michael, 1975). It is an index of the internal consistency of items, that is, tendency to correlate positively with one another. Co-efficient alpha may be thought as the average or mean of all possible split-half coefficient. Thus both K-R$_{20}$ and coefficient alpha can be thought of as average of all possible split-half coefficients for groups tested (Linn & Miller, 2011). The difference between the two is that the K-R$_{20}$ is used for estimating reliability of tests containing dichotomous items (that is, Yes-No, True-False) whereas the coefficient alpha can be used not only for dichotomous items but for tests containing interval and ratio items such as five-point rating scales. K-R$_{20}$ is not suited for evaluating internal consistency in many cases especially in those types of tests for which there are no 'right' or 'wrong' answer as we find in many personality and attitude scale where the testtakers express their responses in terms of so many response options like 'strongly agree', 'agree', neutral, 'disagree' and 'strongly disagree'. None of these responses is correct and none is incorrect. The response of the testtaker indicates only his standing on the continuum between 'strongly agree' and 'strongly disagree'. To meet with such situation, Cronbach developed a method of estimating internal consistency for those tests where their items are not scored as 0 or 1. This method is called as coefficient alpha and the formula for this is as under:

$$ra = \left(\frac{n}{n-1}\right)\left(\frac{\sigma_t^2 - \sum \sigma_i^2}{\sigma_t^2}\right) \quad (5.9)$$

where,

r_a = coefficient alpha
n = number of items
= variance of the total score
= variance of the individual item
Σ = This summation sign indicates that one is to sum individual item variances.

It we pay attention to coefficient alpha, it will be obvious that it extends the K-R method to the types of tests with items that are not scored as 0 or 1. In computing coefficient alpha the sources of error variance are item sampling and test heterogeneity. Unlike Pearson r whose value ranges from −1 to +1, coefficient alpha typically ranges from 0 to 1. This is because coefficient alpha, like other coefficient of reliability, tends to answer questions how similar sets of given data are. Here similarity is assessed on a scale from 0 (absolutely no similarity) to 1 (perfectly identical). Since the negative value of alpha coefficient is theoretically impossible, under the rare circumstance the alpha coefficient value may be reported as zero (Hensen, 2001). There is also a myth about alpha coefficient that larger its value, the better it is. In fact, value of alpha coefficient above .90 may be too high indicating redundancy in items (Streiner, 2003).

One common point in all the measures of internal consistency is that they evaluate the extent to which the different items on a test measure the same ability or trait. All these measures would yield low estimates of reliability if the test is designed to assess several traits, that is, the test is heterogeneous.

- **Inter-scorer reliability:**

In some tests under some circumstances, the score is much influenced by the scorer than anything else. A test can have homogeneous items and yield heterogeneous scores and even then, it may not be reliable if the person scoring the test makes mistakes. Certainly, projective tests, creativity test and tests of moral development fall under this category. If the scorer is a major factor in the reliability of these tests, a report of interscorer reliability is imperative.

Perhaps the simplest way to determine the degree of consistency among scorers in scoring a test is to calculate coefficient of correlation. A test is independently scored by two or more scorers and the scores for pairs of scorers are then correlated. Such correlation coefficient is referred to as a *coefficient of inter-scorer reliability*. It is in practice that the test manuals report the training and experience required from the scorers and then, list inter-scorer reliability coefficients.

Interscorer reliability coefficient is regarded as supplement to other reliability estimates and never replaces them. In such reliability the scorer differences are the source of error variance. Table 5.1: presents the brief summary of methods of estimating reliability along with its details.

Table 5.1: Brief Summary of Methods of Estimating Reliability

Method	No. of Forms	No. of Sessions	Sources of Error Variance	Statistical Procedures
Test-Retest	1	2	Time sampling	Pearson r or Spearman rho
Alternate-forms (immediate)	2	1	Time sampling	Person r or Spearman rho
Alternate form (delayed)	2	2	Item sampling and time sampling	
Split-half	1	1	Item sampling, Nature of split	Pearson r
Kuder-Richardson	1	1	Level of difficulty values, Item sampling	Pearson r
Coefficient Alpha	1	1	Item sampling, Test heterogeneity	Pearson
Interscorer	1	1	Scorer differences	Pearson r or Spearman rho

Reliability of a Difference Score

Sometimes psychological testing implies a *difference score* created by subtracting one test score from other. A difference between performance on two points might be calculated. For example, a group of students' performance might be compared before and after a training programme. Likewise, a difference between measures of two different abilities such as whether a student is doing better in Psychology or in Education, might be a point of interest.

Such difference score creates many problems in comparison to single score and this makes difficult to work with them. Let us elaborate this point: As we know, any observed score is composed of true score and error score. In difference score, error score is expected to be larger than either observed score or true score because such error score absorbs error from both of the scores on the basis of which difference score is created. Apart from this, true score might be smaller than error score because whatever is common in both measures tends to be cancelled out when the difference score is created. Due to these two factors, the reliability of the difference score tends to be lower than the reliability of either score on which it is based.

As we know, the most convenient way to find the difference scores is to create z or sigma score for each measure and then, find the difference between them. The reliability of scores that convey the difference between two z scores (standard score) is given by the following formula.

$$r_{DD} = \frac{\frac{r_{xx} - r_{yy}}{2} - r_{xy}}{1 - r_{xy}} \quad (5.10)$$

where,

r_{DD} = reliability of the difference between scores on X and scores on Y.
r_{xx} = reliability of X
r_{yy} = reliability of Y
r_{xy} = correlation between X and Y

Suppose that the correlation between two measures (X and Y) is .65 and the reliabilities of the two measures are .90 and .80 respectively. The reliability of the difference between these two measures by formula 5.10 would be :

$$r_{DD} = \frac{\frac{(.90 + .80)}{2} - .65}{1 - .65} = \frac{.20}{.35} = .571$$

The example clearly shows that the reliability of the difference score between two test reliabilities (.90 and .80) is only .571. The circumstance in which the reliability of the difference score is lower than the average reliabilities of the two initial tests is not *unusual*. The reality is that it occurs in all situations except when the correlation between two tests is zero.

Index of Reliability

Index of reliability is statistically defined as the correlation coefficient between obtained scores and their true scores. This statistic indicates the extent to which one can depend upon the obtained scores as a measure of true score because it gives the maximum correlation, which the test is capable to yield in its present form. Index of reliability is calculated by finding the square root of the reliability coefficient of the test. The formula is:

$$r_1 = \sqrt{r_{tt}} \qquad (5.11)$$

Where r_1 = index of reliability and r_{tt} = reliability coefficient of the test. Suppose the reliability coefficient of a test is .60, then by formula 5.11, the index of reliability = $\sqrt{.60}$ = .77, which is the maximum correlation the test can yield in its present form.

Which type of Reliability is More Appropriate?

There are different methods of estimating reliability such as test-retest, spit-half, Kuder-Richardson, coefficient alpha and interscorer reliability. Now the question arises: which method is the best? When should the test developer use one method but not the other?

For knowing the correct answers of these questions, we must assess the nature and purpose of the test in hand. Where tests must be administered to the testtakers more than once, it would be reasonable to expect that the test must display reliability across time and then in such case, test-retest reliability would be the most appropriate method. Where the purpose is to assess factority purity, alpha coefficient is the most appropriate method. Thus when a test intends to assess a single factor, alpha coefficient would be the better choice. But when the test contains multifactors as we find in case of intelligence test, coefficient alpha would not be appropriate. Split-half reliability would be a good choice when the items of the test are carefully ordered according to the indices of the difficulty value. Likewise, inter-scorer reliability would be the most appropriate choice for any test that contain subjectivity of scoring.

Conditions affecting Reliability coefficients

The most common conditions that affect the reliability of the test scores are as under: A discussion these common factors is as under:

- **Homogeneity vs. heterogeneity of test items**

Whether test items are homogeneous or heterogeneous have an impact upon the reliability of the test. A homogeneous test is a test, which contains items that measure a single trait or ability. In other words, homogeneity refers to the extent to which items in the test are unifactorial. On the other hand, a heterogeneous test is one that measures different factors. In other words, such test contains items, which measure more than one factor. For homogeneous test, it is quite reasonable to expect that internal consistency will be high. By contrast, if the test has homogeneous items, the most appropriate reliability is the test-retest reliability and if internal consistency reliability is calculated, it will be adversely affected and will be low.

- **Dynamic vs. static characteristics being measured**

What is being measured is dynamic or static is also an important consideration in reliability. A dynamic characteristic is one which changes over time as a function of situational and cognitive experiences. For example, amount of anxiety experienced by the person is a dynamic characteristic, which varies over time due to various situational and cognitive factors. A test that intends to assess such dynamic characteristics, will have low test-retest reliability. However, its internal consistency reliability will not be affected and therefore, it will be its best measure of reliability. On the other hand, when trait or ability is static, that is, not changing over time, the test is said to assess static characteristic (such as intelligence) and then either test-retest or alternate form reliability would be appropriate. Other estimates of reliability will be unduly affected by such static characteristics.

- **Number of items on the test or length of the test**

The reliability coefficient of the test is affected by the number of items, which determines the length of the test. The more items are on the test (that is, the more lengthy the test is), higher is the reliability coefficient. The condition is that added new items must be homogeneous, that is, must measure the same trait or attribute. Partly this happens because the increase in the number of items is accompanied by an increase in the potential variability of the testtakers within the group. If there is a single item test, there cannot be much variability since scores can be either 0 or 1. Another reason of longer test being more reliable one is that such test, also provides a better and consistent sample of the testtakers knowledge or ability than shorter ones.

If the researcher can assume that the ability of the test items, traits or abilities measured by those items and the nature of the testtakers remain same, the increase in reliability as a function of increase in test length or number of items may be given by Spearman Brown formula as under:

$$r_{nn} = \frac{n(r_{tt})}{1+(n-1)_{rtt}} \quad (5.11)$$

where,

r_{nn} = reliability coefficient of lengthened test
n = number of times test has been increased or decreased
r_{tt} = reliability coefficient of original test.

Suppose an intelligence of 50 items has reliability coefficient 0.65. If the this test is increased two times (100 items now), its reliability coefficient by formula 5.11 will be

$$r_{nn} = \frac{2(.65)}{1+(2-1).65} = \frac{1.30}{1+(1).65} = \frac{1.30}{1+.65} = \frac{1.30}{1.65} = .788$$

Thus lengthening the test two times increases the reliability coefficient of the test from .65 to .788.

Likewise, if its length is increased four times (200 items, its reliability will be enhanced from .65 to .88 by the same formula, that is, 5.11.

One can also rearrange the formula 5.11 a little bit and solve for n for estimating by what factor test has to be lengthened in order to achieve a desired reliability coefficient. One can use the following formula to determine how much one requires to lengthen the test (n):

$$n = \frac{r_n(1-r)}{r(1-r_n)} \qquad (5.12)$$

where,
   n = number of times test be lengthened
   r_n = the desired reliability coefficient
   r = the present reliability coefficient

Suppose we have a test with r = .70 and we want the r to be at least .90. We can use formula 5.12 to know the number of times the present test should be lengthened.

$$n = \frac{.90(1-.70)}{.70(1-.90)} = \frac{.270}{.07} = 3.85$$

Thus n = 3.85 which means that for increasing reliability coefficient up to .90, we have to make test 3.85 times as long.

How much the researcher can lengthen a test is, of course, limited by some practical considerations. For example, such lengthening is limited by factors of fatigue and boredom experienced by testtakers, amount of time available for testing and sometimes by the inability of the test developers to construct good test. But within these limits, reliability is enhanced if the test is lengthened.

It should also be carefully noted that there is a point of diminishing returns in lengthening a test. When the reliability of the test is already moderately high, a considerable increase in length of test produces only a modest increase in reliability. For example, if a 20-item test has reliability coefficient of .50 and it is lengthened five times (now 100 items), there occurs a substantial increase in reliability from .50 to .83. But once again, if the test is doubled in length (now, 200 items), only a modest increase in reliability from .83 to .91 occurs. Again, if the test is doubled (400 items) the reliability is enhanced from .91 to .95 a very poor increase in reliability.

One special kind of lengthening is often represented by increasing the number of raters who rate a person or something produced by the person. If several raters with equal competence or equal familiarity with the ratees have rated the persons, then a pooling of their ratings will yield a composite that is more reliable. For example, if a part of raters judge a sample of 10 handwritings and show a correlation of .45, then a pooled ratings of four raters could be expected to correlate with four others as calculated by formula 5.11.

$$r_{nn} = \frac{4(.45)}{1+(4-1).45} = \frac{1.8}{2.35} = .77$$

Psychologists and educators have also devised means for estimating the loss (or decrease) in reliability if the test is shortened. Suppose a 100-item test has a reliability of .80 and the researcher wishes to estimate the reliability of a 30-item test made up of a sample of items taken from this longer 100-item test. Now, the length of this new test is 30/100 = .30 times the length of the original test. The estimate of reliability of the new shorter test, by formula 5.11 would be

$$r_{nn} = \frac{.30(.80)}{1+(.30-1).80} = \frac{.24}{1+(+.70).80} = \frac{.24}{1+(-.56)} = \frac{.24}{.44} = .545$$

This procedure is very much helpful in judging whether shortened test has sufficient degree of reliability.

Ebel (1972) has shown that when a test is doubled in length, its true variance increases four times whereas its error variance increases only two times. However, longer test is no guarantee of high reliability. The Spearman-Brown formula indicates that reliability will increase as long as items added to the test are equivalent to those in original test. In fact, if poor items (such as ambiguous or vague items) are added, it will lower reliability coefficient.

- **Range of test scores or variability of the group tested**

    Reliability coefficient is influenced by the spread of scores or the true variance. In general, the larger the true variance or the spread of scores, greater will be the reliability. If the spread of scores (or variance) in either variable in a correlational analysis is restricted, then the resulting correlation coefficient tends to be lower. On the other hand, if there exists greater variability in the scores of either variable in the correlational analysis, the resulting correlation coefficient tends to be higher. Let us see what happens when the variability in scores is zero. In such extreme case all the testtakers receive the same score and then, naturally the standard deviation is zero. When standard deviation is zero, the variance is also zero. Each testtaker is then at the mean of his group and has a z-score of zero. Since reliability coefficients (or correlations) are defined as the average product of z-scores and here, the product of z scores is zero. Therefore, reliability coefficient is also zero. However, when there occurs some spread of scores or variability in the scores, the chance of some correlation and therefore of reliability is possible.

- **Difficulty level of the test**

    When the test is very easy (having items showing high percentage of giving correct responses) or very hard (having items showing low percentage of giving correct responses), it remains incapable of measuring individual differences since the testtakers tend to respond uniformly. The difficulty level of a test may be defined as M/n, where M is the mean of the scores in a distribution and n is the number of items. As the mean comes nearer to the number of items on the test, the test becomes progressively easier. On a maximally easier test, the mean is equal to the number of items. In such maximally easier test, everyone responded correctly to all items and therefore, there would no variability among the testtakers. In such case variance will be zero and therefore, the reliability will also be zero. On the other hand, difficult tests encourage *guessing* that produces *random error* and this enhances variance. Consequently, it contributes to unreliability. When the tests are comparatively easier, guessing is reduced and therefore, that source of error is reduced.

- **Objectivity in scoring: scorer reliability**

    Scorer reliability refers to the extent to which different observers or raters agree with one another in scoring the same set of papers. Higher the degree of agreement among observes, higher will be the scorer reliability. To the extent that there is subjectivity in the scoring, reliability coefficient are likely to be low. The subjectivity in scoring means that inconsistencies are allowed to create some random errors, which in turn, tend to lower the reliability coefficient.

- **Test-retest interval**

    When there is a longer interval between two administrations of a test, the reliability coefficient is likely to be lowered. This happens mainly because of two reasons. *First*, longer interval tends to increase the probability for many errors that make come through changes in test administration, environment or some personal situations. *Second*, longer interval provides more opportunities for the testtakers to bring changes in terms of factors being assessed. Such changes ultimately cause a change in the testtakers' true scores.

- **Test administration**

    There are three major factors related to test administration that affect reliability of the test. *First*, the testtakers must follow the instructions carefully. This ensures a decrease in errors that usually results when the testtakers are not following the instructions uniformly and properly. *Second*, there should be maximum constancy between two administrations of the tests. If somehow the conditions for two administrations of the test differ, many errors are likely to be introduced. *Third*, good and effective testing practices should be adopted. This tends to decrease the contamination of testtakers' scores with errors.

    Thus reliability of the test is affected by many factors and test developers must keep those factors in view while determining the reliability of the test.

The Concept of Validity

No matter how reliable a test is, it is useless if it is not valid. The term validity is used in conjunction with the meaningfulness of test score, that is, to what the test score truly means. In simple words, validity of the test scores indicate an estimate of how well a test measures what it intends to measure in a particular context with a particular population of people. According to Anastasi and Urbina (1997). "the validity of a test concerns what the test measures and how well it does so. It tells us what can be inferred from test scores." Paraphrasing the definition of validity given by the influential *Standards for Educational and Psychological Testing* (AERA, APA &

NCME, 1985, 1999), it can be said that a test is considered valid to the extent the inferences drawn from it are appropriate, meaningful and useful. Thus a test does not have validity in any absolute sense rather the scores produced by the test are valid for some uses and not valid for others. A common misconception is that a test *is* or *is not* valid. In fact, a test is not valid *per se*. It is valid for a particular purpose and for a particular group. So the question is not being 'valid or invalid' but being 'valid for what and for whom'. A valid test of geography is not very likely to be a valid for personality test.

Validity, like reliability is a matter of degree and not an *all-or-none* property. Any test cannot be said to be wholly valid or not valid at all for the purpose it was constructed. Likewise, validity is *not* a fixed property of the test scores. A test today may be valid but it may not remain valid after thirty or forty years later because many new concepts and meanings may emerge during such a long time. Thus validation is not a fixed process rather an unending process. *Validation* means the process of gathering and evaluating validity evidences. It is the responsibility of the test developer to supply validity evidence in the test manual. Sometimes the test users may plan to conduct their own *validation studies* with their own set of testtakers. Such local validation studies give insights regarding a particular set of testtakers as compared to sample described in the manual.

The validity of the test is influenced by systematic errors of measurement. In other words, validity is the measure of systematic or constant whereas reliability is the measure of variable or random errors. Random errors usually result from uncontrolled and uncorrelated factors such as fluctuating mood, motivation, attention of the persons, noise and other similar distractors. Systematic errors result from defects in test construction, limited time allowed for completion of test essentially in ability test, personal modes of responses, etc. Systematic errors generally act in one direction, causing either over estimation or underestimation of the scores in all cases systematically.

Categories of Validity

There are three important categories of validity, popularly called or *trinitarian model of validity*. The three categories are:
- Content Validity
- Criterion-related validity
- Construct validity

These three approaches of assessing validity associated respectively which *content validity*, *criterion-related validity* and *construct validity* tend to:
(a) scrutinize the content of the test
(b) relate the scores obtained on the test with other test scores or criterion score
(c) provide a comprehensive analysis of
 (i) how scores on the test be better understood within a theoretical framework for understanding the construct, which the test is intended to measure.
 (ii) how test scores relate to other test scores to which it should relate and how do they not relate to other test scores to which they should not relate.

There three approaches to validity should not be thought as mutually exclusive. In fact, each should be thought as one type of evidence which with others, contributes to a judgement concerning the validity of the test.

The above discussed trinitarian model of validity has been criticised (Landy, 1986; Messick, 1995; Cronbach, 1988; Shepard, 1993). These researchers condemned trinitarian model of validity as being fragmented and incomplete and have advocated for a *unitary view of validity*, which takes into account everything from the implication and interpretation of test scores in terms of societal values to the consequences of the test use. According to this unified view of validity, test validity (or even test validation) is a minsomer because what is validated are the implications and interpretations of test scores and uses of these scores for particular applied purpose and not the test itself. When a test is constructed and it is not administered, validation is not an issue. When it is administered and test scores are obtained but kept in almirah, again validation is not an issue. But when those test scores are used for a purpose, the questions arises about the appropriateness of the scores for that use. Here inferences in need of validation are of two types: *interpretative inference* and *action inference*. Validation of interpretative information involves providing multiple lines of evidences in support of particular interpretation of test scores and at the same time demonstrating that other interpretations are less appropriate. Validation of action inferences involves both evidences regarding the meaning of score as well as evidences regarding appropriateness and usefulness of test scores for particular applied purposes. In a nutshell, Messick (1995) as well as others have expanded the boundaries of validity beyond meaning of its score to include implications of some interpretation and utility, relevance as well as some societal values to the consequence of test use. A discussion of the above three validity follows –

Content validity :

Content validity or content-related validity is defined as an assessment of whether a test contains appropriate content and requires that the appropriate process have been applied to that content (Thorndike & Thorndike-Chirst, 2015). Content validity is determined by the extent to which items on a test becomes representative of the universe of behaviour the test intended to sample. If the sample (the set of items) is representative of the population (all possible items), then the test is said to possess the content validity. Content validity requires both *itemvalidity* and *samplingvalidity*. By item validity is meant whether the items of the test represent measurement in the intended content area. Sampling validity refers to how well the test samples the total content area. A test intended to assess knowledge in chemistry might have good item validity because all the items do indeed deal with the facts of chemistry but might have poor item validity because all the items deal only with organic chemistry. Content validity is of prime importance for achievement tests.

There are some statistical methods to determine the overall content validity of the test (Gregory, 2004, Lawshe, 1975). One such popular statistical method is based upon interrater agreement for determining content validity. Let us take an example to illustrate it. Suppose two experts (Expert 1 and Expert 2) evaluate the individual items on a four-point scale ranging from 1 to 4. Out of these four, 1 stands for 'very weak relevance', 2 stands for 'weak relevance', 3 stands for 'strong relevance' and 4 stands for 'very strong relevance'. Ratings of each expert on each item may be grouped, into two categories such as 'weak relevance' (1 & 2) and 'strong relevance' (3 & 4). Subsequently, for each item a joint rating of the two experts may be arranged into 2 × 2 table as depicted in Table 5.2. This table shows a hypothetical data on 50-item test produced by two experts. A close

Table 5.2: Inter expert agreement model for content Validity for a 50-item test

		Expert 1	
		Weak relevance	Strong relevance
Expert 2	Weak relevance	A2 items	B3 items
	Strong Relevance	C5 items	D40 items

scruity of Table 5.2 shows that of the 50 items, 40-items are such, which show strong relevance between the two expert (cell D). The cell B and C reflects disagreement and cell A indicates that 2 items shows agreement between the two experts regarding the fact that items don't belong on the test due to their weak relevance. A coefficient of content validity may be calculated by formula 5.13.

$$\text{Coefficient of content validity} = \frac{D}{(A+B+C+D)} \qquad (5.13)$$

$$= \frac{40}{(2+3+5+40)} = \frac{40}{50} = .80$$

Thus coefficient of content validity is .80, which is high and satisfactory. If more than two experts are used for judging content validity, the same computational procedures could be used with all possible pair-wise combination of experts and then, average coefficient may be calculated.

Several other methods of computing content validity has been proposed (James et al., 1984; Lindell et al., 1999; Tinsley & Weiss, 1975; Lawshe, 1975). it is not possible to deal with all these methods. However, as one illustration, Lawshe's method of quantifying content validity will be discussed here.

Lawshe (1975) attempted to quantify content validity on the basis of agreement among experts or raters regarding how essential a particular item is. Lawshe proposed that expert must respond to the following question for each item :

Is the skill or knowledge assessed by the item ... to the performance of the job?
(a) essential (b) useful but not essential (c) not necessary

For each item, the number of experts or raters who stated that the item is essential is noted. Lawshe proposed that if more than half experts stated that an item is essential, that item is said to have content validity. When a large number of experts agree that a particular item is essential, a stronger evidence for higher degree of content validity exists. Lawshe has developed a formula called *Content Validity Ratio* (CVR) for quantifying content validity.

$$\text{CVR} = \frac{n_e - \frac{N}{2}}{\frac{N}{2}} \quad (5.14)$$

where,

CVR = content validity ratio
n_e = Number of experts stating 'essential'
N = Total number of experts

The CVR may be *negative*, *positive* and *zero*. When fewer than half the experts state 'essential', the CVR is negative. When exactly half of the experts indicate 'essential', CVR is zero and when more than half the expert indicates 'essential', the CVR is positive. Let us take an example to illustrate these three CVRs. Suppose there are 12 experts who rated the items of the test. Item No. 5 was rated 'essential' by 4 experts, item No. 6 was rated 'essential' by 6 experts and item No. 7 was rated 'essential' by 10 experts. Their CVR, by formula 5.13, would be:

$$\text{CVR of item No. 5} = \frac{4 - 12/2}{12/2} = \frac{4-6}{6} = \frac{-2}{6} = -.33 \text{ (Negative CVR)}$$

$$\text{CVR of item No. 6} = \frac{6 - 12/2}{12/2} = \frac{0}{6} = 0 \text{ (zero CVR)}$$

$$\text{CVR of item No. 7} = \frac{10 - 12/2}{12/2} = \frac{10-6}{6} = \frac{4}{6} = .66 \text{ (positive CVR)}$$

In this way CVR is calculated for each item. The obtained CVRs if not significant at .05 level, are not retained. Lawshe has prepared a table which shows the minimum values of CVR at .05 level for different number of expert or panelists. For example, if the number of experts is 5, the minimum value at .05 level is .99, for number of experts being 6 and 7, the same minimum value of .99 is needed. However, if number of panelists is 8, the minimum value at .05 is .75, for 9, it is .78 for 10 it is .62, for 11, it is .49 and so on for 40, it is .29. In the present example, only CVR of item no. 7 is significant because it exceeds the minimum value of CVR (that is, .56). Such item would be retained.

Face validity is another term frequently encountered in psychological testing. Face validity s *not* a form of validity. A test is said to have face validity if it looks valid to testtakers and others. Thus face validity relate more to what the test appears to measure than to what the test actually measure. Thus face validity is really a matter of social acceptability. In fact, if a test definitely appears to measure what it purports to measure 'on the face of it', it could be said that the test has higher degree of face validity.

In reality, face validity of a test is important because if a test lacks face validity in the eye of testtakers, negative consequences may occur. For example, in such situation the testtakers may show a poor attitude towards the test and little cooperation or motivation to do their best on the test.

- **Criterion-related validity**

As its name implies, criterion-related validity is one where a test is shown to be effective in estimating the testtaker's performance by correlating his scores with scores on some outcome measure called *criterion*. In fact, the term criterion-related validity was introduced in 1966 edition of *Standards for Educational and Psychological Tests and Manuals*. In fact, criterion-related validity means the correlation of the present test scores with the scores on some external criterion. There is no hard-and-fast rules for what constitutes a criterion. It may be a test score, a specific behaviour of group, an amount of time, training cost, a rating, a psychiatric diagnosis, an index of absenteism and so on.

As stated above, criterion is any outcome measure against which the test is correlated. Now one important question here arises is that what are the characteristics of a good criterion? Researchers have outlined some important characteristics of a good criteria. Some of these are as under :

- A good criterion must be *relevant* or appropriate. It means that the criterion must be pertinent or applicable to the matter at hand. For example, if a test is intending to assess knowledge in the field of psychology, and if the criterion for this test is classroom performance of testtakers in the subject psychology, it would constitute an example of relevant criterion. In fact, the *Standards for Educational and Psychological Testing* sourcebook (AERA, APA & NCME, 1985) very clearly has pointed out that all criterion measures must be described accurately and the rationale for choosing them as relevant criteria should be explicitly made clear.
- The criterion must itself be *reliable*. A unreliable criterion means it is unpredictable, regardless of the merit of the test. In fact, both the reliability of the criterion as well as that of the test determine the theoretical upper limit of the validity coefficient and the validity coefficient is always less than or equal to the square root of the reliability of the test multiplied by the reliability of the criterion. In terms of equation, this can be expressed as :

$$r_{xy} = \sqrt{(r_{xx})(r_{yy})} \qquad (5.14)$$

where,

r_{xy} = The validity coefficient of the test
r_{xx} = The reliability of the test
r_{yy} = The reliability of the criterion

The equation 5.14 clearly gives hints that to the extent the reliability of either test or the criterion (or both) is low, the validity coefficient tends to diminish

- The criterion must also be *valid* for the purpose for which it is being used. If a test (say X) is being used as a criterion to validate another test, a clear evidence should exist that the test X is valid.
- A criterion must also be *uncontaminated*. A criterion is said to be contaminated when it possesses its artificial commonality with the test. When a test is correlated against a criterion and the criterion possess some such items, which are common to the test, the correlation between these two will be artificially inflated. A criterion can also be contaminated when the criterion consists of ratings from the experts who somehow have pre-knowledge about the testtakers' scores. When experts have pre-knowledge about the scores of the testtakers, their ratings consciously or unconsciously might be influenced by such knowledge, thus contaminating the criterion. When the test is validated against such ratings of the experts, the experts must not be given any knowledge about the test scores. The central idea is that the criterion measure should be free from bias. By this, it is meant that the criterion should be such on which each individual has the same opportunity to make a good score or on which each equally capable person gets the same score regardless of the group to which he or she belongs.
- The criterion must be *conveniently available*. How long one has to wait to get a criterion score for each testtaker? What will be its cost? etc. In fact, these are the practical problems which must be satisfactorily met. Any choice of the criterion measure must at least address these questions. (Thorndike & Thorndike-Christ, 2015).

Criterion-related validity contains two types of validity: *Concurrent Validity* and *Predictive Validity*. In concurrent validity the test scores and criterion scores are obtained simultaneously and subsequently, they are correlated. The resulting coefficient is the coefficient of concurrent validity. For example if a newly constructed test is administered on a sample upon which an already constructed intelligence test which is reliable and valid, is also administered. This already constructed test acts as a criterion. The correlation between the two sets of scores becomes the evidence for the concurrent validity. Higher the correlation coefficient, higher is the concurrent validity of the test. Concurrent validity is much desirable for achievement tests and diagnostic clinical tests. An evaluation of concurrent validity indicates the extent to which the test scores correctly estimates the person's present position on the relevant criterion.

In addition to establishing correlation between test scores and criterion scores for providing evidence to concurrent validity, such validity can also be estimated through the *method of discrimination*. When the test scores can be meaningfully and satisfactorily used to discriminate between persons who possess a certain characteristic and those who don't possess such characteristic or between those who possess more of certain characteristics than those of who possess less of certain characteristic. For example, a test of extroversion would have concurrent validity if scores resulting from it, could be used to classify correctly extroverts and non-extrovert persons.

Predictive validity (also known as *empirical validity* or *statistical validity*) refer to the degree to which a test can predict how well a person will do in future situation. An arithmetic aptitude test will be said to have predictive validity if scores on it accurately predict which student will do well in arithmetic and which student will not. Thus in predictive validity the correlation is computed between the test scores and the criterion measure to be obtained sometime in future. Predictive validity is especially useful for entrance examination test and employment tests. In estimating predictive validity, the test which is used to predict success is called as *predictor* and the behaviour which is predicted is referred to as the *criterion*. In estimating predictive validity of a test, the first step is to identify a carefully defined criterion, which must be a valid measure of the behaviour to be predicted. For example, if the researcher wishes to establish the predictive validity of an arithmetic test, final examination scores on the completion of course in arithmetic would be a valid criterion but the number of days absent during the course would not be a valid criterion. Once the criterion has been clearly identified, the procedure for determining predictive validity may be set as under:

- Administer the predictor variable or the test
- Wait until the behaviour that is to be predicted (criterion variable) occurs
- Obtain measures on the criterion variable
- Correlate the scores obtained on the test and the criterion variable.

If the obtained correlation coefficient is high, the test is said to demonstrate high predictive validity. Sometimes a combination of predictors is used to predict a criterion. In such situation prediction equation may be developed. An individual's scores on each of a number of tests are inserted into equation and his future performance is, then, easily predicted.

One very simple way to estimate the relationship between a predictor score (test score) and a criterion score is to construct an expectancy table, which is basically a two-way table that lists predictor scores on the left side and the criterion score across the top. Such table displays the established relationships between the test scores and expected outcomes on a relevant criterion. Entries in the table represent the number or percentage of individuals at each intersection. Table 5.3 represents a hypothetical example of what the results might be if one attempted to validate the predictive validity of I.Sc. entrance test. Suppose that a total of 600 students appeared in the test. The maximum score for a student on the test was 100 marks. Further suppose that pass marks was a score of 80 or above. The left-hand column of Table 5.3 indicates the number of students who passed (secured a score of 80 or above) and the number who did not pass. The symbol n indicates the number of

Table 5.3: A simplified Expectancy Table

	Division at the End of I.Sc. Exam		
Entrance Test	1st Division n (%)	2nd Division n (%)	3rd Division n (%)
Pass (n = 400)	100 (25)	180 (45)	120 (30)
Fail (n = 200)	30 (10)	50 (25)	130 (65)

students in each category and (%) indicates the percentage of the total that figure represents. Table 5.3 clearly indicates that only 25% of the students in pass group got 1st division whereas 10% of the fail group got first division. Likewise, 30% of the pass group got 3rd division whereas 65% of this group got 3rd division. These results would provide some evidence of the predictive validity of the Entrance test and tended to indicate that as a group who passed did better in 'the final I.Sc. Examination'. The ratio of the number of successful to the total number of students appeared is called as *success ratio*. In the example the success ratio is .67.

If one makes a comparative study of concurrent validity and predictive validity, the predictive validity is often found to be lower than concurrent validity. This happens probability due to the fact that the degree of association between the test scores and criterion scores decreases over time. While discussing both concurrent validity and predictive validity, it has been said that if the resulting coefficient is high, the test is said to possess good validity. Now one question is: *how high is high*? This is not easy to answer. There is no such magic number that a coefficient must reach in order to be called as 'high'. In general, it is a relative matter. If there is only one test available to predict a given criterion, a coefficient of .50 might be acceptable as 'high'. On the other hand, if there are more than one test available with higher coefficient, a coefficient a coefficient of .50 might be inadequate.

- **Construct validity**

Sometimes we ask question about an educational and psychological test like: What does the score of the test tells us about an individual? Does this score correspond to some meaningful trait or *construct* that may help us in understanding the behaviour of that person? For this, the term *construct validity* is used where we don't ask, "How well does the test score predict future performance?" nor we ask, "How well does the test scores correlate with the given criterion measure?"

Construct validity refers to the degree to which a test measures an intended hypothetical construct. In other words, construct validity is concerned with a judgement about the appropriateness of inference drawn from test scores regarding the standing of the person on a variable called *construct*. A construct is a scientific idea developed or hypothesized to explain behaviour. Therefore, it is a non-observable trait such as intelligence, job stratification, etc. An intelligence is construct, which may be used to describe why a person performs better than others. Likewise, job satisfaction may be used to explain why one employee on the job is more motivated and punctual than others. Some other important examples of construct are self-esteem, depression, aptitude, social adjustment, love, curiosity, creativity, etc.

Computing construct validity of a test is by no means an easy task. The process basically involves testing hypothesis deduced from a theory concerning the construct. For example, let us suppose that we are going to estimate construct validity of a creativity test. If a theory of creativity hypothesized that one highly creative people will be giving more alternative solutions to the given problem than the low creative people and if the testtakers who score high on the creativity test subsequently come up with more alternative solutions to the given set of tasks, this would be a support for the construct validity of the test. On the other hand, if the testtakers obtaining high scores on the test fail to produce more alternative solutions it would not necessarily mean that the test is not measuring creativity. The hypothesis related to the behaviour of more-creative persons might by wrong. Therefore, several independent studies are needed to establish the construct validity of a test. Thus construct validity is established through a series of activities where the investigator simultaneously defines construct and also develops ways to measure it. In fact, construct validity involves assembling evidences about what a test means and over a series of studies, the meaning of a test gradually begins to take shape. Therefore, gathering evidences for construct validity is an on-going process. According to Cronbach and Meehl, 1955), the process of collecting construct validity evidence should start when "no criterion or universe of content" could be accepted as entirely adequate to define the quality to be assessed.

There are some considerations that must be kept in view while thinking about a construct. Some of these are as under :
- The hypothesized construct must be measurable
- A distinction should be made between hypothesized construct and other constructs that may appear similar
- Evidences should be obtained from different sources to support the construct
- Evidences should be obtained to show that constructs don't correlate with irrelevant factors. It means construct should have divergent validity
- A construct, if becomes necessary, must be modified to conform with additional informations. With the accumulation of new evidences, the researcher should modify the nature of the construct.

Interestingly, construct validity is viewed as the unifying concept for all validity evidence (AERA, APA & NCME, 1999). All types of evidences for validity, including evidences from content validity and criterion-related validity, come under the umbrella of construct validity (Cohen & Swerdlik, 2005).

Researchers are of view that several procedures may be used to provide different kinds of *evidences* that a test has construct validity. Most studies of construct validity fall into one of the

following categories providing evidence for construct validity of the test:
- Analysis for demonstrating that test items or subtests are homogenous and therefore, measuring a single construct
- Test scores increase or decrease due to changes such as changes in age, passage of time or an due to experimental manipulation as predicted according to a theory
- Pretest scores (score before any training programme or similar events) differ from the poststest scores (after the events)
- Individuals belonging to different cultures or groups show differences in test scores as predicted by the theory
- Test scores should correlate with scores on other tests as predicted from a theory which covers manifestation of construct under consideration. However, they must not correlate with other variables with which they are not expected to correlate according to the things
- Factor analysis of test scores in relation to some other sources of information.

A brief discussion of these evidences is presented below :

Evidence from the Homogeneity of test:

Homogeneity is one of the first steps in certifying that the test has construct validity. If a test measures a single construct then its items (or subtests) are likely to be homogeneous (also called as internally consistent). Although in most cases, homogeneity is built into the test during the stage of test development, the goal of homogeneity can be achieved by correlation each item with total score and select items that show high correlated with total score. If all test items show significant correlation coefficients with the total scores and high scorers on the test tend to pass each item more than low scorers pass, then each item is probably measuring the same construct as the total test. Each item is, thus, contributing to test homogeneity.

Test developers can improve the homogeneity of a test consisting of dichotomous items (true-false item) by eliminating those items, which tend to yield low correlation with total score. Coefficient alpha may be taken as the index of homogeneity of a test that consists of multiple-choice items as we find in attitude and opinion questionnaires (Novick & Lewis, 1967).

Although homogeneity of test is an important evidence for construct validity of the test, alone it is a very weak evidence. In fact, the homogeneity is not the 'be-all and end-all' evidence of construct validity. In fact, knowing that a test is homogeneous provides no information about how the construct which is being measured, relates to other constructs. Therefore, it is essential to report some other evidences too along with the evidence of homogeneity of test as the evidence of construct validity of the test.

• Evidence of appropriate changes with age

There are some such constructs, which are expected to change over time. For example, construct of vocabulary knowledge tends to increase from early childhood to later childhood and from later childhood to adolescence and so on to the adulthood and the old age. For any new test of vocabulary knowledge, the most important piece of evidence of construct validity would be that older individuals would score higher on the test of vocabulary than younger individuals assuming that their health and educations are kept constant. However, there are constructs, which are not affected by the lapse of time. For example, it is not clear whether with lapse of time, construct of extroversion will increase or decrease or remain stable. Therefore, developmental changes would not be an evidence of construct validity of such scale. Evidence of developmental changes or change over time, like the evidence of test of homogeneity, does not itself provide evidence for the fact that how the construct being assessed relate to other constructs. Therefore evidence of appropriate developmental changes alone is not a very powerful evidence of construct validity.

- **Evidence from pretest/posttest differences**

Evidence that scores change as a result of some intervention or experience between pretest and posttest provide evidence for the construct validity of the test. In other words, theory-consistent intervention effects showing changes in appropriate direction provides evidence for construct validity. One investigator cited in Roach et al. (1981) compared the scores on Marital Satisfaction scale before and after sex therapy treatment programme. Scores showed a significant change between pretest score and posttest scores suggesting evidence for construct validity of the test. Likewise, it is expected that the spatial orientation ability of the elderly persons will tend to enhance when they receive some cognitive training especially to enhance such ability. If the researcher has constructed spatial orientation test and wants to compute construct validity of the test, he can administered this test to a group of elderly persons and obtain the pretest scores. Subsequently, these persons may be given cognitive training to enhance such spatial orientation ability for any specified period of time and again the same test maybe administered in order to get posttest scores. If the posttest scores are significantly higher than the pretest scores, he gets an evidence for the construct validity of the test.

- **Evidence from group differences**

Another way to provide evidence for construct validity of the test is to demonstrate that scores on the test differ in a predictable way as a function of membership of different groups or cultures. If the individuals thought to be high on construct measured by the test, obtain high scores and the individuals thought to be low on the construct, obtain low scores, it becomes an evidence for the construct validity of the test. The rationale behind this is that if a test is a valid measure of the concerned construct then test scores from groups of people who are presumed to differ with respect to that construct should have correspondingly different test scores. This method is also called as *known-group validity* because the test is given to two contrast groups of people who are known to be different on the trait in question (Hattie & Cooksey, 1984). Let us take an example: suppose a researcher is constructing a depression scale where higher test scores indicate higher degree of depression. The test is administered to the psychiatrically hospitalized persons for depression as well as to those of a normal group of people. If the psychiatrically hospitalized persons score significantly higher than the normal persons, it can be presumed that the scale has construct validity. Roach et al. (1981) conducted one study in which they have attempted to establish the construct validity of the *marriage stratification scale,* which measures the construct of marriage satisfaction. They took two groups of married couples – one group was satisfied as shown by ratings of pears and counsellors and the other groups was not satisfied. Mean scores of the two groups on the scale was subjected to the test of significance of difference that yielded a significant *t* value suggesting that the two groups differed. This provided evidence for the construct validity of the Marriage satisfaction scale.

- **Evidence from Convergent and Discriminant Validation:**

A convergent validity is said to be demonstrated by the test scores correlated highly with other variables with which it shares an overlap of construct. For example, a test designed to measure of general anxiety is expected to correlate with a measure of general anxiety, and if it occurs, it becomes an evidence of convergent validity. A discriminant validity is said to be demonstrated when a test does not correlate with variables or measures of construct from which it differs or theoretically not expected to correlate. For example, a test of intelligence is not expected to correlate with the measure of social interest and if the scores of those two tests yield a very low correlation, it becomes evidence for discriminant validity.

Campbell and Fiske (1959) have provided an explicit method of simultaneously confirming the convergent and discriminant validity of a psychological test. Both convergent validity and discriminant validity are good evidences for construct validity of the tests (Cohen & Swedlik, 2005). The procedure outlined by Campbell and Fiske is called the analysis of *Multtrait Multimethod Matrix of Correlation* (MTMM), which requires that several different methods be used to measure each of several different traits. *Multitrait* means 'two or more traits' and '*multimethod*' means two or more methods. The multtrait multimethod matrix (MTMM) of correlation is the matrix or table, which results from correlating traits within and between methods. Thus the design of Campbell and Fiske demands assessment of two or more traits by two or more methods Table 5.4 presents a hypothetical example of MTMM approach. In Table 5.4 three traits (A, B and C) are measured by three methods (Methods 1, 2 and 3). The three traits are sociability (A), Ascendance (B) and Trustworthiness (C) and three methods are self-ratings (method 2), Peer ratings (method 2) and observation (method 3). Therefore, A2 means trait A (Sociability) is being measured by Method 1 (self-ratings).

Table 5.4: Multitrait-Multimethod Matrix with hypothetical correlations (Based on suggestion by Campbell & Fiske 1959)

	Traits	Method 1			Method 2			Method 3		
		A_1	B_1	C_1	A_2	B_2	C_2	A_3	B_3	C_3
Method 1	A_1	(92)								
	B_1	53	(93)							
	C_1	48	38	(85)						
Method 2	A_2	.60	24	18	(94)					
	B_2	20	.61	11	66	(84)				
	C_2	.12	.11	57	60	59	(89)			
Method 3	A_3	58	25	12	64	45	34	(90)		
	B_3	20	60	13	20	58	36	68	(93)	
	C_3	.10	11	48	35	33	64	60	61	(90)

In the example shown in Table 5.4 all values of correlations (points omitted) are hypothetical and shows that nine tests are being studied (that is, three traits each being measured by three different methods). Each of these tests is administered twice to the same groups of teststakers and scores on all pairs of tests are correlated. The result is *multitrait multimethod matrix*, which provides a very important source of data relating to not only to convergent validity and divergent validity but also to reliability of the tests. Let us analyse this.

The cells of the matrix showing correlations along main diagonals given in parenthesis are *reliability diagonals*. They represent correlation between two assessments for the same trait using the same measurement method. Therefore, they reflect the reliability of the test and the researcher would like these values to be as high as possible and they should be the highest value in the matrix. In the language of MTMM analysis, these are *monotrait monomethod correlations*.

The correlations in boldface along three shorter diagonals provide evidence for convergent validity. There are called *validity diagonals* and are important for evidence of construct validity. These correlations should be strong and positive as shown in the table. To the extent that the different methods of assessing the same trait yield high correlations, the construct and its measures tend to demonstrate convergent validity. Thus it can be said here that the different methods of measurement converge on the trait.

Table also contain correlations between different traits measured by the same method, that is, *hetrotrait monomethod correlations* and correlations between different traits measured by different methods, that is, *hetrotrait hetromethod correlation*. Hetrotrait monomethod correlations have been shown by solid triangles and hetrotrait hetromethod correlations have been shown by dotted triangles. Hetrotrait monomethod correlations represent some similarities due to both natural relationship between the constructs and the very fact that they have been measured by the same method. Such correlations should be similar than the correlations showing reliabilities. Hetrotrait hetromethod correlation should be the lowest of all correlations in the matrix and they provide evidence for discriminant validity of the test. Such correlations are lowest in the matrix because they include only trait correlations.

In summary, it can be said that Campbell and Fiske (1959) have clearly combined the need to collect evidence of reliability, discriminant and convergent evidence of validity into one study. They have called it *multitrait-multimethod design* for investigating construct validity. The four types of correlations coefficient of the matrix are *coefficient in parenthesis, coefficients in triangles with solid lives coefficients in triangles with broken lines* and *coefficient along three diagonal sets of bold face numbers*. The coefficient in parentheses are indicative of reliability of the test; the coefficients in triangles with sold lines are indicative of convergent evidence of validity and the coefficients in the triangles with broken lines represent discriminant evidence of validity. Three diagonal sets of bold face numbers reflect validity coefficients because here scores obtained for the same trait by different methods are correlated (monotrait-hetromethod).

- **Evidence from factor analysis:**

Factor analysis is one another evidence for the construct validity. In fact, both convergent and discriminant evidence of construct validity can be provided with the help of factor analysis. Now, what is *factor analysis*?

Factor analysis refers to a set of mathematical procedures designed to identify factors or some specific variables, attributes or characteristics on which the individuals tend to differ. In simple words, factor analysis is an advanced procedure based on the concept of correlation that helps the researcher explain why two tests are correlated (Murphy & Davidshofer, 1994). Basically, there are two forms of factor analysis: *confirmatory* and *exploratory*. In confirmatory factor analysis the purpose is to confirm that the test scores and variables fit a pattern predicted by the theory. In exploratory factor analysis the purpose is to summarize the interrelationship among several variables in a concise and accurate way. In psychometric research, factor analysis is frequently used in the second form where it is basically employed as data reduction technique in which several sets of scores and correlations between them are analyzed in detail. In such studies the goal of factor analysis is to identify factor or factors in common among test scores on subtests within a particular test or the factors in common between scores on various tests under study. For example, if 20 tests have been administered to a sample of psychiatric patients, the first step in factor analysis is to compute the correlations between the scores on the $20(20-1)/2 = 190$ possible pairs of tests. The resulting table is called *correlation matrix*. Subsequently, the factor analyzer searches the pattern of intercorrelations carefully and identify a small number of factors and name those factors. Naming factors that came out from factor analysis depends upon their knowledge, judgement and verbal abstraction ability than upon mathematical expertise of the factor analyzer. There is no hand-and-fast rules and he utilizes his own judgement in deciding what factor name would be appropriate in communicating the meaning of the factor. For example, if four tests, namely, test of comprehension, arithmetic test, similarities test and information tests are administered to group of people and the resulting 6 correlations ($4(4-1)/2 = 6$ are analyzed and found that the correlations among information test, similarities test and test of comprehension are high, one can conclude that a common factor (may be named as verbal comprehension) exist. In this way, factors are identified and named.

After naming the factor, a table of factor loadings is produced. Infact, factor loading is a sort of metaphor. Each test is considered as a vehicle, which carries a certain amount of one or more abilities. Loading a factor in a test provide information about the extent to which a factor tends to determine the scores of the test. In reality, a factor loading is a correlation between an individual test and a single factor. Factor loadings tend to vary between +1.0 to −1.0. Table of factor loadings is the final outcome of the factor analysis. This table helps the researcher in describing the factorial composition of a test, and thus, provides information regarding the construct validity of the test. Let us take an example to illustrate this point. Suppose a new test intending to measure intelligence can be factor analyzed with other known measures of intelligence as well as with other kinds of measures such as measures of depression, anxiety, self-esteem, perfectionism, leadership, etc. High factor ladings (that is high correlation of test with a factor) by the new test on intelligence factor would provide evidence for convergent evidence of construct validity and the moderate to low factor loadings by the new test with respect to the measures would provide evidence for discriminant evidence of construct validity.

Sources of Evidence of Validity

Test validity is a function of how the scores are used. Validity is no longer thought as a characteristic of test. The *Standards for Educational and Psychological Testing* (AERA, APA & NCME, 1999, 2014) points out that validity refers to whether there is evidence supporting the use and interpretation of test scores for their intended purpose.

There are two views of validity: *traditional view of validity* and *current views of validity*. These two views are discussed below:

Traditional view of validity:

Traditional view of validity is often described as an evaluation of whether the test measures what it intends to measure. According to this view, the test developer would develop a test of assessing extroversion based on some well-researched theory of personality. He would then evaluate how well the test was measuring what it was designed to measure by comparing test takers' obtained score on this new test with their scores on different test that is designed to assess the same trait, that is, extraversion. If the obtained scores on the two tests are similar, the test developer may conclude that the test was valid because there was evidence to support the fact that test was measuring what it was designed to measure. When any test developer claims that a test is valid, test users are likely to assume that a valid test is good for measuring almost anything. But this is not the reality and there may be some unintend results. One common misconception that persists today is that when we talk about validity, we are talking about evaluating a characteristic of the test implying that a test can be judged as being valid or invalid. The important point is that a test can measure exactly what it intended to measure and yet not be valid to use for a particular purpose.

Current views of validity:

The current view of validity focuses on the correct interpretations that could be made from the test scores. A test can measure what it is intended to measure and yet not be valid to use for a particular purpose. In the past, three forms of validity were usually referred: content, criterion-related and construct. Today, validity is viewed as a unitary or single concept by *Standards* (AERA, APA & NCME, 1985, 1999). The emphasis is being given on the importance of evaluating the interpretation of test scores and in providing evidence for a sound scientific basis for the proposed score interpretation (Goodwin & Leech, 2003) Today the *Standards* have described five independent sources evidence of validity as under:

- **Evidence based on the content of the test:**

All psychological and educational tests are designed to measure something and items of the test, their format, wording and processes required to testtakers (together called content of the test) must relate to that 'something', which the test intends to measure. Previously, it was called *content validity*. This source of evidence of validity focuses upon logically examining and evaluating the content of a test for determining the extent to which the content is representative of the concepts that the test is designed to measure.

- **Evidence based upon the internal structure:**

This type of evidence for validity emphasizes upon whether the conceptual frameworks that were used in test development could be demonstrated using appropriate analytical techniques such as factor analysis. For example, suppose a test was designed to measure a single concept such as neuroticism. If the analysis of the test results shows that only one factor or concept (here neuroticism) accounts for a majority of information the test takers have provided, it becomes an evidence for validity of the test. On the other hand, if the analysis suggests that the scores were affected by more than one underlying concept or factor, the validity of the test based upon its underlying single concept structure would be questioned.

- **Evidence based on response processes:**

This is another source of evidence of validity where the researcher observes the testtakers as they respond to the test and/or interview them when they answer the test items. These observations and interviews are used to understand the mental processes of the testatakers while responding the test items. For example, if a test of reasoning is being administered to the testtakers, as it is expected, they should respond mentally processing the test information rather than responding with some memorized facts.

- **Evidence based on relations with other variables:**

    This evidence of validity is based upon finding the relations of the present test with some other variables or tests that claim to measure the same attribute or construct. These other variables are called *criteria* and it has traditionally been referred to as *criterion-related validity*. The criterion-related validity is also a facet of construct validity and their source of validity evidence typically involves correlation between the test scores and another measure of the same trait, attribute or ability. Likewise, the researcher would want to know that the test scores are not related to other measures to which he would not expect them to relate. For example, a test of mechanical aptitude should correlate with another measure of mechanical aptitude while should not correlate with a vocabulary test. Two types of research designs are used to demonstrate criterion-related evidence of validity: predictive and concurrent. The predictive method yields predictive validity and concurrent method yields concurrent validity. In predictive method the test is administered to a group of the testtakers, researcher waits for a specified period and then, obtains scores on the criterion matured after the specified period. Subsequently, the two sets of scores are correlated. For example, if the testtakers who obtained high scores on a test given to them and also obtain high scores later on the criterion measure after the lapse of specified period, whereas the testtakers who scored low on the test also received lower scores on the criterion measure, it is said that the test has the evidence of predictive validity. In concurrent evidence of validity the emphasis is on determining whether the scores on the test are related to a criterion, which is available at the same time as the test is given. Since in concurrent validity the criterion is available in present time, the researcher is not required to wait for some specified time. For example, if scores on the test of mechanical aptitude are correlated with already scores on another test of mechanical aptitude already available in the present time, it would provide concurrent evidence of validity of the test.

    Evidence based on relations with other variables also relate to what is called *construct validity*. The construct validity involves gathering evidences showing that a test is based on sound psychological theory. The process of establishing construct validity for a test involves accumulating evidence that the test scores relate to some observable behaviours in a way predicted by the theory underlying the construct. Two basic sources of evidence for demonstrating construct validity are: *convergent validity* and *discriminant validity*. Convergent evidence of validity demonstrates that other or similar constructs that theoretically should be related to construct assessed by the test are, in reality, related. Discriminant evidence of validity shows that constructs that should not be related to the construct measured by the test, are, in reality, not related.

- **Evidence based on consequences of testing:**

    Whenever the researcher makes a psychological measurement based on the educational and psychological test, there occurs some intended or unintended consequences. For example, suppose the test is intended to select personnel for a certain job. The intended consequences of this test would be to obtain accurate information from the testtakers relevant for better performance on the job. However, the test may be biased in favour of male testtakers because it contains such items over which generally males have superiority or the test may favour one certain group of testtakers over the other. This is an unintended consequences of the test. Therefore, the test users must be aware of the fact that it is important to make distinction between intended and unintended consequences of testing commonly associated with the validity of the test itself.

Interpreting Reliability and Validity Coefficient

The degree of reliability and validity are commonly expressed in terms of correlation coefficient. A high coefficient indicates high reliability or validity. If a test were perfectly reliable and valid, all coefficients of correlation would be 1.00. This would mean that the person's scores on the test perfectly reflected his true status with respect to the variable being measured. However, the reality is that no test is perfectly reliable and valid. Scores on the test are affected by systematic errors as well as by the errors of measurement resulting from so many sources. High reliability indicates that errors of measurement or random errors have been reduced whereas the high validity coefficient indicates that the systematic errors have been reduced. As we know, systematic errors affect scores in a systematic way (lowering or enhancing them all) whereas the random errors affect scores in a random way, that is, some scores may be decreased where as others may be increased. Researchers have demonstrated that systematic and random errors can be caused by the characteristics of the test itself (test being inappropriate for the testtakers), the conditions of administration (instructions may not be properly understood) or by the current mental status of the testakers taking the test (testtakers being fatigued or unmotivated, etc). High reliability and validity coefficient indicates that these various sources of errors have been eliminated to a greater extent.

Although the reliability and validity of the test are expressed in terms of correlation coefficient, what does this correlation coefficient do is not clear to most of the students. Some students wrongly but understandably think that a correlation coefficient of .50 means that two sets of the given score are 50% related. This is not true. In psychological and educational testing a correlation coefficient squared (called coefficient of determination) indicates the amount of common variance shared by the two sets of scores. Whenever a test s administered to a group of testtakers, everyone will not be getting the same test scores and therefore, a range of scores will be obtained and score variance will be the result. Common variance is defined as the variation of one set of scores, which is attributed to its tendency to vary with the other set of scores. If two sets of scores are perfectly related, the variability in one set of scores has everything to do with the variability in other set of scores. On the other hand, if two sets of scores are not related (that is, when the test is unreliable one), the variability in one set of scores has nothing to do with the variability of other sets of scores. In other words, when there is perfect relationship between two sets of scores, there is 100% common variance. On the other hand, where there is no relationship between two sets of scores, the common variance is zero. The student must keep in mind that the percent of common variance is generally *less* than the numerical value of coefficient of correlation. For example, if the coefficient of correlation between two sets of scores is .84, then the common variance is equal to $(.84)^2 = .706$. This squared value is technically known as coefficient of *determination*. A value of .706 yields 70% common variance between two test of scores. Likewise, a coefficient of .50 may look attractive at first, but its common variance is only .25 which means that the two sets of scores have only 25% common variance. Similarly, a correlation coefficient of 00, when squared, will yield zero indicating zero common variance between two scores. A correlation coefficient of 1.00 when squared will yield 1.00, indicating 100% common variance between two sets of scores.

Now one important question here arises is how high a correlation coefficient needs to be for being called as 'good' or acceptable. There is no fixed rule that a coefficient must be .80, for example, to be called as 'good' or 'acceptable'. Infact, the interpretation of coefficient depends upon for what purpose it is going to be used. When the purpose is to use the correlation coefficient in research studies investigating relationship between variables, a correlation coefficient .45, for example, would be considered useful. But this coefficient would not be considered good for reliability of the test, such value may not be useful for validity especially, for predictive validity and is terrible for the reliability. A coefficient of .60 may be considered as useful in a study of possible predictors but this will be considered as unsatisfactory for the estimate of reliability. In fact, what constitutes an acceptable level of reliability is to some extent, determined by the type of test. For achievement and aptitude tests, a reliability coefficient .90 is considered good. Personality tests don't generally report such high reliability coefficients and therefore, reliabilities in the range of .70 to .80 is considered as acceptable range. Attitude scales reliabilities usually fall in the .60 to .80 range when tests are developed in new area such as curiosity, surgency etc., one has to keep oneself satisfied with low reliabilities.

Sometimes the test is composed of many subtests, then the researcher should assess the reliabilities of each subtest and not only the reliability of total test. The reliability of the given subtest is usually lower than the reliability of the total test because reliability, as we know, is the function of length of the test. A close examination of the reliabilities of the subtest becomes necessary especially when one or more subtests may be independently used by the researchers.

Relation between Reliability and Validity

Reliability and validity are two related dimensions of the test efficiency. Statistically, validity is defined as the correlation of the test with some other test or criterion measured whereas reliability is defined as the self-correlation of the test. A test which is not correlating well with itself, is not expected to correlate with others. It means that the test having poor reliability is not expected to yield high validity. Here validity is said to be dependent upon reliability and this is true especially when the test is homogeneous in nature. If a test is heterogeneous, validity especially predictive validity may be high without the underlying high reliability. A heterogeneous test measures different factor within its content and therefore has a better chance of sharing common factors with factorially complex criterion.

Reliability is thus necessary but not sufficient condition for validity. The validity of the test may be higher than the reliability but not higher than the index of reliability because the index of reliability provides the maximum limit of correlation than the test can yield with the true measures. The maximum validity coefficient between two variables is equal to the square root of the product of their reliabilities or $R_{12max} = \sqrt{R_{11} \times R_{12}}$ where R_{11} and R_{12} are the reliabilities of the two variables and R_{12max} is the maximum validity coefficient. In a nutshell, we can have reliability without validity. However it is logically impossible and very difficult to demonstrate that an unreliable test it valid.

Aiming to have both high reliability and high validity of the same test is sometimes not appreciable because in such case the researcher is said to work at cross purposes. Major requirement of high reliability and high validity are opposite of each other. Maximal reliability requires items of equal difficulty and high intercorrelations between items whereas maximal validity requires item of varying difficulty values and low intercorrelations among items (Guiford, 1956). Obviously then, to aim at high reliability and high reliability of the same test would then be as working at cross purpose. An in-between way suggested by Tucker (1946) is that if inter-item correlations range from .10 to .60, (moderate correlations) one can expect to have both reliability and validity of the test to a much satisfactory level. In other words, items within these range of correlations should provide tests of both high reliability and validity.

Factors Affecting Validity

There are several factors that affect the validity coefficient of the test. Major such factors are as under :

- **Vague directions:**

Directions that fail to clearly indicate to the testtakers how to respond with items and how to record the responses tend to reduce validity.

- **Improper arrangement of items:**

Usually in any test the ideal arrangement of items is one where easy items are placed in the beginning and the difficult items are placed later. However, when this ideal arrangement is not followed and difficult items are placed in the beginning in the test, this may cause testtakers to spend too much time and prevent them from reaching items that they could easily answer. Such improper arrangement influence the validity of the test by lowering the motivation of the testtakers.

- **Identifiable patterns of answers:**

    If correct answers of items are placed in some systematic pattern such as A, B, A, B, C, D, CD, A, B, and so on, it enables testtakers to guess the answers to some items more easily and this lowers the validity of the test.

- **Reliability of the test:**

    The *maximum* validity coefficient of a test is directly related to the reliability of the test. The formula 5.14 expresses this relationship.

    $$\text{Maximum validity} = \sqrt{Reliability} \qquad (5.14)$$

    If the reliability of the test is .80, its maximum validity would be = .89, that is, the validity of such test cannot be higher than .89. It is possible (but not likely) that a test will correlate more highly with an external criterion (validity) than it does with itself (reliability).

    It is a matter of common belief that the validity especially the predictive validity of the test is directly proportional to the reliability, that is the more reliable the test is, the more valid it is. When the reliability increases, it means the proportion of true variance in the test increases and in this situation, other things being equal, the commonality should increase the loadings on the common factors. If one or more of the common factors are shared by the criterion predicted, the predictive validity is likely to increase. However, if the test and the criterion share no common factors, no amount of increase in reliability will enhance the validity of the test.

- **Length of the test:**

    As we know, the homogeneous lengthening of a homogeneous test increases the reliability of the test. So also, such lengthening also increases the validity of the test. By homogeneous lengthening of the test is meant increase or addition of such new items in the test which possess identical features as those of original items. Using Spearman-Brown formula, we can calculate increase in validity of the test when the homogeneous test is lengthened n times (such as 2 times, 3 times, 4 times or any number of times). The formula is as under:

    $$r_{y(nx)} = \frac{(n)(r_{yx})}{\sqrt{n + n(n-1)r_{xx}}} \qquad (5.15)$$

    where,

    $r_{y(nx)}$ = validity of the test after lengthening n times
    r_{yx} = validity coefficient of the test
    r_{xx} = reliability of the test
    n = number of times test has been lengthened or repeated on the same sample

    Suppose, a test has a validity of .60 and reliability of .50 and it is enhanced four times of its present length. Its validity after applying formula 5.15 would be

$$r_{y(nx)} = \frac{4(.60)}{\sqrt{4+4(4-1).50}} = \frac{2.4}{3.46} = .694$$

Certainly, enhancing the length of the test would increase the validity of the test from .60 to .694.

If the research is interested in knowing how much homogeneous lengthening is needed in order to achieve a desired level of validity, he could do this by formula 5.14 which is as under:

$$n = \frac{r_{y(nx)}^2 (1 - r_{xx})}{r_{xy}^2 - r_{y(nx)}^2 (r_{xx})} \qquad (5.15)$$

where,

$r_{y(nx)}$ = desired level of validity

The rest of the symbols are defined as they have been defined in formula 5.14.

Suppose a test has reliability coefficient of .50 and validity coefficient of .60. If the researcher wants to enhance the validity to a level of .70, how many times should the present test be lengthened? The answer can be given by using formula 5.15 as under

$$n = \frac{(.70)^2 (1 + .50)}{(.60)^2 - (.70)^2 (.50)} = \frac{(.49)(.50)}{.36 - (.49)(.50)} = \frac{.245}{.115} = 2.13$$

Thus the present test should be lengthened about two times for yielding the validity coefficient of .70.

Sometimes the researcher lengthened both the test as well as the criterion for enhancing the validity of the test. In such case formula 5.15 will not serve the purpose. To calculate enhanced validity coefficient when both the test as well as the criterion have been lengthened n times, formula 5.16 will be appropriate one.

$$r_{c(nxmc)} = \frac{r_{cx}}{\sqrt{\left(\frac{1-r_{xx}}{n} + r_{xx}\right)\left(\frac{1-r_{xxc}}{m} + r_{xxc}\right)}} \qquad (5.16)$$

where,

$r_{c(nxmc)}$ = correlation between test lengthened n times and the criterion lengthened n times

r_{cx} = correlation between the test and the criterion
r_{xx} = reliability of the test
r_{xxc} = reliability of the criterion
n = number of times test has been lengthened
m = number of times criterion has been lengthened

Suppose the correlation between the test and the criterion is .60 and reliability of the test is

.50 and the reliability of the criterion is .58 and both test and the criterion have been lengthened two times each. In such situation the enhanced validity coefficient of the test by formula 5.16 would be calculated as under :

$$r_{c(n \times mc)} = \frac{.60}{\sqrt{\left(\frac{1-.50}{3}+.50\right)\left(\frac{1-.58}{3}+.58\right)}} = \frac{.60}{\sqrt{(.167+.50)(.133+.58)}}$$

$$= \frac{.60}{\sqrt{(.667)(.720)}} = \frac{.60}{\sqrt{.84}} = .87$$

Thus three times lengthening both the test as well as the criterion results into enhancement of validity of the test and criterion from .60 to .87.

- **Relation of validity to errors of measurements:**

The validity coefficient of the test are low to the extent that either predictors (test) or criteria are unreliable or are composed of errors of measurement. The correlations assume consistency from one variable to other and errors of measurements are, by their nature, inconsistent. If errors of measurements are eliminated, the resulting correlations would be between true scores. These errors of measurements *attenuate* or lower correlation coefficient.

Researcher have pointed out that a correction for attenuation may be introduced in the criterion only or in both predictor (test) and criterion. The formula for correcting the criterion for attenuation is.

$$r = \frac{validity\ coefficient}{Reliability\ of\ the\ predictor} \quad (5.17)$$

For example, if the test that has the predictive validity of .50 and the reliability coefficient is .70, its predictive validity by correlating the test scores with an errorless criterion would be:

$$r = \frac{.50}{\sqrt{.70}} = \frac{.50}{.83} = .60$$

It is also possible to correct both predictor and criterion for attenuation. This would provide a maximum validity coefficient because the effects of errors of measurement would be eliminated from both predictor and criterion. The formula for correcting both the criterion and the predictor scores for attenuation is :

$$r = \frac{validity\ coefficient}{\sqrt{Predictor\ reliability \times criterion\ reliability}}$$

Suppose the validity coefficient of a test is .50 and the reliability of predictor is .70 and the reliability of the criterion is .60, the maximum correlation between the predictor and the criterion would be:

$$r = \frac{.50}{\sqrt{.70 \times .06}} = \frac{.50}{.65} = .77$$

This is a correlation between perfectly reliable (errorless) predictor and a perfectly reliable criterion. When the validity was found by only correcting the criterion for attenuation it was only .60 but when such correction for attenuation was introduced in both the predictor as well as the criterion, there was a significant increase in the validity coefficient of the test from .60 to .77.

Summary and Review

- Reliability means consistency of test scores. There may be internal consistency or external consistency such as temporal stability or consistency.
- The logical or technical meaning of reliability is based upon the classical test theory or true score model. Logically reliability is defined as the proportion of the true variance in total variance, which consists of true variance plus error variance.
- There are various sources of inconsistency or error variance. Important such sources are test construction, test administration, test scoring and interpretation as well as some other miscellaneous factors.
- There are two modes of expressing reliability of the test: coefficient of correlation and standard error of measurement.
- There are several methods of estimating reliability. Important ones are: test-retest reliability, parallel-form and alternate-form reliability, internal consistency reliability, and inter-scorer reliability. In internal consistency reliability three important methods have been considered. They are split-half method, Kuder-Richardson formulas and coefficient alpha.
- Reliability of difference score can be conveniently calculated with reference to sigma score or z score.
- Index of reliability is defined as the square root of the reliability coefficient of test and it indicates the maximum correlation, which the test is capable to yield in its present form.
- There are various conditions that affect reliability of the test. Some of the important ones are: homogeneity vs heterogeneity of the test, dynamic vs. static characteristic of the test, length of the test, range of test scores, difficulty level of test and objectivity in scoring.
- The validity of the test scores refers to the estimate of how well the test measures what it intends to measure in a particular context with a particular population of people.
- According to trinitarian model, there are three categories of validity: content validity, criterion-related reliability and construct validity. Concurrent validity and predictive validity are the two important categories of criterion-related validity.

- Face validity is not a form of validity in technical sense. A test is said to be have face validity if it looks valid to the testtakers and others.
- Reliability and validity coefficient need to be interpreted carefully. True meaning of the coefficient of correlation is derived from common variance, which is reflected by the coefficient of determination.
- Reliability and validity are related concepts. However, reliability is necessary but not a sufficient condition for validity.
- There are several factors that affect the validity of the test. Important such factors are vague directions, improper arrangement of items, reliability of the test, length of the test, errors of measurement, etc.

Review Questions

1. Define reliability. Discuss the various sources of error variance.
2. Give the technical meaning of reliability. In this connection, discuss the concept of true variance, error variance and obtained variance.
3. Discuss the method of test-retest reliability. Make a comparative study of this method and method of parallel-form reliability.
4. Make distinction between internal consistency and external consistency of the test. Discuss the various methods of estimating internal consistency reliability of a test.
5. Discuss the various factors that affect reliability of the test.
6. Define validity. Discuss the trinitarian model of validity.
7. Citing examples make distinction between content validity and face validity.
8. What do you mean by construct validity. Discuss the important types of evidences put in support of construct validity.
9. Citing examples discuss the concept of concurrent validity and predictive validity.
10. Discuss the factors that influence validity of the test.
11. Write short notes on the following:
    (a) coefficient alpha
    (b) K-R formula so and 20 and 21
    (c) Relation of reliability and validity
    (d) index of reliability.

■■■

Chapter – 6

Norms and Transformation of Scores

Learning objectives :
- Learning and Nature of Raw scores
- Nature of norms
- Types of norms
    - Age norms
    - Grade norms
    - Percentile norms
    - Standard Score norms
- Transformation of Raw Scores
- Normalizing Standard Scores
- National Norms Vs. Local norms
- Summary and Review
- Review Questions

Key Terms:

Age norms, Raw score, Grade norms, Percentile norms, Standard Score norms, Z-score, T-score, Stanine, C-scale, Standardized score, Normalized Standard Score, Percentile, Percentile rank.

One important purpose of testing is to derive some meaningful conclusion from the test scores. For this, psychologists and educationists have done appreciable efforts. They have been able to develop some normating criteria against which the scores are compared for some meaningful inferences. The present chapter is concerned with the explanation of all such criteria and ways needed to facilitate the interpretation of test scores.

Meaning and Nature of raw score

Raw score provides the most elementary level of information about a psychological and educational test. For example, in intelligence testing and personality testing often performance of testtakers is expressed in terms of some numbers such as a score of 50 or a score 90 or something else. Standing alone, the number has no meaning at all and is completely uninterpretable. At the superficial level, we even don't know whether the number 50 represents a perfect score of 50 out of 50 or a very low percentage of possible score such as 50 out of 200.

In fact, it is difficult to derive any meaning from the raw scores without making any comparison to some frame of reference of one variety or another. For example, we may want to know how other have done on this test and whether the obtained score is high or low in comparison to a representative group of subjects. In other words, a raw score becomes meaningful only when it is compared to norms, which are an independently frame of reference derived from a standardization sample or a norming group, which consists of a sample of persons who are representative of the population for whom the test is intended.

Most of the raw scores on the psychological and educational tests are interpreted by consulting norms. Such tests are popularly called as *norm-reference tests*. Besides these, there are *criterion-referenced tests* that help in determining whether a testtaker can attain an objectively defined criterion such as multiplying 3-digit problems with 90 percent accuracy. For criterion-referenced tests, norms are *not* essential. Any testtaker's raw score is interpreted by making comparison to the performance of the standardization sample. Such comparison answers questions like this : Does the raw score coincide with the mean performance of the standardization group? Is it above average or below average or fall near the upper or the lower end of the distribution?

Nature of Norms

Norms provide the average test performance of the standardization sample. By standardization sample is meant a sample of the persons who are taken to be representative of the population for which test is meant. For representing the population adequately, it is essential that sample must be cross-sectional in nature, that is, it must represent all sections of the population. For comparing the raw scores with the performance of the standardization sample, they are usually converted into derived score. Psychometricians have given two reasons for such conversions. *First*, the derived scores indicate the persons' relative standing in reference to other individuals. *Second*, derived scores provide comparable measures that permit a direct comparison of the person's performance on different tests. Thus the person's relative performance in many different tests can be compared with help of the derived scores. Let us take an example. Suppose Sohan has got a score of 40 on an arithmetic test and a score 50 on a spelling test. On the basis of the comparison of these raw scores, nothing can be concluded. We cannot say on which of the two tests, his performance is better or worse. Moreover, these raw scores are usually expressed in different units, which make a direct comparison of such scores impossible. However, when these raw scores are converted into derived score, such derived score can be expressed in the same units and are related to the same or closely similar normative samples for different tests.

The major characteristics of norms may be now summarized as under :
- Norms are based upon the performance of the standardization sample, which is representative of the population

- Norms are average test performance of the sample
- Norms *don't* provide any *standard* or *criterion* rather they help in classifying the testtakers from low to high, across of continuum of trait, ability or achievement. This is done by converting the raw scores into derived scores.
- Norms are not the measure of what ought to be, that is not of a good or standard, but a measure of what is the status quo.

Types of Norms

There are many types of the norms. Of which, the most popular ones are: *Age norms, grade norms, percentile norms* and *standard score norms*. A discussion follows :

Age norms:

Age norms or age-equivalent scores are norms that display the level of the test performance for each separate age group in the standardization sample. In other words, age norms display the average performance of different samples of testtakers who lie at the different stages at the time the test was administered. Age norms are most suited to traits that exhibit continuous and relatively uniform growth with age. During childhood, we can observe continuous growth in height and weight as well as in wide range of perceptual, motor and cognitive abilities. The purpose of age norms is to facilitate comparisons of the same aged persons. In age norms the performance of the testtaker is interpreted in the standardization sample of the same age. The age span for the normative age group varies depending upon the degree to which performance on the test is age-dependent. For example, the characteristics that quickly change during childhood with advancement of age (such as intelligence) need narrowly defined age span such as 6 years 2 months (6-2), 6 years 4 months (6-4), 6 years 6 months (6-6). On the other hand, the characteristics that change slowly with enhancement in age needs a relatively bigger age span such as 5 years to 7 years, 8 years to 10 years so on.

One of the chief limitations of age-norms is the concern about equality of age units. It is a matter of common observation that as a child reaches adolescence or adulthood, the age more or less ceases as a unit of expressing the performance of the individual. Therefore, age norms are most appropriate and meaningful for infancy and childhood. Psychometricians have suggested that age norms should not be used for the cognitive abilities beyond childhood because after that the pattern of growth of these abilities depend too heavily upon formal school experiences.

- **Grade norms**

Conceptually, *grade norms or grade equivalents* are similar to age norms. A grade norm is one which displays the level of test performance for each separate grade in the representative sample. For developing grade norms, the test is administered to representative sample of children over a range of consecutive grade levels (such as grade first to grade eight). Subsequently, average score (mean or median) for children at each grade level is computed. Thus, if the average number of arithmetical problems sorted correctly on an arithmetic test by the fifth-graders in the normative sample is 25, then a raw score of 25 corresponds to a grade-equivalent of 5.

Grade norms are rarely used with ability tests. However, they are very popular in school setting when one wants to report the achievement levels of school children. However, grade norms have some limitations. *First*, since content of instruction varies somewhat from grade to grade, grade norms are used only for common subjects or common consent taught throughout the grade levels covered by the test. *Second*, grade norms are sometimes misinterpreted. For example, suppose fifth-grade child obtains a grade-equivalent of eight in the arithmetic test. Does this mean that a student of fifth grade has the same arithmetic ability as the average eight-graders? The answer is no. He certainly showed a superior performance in fifth-grade arithmetic but it could not be established that he has the prerequisites of eight-grade arithmetic. Accurately interpreted, all such findings indicate that this student and the hypothetical average eight-grader solved the same number of problems on the test. Grade norms don't provide information as to the content or types of items a student could or could not answered correctly. *Third*, grade norms sometimes are mistakenly regarded as performance standard. For example, a teacher teaching in class seven may assume that all students of this class should fall at or very close to grade-equivalent of seven on the achievement test. This may not come true due to the fact that there exists individual differences among students.

- **Percentile norms**

Percentile norms are the data from a standardization sample converted to percentile form. What is percentile? Percentiles are the *specific scores* or point within a distribution. In fact, percentile are scores expressed in terms of the percentage of persons in the standardization sample who fall below a given raw score. In other words, percentile are the expression of the percentage of people whose score on the test falls below a particular raw score. Let us take an example. Suppose that 35 percent of the persons obtain fewer than a score of 20 on a arithmetical reasoning test, then a raw score of 20 corresponds to the 35th percentile (P_{35}). In this example, percentile score is 20 and percentile rank is 35. A percentile rank answers the question : what percent of scores fall below a particular given score? Thus percentile indicates the ranking that tells about the relative position of person in the standardization sample.

The median in the distribution indicates 50th percentile (P_{50}). Percentile above 50 indicate better performance and percentile below 50 indicates poor performance on the test. Likewise, 25th and 75th percentiles are called Q_1 and Q_3 respectively and Q_2 being the 50th percentile. Percentile should not be confused with the percentage score. A percentile is a *converted score*, which indicates the percentage of testtakers. A percentage score refers to the raw score (that is, the number of items answered correctly) multiplied by 100 and divided by the total number of items.

Percentile norms are more popular because they can be easily understood even by non-technical or untrained persons. They are suitable for any type of test and can be used equally well with adults as well as children.

Major drawback of percentile norms is that of marked *inequality* of their units especially at the extreme of distribution. If the distribution is normal one, then raw score differences near the centre or median are exaggerated in percentile transformation whereas the raw score differences near the ends of the distribution are shrunk to a greater extent. (see Figure 6.1). Thus the real difference between raw scores are minimized near the ends of the

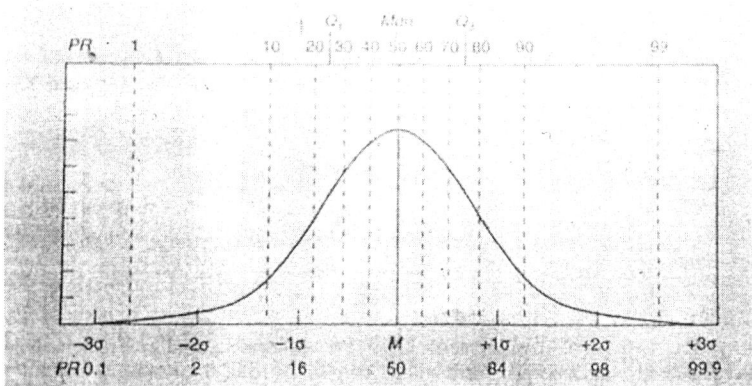

Figure 6.1 Percentile Transformation in a Normal Distribution

distribution and exaggerated in the middle of the distribution. This problem becomes more serious with highly skewed data. As we know, in a normal distribution the highest frequency of raw scores occurs at the centre of the distribution. This being the case, the differences between all those scores that cluster at the centre of the distribution, might be quite small, yet even the small differences will appear as big differences in percentile. The reverse is true at the extremes of the distribution where the differences between raw scores may be bigger ones, though this is not supported by the small differences in percentiles.

- **Standard Score norms**

Since units of a score system based on percentile ranks are clearly unequal, the researcher is forced to look for some other unit that have the same meaning throughout the whole range of distribution. Standard score norms have been developed to serve this purpose. Standard score norms are norms, which are expressed in terms of some standard scores. The standards about the standard scores is that it has a fixed mean and a fixed standard deviation. A standard score uses the standard deviation of the total raw score as one of the fundamental unit of measurement. Standard score expresses the distance of score from mean of the distribution in terms of standard deviation units. A standard score not only expresses the magnitude of deviation from mean but it also expresses direction of departure (positive or negative). For example, a raw score that is exactly one standard deviation unit above the mean converts to a standard score of +1.00. Likewise, a raw score that is exactly one standard deviation unit below the mean converts to a standard score of –1.00. This type of standard score is more popularly known as z-score or sigma score or *zero plus or minus one scale*. This is called so because it has a mean set at zero and a standard deviation set at zero. The formula for z-score is as under :

$$z = \frac{X - M}{SD} \quad (6.1)$$

where,
        z = sigma score or z score
        X = Person's obtained score
        M = Mean of the normative group
        SD = Standard deviation

Suppose, a testtaker has obtained a score of 20 on a test and mean of the group of testtakers is 14 and standard deviation is 5, the his z-score by formula 6.1 would be

$$\frac{20 - 161}{5} = \frac{6}{5} = 1.20$$

Since z-score is a linearly transformed score, it retains the exact numerical relations of the original raw scores and thus it possesses the desirable property of retaining the relative magnitude of distance between the successive values in the original raw scores. Let us take an example. Suppose four persons A, B, C and D have obtained a score of 45, 50, 60 and 70 on a test with mean of 40 and standard deviation of 10. The first two scores differ by 5 raw score points and the last two scores differ by 10 raw score points. When these raw scores are converted into z-score by formula 6.1, then results are +.50σ, +1.00σ, +2.00σ, and +1.00σ, respectively. It is clear that the first two standard scores differ by 0.50 whereas the last two standard scores differ by 1.00 standard scores that is, twice the difference of the first pair. This illustrates what is called linear transformation of raw scores into z-score. When each of the raw score is transformed into a z-score, the resulting collection of standard scores always has a mean of zero and standard deviation of standard scores is 1.00. Since variance is equal to standard deviation squared, its variance is also 1.00.

    z-sores are quite satisfactory except for two things as under :
    (a)   z-sores require the use of plus and minus signs, which may by overlooked or miscopied.
    (b)   z-sores are obtained in plus or minus signs, which may be misplaced.
People in general, dislike to think of themselves as negative or fractional qualities.

    Usually we transform the raw scores into z-sores for depicting results on different tests according to common scale. If the two distributions of test scores possess the same form, we can make direct comparison of the raw scores by transforming them to standard scores. If the distributions don't have the same form, that is, one is normal and another is skewed, standard score comparisons can be very misleading.

Transformation of Raw Scores

Every researcher wants to make a sense out of the obtained raw score. For this, he transforms the raw scores into more interpretable and useful forms of information.

Standard score is, in fact, a raw score converted from one scale to another such scale, which has arbitrarily set mean and standard deviation. Raw scores are converted into standard scores because standard scores are more quickly interpretable than raw scores. When a raw score has been converted into standard score, the position of the testtaker's performance relative to other testtakers becomes readily apparent.

Standard scores may be obtained by either linear or nonlinear transformations of the raw scores. A linear transformation or conversion is defined as transformation where the standard scores retain the exact numerical relations of the raw scores because they are found by subtracting a constant (mean) from each raw score and dividing it by another constant (standard deviation). All features of the original distribution of raw scores are found in the distribution of these standard scores. Due to this reason any statistical computations that can be carried out with original raw scores can also be carried out with linearly transformed standard scores. z-score or also called as sigma score and percentage are very good examples of linerly transformed standard score.[5] One requirement of linear transformation of raw scores into standard score (or z-score) is that the form or shape of the distributions must be approximately same. If two distributions of the test scores of testtakers possess the same form (both normal or both skewed to more or less equal extent), a direct comparison on raw scores can be done by transforming them to standard scores. Let us take an example to illustrate it. Suppose Mohan has obtained a raw score of 80 on a spelling test for which the normative sample has a mean of 68 and standard deviation of 10. Further, he obtained a score of 70 on vocabulary test for which the normative sample has a mean of 60 and standard deviation of 12. In which skill area, Mohan has greater ability, spelling or vocabulary. If the normative samples for both test scores are of the same form (say normal), we can compare spelling ability and vocabulary ability by converting each score to standard scores. The spelling standard score for Mohan is $(80-68)/10 = 1.20$ whereas the vocabulary standard score for Mohan is $(70-60)/12 = .830$. It can be said that relative to the normative samples, Mohan has greater ability for spelling than for vocabulary.

On the other hand, when there are dissimilarly shaped distributions (one normal and other skewed) and the researcher wants to achieve comparatibility of scores, he resorts to what is called nonlinear or area transformation so that the score can be shown to fit to any specified type of distribution curve. Percentile scores, percentile rank, stanine are the examples of nonlinear transformations. Here percentile rank are computed to assign z-scores to raw score, based on the percentage of the normal distribution that falls below the z-score. Subsequently, these z-scores are

---

A discussion of z-score has already been presented in previous section under the subhead 'standard score norms'.

transformed with an arbitrary standard deviation and mean to a desired scale. The resulting set of distribution will tend to form a normal distribution, irrespective of the shape of distribution of raw scores. Thus in non-linear transformation the resulting standard scores does not necessarily have a direct numerical relationship to the original raw scores. As a result of such transformation, the original distribution is said to have been *normalized*. In a nutshell, in nonlinear transformation both unit of measurement and unit of reference are changed. Such transformations rely on the normal and tend to enhance the differences between the individuals in the middle of the distribution and compress the individual differences between individuals at the extreme of the distribution.

Besides z-scores, there are many other varieties of standard scores such as *T scores*, *stanine*, etc. Many psychologists and educators have praised the concept of standard score (z-scores) except for decimal fractions and positive and negative signs. To deal with these problems, they have devised a number of variations of standard score that are collectively known as *standardized scores*.

The students may ask what is the *difference* between standard score and standardized score? Conceptually, standardized scores are identical to standard score because both kinds of score contain exactly the same information. The shape of the distribution of the scores is not affected and the relationship between standard score and standardized score is always described by a straight line. However, the difference between standardized score and standard score is that the standardized score is always expressed as positive whole numbers with *no decimal fractions or negative signs*. Standardized scores can be tailored to produce any mean and standard deviation. The mean of the standardized score can be set at any convenient value such as 100 or 500 and standard deviation at 15, 20 or 100. If the preselected mean is at least *five times* as large as standard deviation, it is considered better for the good health of any standardized score (Gregory, 2004).

One most popular standardized score is T score proposed by McCall (1992, 1939) and named it in honour of his professor E.T. Thorndike. T score scale has a mean set at 50 and a standard deviation set at 10. Accordingly, it is also sometimes called as *fifty plus or minus ten scale*. For transforming a raw score to T-score, the formula 6.2 may be used :

$$T = \frac{10(X - M)}{SD} + 50 \qquad (6.2)$$

In this formula the term X-M/SD is equivalent to z (cf. formula 6.1) so formula 6.2 may be simplied as

$$T = 10z + 50 \qquad (6.3)$$

T score system consists of a scale that ranges from 5 standard deviation below the mean to 5 standard deviation above the mean. Thus a raw score that fall exactly at 5 standard deviations below mean would be equal to a T score of 0, and a raw score that fall at the mean would be equal to a T of 50. Likewise, a raw score falling at 5 standard deviations above the mean would be equal to a T of 100. Stated very simply, on this scale a score of 50 corresponds to the mean and a score of 60 to one SD above the mean, a score of 70, two SD above the mean and so on. (see Figure 6.2) T score scales are most popular with personality tests and we find that each clinical scale of famous MMPI is lashed with an average score of 50 and standard deviation of 10 for the normative sample.

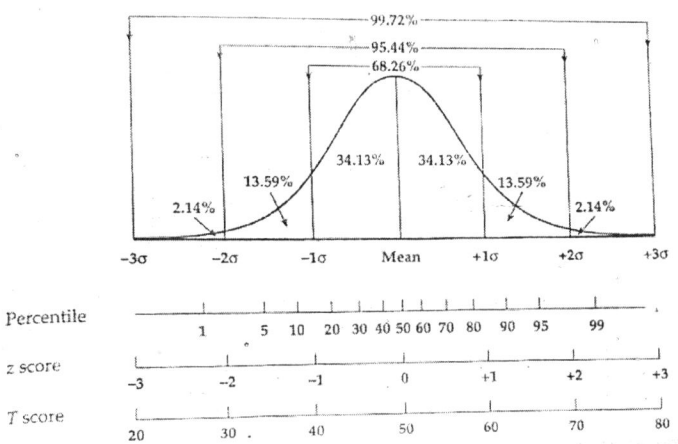

Figure 6.2   Equivalences among Common Raw Scores Transformation in a Normal Distribution.

Stanine scale developed by the United States Air Force during World War II is derived from the phrase *Standard nine-point* scale. This is a normalized standard score that has become popular for educational tests. The scale has a mean of 5 and standard deviation of approximately 2. Thus in stanine scale all raw scores are converted into a single-digit system of scores ranging from 1 to 9. The transformation from raw scores to stanine is very simple. Here first the scores are arranged from lowest to highest and the bottom 4 percent of scores converts to a stanine of 1, the next 7 percent of scores converts to 2 the next 12 percent to 3, etc. (see Table 6.1). It is still more clarified from Figure 6.3. The biggest advantage of stanine score is that it eases computational problems due to restriction of single digit number.

Table 6.1: Distribution of Percentages for Stanine Conversion

Percentage	4	7	12	17	20	17	12	7	4
Stanine	1	2	3	4	5	6	7	8	9

Some statisticians have provided several variations of stanine scale. For example, Canfield (1951) proposed *Sten scale* consisting of 10-unit, with 5 units above and 5 units below the mean.

Guilford and Fruchter (1981) have proposed another variant of stanine called *C scale*, which consists of 11 units with the standard deviation of 2.

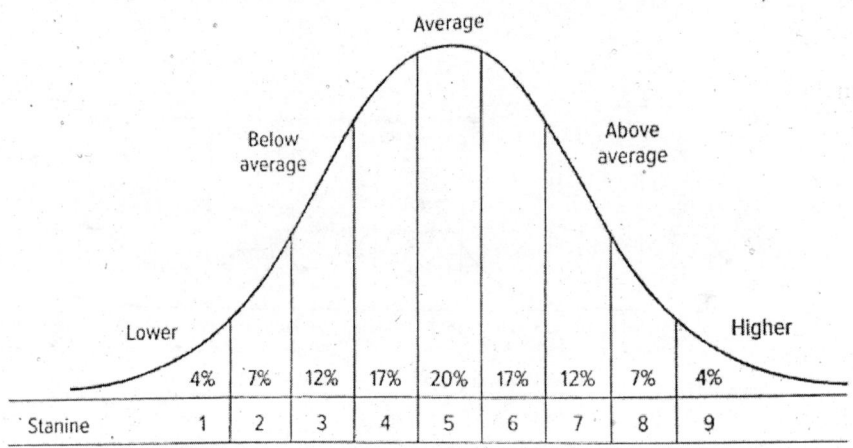

Figure 6.3: Stanines and the Normal Curve

Psychologists and educators prefer to work with normal distributions because statistical properties of the normal curve are obvious and standard scores from such distributions can be easily and directly compared. But sometimes the situation becomes otherwise and they have to deal with skewed distributions. For example, what should be done if the test under development is yielding a nonnormal distribution such as a skewed distribution.

In such situation test developers try to normalize the distribution. Technically, normalizing a distribution involves stretching the skewed curve into a shape of normal curve creating a scale of standard scores called a *normalized standard score scale*. Normalized standard sores are standard scores expressed in a distribution that has been transformed to fit a normal curve (Anastasi & Urbina, 1997). The computation of a normalized standard score is easily done. First, we use the percentile for each raw score to determine its corresponding standard score. If it is done for each and every score in a normal distribution, the resulting distribution of standard scores will be those of normalized standard scores.

The conversion of percentile to normalized standard score may seem a simple and ideal solution to nonnormal distributions but it involves some technical objectives and serious drawbacks. Since normalized standard scores are a nonlinear transformation of the raw scores, the mathematical properties of the raw scores may not be true for the normalized standard score that is above the mean. Moreover, such transformation should be carried out only when the sample is *large* and representative and when there is reason to believe that nonnormal distribution of raw scores has occurred due to inappropriate difficulty level of the test items such as there are too many difficult or too many easy items in the test. It should be made clear that when the original distribution of raw scores approximates that of normal distribution, the linearly derived standard scores and the normalized standard scores will be very similar. Despite the fact that these two types of scores are derived in a different way, the resulting scores will be nearly identical. Obviously, then the process of normalizing the distribution of scores which already approximates normal distribution, will produce little or no change.

Keeping in view what has been said above, it is most desirable for the test developers to obtain a normal distribution of the raw scores by making some adjustment with the difficulty level of the items of the test rather than take pains to normalize the markedly a nonnormal distribution.

National Norms vs. Local Norms

National norms as the name implies, are norms, which are derived from a normative sample taken to be representative of large number of people belonging to different variables of interest such as gender, age, socio-economic status, geographical location as well as the different types of communities within the different parts of the country such as south, north, east and west. When the national norms are being developed, some questions are raised. These questions relate to what the test is designed for and what the test is intended to do. For example, if the test is to be used in the schools, norms might be developed for students of each grade and several factors related to the representativeness of school from which members of the norming sample would be drawn, must be kept in view. For example, some popular questions relating to selection of schools might be: Is school privately owned or owned and controlled by the government? Is the school located in urban area a rural area? is the school running in rented house or it has its own building? Has the school facility of library? What is the teacher-students ratio in school? Does the school have transportation facility? etc. Based on such questions, schools are selected from all over the country and then, a required number of sample, usually a larger sample, is selected for developing the national norms.

Usually different psychological and educational tests claim to have a nationally representative sample. But a close scruity of the description of sample taken generally reveals that sample differs in many different ways even from those similar tests claiming to measure the same thing. Therefore, a good suggestion is that test users must check the manual of the test under consideration to examine how comparable the two tests really are.

If the researcher wants to compare the findings on two tests that measure the same trait or ability, an equivalency table for the scores on the two tests must be developed. Such equivalency table is technically thrown as *national anchor norms*, which provides some stability to test scores by anchoring them to other test scores. The method of developing such norms starts with the computation of percentile norms for each of the tests to be compared. With the help of such *equipercentile method*, the equivalency of scores on different tests is calculated with reference to the corresponding percentile scores. For example, suppose the 90th percentile corresponds a score of 60 on Test A, and 90th percentile corresponds to a score of 40 on Test B, we can say the score of 60 on Test A is equivalent to a score of 40 on Test B. For calculating national anchor norms, it is essential that both tests must be administered on the same sample. Thus national anchor norms provide a good platform for judging the equivalency of scores on various tests. However, because of some technical considerations, these equivalences cannot be treated as precise equalities (Angoff, 1971).

Local norms are norms, which provide normative information with respect to performance on test by local population. The group involved in developing such norms are narrowly defined within a particular setting. For example, the principal of the college may develop norms on its own student population. The management of a particular company may develop local norms on applicants for a certain job. These local norms are considered more appropriate than national norms for some special purposes such as college academic achievement, predicting about the individual's progress over time, predicting about subsequent job performance, etc.

Sometimes it is often helpful and important to develop separately *subgroup norms* despite the fact that norms are available for a broadly defined population. In developing subgroup norms the normative sample is segmented by any of the criteria such as education, age, socio-economic level, sex, geographical region, community type, etc. Such segmentation becomes true whenever recognizable subgroups produce appreciably different score on a particular test.

Major Cautions in Interpreting Test Scores

When one is going to interpret the test scores, one should keep in mind some general cautions that apply to the interpretation of any test scores. Some of the important such cautions are as under :

1. **Test score should be interpreted in terms of the specific test from which the scores have been derived.**

No two tests measure exactly the same thing. Psychological and educational tests are prone to wide variation and differences are rarely displayed in the test title. For example, one biology test may be limited to simple knowledge of vertebrates whereas other biology test may contain a number of problems of invertebrates only. Likewise, one mathematics test may be limited to simple computational skills whereas the other mathematics test may contain a number of arithmetical reasoning problems. Obviously then, with such variation it will be misleading to interpret the test score of a student as representing some general achievement in the concerned area. Such variations are also found in intelligence tests, creativity tests and aptitude tests. Therefore, these tests scores should also be interpreted very cautiously.

2. **Test score should be interpreted as a band of scores rather than as a specific value :**

Any obtained tests score is not completely free from error. Therefore, during the interpretation of test score, scope for allowance of such errors be made. One best way to deal with test score, then, is to consider the test score as a band of scores such as one standard error of measurement above or below the obtained score. Suppose the test score is 65 and standard error is 3, then the test performance should be interpreted as band ranging from score 62 to 68. This will help us in making interpretations that are more accurate than the test results would allow.

3. A test score must be verified by other background evidences :

When interpreting test score, it is very difficult to say that all important basic assumptions of testing have been properly met. Therefore, errors may creep into the interpretation. For example, it is very difficult to ensure that during testing maximum motivation among the testtakers and equal educational opportunity have been maintained. Likewise, it is difficult to say that the conditions of testing such as administration, scoring, etc. have been properly met. Consequently, in addition to the predictable error of measurement that could be controlled through standard error bands, a test score may incorporate some indeterminate amount of errors due to these unmet assumptions or uncontrolled conditions. Therefore, it is essential to protect the interpretation of test scores against these errors. Consequently, it is suggested that test scores must be coordinated by background facts as well as they must be verified by constant comparison with some other available data. (Cronbach, 1970).

4. A test score should be interpreted in the context of all testtakers' relevant characteristics.

A test score must be interpreted in light of the several relevant characteristics of the testtakers. Among these relevant characteristics are socio-cultural background, educational experiences, emotional adjustment, language background, improper motivation or similar factors. Thus if a testtaker gets a poor test score on test, rather than interpreting it as a sign of poor performance, one should interpret this score in terms of several other factors mentioned above that might have interfered with the testtaker's performance.

5. A test score should be interpreted in light of decision to be made :

Any test score should not be considered as simply being high or low in general. Rather it must be interpreted in the light of the decisions likely to be taken. For example, a score of 30 (out of a maximum of 100) on a mechanical reasoning test may be considered as good for taking admission into training course of some mechanical supervisors but it will not be considered good for appointment of the applicant to the post of a mechanical supervisor.

These are some important cautionary principles, which must be taken into account if the misinterpretation and misuse of the test scores are to be prevented.

Summary and Review

- Raw scores are the obtained total score on the test. Strictly speaking, a raw score in itself has so meaning unless it is compared with some norms.
- Norms are defined as the average test performance of the normative sample or standardization sample.
- There are four important types of norms: age norms, grade norms, percentile norms and

- standard score norms.
- Age norms are norms that display the average performance on the test by each separate age group in the normative sample.
- Grade norms are norms that displays the level of test performance for each separate grade in the standardization sample.
- Percentile norms are those norms, which are based upon percentile.
- Standard scores norms are those norms, which are expressed in terms of various standard scores. z-score is one important type of linearly transformed standard score.
- Raw scores are transformed into standard scores in both linear way as well as in nonlinear way.
- Standard scores may be normalized and a normalized standard score scale may be built. When the test developers get skewed distribution rather than normal distribution on the test, they are required to normalize it, that is, they are require to transform in such a way that it may fit to normal curve.
- Both national norms as well as the local norms are important. But psychologists and educators, in general, prefer local norms to national norms.
- There are some vital principles that must be kept in view while interpreting the test scores.

Review Questions

1. Define norms. Discuss its various types
2. What is age norms? Discuss its advantages and disadvantages.
3. Make a comparatve study of age norms and grade norms.
4. What is percentile norms? Discuss its major limitations.
5. What do you mean by standard score? Also explain the concept of standard score norms in this connection.
6. Citing examples make distinction between linear transformation and nonlinear transformation of raw scores into standard score.
7. Distinguish between :
    (a) Standard score and Standardized score
    (b) Z-score and T score
    (c) Age norms and Grade norms
    (d) Percentile and Percentile rank
    (e) National norms vs Local norms
8. What kind of raw scores are subjected to normalization procedure? Discuss the limitations of the normalizing the standard scores.

Chapter – 7

Taxonomy of Educational Objectives

Learning objectives :

- Dimensions of Educational objectives
  - Process and product objectives
  - Behavioural and Implicit Objectives
  - Intermediate and ultimate objectives
  - Restricted and Inclusive objectives
- Taxonomy of Educational objectives
  - Bloom's Cognitive Educational objectives
  - Krathwohl's Affective Educational objectives
  - Simpson's Psychomotor Educational objectives
- Summary and Review
- Review Questions

Key Terms :

Educational objectives, Process objectives, Productive objectives, Behavioural objectives, Implicit objectives, Intermediate objective, Ultimate objective, Restricted objective, Inclusive objective, Cognitive objective, Affective objective, Psychomotor objective, Characterisation, Internalisation.

There are three interacting aspects of educational processes: educational objectives, learning experiences and evaluation procedures. Educational objectives are goals established either explicitly or implicitly. Learning experiences are subsequently designed to promote the attainment of these educational objectives. Finally, evaluation is done to determine the extent to which objectives have been attained. The results of the evaluation may affect the educational objectives. Thus there is an interrelationship among roles of educational objectives, instruction (or learning) and evaluation. The objectives should provide direction to the curricular methods and contents. The evaluator tries to determine the degree to which objectives have been attained both individually and collectively. Thus the results of the evaluation provides a feedback that may introduce some changes in either objectives or instruction, or both Educational objectives can be explained as being process or product, behaviour or implicit, immediate or ultimate and restricted or inclusive. In the present chapter, attempt has been made to present a description about dimensions of educational objectives as well as a taxonomy of educational objectives. The stated purpose of such taxonomy is to facilitate a communication.

Dimensions of Educational Objectives

There are several dimensions of educational objectives. Important ones are : *process and product objectives, behavioural and implicit objectives, immediate and ultimate objectives* as well as *restricted and inclusive objectives*. A brief discussion of these dimensions of educational objectives is presented below:

- **Process and product objectives**

Process objectives are those educational objectives, which describe *who* is responsible for the given activity and *when* it will be achieved. There are three parts of this dimension: 'who', 'activity' and when. The *who* part of the process objective refers to the teacher or students but sometimes it may also refer to the parent or administrator. The *activity* part refers to the description of what should take place. For example, students may be required to memorize three poems or the teacher may order for good preparation of N.C.C. parades, etc. The *when* part of the process objective refers to the specific time by which the activity should be completed.

Process objectives, thus, involves specifying what someone must do by a fixed time to monitor a programme and find out whether it is running on correct schedule. Such objectives also provide help to the evaluator for determining the prior sequencing the activities. They also help ensure that all necessary steps in the programme have been properly sequenced. Besides, they also provide evaluator with a target date of carefully completing each activity.

One limitation of process objective is that they fail to indicate the *effectiveness* of the activity. In fact, effectiveness depends on what is called *product objectives*.

A *product objective* is defined as objective, which indicates what the student must do or know as a consequence of instruction. Giving an assignment from the textbook to the student is a process objective but what he is expected to know as a consequence of this assignment is the product objective. For example, at school level physical education teacher's product objective may be to develop skills for 'good sports' among his students.

It is argued that all process objectives ultimately are related to some product objectives. For example, doing various types of exercises in physical education is related to producing a 'healthy product' that is, good physical and mental health among the students. Those subjects in which process objectives are not closely related to product objectives are generally given low priority in the curriculum.

- **Behavioural and Implicit objectives**

Behavioural objectives are those objectives, which specify some terminal behaviour (or also called as observable task) that the student is expected to perform for demonstrating that the goal has been achieved. Behavioural objectives have three elements:

(a) Such objectives have an overt activity or terminal behaviour to be performed by the student.
(b) Such objectives contain specification of conditions under which the terminal behaviour is to be manifested.
(c) Such objectives specify a minimum level of acceptable performance.

Let us take an example to illustrate behavioural objectives:
1. *Recall* at least 12 names of cities out of 20 names without using notes within one minute
2. *Write* any poem that you have memorized within last two days from memory without a single error within ten minutes.

In each of these two example the italicized verbs indicate an observable response and is, therefore, a terminal behaviour or observable task. The conditions under which these behaviours are to be demonstrated have also been mentioned (without using notes, and within the minute from memory) as well as the minimum level required for acceptances, (12 out of 20 names), and (without a single error). Under some conditions, it may not be necessary to specify these various conditions.

Implicit objectives, on the other hand, are those objectives, which imply a nonobservable state of *knowing, understanding, appreciating, interpreting, clarifying* and so on. As we know, all behavioural objectives involve student product but not all product objectives are behavioural. Some imply an inner or nonobservable responses, which are called as implicit objectives. For example :
- Understands the distinction between a psychologist and a psychiatrist
- Appreciates the role of vitamins and minerals in maintaining good health.

The verbs 'understands' and 'appreciates' indicate something unobservable and therefore, indicate implicit objectives. Implicit objectives may provide guidelines to teachers that some goals are important although their measurement may be a difficult task. Sometimes implicit objectives can form the basis for behavioural objectives.

- **Immediate and Ultimate objectives**

Immediate objectives are those objectives over which teachers have some control and they can be managed and taught to the students. Some school activities such as marriage, employment and voting behaviour are designed to prepare students for events. Teachers have to wait for a considerable period of time for assessing their effectiveness. These are, in fact, ultimate objectives over which teachers have little control because what will students do after being adults, are beyond the control of the teachers. Although teachers cannot check directly on the effectiveness in teaching ultimate objectives, they must emphasize that students have acquired these behaviours within themselves. Thus it may not be possible to measure 'intelligent voting' but certainly, it is possible to make sure that the students understand the immediate issues relating to voting, such as different types of political parties, advantages and disadvantages of forming government by a particular party, which party can go hand-in-hand with a particular party, etc.

- **Restricted and Inclusive objectives**

    Restricted objectives are those objectives, which are so restricted that they may be treated as one single item. Inclusive objectives are those objectives, which are so broadly stated that they allow for the inclusion of several behaviours from the same designated universe. Some examples from restricted objectives and inclusive objectives are as under :

    Examples from restricted objectives :
    (a) Multiply: $2 \times 2 \times 2 \times 2 \times 2 \times 2$
    (b) Name two major proponents of the third force in psychology
    Examples from inclusive objectives :
    (a) Find out the solution of 2^6
    (b) Point out the major contributions of Maslow and Rogers.

    The examples clearly indicate restricted objectives are in fact so restricted that they can be treated as one single item. A student may be asked to find out the answer of $2 \times 2 \times 2$ on a separate sheet of paper. However, when all the multiplication objectives are considered together, they completely describe the inclusive objectives: find solution of 2^6. This explanation also applies more or less to the second example.

    It is obvious, then, that inclusive objectives are like general conclusions. They may completely describe everything that a student is expected to do or they may contain some selected samples of behaviour. One important point to be made is that only inclusive objectives need to be written by the teacher if the restricted counterparts are completely enumerated as in the first example because this example incorporates all elements common to the restricted objectives. But if inclusive objectives can be measured by many different restricted objectives, it becomes necessary then to write the inclusive objectives and then, indicate about the restricted samples to be used.

Taxonomy of Educational Objectives

In science, taxonomy is defined as an orderly and systematic classification of plants and animals according to their some presumed natural relationship. This term has been used in education for describing a comprehensive classification scheme for educational or instructional objectives. There are different taxonomies of educational objectives and each taxonomy classify all objectives into a hierarchy of categories based on presumed complexity. Each succeeding category involves behaviour thought to be comparatively more complex than the previous one and each is considered to be prerequisite to the next. In other words, performance of objectives at higher level indicates ability to perform satisfactorily at lower level. The stated purpose of the taxonomies is to facilitate communication. The identification and description of categories of behavioural outcomes permits the educators to use the same terms to describe the same behaviours in more or less the same way.

Educators have proposed three important taxonomies of educational objectives, according to three domains of learning. Three domains are : *the cognitive domain*, *the affective domain* and *the psychomotor domain*. Thus there are *cognitive objectives*, *affective objectives* and *psychomotor objectives*. *Cognitive objectives* are those objectives which emphasize the attainment, retention and development of knowledge and intellect. In simple words, such objectives broadly include all cognitive processes. The *affective objectives* are those objectives, which encompasses the behaviours characterized by feelings, emotions interest and values. Affect is either positive (or favourable) or negative (or unfavourable). Thus students may be favourably or unfavourably impressed by their teacher, curriculum or school. The *psychomotor objectives* are concerned with physical abilities such as motor skills, manipulation, neuromuscular coordination or so on. Running, speaking, playing handwriting, etc. are classified as psychomotor activities.

The categorization of behaviours into domains does not mean that the domains are completely independent of each other. Reality is that they are interdependent. For example, when a student gives a talk in the classroom, all three domains are involved though they may not be involved equally. During the speech by the student, if the teacher mainly emphasizes upon the student's gesture, movement and voice inflection, he is emphasizing upon psychomotor objectives. If he is emphasizing upon the organisation of content of the speech, he is emphasizing cognitive objectives. Likewise, if he is concerned with pleasantness or unpleasantness of the task, he is emphasizing upon affective objectives.

- **The Cognitive domain: Bloom's taxonomy of Cognitive Educational objective**

The cognitive domain incorporates those objectives that deal with the recall or recognition of the learned material as well as development of intellectual abilities and skills. The largest proportion of educational objectives fall into the cognitive. Bloom et al., (1956) have developed a taxonomy of cognitive objectives. This taxonomy has been most popular and has definitely made educators aware of the wide range of abilities involved in learning. They have divided cognitive objectives into six hierarchical categories from simple to complex as illustrated in Figure 7.1. These six ascending levels are : *knowledge, comprehension, application, analysis, synthesis* and *evaluation*. Each of these six major categories represents a different kind of behaviour involved in each category. The rationale for the hierarchy of these six categories is based upon the assumption that each level is the extension of all previous levels. For example, to achieve an objective in application category requires (at least in theory) that some goals in comprehension category be achieved, which in turn, can be achieved only if certain information in the knowledge category is acquired. A brief description of these six categories of cognitive objectives is as under :

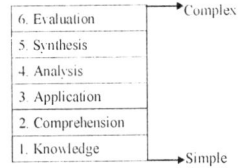

Figure 7.1: Ascending levels of Bloom's Cognitive objects

1. Knowledge :

According to Bloom, the simplex level of complexity is knowledge. This category includes memorization behaviours, that is, recall or recognition of previously learnt information. Knowledge represents the *lowest* level of learning outcomes in the cognitive objective. Examples of objectives within the knowledge category are the knowledge of dates, persons, specific facts and terminology. For example :

In which year Wilhelm Wundt established the first laboratory as an experimental psychologist:
- (a) 1879
- (b) 1869
- (c) 1995
- (d) 1885

Answering this question requires no reasoning and only a simple correct knowledge of year is enough.

2. Comprehension :

It is the second level of category in cognitive objective. Behaviours in this category are those, which show understanding and not just memorization. Translation, interpretation and extrapolation are common behaviours at this level. Comprehension is evidenced by the accuracy with which communication is paraphrased or rendered from the language. This type of objective requires students to restate the fact or problem in their own words, to give an example of a principle or concept or to point out some implication or consequences. For example :

Which of the following is the best definition of the term learning?
- (a) Changes in behaviour due to practice
- (b) Changes in behaviour due to practice and experience
- (c) Changes in behaviour due to experience
- (d) Changes in behaviour due to practice and experience for making healthy adjustment with environment

This problem required students to judge the best of four definitions, which vary in correctness and completeness. Had three of these choices been totally incorrect, this item would be functioning only at the knowledge level of taxonomy.

3. Application :

Application is the third level of complexity. It refers to being able to use or apply an abstract concept to a specific situation In other words, application refers to use or to apply learned material in new concrete situation. Learning outcomes in this area require a higher level of understanding than those under comprehension. For example :

If a person is planning to sunbathe, at what time of the day is the most likely to receive a severe sunheat? He is likely to experience a severe sunheat in the middle of the day, that is, in between 11 A.M. to 1 P.M. because:
- (a) The ultraviolet of the sunlight is mainly responsible for sunheat.

(b) When rays of sun fall directly, that is, straight down on a surface, more energy is received by that surface when the rays of sun fall obliquely on that surface.
(c) The noon sun will produce more heat than the morning or afternoon sun.
(d) When the sun is directly overhead, the rays of the sun pass through less absorbing atmosphere than when the sun is lower in the sky.

The examples of the item illustrate that mere parroting of the definitions of the textbook would not be much helpful to a student in arriving at correct answer in this applied setting.

4. Analysis :

This is the fourth level of complexity and it refers to the ability of the students to break down information into its constituent parts so that its organisation structure may be clearly understood. This may include the identification of the different parts, a careful analysis of the relationship between parts and recognition of the organisational principles involved. Learning outcomes at this category represent a higher intellectual level than comprehension and application because they demand an understanding of both the content and structural forming content. For example :

The general components of attitudes are :
(a) cognitive component
(b) behavioural component
(c) affective component
(d) conative component

The students are required here to break down the components of attitudes and recognize them.

5. Synthesis :

Synthesis refers to the ability to put parts together to form a new whole pattern. When a student is asked to produce a story, composition, hypothesis or a theory on his own run, he is said to synthesize the knowledge. At this level, student produces something new and unique instead of breaking knowledge into simple elements. Thus synthesis involves a production of unique communication (theme or speech) or a plan of operations (research proposal). Learning outcomes in this category stress creative behaviours giving major emphasis on new pattern or structures. For example :

Develop a original plan to reduce inflation in India Formulate a theory of learning to apply it to classroom teaching.

In these examples, synthesis exercises are reflected where students are required to combine the ideas in such a way as to formulate a new communication or structure.

6. Evaluation :

Evaluation is concerned with the ability to judge the value of material (such as poem, novel, research report) for a given purpose. These judgements are based on internal criterion or on external criterion. Learning outcomes in this category are the *highest* in the cognitive hierarchy because they contain elements of all of the categories discussed so for plus value judgement based on some clearly obtained criteria. Examples are :

    Ability to criticise a theory or plan

    Ability to indicate logical fallacies in arguments.

Three taxonomy categories, namely, analysis, synthesis and evaluation have proved to be of much less value for curriculum development as well as for that of educational evaluation than those of knowledge, comprehension and application categories. The logic of hierarchical continuum of complexity is also less compelling beyond application level (Michael, 1968)

Although Bloom's six-level classification of cognitive domain is the most extensive attempt, it is not the only useful scheme available. Ebel (1956) formulated six ascending levels and attached to the some ideal percentages he recommended for good achievement test as under:

Content details	0 percent
Vocabulary	less than 20 percent
Facts	less than 20 percent
Generalizations	More than 10 percent
Understanding	More than 10 percent
Applications	More than 10 percent

Beside these, Michael (1968) have suggested the structure of intellect proposed by Guilford (1967) as a very comprehensive basis of achievement examinations. Scriven (1967) have also suggested a comprehensive system of educational objectives which encompasses cognitive, social, attitudinal and psychomotor dimensions.

The affective domain: Krathwohl's Taxonomy of Affective Educational objectives:

The taxonomy of affective objectives has been developed by Krathwohl et al., (1964). Objectives in domain relate to feelings, attitudes, interests and values. There are five levels describing the affective domain of educational objectives: *receiving* (attending) *responding*, *valuing*, *organization* and *characterization*. Thus the categories represent a hierarchy of acceptance, which ranges from willingness to receive or attend to characterisation by a value. The degree of *internalization* is supposed to be the unifying hierarchical factor underlying the affective taxonomy. The shallowest degree of internalization of feeling is represented by *awareness* in receiving category and the deepest is represented by *characterization*. Often the achievement of affective objectives is determined through administration of self-report measures.

A discussion of the five categorises of affective objectives is presented below :

1. Receiving (1.0) :

At this level the person is sensitized to the behaviour of interest to the extent that he is willing to receive or to attend (pay attention) do it. For example, a student listens attentively when a teacher reads a poem or poetry in the classroom. there are three subcategory under receiving

(1.0): *awareness* (level 1.1), *willingness to receive* (level 1.2) and *controlled* or *selected attention* (1.3).

The lowest subcategory under receiving is awareness (1.1) that presumes some minimum consciousness or knowledge of the behaviour of interest. Thus here the student becomes aware of the stimulus and takes it into account.

The next level is *willingness to receive* (1.2). Here instead of simply being aware, the student shows willingness to attend or to at least not to actively avoid some stimulus. Evidence for showing willingness to receive is usually expressed by the student's indication of whether he likes, dislikes or is indifferent to some activity. Here the concern is not that how much a student likes or dislikes the activity but whether or not he has some minimal willingness to consider it.

The next level under receiving category is *controlled* or *sustained attentions* (1.3). At this level, the student is able to differentiate or attend to certain portions and disregard others. Thus he expresses preference for some activity and actively select it from others. He is fully capable of controlling his own attention to attend to some stimuli that are liked. For example, he may be able to attend to background music in the movies despite competing and distractive stimuli.

2. Responding (2.0)

Responding lies at the level 2.0 where a student does more than something to attend to the stimuli. In fact, he responds to them enthusiastically. Thus in responding there occurs active participation on the part of the student. Thus he not only attends to stimuli rather reacts to it in some way.

There are three subcategories or levels under the category of responding: *acquiescence in responding* (2.1), *willingness to respond* (2.2) and *satisfaction in response* (2.3).

The lowest sublevel is *acquiescence in responding* (2.1). The two important key words that describe this level are *obedience* and *compliance*. A student obeys his teacher, is compliant and does submissively whatever told. A student who visits the library because the teacher requires so, is functioning at this level.

The next sublevel under responding category is *willingness to respond* (2.2). At this level a student engages in an activity due to his genuine interest and not because his teacher has suggested or requires it. In a nutshell, a student here responds voluntarily and not out of obedience or compliance. He is under no obligation nor he faces any threat for punishment for not responding. A student who visits the library voluntarily and read books or studies some subject in greater depth than is required, is responding at this level.

The highest subcategory or sublevel under responding is *satisfaction in response* (2.3). At this level student enjoys what he is doing and gains satisfaction. For example, here student not only be visiting the library rather he enjoys it and have satisfaction Stratification in response can be assessed by asking students to indicate it on three or five point scale or by using student's gesture, comments, grimaces, laughter, etc.

3. Valuing (3.0):

The third level in affective objective is valuing (3.0), which is concerned with worth or values a student attaches to a particular behaviour or phenomenon. Valuing also implies consistency in responding to general class of objects to the extent where others recognize this general pattern. Again this category or level has three subcategories or sublevels: *acceptance of a value* (3.1), *preference for a value* (3.2) and *commitment* (3.3).

Acceptance of a value lies at the lowest sublevel (3.1) in valuing category. At this level the student has a tentative commitment towards some belief or attitude, which he considers worthwhile. He questions the belief and may switch rather than fight with it. He regards the object of his commitment as more valuable than a mere source of pleasure or satisfaction.

The next sublevel (3.2) is the *preference for a value*. At this level, student actively pursue his valued object or belief and is willing to devote time and effort in pursuing values. Here the degree of commitment is greater than what is found in 3.1. The difference between 3.1 and 3.2 is largely that of degree. Here at this level the student is more involved and demonstrates the sense of involvement by actively contributing to the value object or phenomenon. For example, he may work for achieving distinction in chemistry to whom he placed great value.

The next sublevel in valuing is called commitment (3.3). At this sublevel, the student is strongly certain that his beliefs are correct. He shows a high sense of loyalty to beliefs and is willing to work hard for the goals. In fact, he has a strong devotion to a cause or belief that demands a continued and sustained effort to achieve some goal. It means that the committed student would rather fight than switch.

4. Organisation (4.0):

Organisation is the fourth level (4.0) in affective domain, which is concerned with bringing together different values resolving conflicts among them and starting building the internally consistent value system. In other words, here the emphasis is on comparing, relating and synthesizing values. There are two sublevels of organisation: *conceptualization of a value* (4.1) and *organisation of a value system* (4.2).

The conceptualization of a value (level 4.1) incorporates an abstract or symbolic set of interrelated values. These values are conceptualized by analyzing their interrelationship as well as by drawing generalisations, which represent that value system. For example, a student who is interested in knowing the factors common to different branches of science is working at this sublevel.

At the second sublevel of organisation is organisation of a value system (4.2). At this level, all the values conceptualized at 4.1 level are ordered so as to make them internally consistent and compatible with each other.

5. Characterization by a value or value complex (5.0):

This is the highest level of affective objectives (5.0). Persons at this level not only have organised value system but are fully capable of behaving in accordance with a consistent philosophy of life. They have controlled their behaviour for a sufficiently long times and have developed a characteristic lifestyle. Here the behaviour is pervasive, consistent and predictable. Persons well know who they are and what they stand for.

There are two sublevels in this category: *generalised set* (level 5.1) and *characterisation* (level 5.2). By generalised set is meant that when the person faces new problems, he is predisposed to respond in accordance with his value system. In fact, his values tend to determine what he perceives as very important, what approaches he will consider to remedy problems and how sincerely he pursues a given course of action.

At level 5.2 is the *characterisation* which is the highest level in affective objective. Characterisation involves development of a philosophy of life that is all encompassing and that gives character to those who are capable of reaching at this level. Such philosophy is internally consistent and the persons act in accordance with this philosophy. There are few persons who have reached at this level and those who have arrived at this level are recognized as competent persons. This is what Maslow has called the stage of self-actualization. A few examples of such persons are Mahatma Gandhi, Jesus Christ, Mother Teresa, etc.

In general, students are not expected to advance much beyond levels 2 or level 3 as a result of their academic activities in the classroom. They can arrive at level 4 only if teachers help students clarify and organise their own value systems. Level 5 will be attained by a few students, perhaps by those who are capable of arriving at the stage of self-actualisation as Maslow has preferred to call it. In fact, the development of level 5 especially characterisation (level 5.2) much depends on the acquisition of more immediate values and beliefs, which the teachers can help students acquire.

Psychomotor domain: Simpson's Taxonomy of Psychomotor educational objectives

Psychomotor domain, as the name implies, incorporates behaviours that heavily depend upon the muscular activities such as jumping, running, speaking, typing, hammering, drilling and so on. All students engage in psychomotor activities to different degrees. At the elementary level, the psychomotor activities like eating, drinking, walking are much more emphasized. During elementary grades the physical activities like catching, throwing, developing eye-hand coordination are emphasized. Several serious attempt has been made at developing a taxonomy for the psychomotor domain. Two most popular such attempt were those of Simpson (1972) and Harrow (1972). As an example, we shall discuss here only that of Simpson (1972).

In Simpson's classification of educational objectives based on psychomotor domain, there are seven categories as under :

1. Perception :

This is the first level of psychomotor adjectives where the student uses organ to obtain various cues that guide motor activities. This category ranges from sensory stimulation through cue selection to translation of cue perception into action in a performance.

2. Set :

Set means readiness to take a particular type of action. This category includes three types of sets such as mental set, physical set and emotional set. Perception of cues serve as an important prerequisite for these types of set.

3. Guided response :

Guided response, as name implies, incorporates the early stages in learning a complex skill. This includes imitation as well as trial and error. Here performance is assessed by a suitable set of criteria or by a competent instructor.

4. Mechanism :

Mechanism is concerned with performance acts where the learnt response becomes habitual and the various patterns of movements can be performed with some confidence and proficiency.

5. Complex and response :

Complex overt response refers to skillful performance of motor acts having complex movement patterns. Here proficiency is reflected by smooth and accurate performance requiring lest time and energy. Learning outcomes at this level includes highly coordinated motor activities.

6. Adaptation :

Adaptation refers to the well-developed skills that can be modified to fit the demands of the problem situation.

7. Origination :

Origination means creating new movement patterns to fit a particular situation or specific problem. At this level, creativity based upon highly developed skills is emphasized.

Thus we find that there are three taxonomies of educational objectives: taxonomy of cognitive objectives, taxonomy of affective objectives and taxonomy of psychomotor objectives. Of these three, taxonomies of cognitive and affective objectives have been generally accepted. While several taxonomies of psychomotor domain have been developed, none of them have achieved degree of acceptance of other two. Of the cognitive and affective taxonomy, the affective taxonomy has not had that impact on education that cognitive taxonomy had. Perhaps this is partially due to the unique measurement problems associated with assessment of affective domain.

Summary and Review

- There are several dimensions of educational objectives such as process and product objectives, behavioural and implicit objectives, immediate and ultimate objectives, as well as restricted and ultimate objectives.
- Based on three domains of learning (cognitive, affectives as well and psychomotor), there are three types of educational objectives: cognitive educational objectives, affectives

- educational objectives and psychomotor educational objectives.
- Taxonomy of cognitive educational objectives was developed by Bloom and his colleagues in 1956. It contains six categories, which from order of simple to complexity are: knowledge, comprehension, Application, Analysis, Synthesis and Evaluation.
- Taxonomy of affective educational objectives was developed by Krathwohl and his associates in 1964 and this incorporates five categories, which in order of degree of increasing internationalisation by a value.
- Taxonomy of psychomotor educational objectives has been developed by several experts. Of these, the taxonomy developed by Simpson in 1972 have been very popular. This taxonomy includes seven categories: Perception, Set, Guided response, Mechanism, Complex overt response, Adaptation, and Origination.

Review Questions

1. What do you mean by educational objectives? Discuss the different dimensions of educational objectives.
2. Citing relevant examples, discuss Bloom's taxonomy of cognitive educational objectives.
3. Outline a plan for presenting a taxonomy of affective educational objective.
4. Discuss the taxonomy of psychomotor educational objectives.

Chapter – 8

Classroom Instructional goal and objectives : Educational Decision-making

- Nature and Purpose of classroom instructional goals and objectives
- Instructional objectives as learning outcomes
- Criteria for selecting appropriate instructional objectives
- Some important ways of stating instructional objectives
- Classroom Instructional Decisions
- Summary and Review
- Review Questions

Key Terms:

Affective measures; Instructional goals; Instructional objectives; Learning outcomes; Oral test, Product evaluation, Paper-and-Pencil test; Performance tests.

Instructional goals and objectives play a vital role in teaching. Defining desired outcomes is the first step in goal teaching. In fact instructional goals, objectives and learning outcomes serve as guides for both teaching and learning, communicate the purpose of instruction to others and provide guidelines for assessing learning outcomes.

In educational set up, teachers, students and parents are required to take a number of decisions that may influence educational experiences. For example, teachers may decide how to tailor the curriculum for meeting the needs of students as well as he may decide how to evaluate student's progress. Students are required to take decision about selecting the course and vocational planning. Parents also take some decision, which sharpen the educational path of their children.

In the present chapter an attempt has been made to highlight all the important facts and necessaries of instructional goals and objectives as well as ways of making good educational decision-making so that the students may be on right path to their educations and teachers may also be able to guide them satisfactorily and import value-based education to them.

Nature and Purpose of Classroom instructional goals and objectives

In classroom teaching-learning, instructional goals and objectives play a vital role both in instruction process as well as in assessment process. Objectives are generally written at varying level of specificity. At the one end of continuum, there are very broad or global statements of long-term outcomes. These are commonly called as *general objectives* or *goal*. At the other end of the continuum, there are precise statements of more short-term outcomes, each representing only one behaviour or result. These are called specific objectives. According to dictionary meaning, there is virtually no difference between a goal and an objective. A goal is defined as the end towards which effort is directed whereas an objective is defined as something toward which effort is directed.

Specific objectives are concerned with the most discrete intended outcomes. They are the measurable outcomes upon which day-to-day activities like instructions are based. Therefore, specific objectives are related to instructional objectives. Instructional objectives are those objectives that deal with changes in student performance or behaviour, which indicate that learning has occurred as a result of instruction. They are also called as *classroom objectives*, *performance objectives*, *behavioural objectives* and *learning objectives* (Gay, 1980). Some instructional objectives are considered to be required as minimum essentials, that is, objectives which must be acquired by all students irrespective of their background or ability. There are called *mastery objectives*. Various mastery objectives may be identified at different grade levels. In mastery system, amount learnt, as shown by objectives achieved, remains constant but the time required to achieve objectives varies among students. This happens due to differential aptitudes and abilities among students. Some need more time than others. But if students are given sufficient time and appropriate instruction, about 95% of the students would be able to achieve objectives and master what is taught. Bloom (1965) has suggested that all instruction should be based on mastery learning model. The instructional goals and objectives have some specific purposes. A brief discussion of these purposes is as under :

1. Instructional goals and objectives provide direction for instructional process. They do so by clarifying intended learning outcomes.
2. Instructional goals and objectives convey instructional intent to students, parents teachers as well as to the public.
3. Instructional goals and objectives tend to provide a basis for assessing the learning of students by specifying performance to be measured.

Instructional objectives as learning outcomes

When instructional objectives are viewed in terms of learning outcomes, we are concerned with the *product* of learning rather than the *process* of learning. However, it does not mean that the process of learning is unimportant. In facts, the process of learning such as outlining, drafting, revision, etc. are quite important. But the instructional objectives especially long term instructional objectives tend to emphasize the product of the learning. In fact, there are *three* basic points regarding the role of instructional objectives that must be kept in view :

- Objectives tend to establish direction. When objectives are mentioned in terms of some learning outcomes, they go beyond knowledge of specific course content. The course content is listed under process and it is through this content that objectives are attained.
- Products show varying degree of dependence on the course content. For example,

ability to write a good research report (product) is dependent upon the knowledge of steps to write reports as well as how to write the reports.

- Objectives vary in complexity. There are some learning outcomes, which are very simple whereas there are learning outcomes, that are difficult and complex. For example, knowledge about components of attitudes is simple but the formation of attitude is a complex learning outcome. Simple learning outcomes can be measured through short-answer or fixed response paper-and-pencil test. It is difficult to come to a conclusion regarding 'formation of attitude' in single course and it requires analysis of the behaviour of the students.

There are several types of learning outcome, which are the expected outcomes of the objectives. Any such classification is although arbitrary, it serves several useful purposes. For example, such classification provides hints to the types of learning outcomes that should be attended to and it directs attention towards various types of changes in students' performance in several areas. The following are the major types of learning outcomes that specify the important areas in which instructional objectives might be easily classified.

1. Knowledge
   - Specific fact
   - Concepts and Principles
   - Methods and Procedures

2. Comprehension
   - Concepts and principles
   - Methods and Procedure
   - Numerical data
   - Problem situation

3. Application
   - Concept and Principles
   - Methods and Procedures
   - Factual information

4. Thinking skills
   - Scientific thinking
   - Evaluative thinking

5. General skills
   - Communication skills
   - Social skills
   - Performance skills
   - Laboratory skills

6. Attitude
   - Social attitudes
   - Scientific attitudes

7. Interests
   - Vocational interest
   - Personal interest
   - Educational interest

8. Appreciations
   - Scientific achievements
   - Social achievements
   - Achievement in music, art and literature

9. Adjustments
   - Social adjustment
   - Emotional adjustment
   - Home adjustment

The specific areas under each category are only suggestive.

Criteria for Selecting Appropriate Instructional objectives

In selecting instructional objectives, the role of teacher, curriculum committees, school Boards do play a significant role. Now the question is which objective should be given priority at various grade levels and in various subject-matter areas. How do we identify those instructional objectives, which will prove to be most useful for teaching and assessment purposes? Fortunately; there are some criteria for a judging the suitability and appropriateness of the instructional objectives. Some of these criteria are as under :

- The instructional objectives should be in harmony with the content, values and general goals of the school and Education Board:

The instructional objectives framed by the individual teachers should be in harmony with the general goals, values and content standards outlined by school and Education Board of the Government. For example, if school and Education Board give emphasis upon moral education and sex education among others, the teacher should include instructional objectives relating to these two categories in their list. Instructional objectives inconsistence with these valued outcomes should be excluded from the list.

- The instructional objectives should be in tune with sound principles of learning :

Instruction objectives framed by the individual teachers should respect the various psychological principles of learning because these objectives basically indicate the desired outcomes of a series of learning experiences. Such objectives should be in tune with the principle of readiness, principle of motivation, principle of transfer, principle of retention, etc. In other words, the instructional objectives should fit with the requirements age and experiential backgrounds of the students (principle of readiness); they should include such learning outcomes that may easily be applied to different situations (principle of transfer); they should reflect such learning outcomes that are of permanent nature (principle of retention) and they should address to the needs and interests of the students (principle of motivation).

- Instructional objectives should include all important learning outcomes of the course

Any instructional objective in order to be called as a *good objective* must include all important learning outcomes of the course. In other words, instructional objectives derived from knowledge, comprehension, application, thinking skills, attitudes appreciations etc. must be properly maintained in balance. The common practice is to maintain knowledge objectives but objectives derived from the areas of comprehension, application, thinking skills, attitudes, methods of instructed are generally ignored.

- Instructional objectives should honour the abilities of the students as well as the various facilities and the time available :

Individual teachers should set instructional objectives in such a way that they should honour the beginning skills and abilities of students, time available for achieving those objectives as well as the resources available for attaining those objectives. Moreover, they should also keep in view the developmental levels of the students. Instructional objectives honouring these things can be considered as good objectives.

Some Important ways of stating Instructional objectives

Preparing a list of instructional objectives is not an easy task. Teachers or educators who are preparing instructional objectives have to keep several things in view. Of these, there are two things which are important :
- List of instructional objectives being prepared for a course study should be as complete as possible.
- Instructional objectives should be stated in such way that they already reflect the learning outcomes that is expected from the objectives.

That are two simple ways of stating instructional objectives (Gronlund, 2000) :

(a) State the general objectives of instruction as intended learning outcomes
(b) Under each objective, a sample of specific performance, which the students are expected to achieve, should be mentioned.

As a consequence of these two steps, statement of general objectives and specific learning outcomes will take the following shape.

1. Understand concepts
1.1 Defines concept well
1.2 Makes distinction among different subconcepts
1.3 Identifies examples to illustrate the concept
1.4 States some hypotheses relating to the concept.

In this example the expected learning outcome is concerned with understanding and general objective starts with the verb understands. There are some facts, which must be kept in view about the specific learning outcomes stated within the general principles. *First*, each learning outcome begins with a verb (such as defines, makes, identities, states) that can be easily measured by any outside observer. In other words, these verbs provide specific learning outcomes in form of observable performance of the student. *Second*, the list of specific learning outcomes should not be considered as fixed and complete rather they are only a representative sample of many specific ways through which understanding of concept is displayed. Finally, specific learning outcomes remain independent of specific course content because such outcomes specify the types of performance by students acceptable as evidence of understanding.

One important problem in stating general instructional objectives is that these objectives must have the proper level of generality. In other words, these objectives must be so general that they provide direction for instruction but they should not be so specific that the instruction is reduced to a training programme. If the objectives have proper level of generality, the teacher has greater freedom of choice regarding a particular method of instruction out of many available. For example, the general instructional objective of *understands concept* may be achieved through lecture, demonstration, discussion, different textbooks or combination of these methods. The following ways of stating general instructional objective enjoy proper level of generality.

- *Interprets* graphic materials
- *Writes* a well-organised theme
- *Demonstrates* conversational skills
- *Applies* principles to novel situations
- *Sings* well on the tune of music
- *Demonstrates* scientific aptitude

In the above example, the verb of the each statement are general enough to incorporate a range of specific learning outcomes. Sometimes the specific learning outcomes for my general objective becomes easier to write and conveys instructional intent in a better way provided each statement begins with *action verb* that indicates some observable responses. Example of such action verbs are *sings, communicates, identifies*, etc. This type of statement clearly displays that each specific learning outcome is directly relevant to the general instructional objective it is defining.

Sometimes the teacher who is to prepare general instructional objective is faced with the problem of deciding how many specific learning outcomes be included under a general instructional objective. The answer is that any fixed number of specific learning outcomes cannot be specified. Since it is not possible to include all possible learning outcomes under a general instructional objective, the included learning outcomes must be representative one. In general, four or five learning outcomes under a general instructional objective are more common. As a general rule, enough learning outcomes should be listed for each objective to demonstrate the typical performance of the student. The following general objectives and specific learning outcomes illustrate a satisfactory level of specificity for the intended learning outcomes:

1. Demonstrates skills in critical thinking
1.1 Differentiate between realistic think and artistic thinking
1.2 Identifies fallacious reasoning in written material
1.3 Identifies the limitations of given data
1.4 Explain the meaning of a concept satisfactorily
1.5 Distinguishes between two similar things clearly

One should keep in mind that in writing the specific learning outcomes, the teacher tries to describe what types of student performance is going to represent each general objective and not that what specific content the students are to learn.

While summarizing the whole things, it can be said that one most acceptable principle is that the final list of instructional objectives for any course must include all important learning outcomes and must mention clearly how students are expected to perform at the end of the learning course. According to Gronlund, (2000), in stating any instructional objectives, there are some steps, which must be followed. These steps are as under :

A. Stating the general instructional objectives :

This general step covers the following substeps as under :

- To state each general objective as an intended learning outcome
- To begin each general objective as verb (such as interprets, knows etc.)
- To include under each general objectives only one general learning outcome (avoid 'knows and understand' because it contains not one but two general learning outcomes)
- To include a readily definable domain of responses under each general objective
- To keep each general objective free from course content
- To minimize the overlapping of one objective with other objective.

B. Stating specific learning outcomes

This second step includes the following seven substeps as under :
- To include within each general instructional objective a representative sample of specific learning outcomes, which the students are expected to demonstrate.
- To begin each specific learning outcome with an action verb such as *describe*, *explain*, *point out* etc.
- To ensure that each specific learning outcome is relevant and important to the general objectives it describes.
- To mention enough specific learning outcomes for properly and adequately describing the performance of the students who have achieved the objectives.
- To keep each specific learning objective free of the course content so that the list, according to the requirement, can be used with various courses.
- To make reference materials available for those learning outcomes, which are complex and therefore, difficult to define.
- To add another level of specificity to the list of outcomes, if the situation so warrants.

These are important steps involved in stating general objectives.

Classroom Instructional Decisions

One of the very important tasks for a teacher is to ensure that students must achieve instructional objectives in the classroom. A classroom teacher generally employs a wide range of instructional objectives for a particular class. These instructional objectives may be *cognitive objectives* or *affective objective*. The cognitive objectives may include developing a knowledge in a particular area or the development of some cognitive skills such as reading, writing, listening, etc. Affective objectives may involve the development of attitudes, interests, values, etc. A teacher must monitor the progress of students for making good decisions about where to begin teaching, when to go to the next unit of instruction, whether a student needs particular help for mastering the learning task, whether a particular unit needs to reteaching, etc. In fact, quality of these decisions affect the effectiveness of the classroom instructional program. Any effective teaching requires that the classroom progress of the students must be monitored through the process of both informal assessment and formal assessments. Observing students while they are doing the tasks, asking questions, asking students to verbalize the working while solving a problem are some of the examples of informal assessments. Such assessments are helpful in assessing the understanding of the students presented and consequently, help teachers in taking better decisions in planning instructions. If teacher finds that the students are not learning according to the expectation, he takes a decision about making some alternation in the present instructional methods. In fact, such assessment techniques guide the course of instruction and therefore, they constitute the example of *formative assessment*. Formal assessment includes assessment by means of various types of psychological tests. Such assessment usually takes place after the course of instruction is over and therefore, they are called *summative assessment*, whose basic purpose is to know about what students have learned after finishing the course. The tests used in summative assessment are both standardized tests and teacher-made tests. Results of these tests usually form the basis of taking decision about students' grade and changes in placement. Of these two methods, that is, formal assessment and informal assessment, the teacher prefer formal assessment to informal assessment in making assessment of the students' progress.

Types of Assessment Instruments

For assessing the effectiveness of instructional objectives, teachers use varieties of assessment instruments. Some of the popular types of assessment instruments used by the teachers are :
- Teacher-made assessment instruments
- Standardised achievement tests
- Assessment material packaged with curricular materials

A discussion of these assessment tools is as under :
- Teacher-made assessment instruments :

For taking day-to-day instructional decisions, classroom teachers usually develop their own tests or assessment procedures. Researchers have pointed out that popular such teacher-made assessment instruments are : *oral tests, paper-and-pencil tests, product evaluations, performance tests* and *some affective measures*. A brief description of these five tests is as under :

1. Oral tests :

This is a popular measure of assessing knowledge objectives in particular course. Besides, there are some other objectives such as objectives in language classes, that can be appraised with the help of oral tests where teachers ask some questions, which are answered by students orally. Oral tests have advantage of providing an opportunity to appraise the integration of ideas by removing the effect of skill in written expression. However, they have some disadvantages. For example, oral tests are more time consuming for the teacher and evaluation of performance also tends to be more subjective. Moreover, oral tests are not much helpful in identifying the underlying difficulty of students in reading.

2. Paper-and-Pencil tests :

Paper-and-pencil tests constructed by the teachers consists of items that ask question about the knowledge of students for a particular area. Not only that, such tests also assess the students' capacity to use the knowledge to solve problems or some general educational skills such as reading, writing, etc. But such teacher made paper-and-pencil tests have proved to be a poor measuring instruments due to some reasons. *First*, such tests tend to have poor psychometric properties like reliability and validity. *Second*, items of such tests often don't match the goals of the class. As a consequence, teacher made paper-and-pencil tests seldom tend to provide relevant information regarding the effectiveness of instructional objectives. Thus instructional decisions based on the information obtained from paper-and-pencil tests become questionable.

3. Product evaluations :

Product evaluations are needed in those instructional objectives, which require student to produce a product that meets certain standards of acceptability. Examples of product are writing a poem, drama or business letter, producing a wooden horse toy in wood shop class, etc. Since paper-and-pencil tests are not adequate for such product evaluations, teachers prefer to evaluate the products themselves. This is called product evaluation. Despite the fact that the product evaluation seems straightforward, the major limitation of product evaluation is that it is very difficult to properly assess the aspects of product relevant in their quality as well as in terms of establishing some standards showing the proper degree of adequacy or excellence in the product.

4. Performance Tests :

Performance tests are those tests, which required students to perform something or to carry out a procedure such that don't leave a very concrete and tangible product that can be assessed. Such instructional objectives may include playing a musical instrument, preparing a oral report, etc. These instructional objectives can be assessed by assigning some appropriate tasks to the students and making a strict observation and rating of the performance as it occurs. Such evaluation is known as *performance assessment*. The major problems with such assessment lie in recognizing the most critical or salient aspect of the performance that should be observed as well as in correctly applying the appropriate criteria for discriminating various degrees of competency in the performance.

5. Affective measures :

There are some instructional programmes particularly at primary and secondary school levels, which have affective objectives, apart from cognitive objectives. As these affective objectives are concerned with personal and social attributes that the students are expected to develop, their assessment is difficult. This difficulty lies due to the fact that such attributes lie within the students personality and therefore, cannot be observed directly. Examples of such attributes are self-concept, values, interest, attitude, sense of cooperation with others, etc. Educators, however, are of view that an inference about such inner feelings or motivations or attributes can better be made from observing the behaviour of the students. For example, if we find that a student is always interested in discovery of truth and does behaviour accordingly, it can be inferred that he is displaying a particular kind of value called theoretical value as named by Allport, Vernon and Lindzey (1960).

Summary and Review

- In classroom teaching-learning instructional goals and objectives are very important. Objectives may be general or specific. These objectives serve various purposes.
- When instructional objectives are viewed as learning outcomes, the teacher is concerned with the product of learning rather than the process of learning. Learning outcomes are defined as the expected outcomes of objectives. There are several types of learning outcomes.
- Some criteria have been laid out for selecting appropriate instructional objectives. Four such criteria have been mainly emphasized.
- Educators and psychologists have outlined some important ways of stating instructional objectives.
- There are several types of assessment instruments, which if applied properly, enable teacher take effective instructional decisions.

Review Questions

1. Discuss the nature and purposes of classroom instructional goals and objectives.
2. Describe instructional objectives as learning outcomes.
3. Explain the major criteria for selecting appropriate instructional objectives.
4. Discuss some important ways of stating instructional objectives.
5. Discuss the different types of assessment instruments commonly employed by a teacher in making classroom instructional objectives.

■■■

Chapter – 9

Understanding the various domains of Curriculum

- Meaning of curriculum
- Nature of curriculum
- Types of curriculum
- Characteristics of good curriculum
- Objectives of curriculum
- Major issues involved in curriculum development
- Basic Principles of curriculum development
- Curriculum and syllabus
- Foundations of curriculum
- Functions of curriculum
- Models of curriculum development
- Summary and Review
- Review Questions

Key Terms:

Curriculum, Core curriculum, Overt curriculum, Null curriculum, Learner-centered curriculum, Activity-centered curriculum, Naturalism, Idealism, Existentialism, Deductive Model, Inductive Model, Tyler rationale.

Curriculum is one of the most important part of any educational innovative programme. In fact, it is the heart of any learning institution such as school or university. The design and goals of ny curriculum clearly reflects the inherent educational philosophy of the educators who framed it. Curriculum should be grounded on the latest concepts and approaches to the learning. In India as well as in foreign countries, the development of curriculum is ordinarily done by the renowned educators and by highly trusted organizations. This is done keeping in view the basic fact that the curriculum provides the basic path to the acquisition of knowledge skills and abilities. In India this task is very profoundly done by NCERT (National Council of Educational Research and Training) SCERT (State Council of Educational Research and Training) and UGC (University Grants Commission) and by the other reputed educational institutions. In the present chapter, various domains of curriculum will be discussed in detail.

Meaning of Curriculum

The term curriculum has been derived from a Latin word 'currere', which means a race or the course of a race. Therefore, etymologically, the curriculum is a run-way course on which a person runs to reach a goal. In other words, it is a track around which the competitors do a race for the entertainment of others. The first known use of the word curriculum appears in the *Professio Regia* in 1576 by a Paris Professor Petrus Ramus. In 1633 the University of Glasgow referred its course of study as a curriculum and this was the first use of the word curriculum in English. By 19th century several universities of world routinely referred their complete course of study as curriculum.

Although there is no well agreed upon definition of curriculum, curriculum is a wider agreement among the educational professionals as well as the government officials regarding what the learners should take on during some specific periods of time. Also the curriculum defines *what, why, when, where, how* and with *whom* to learn? In other words, curriculum is the total or aggregate of courses of study provided in a learning environment. A very popular authority in the field of curriculum named Oliva (1997) has explained the meaning of the word curriculum having a multi-dimensional approach in view. According to Oliva (1997), curriculum is

- that which is taught in school
- a set of subjects
- a programme of studies
- a sequence of course
- a course of study
- a content
- a series of experiences undergone by learners in school
- a set of materials
- everything that occurs within the school, including extra-class activities, guidance and interpersonal relationship.

A dictionary of Education edited by Wallace (2015) defines curriculum as, "the content and specification of a course or programme of study... or in a wider sense, the totality of the specified learning opportunities available in one educational institution ... or in its very widest sense the proramme of learning applying to all pupils in the nation." In simple words, curriculum includes all the learners' experience, within or outside school, included in the programmes that have been developed to help them emotionally, developmentally, socially, morally as well as spiritually. In 1964-1966 the popular Kothari Commission also viewed that school curriculum is the totality of learning experiences that a school ordinarily provides for the pupils through all the manifold activities in the school or outside the school and are done under supervision of some one.

In a nutshell, it can be said that curriculum is such blue print or plan of any educational institution that includes the totality of experiences earned by learners through the prescribed different types of activities done in the institution or outside the institution. It is one of the means to achieve the end or goal of education. It is a tool in the hands of the teacher to mould his pupils in accordance with the ideals of his educational institution. Cunningham has rightly said that curriculum is a tool in the hands of the artist (the teacher) to mould his material (the pupil) according to his ideal (objective) in his studio (school).

Types of Curriculum

There are many types of curriculum. Some of the popular types of curriculum are discussed below :

- **Common Curriculum :**

A common curriculum is one that is meant for all learners in a school or to all learners for all schools in a region or even in a country. An important feature of such curriculum is that all the learners develop the same learning experiences because they cover the same topics. In India the curriculum released by Central Board of Secondary Education for different classes is an example of common curriculum because this is followed by all the maintained schools.

- **Core Curriculum :**

Core refers to the heart of the experiences every learner must go through. A curriculum which is compulsory and common to all is called core curriculum. This types type of curriculum is generally is used when other parts of the curriculum are optional so that it can be identified as core curriculum content. Traditionally, core curriculum incorporates all required content areas in the school programme. However, more recently, this term is often used to refer to a type of course such as general education, united studies, common learning, social living and some integral programmes. As a consequence, such curriculum provides experiences needed by all the learners and the experiences cut across subject lines. In other words, the term core curriculum is, in some way, applied to all or part of the total curriculum, which is required of all learners at a given level. Thus core curriculum is centered around certain essential learning outcomes common for all learners. One obvious advantage of core curriculum is that it utilizes the problems of personal and social development, which is common to all learners.

- **Overt Curriculum :**

Also known as explicit or written curriculum, it is that curriculum, which is written as part of some formal instruction or education. It may include a text, films and other supporting teaching materials, which are overtly selected to support a programme of instruction and reviewed by teachers and curriculum experts. Thus overt curriculum is confined to some written directions.

- **Hidden Curriculum :**

Also known as covert curriculum, it refers to the fact that the learners during their exposure to the formal curriculum, acquired or learn knowledge, skills or attitudes, which were not the part of the formal curricular. In other words, hidden curriculum consists of contents or things, which students learn not because they had been included in planning of those responsible for the development of curriculum but because of the way in which the school has planned and worked (Kelly, 2009). Thus incidental learning is involved in hidden curriculum. It may include both positive or negative messages. For example, in a disciplined message the students must raise their hands before asking questions in the class, they must enter the classroom in time and maintain silence, they must obey the teachers, they must engage in endless competition for grades and the like reflect the hidden curriculum of any educational institution.

- **Null Curriculum :**

Null curriculum (or excluded curriculum) includes something which is not taught providing a clear message that these things are not relevant for learning experiences or for society. Such curriculum is based upon the assumption that what the schools don't teach may be as important as they do teach. Eisner (1994) has coined the term null curriculum. Some people are entitled to take decision as what is to be included and what is to be excluded from the overt curriculum. As we know, it is physically impossible to teach everything in schools. Many topics and areas of the subjects are dropped as parts of overt curriculum. For example, we teach about certain culture and histories but not about others. Likewise, when the concept of evolution is dropped from Biology curriculum, it constitutes the example of null curriculum. Both our inclusions and omission send some messages to the students. In fact, what the curriculum neglects is as important as what it teaches.

- **Learner-centered Curriculum :**

Also known as child-centered curriculum, it aims at providing all-round development of the learners. Such development includes intellectual, emotional, physical, social, as well as spiritual development. In this curriculum, the learner is taken to be the centre of all possible kinds of curricular and co-curricular activities. It facilitates the mind of the learners because it fulfills their psychological and mental requirement. Thus in such curriculum the child occupies a very important and pivotal place. That is why, the education based upon such curriculum is known as *Paedocentric* or *child-centered education*. The National policy of Education that was framed in 1986 and subsequently, modified in 1992, has emphasized upon the development of child-centered curriculum.

In the learner-centered curriculum, the children are to grow on their own but the teachers are required to guide them. Here the learners are provided with all kinds of learning experiences keeping in view their interest, abilities, and skills. The children's experiences become the basis of teaching and tools of various curricular and co-curricular activities. One major advantage of learner-centered curriculum is that it provides for the active participation of the learners in the learning process.

- **Activity Curriculum :**

This is also very popular curriculum in the Indian context. In this type of curriculum the study theme or the subject-matter is translated in terms of activities and the learners gain knowledge as a result of the successful completion of those activities. Thus an activity-centered curriculum is basically an attempt to treat learning as an active process. Such curriculum discards centered on the activities of children according to their interest. In schools or colleges, the activities are organised on the basis of interest, attitudes and needs of the learners. In this curriculum the spirit of the project work, field work and experiment prevail the school. One of the major advantages of activity curriculum is that since it promotes team work and cooperative activities among learners, it provides physical fitness, alertness, initiativeness and emotional satisfaction among them. Most of the educationists like Montessori, Froebel, Pestalozzi, Comenius, Rousseau and so on have supported activity curriculum in their viewpoints. Thus activity-centered curriculum is based upon the project method or problem solving method of teaching and most frequently needed in subjects like physical education, geography history, etc. The Wardha Scheme of Education has much emphasized upon such curriculum.

- **Negotiated Curriculum :**

A curriculum which has been agreed through consultation between teacher and the learners themselves is called as negotiated curriculum. In such curriculum, the needs and interest of the learners are given top priority. What is to be covered and achieved is completely chosen by the learners. In simple words, the negotiated curriculum is related to the learner-centered education where curriculum revolves round the needs and interests of the learners.

- **Spiral Curriculum :**

This curriculum is based upon ipsative approach to learning. When the structure of the curriculum is such that its topics are revisited each time in a more detail or depth as the learner acquires more knowledge and skills it constitutes what is called as spiral curriculum. For example, a topic entitled Learning and Motivation' appears at the intermediate stage, then it is also repeated at Bachelor stage and subsequently, also taught at Master stage. At each subsequent stage, the topic is covered in more detailed with enhancement of knowledge among the learners. Thus curriculum for this topic becomes an example of spiral curriculum.

- **Concomitant Curriculum :**

This type of curriculum includes those experiences, which are emphasized by family at home. This type of curriculum includes topics related to religious beliefs, values ethics and morals. Such curriculum may be received at some renowned temples, church, Masjids, Gurudwaras, etc.

- **Electronic Curriculum :**

Electronic curriculum incorporates those activities and lessons that are learned through the searching Internet or through using e-forms of communication. Such curriculum may be formal or informal and the internet lessons and topics may be good or bad, correct or incorrect depending upon one's view. Learners who use Internet, personal e-mails and sites like Facebook, Twitter and Youtube are frequently surrounded by various types of messages and information. Some of these information and messages may be incorrect, outdated, biased or even perverse, too.

- **Subject-centered Curriculum :**

    Subject-centered curriculum is perhaps the oldest and a very popular curriculum for teachers, students and guardians. In this curriculum, focus is given on the content of the subjects to be taught. Such curriculum mostly corresponds to the textbook, written for the specific subject. Here usually the schools divide the hours of teaching to the different subjects such as mathematics, science, literature, geography, history, etc. Henry Morrison and William Harris are the popular supporters of the subject-centered curriculum. The subject-centered curriculum has one obvious advantage that it is most easily understood by the teachers and learners and therefore, it is also easily delivered. Apart from this, several books and supporting instructional materials are commercially available. However, the demerit is that the curriculum emphasizes upon the content so much that it completely neglects the learners' interests and natural tendencies.

- **Problem-centered Curriculum :**

    Problem-centered curriculum relies on social problems, needs, abilities and interest of the learners. Such problems generally relate to contemporary life problems, life situations etc. Here the content of the curriculum cut across the boundaries of the subjects. These problems are selected by either teachers or students on the basis of some defined criteria. The obtained information is organized, analyzed and interpreted. The motivating force behind the development of problem-centered curriculum is the Herbart Spence's curriculum writing and his frequent emphasis upon the activities that sustain and enhance human lives by maintaining the person's social and political relations.

- **Collateral Curriculum :**

    This type of curriculum is intentionally designed by the curriculum developers to afford the learners an opportunity to learn empowering concepts, principles and ideas which lie outside the subject to be taught. Teacher emphasizes upon learning outcomes of the collateral curriculum. However, its knowledge is neither specified in the instructional objectives nor is it evaluated. Thus collateral curriculum is a sort of planned hidden curriculum.

Characteristics of a good Curriculum

The major characteristics of a good curriculum are as under :
- A good curriculum must have scope for continuous monitoring and evaluation.
- An effective curriculum must be responsive to the needs of the society and community. In fact, it establishes a good cooperation with the programmes of the societies.
- A good curriculum must be based upon the needs of the learners. In fact, any good curriculum reflects the needs of the society as a whole.
- A good curriculum must be flexible in nature. It should be open to revision and should meet the demands of globalization.
- The nature of curriculum to be effective must be such that the learner is able to take the

best learning experiences throughout.
- A curriculum to be called good and effective must provide instructional equipment including guidance and counselling programme, library and laboratories, student-teacher relationship and some other school-related work experiences.
- A good curriculum is one which can be safely regarded as the product of many minds. In fact, a good curriculum is democratically developed by the efforts of a group of individuals from the different sectors who know better about the needs and interests of the learners.
- A curriculum to be called good must provide a logical continuity of the experiences of the subject-matter.
- A good curriculum must stand of three pillars–*relevance equity and excellence*.

Objectives of Curriculum

There are some explicit objectives of any good curriculum. Some of such objectives are as under :
- A curriculum aims at promoting good human relationship and lays the foundation of some satisfying and useful relations with others.
- A curriculum also intends to develop overall good positive traits like honesty, integrity, judgement, goodwill, cooperation among the learners.
- A curriculum also tends to ignite and inspire full personality development of each learner.
- A curriculum also promotes critical and creative thinking on the part of learners.
- A curriculum also intends to promote democratic sense of citizenship among the learners.
- A curriculum also aims at informing learners about the rubrics of the humanities, the arts, religion and social sciences.
- A curriculum also aims at promoting such scholars whose interest in research, discovery and curiosity may benefit the whole society.

Major Issues involved in Curriculum development

While developing or constructing a curriculum, the authorities responsible for it, usually face some problems. There are some basic questions, which must be taken into account at the time of formulating the curriculum. Some of these questions are as under :
- *What is to be included in curriculum?* This is one of the baic question. What should be included or what should not be included in the curriculum often haunts the mind of curriculum developers. They have to be very careful about it and they must keep in view the primary needs, interest and abilities of the learners for whom the curriculum is meant.
- *Why is curriculum formulated?* This question is also important while developing a curriculum. In fact, curriculum is formulated to meet the basic needs of education. Therefore, both the curriculum developer as well as the learners must have a clear frame of mind regarding the aims of education.

- *Who is to be confronted with the curriculum?* This question is also very much important because after all the curriculum is meant for the learners and it must not be above the abilities and acquired skills of the learners. The curriculum must also cater the needs of the individual learners.
- *How is curriculum to be used?* This question reminds the curriculum developers to the various techniques and methods of teaching-learning process that are commonly used in implementing the curriculum. Which method of teaching would be more beneficial for the given curriculum, has to be kept in view by the curriculum developers. For example, if the curriculum emphasizes upon the experimental work, demonstration method of teaching the curriculum would be more effective. On the other hand, if the curriculum is heavily loaded with various theories or models, lecture method of teaching would be more effective for the learners.
- *When is the curriculum to be used?* This issue is related to the different stages of child development. At the beginning stage, a different curriculum is framed and for the higher stages, a curriculum matching their stage has to be developed. No one curriculum can be enforced for both the stages and therefore, the curriculum developers have to keep the issue of stage matching in their views.
- *Who is to use the curriculum?* Again this issue is vital. There are two important users of the curriculum–teachers who teach and the learners who learn. The curriculum developers have to keep the abilities and skills of both teachers and learners in their mind.
- *How is the outcomes of curriculum be evaluated?* The developers of the curriculum has also to keep in mind how would the outcomes of the proposed curriculum would be evaluated. Whether the outcomes would be evaluated by some objective tests or essay tests or by some other methods. This is related to the success of both the learners as well as the teachers.
- *Who is to formulate the curriculum?* Most of the times curriculum is formulated by the teachers alone. However, it is also developed through the efforts of a group of experts taken from different sectors in society. The formulation of the curriculum is much affected by the expertise of the persons responsible for it.
- *How are the required changes to be made in curriculum?* This is also an important issue to be kept in mind. In any curriculum the required changes must be made from time to time so that it may conform to the needs of the learners. In other of words, the nature of the curriculum should be flexible one and not something static one.

Basic Principles of Curriculum construction

A curriculum is not only a set of contents or subjects but is an entire range of activities and experiences, which the learners are expected to learn so that they may act as the most responsible person of the society and also be able to get full satisfaction in all spheres of his life. Therefore, it is essential that the curriculum should be developed on the basis of some sound principles. A few such important principles are as under :

- **Principle of child-centeredness :**

    Modern approach to curriculum must be paedocentric as well as democratic in the sense that child occupies the most important role in the whole teaching-learning process. The abilities, interests, attitudes and needs of the child should be the key factors at the time of construction of the curriculum.

- **Principle of community centeredness :**

    This principle stresses that both the individual development and social development of the child should get due attention at the time of construction of curriculum because he lives in and for the society. The needs and desires of the child must coform with the needs and desires of the society. The attitudes, values and skills prevalent in the community must be reflected in the curriculum so that the children may easily get knowledge and understanding of their cultural values and norms.

- **Principle of activity centeredness :**

    The curriculum should reflect multifarious activities of the learners. The purposeful activities both within the class and outside the class should be clearly mentioned in the curriculum. According to the demands of this principle, the curriculum must include activities according to the interests, needs and developmental stages of the children. Thus a curriculum while honouring this principle, should provide constructive, creative and project activities.

- **Principle of coordination and integration :**

    As we know, in curriculum the learners are provided with selected experiences through various activities and subjects. This principle strongly demands that all these activities and subjects must be properly integrated and coordinated or correlated so that the learners may be able to develop a wholesome global view of the subjects. Therefore, these activities and subjects should not be placed into tight compartment rather be properly correlated and integrated. This principle is also called as *principle of correlation*.

- **Principle of variety and flexibility :**

    The curriculum should be broad-based so that different types of activities could be included in it and the learners be able to take up subjects and participate in the activities according to their needs, interests, abilities and attitudes. The needs of the learners change from one place to other. For example, the needs of boys and girls as well as the needs of rural areas and urban areas are different. The curriculum should provide opportunities to meet these needs in a convenient way. Therefore, there should be enough flexibility and elasticity in the curriculum to suit these varieties. The report of Secondary Education Commission (1952-53) has also emphasized that the secondary school curriculum should provide variety and flexibility. In a nutshell, the curriculum should be dynamic and should change with changing time.

- **Principle of conservation :**

    This principle holds that the curriculum should be constructed in a such a way as to cultivate a sense of respect for one's culture, which consists of traditions, attitudes, values, skills, conducts that are handed down from one generation to the next in the society. So curriculum must be developed in a way to preserve and transmit our cultural heritage. This is also one of the main function of education.

- **Principle of forward looking :**

    One aim of education is to equip the learners to lead a successful life not only in present time but also in future. Therefore, the curriculum should cater not only the present needs but also the needs of the future life. The curriculum must contain activities, which may enable the child to acquire such abilities and skills that may help him in making effective adjustment in his future life.

- **Principle of creativity :**

    A curriculum must incorporate those provisions, which can develop the creative powers of the learners so that they may behave like a contributory and responsible member of the society. In other words, the curriculum must contain those provisions, which may promote the creative and constructive abilities and capacities among the learners. Thus the curriculum which suits the needs of today and those of future, should also contain some creative subjects.

- **Principle of totality :**

    The curriculum should incorporate the totality of experience that the learners are given through the manifold activities done in the classroom, library, workshop, laboratory, play ground, informal contact, etc. In this way, a good curriculum must include the whole activities done in school or college by the learners. The Secondary Education Commission (1952-53) has rightly commented that curriculum never means only academic activities but it includes 'the totality of experiences' earned by the learners.

- **Principle of values :**

    The curriculum must include activities, which may promote desirable values in learners. Thus any curriculum which inculcates social value, moral value, asthetic value, spiritual value and democratic value must be promoted by the developers of the curriculum.

- **Principle of education for leisure :**

    A curriculum should be developed in such a way that the learners may gainfully and meaningfully utilize their leisure time. Often, it is found that learners waste their leisure time because they have no plan to utilize this time meaningfully. The curriculum should incorporate provisions for meaningfully and constructively utilizing the leisure period by doing some prescribed and interesting activities.

These are some important principles which must be kept in view while developing a curriculum by teachers, authorities and learners.

Curriculum and syllabus

Many people get confused between what is a curriculum and what is a syllabus. Often, they equate curriculum with a syllabus. But this not a reality. If people equate curriculum with syllabus, in fact, they are likely, "to limit their planning to the consideration of the content or the body of knowledge they wish to transmit" (Smith, 2000). In fact, curriculum is a focus of study that consists of different courses which aim at reaching a particular proficiency or qualification. For example, some high schools develop a vocational-prep curriculum that includes specific skill building courses such as electronics, construction trade, computer science, etc. Syllabus is simply an outline and time line of a particular course. It presents a brief overview of course objectives, course expectations, homework deadlines, examination dates, etc. It is often available on the very first day the start of the course. The purpose of the syllabus is to allow the learners to work their schedule for maximizing their own efficiency. In a nutshell, syllabus is a refined detail of the curriculum for a particular subject at a particular stage of learning. The important distinctions between curriculum and syllabus are as under :

1. Curriculum is based upon the philosophy, goals and values of the education whereas syllabus does not take into account these factors.
2. Curriculum is basically concerned at balancing among different subjects whereas syllabus is concerned with balancing of the coverage of content.
3. Curriculum includes both indoor and outdoor activities of the school whereas syllabus is concerned only with the indoor activities.
4. Curriculum is the sum total of various school subjects, activities and learning experience whereas syllabus reflects a list of unelaborated headings.
5. Curriculum is mainly concerned with the content of the subjects whereas the syllabus emphasizes more upon the time frame within which content of the subject may be covered.
6. Curriculum has a variety of roles to play and it is considered as a plan, as a subject-matter or content, as an experience, as a field study and as an objective. The syllabus has a very limited role to play and is of limited importance.
7. Curriculum includes both co-curricular and extracurricular activities whereas syllabus does not have such co-curricular and extra-curricular activities.
8. Curriculum is a wider and inclusive concept, which includes syllabus also whereas syllabus is simply the part of curriculum.

Nature of Curriculum

The educationists have tried to study the nature of curriculum from different angles such as a plan, as a subject-matter, as an experience, as an objective, and as a process. A brief discussion follows :

- **Curriculum as a plan :**

    Here curriculum has been considered as a plan or programme for all learning experiences encountered under the direction of school (Oliva, 1997). So here nature of curriculum reflects a general overall plan of specific materials of instruction or content offered by school for the wholesome development of learners. It is in this sense Bibao et al. (2008) have viewed curriculum as all the experiences in the classroom which are planned and enacted by teacher and also learnt by the students. Thus as a plan curriculum includes strategies for obtaining desired ends.

- **Curriculum as a subject matter :**

    Curriculum can be understood in terms of the subject-matter (Mathematics, Geography, English, Telgu, Hindi) or in terms of content, that is, the way the information is organized and assimilated. The important concept of the curriculum is that of the subjects and subject-matter taught by the teacher and learnt by the learners. Both formal and informal content are retained in a curriculum and on the bases of such contents, learners are able to acquire knowledge as well as understanding and also subsequently are able to develop skills and abilities for overall development.

- **Curriculum as an experience :**

    Curriculum is the of experiences the learners received through the manifold activities done in school, classroom, library, laboratory, play ground as well as in various informal contacts between teachers and pupils. This fact has not only emphasized by the Secondary Education Commission (1952-53) but also by some other educationists like Tanner and Touner (1975), who emphasized that curriculum allows the reconstruction of knowledge and experience systematically gained in the school or colleges.

- **Curriculum as an objective :**

    The educationists also have tried to understand the nature of curriculum in terms of activity-based objectives. From this angle, curriculum is a series of learning experiences, which the learners develop by way of attaining some activity-based objectives. Some experts clearly view curriculum as the series of objectives that the learners must attain through a series of learning experiences.

- **Curriculum as a process :**

    Curriculum as a process is seen when the teacher enters a particular schooling with his ability to think critically and understanding of his role and expectations the learners have of them (Smith, 2000). He then proposes for action, which sets out some principles and features of his educational programme. Subsequently, he encourages dialogue with learners out of which a certain course of thinking and action emerge. Teacher also evaluates the process as well as what he can see of the outcomes.

Foundations of Curriculum

The foundations of curriculum spells out the external boundaries of the knowledge of curriculum and define some valid source of information from which relevant theories, principles and ideas are derived. The foundations of curriculum are usually considered from *philosophical*, *psychological* and *sociological points* of view. A discussion follows :

1. Philosophical Foundation of curriculum :

As we know, the field of education has been influenced by the philosophy, which has a clear impact upon the purpose, methods and curriculum of the school. The philosophical base can be considered as the starting point in curriculum development. In fact, the philosophy is viewed as an all-encompassing aspect of the educational curriculum. Important philosophical views that have a bearing upon curriculum development are as under :

- **Pragmatism :**

The word pragmatism, is derived from a Greek word 'Pragma', which means 'to do', 'to accomplish'. Therefore, action or activity or experience is the key word in the pragmatism, which does not believe in set values. Present is more important than the past and the future and it adopts a flexible viewpoints. The chief propounders of pragmatism are C.S. Pierce (1839-1914) William James 1842-1910) and John Dewey (1859-1952). According to pragmatism, the child's activities are spontaneous activities, purposeful activities and socialized activities and the teacher's job is to keep the learners alive to their purposes, capacities and limitations. This philosophical view emphasizes upon learning by doing, direct experience, correlation and integration of subjects, use of audio-visual aids etc. A curriculum based upon pragmatistic view reflects practical utilitarian subjects, that is, such curriculum is based upon the principle of utility, integration and child's personal needs, interests and experience. Therefore, pragmatism lay emphasis upon broad-field curriculum, experience-centered curriculum, problem-based curriculum and diversified curriculum so to say. In construction of curriculum, the teachers should pay special attention to the principle of utility, principle of integration and principle of natural interest.

- **Naturalism :**

Naturalism or naturalistic philosophy emphasizes that nature alone is the chief source of knowledge and everything that comes to us from the nature is good but it is degenerated in the hands of men. The chief supporters of naturalistic philosophy are Francis Bacon (1562-1626), Rousseau (1712-1778), Herbert Spencer (1820-1930) and Rabindra Nath Tagore (1861-1941). Naturalism stresses upon the fact that child is not an adult and he is not to be judged from the standards of an adult. Education of children should be in accord with the nature of each stage of the child infancy, childhood, and adolescence. The children should be allowed to learn on the basis of his direct experience with the things. Teacher should act as mere observers and he is not to interfere with the activities of children. Therefore, naturalistic philosophy emphasizes upon child-centered methods of teaching and while developing curriculum, it lay emphasis upon learner-centered curriculum in which activity based teachings in natural surroundings are imparted.

- **Idealism :**

As we know, there are two important facets of human life – *spiritual* and *material*. When the emphasis is on the spiritual facet, it constitutes what is called idealism. Accordingly, for the idealism or idealistic philosophy, 'mind' and 'soul' are more important than 'matter' and 'body'. Idealism believes that man is essentially a spiritual being, God is the source of all knowledge, 'spirit' and 'mind' constitute the reality and values are eternal, absolute and unchanging. Man's spiritual nature expresses itself in form of morality, intellectual culture and religion. The chief western exponents of idealism are Plato, Comenius, Kant, Pestalozzi and Froebel and eastern exponents are Swami Dayananda, Rabindra Nath Tagore, Radhakrishnan, Swami Vivekanand and Mahatma Gandhi. The essence of idealistic philosophy is contained in *Upanishads*.

The education, according to idealistic philosophy, is by and large teacher-oriented and not learner-centered. The goal of education is to unfold divine in human personality. According to idealistic philosophy, curriculum must include those subjects and activities, which emphasize higher values of life, curriculum should also lay emphasis upon the prominence humanities and ethics. Religion and art must find a very important place in the curriculum. In fact, curriculum based upon idealistic philosophy reflects cultural heritage and civilization of the whole human race. In a nutshell, idealistic curriculum places great importance upon knowledge-based curriculum and subject-centered curriculum emphasizing intellectual, aesthetic and moral activities for attainment of ideals of life.

- **Existentialism :**

Existentialism is such a philosophical view, which emphasizes that the existence is the ultimate purpose of human life and the centre of existence is human being rather than truth. This viewpoint considers every human being as unique. The major exponents of existentialism are J. Paul Satre (1905-1985) and Karl Jaspers (1883-1969). Existentialism emphasizes the view that man is not God's creation rather he is the product of physical and biological forces. Every human being has the responsibility of creating his values. Existentialistic philosophy does not have faith in a fixed curriculum. The learners should be given freedom to choose their own curriculum. As such, this philosophy emphasizes upon completely individualized curriculum to be composed of human conditions, choices and life-situations. Existentialistic philosophers prefer subjective knowledge of humanistic subjects to the objective knowledge of scientific subjects because they are concerned more with the problem of becoming rather than with the problem of being and more with existence than with essence.

2. Psychological foundation of curriculum :

Psychological foundation comparises various kinds of accumulated knowledge about the needs, interests, motivation, physiological conditions, learning, remembering, reasoning as well as about the developmental stages of personality. Naturally, such vast knowledge tends to force the teachers to develop curriculum keeping in view the demands of these psychological factors. Accordingly, the teachers and other persons responsible for construction of curriculum try to design curriculum keeping in view the needs and interests of the learners, grade the curriculum according to the age of the development of the learners, and keep the curriculum in commensurate with the mental abilities of the learners. The curriculum developers should construct curriculum in such a way that the basic principles and laws of learning, motivation, personality development, interest, attention, transfer of training must be properly respected. Thorndike's laws of learning such as law of exercise, law of readiness and law of effect, Skinner and Pavlov theory of learning, Ebbinghaus theory of forgetting, Interference theory, Perseveration-consolidation theory of forgetting have proved to be very useful for curriculum construction.

3. Sociological foundation of curriculum :

Societal and cultural aspects are also important for curriculum construction. In fact societal and cultural norms, values, folkways, myths, etc. are very important factors that must be kept in view by the teachers while constructing curriculum. In fact, curriculum provides the most important base for healthy education, which is a process that takes place in society, for society and by society. The changing nature of culture has also an impact upon the curriculum. In a nutshell, the sociological foundation of curriculum requires that the curriculum developers should also take into consideration the prevalent needs of the society as well as the educative demands and changing nature of the culture.

In a nutshell, the task of healthy development of curriculum must incorporate the demands of philosophical, psychological and sociological foundations of education particularly those related to curriculum.

Functions of Curriculum

As we know, curriculum is all of the experiences that the learners have in programme of education whose purpose is to achieve the maximum growth of personality. Besides this, the educationists have pointed out the various functions of curriculum. Some of the important ones are as under :

- **Curriculum functions to promote the wholesome development of individuals:**

As it is known that there exists individual difference among the learners with respect to knowledge, abilities, interests, motivations, attitudes, ideals, skills, etc. As a consequence, all learners don't learn equally well at equal pace By adopting a particular design and types of curriculum, the developers of the curriculum try to construct and shape the curriculum in such a way that the maximum number of learners are benefitted more or less equally from the curriculum. So one obvious function of curriculum is to meet the various abilities, interests, needs and motivations of the learners so that they are able to develop their personality. Moreover, any good curriculum includes both curricular and co-curricular activities. The curricular activities function to make mental abilities grow while co-curricular activities directly promote around development necessary for the wholesome and balanced development of personality.

- **Curriculum functions to produce responsible and honourable citizens :**

Any well organised educational programme aims at producing responsible, honourable and useful citizens. The programme does this work by incorporating a good curriculum. When the learners acquired positive and healthy experiences as a result of exposure to the good curriculum, they gradually become the most responsible, useful and honourable citizens of the country. Not only this, such learners also promote the sense of goodness in other citizens.

- **A good curriculum also functions to preserve and transmit the cultural heritage :**

A curriculum helps the society in preserving its culture and transmitting the same to the next generation. Usually the curriculum preserves the culture in literature, art and history and with the help of some suitable methods of curriculum implementation and evaluation, those things are easily transmitted to the learners of the next generation.

- **Curriculum helps in the development of some basic skills in the learners :**

A curriculum also functions in such a way that some basic skills of the learners such as *Three R's*, understanding some language, learning and memorization skills etc., may be properly developed. Without fail, a good curriculum discharges this function effectively.

Besides these major functions of curriculum, there are also some functions, which are done by the curriculum.

- Curriculum promotes physical and mental health of the learners.
- Curriculum helps building a positive attitude towards the life.
- Curriculum inculcates important life values among the learners.
- Curriculum provides a correct knowledge about the world in formal as well as informal ways.
- Curriculum makes the learners broad-minded by providing a liberal outlook towards the life.
- Curriculum produces educationists, scientists, agriculturists and other specialized people by providing a good and appropriate opportunities for developing their latent skills and aptitudes.
- Curriculum provides the teachers to improve their skills of teaching by providing some specialized tasks before them.

Models of Curriculum Development

Curriculum is an important means to obtain the aims of education that is dynamic and go on changing with changes in social environments. As it includes some well selected activities and experiences needed for development of learners according to social requirements, it needs to be developed very carefully.

What is curriculum development? Most of the educationists are of view that curriculum development is the process of *planning, implementing* and *evaluating* curriculum which ultimately results in a curriculum plan. It is an orderly, logical and cohesive construction of knowledge and experience. One very important way of developing curriculum plan is through *modelling*. Models are some objective patterns that serve as a guidelines to action. A curriculum model determines the type of curriculum used, encompasses educational philosophy and provide an approach to teaching and methodology. Using a model for curriculum development can, in fact, result in greater efficiency and productivity (Oliva, 2009). On the basis of a careful examination of the model of curriculum development, the *phases* essential to the process can be easily analyzed.

Various models of curriculum development can be properly grouped into two general categories : (a) *deductive models* (b) *inductive models*.

Deductive models are so called because they proceed from the general to the specific, that is, from the needs of society to specifying instructional objectives. Another feature of deductive models are that they are *linear* because they inculcate certain order or sequence of logical steps from beginning to the end. Still another feature of the models is that they are *prescriptive* since they suggest that what should be done and what many curriculum developers should usually do. Here a discussion of two important deductive models would be undertaken: *Tyler's behavioural model* and Saylor, Alexander and *Lewis's administrative model*. These are also called as *classical model* which are, in fact, means-end model. These classical models assume a desire end (goal and objectives) as well as a means of achieving those ends (learning experiences) and a process (evaluation) for determining to what extent the means have been successfully in achieving the end.

Inductive models of curriculum development are another category. Inductive models are so called because they begin with the development of curriculum materials and subsequently, go to some generalization. Another feature of inductive models are that they are *nonlinear*, which permit curriculum developers to skip components in the model, enter the mode at various points, reverse the order and if preferred by the developers, two or more components of the model may be attended simultaneously. Still another feature is that inductive models are *descriptive* in nature. A descriptive model is one which has a *beginning* (platform), *a process* (*deliberation*) and an *end* (design). By platform is meant the very principles or beliefs that guide the developers of the curriculum. These principles produce *deliberation* regarding the process of taking decision from among alternatives. Such deliberation model is also called as *naturalistic model* (Walker, 1971). Again under the head inductive models, three important models would be discussed. *Taba's Instructional strategiesmodel, Weinstein and Fantini's Humanistic model* and *Eisner's Systematic-Asthetic model*.

What are the major differences between inductive/descriptive/naturalistic model on the one hand and deductive/classical/prescriptive model on the other hand?

- Inductive models are nonlinear whereas the deductive models are linear.
- Inductive models are descriptive whereas the deductive models are prescriptive in nature.
- Inductive models provide a postmodern view of curriculum because they are temporal and naturalistic. On the other hand, deductive models are classical and means-end model because these classical models assume an end (or goal and objectives) as well as a means of achieving the ends and a process (evaluation) for determining to what extent the means have been successful in achieving the end.
- Inductive models provide a unique grassroot approach to the curriculum development as compared to the deductive models.

Tyler: Behavioural Model

Tyler model is one of the most popular models of curriculum development. This model was developed by Tyler (1949) in his famous book entitled *'Basic Principles of curriculum and instruction'*. In this book he suggested four fundamental questions, which he felt were necessary to be answered for the curriculum development. These four questions were.

- What educational objectives should the school seek to attain?
- What educational experiences can be provided that are likely to attain these objectives?
- How can these educational experiences be effectively organized?
- How can these objectives be evaluated?

These four questions are often referred to as *Tyler rationales*, have been built into a model of curriculum development called as Tyler model. These questions may be formulated into a simple four-step process or principles by which a curriculum is planned and developed. In fact, the original work of Tyler (1949) is one of the technical scientific approach and his work equates with the *product model* emphasizing *plans and intensions* in contrast to the *process model* emphasizing *activities and effects* (Neary, 2003). Those four steps or principles are as under :

- Step 1 : Stating educational objectives
- Step 2 : Selection of learning experiences.
- Step 3 : Organisation of learning experiences.
- Step 4 : Evaluation

A discussion follows :

Step 1 : Stating educational objectives

Tyler has pointed out that the curriculum planners identify general objectives by gathering data from three sources: learners, contemporary life outside the school and the subject-matter specialists.

In other words, in this first phase attempt is made to formulate tentative educational objectives through analysis of information collected from the questionnaire, tests results and teacher observation, etc. regarding learners, societal analysis of the aspects of the lives outside the school and the subject matter information from specialists. Tyler posits that the curriculum objectives that don't address the needs, interests and motivation of the learners, the community or society as well as the subject-matter will not be considered as the best curriculum.

Step 2 : Selection of learning experiences :

At this stage, the tentative objectives obtained from three sources are filtered through two screens: the educational philosophy of school and the knowledge of the psychology of learning. As a result of this screening, final versions of *educational objectives* are formulated on these objectives are now known as instructional objectives. Such objectives indicate both behaviour to be developed and area of content to be applied.

Step 3 : Organisation of learning experiences

At this stage, the educational experiences are then selected and organized in such a way as to be helpful in
- developing thinking skills
- developing responsible social attitudes
- creating interest
- developing research skills.

These learning experiences are the result of the interaction between the learner and his or her environment. Therefore, such learning experiences are to, a greater extent, the function of needs, interest, perceptions and past experiences of the learners. Here teacher's role is too limited.

Step 4 : Evaluation

In this final step of curriculum development, it is evaluated that to what extent the learning experiences have been successful in obtaining the educational objectives. In other words, it is determined that to what extent the educational objectives have been realized by the curriculum. In fact, here the statements of objectives not only serve as the basis for selecting and organizing the learning experiences but also provide some standards against which the programme of curriculum is evaluated. Thus, broadly speaking, curriculum evaluation involves the process of matching initial expectations in the form of behavioural objectives with the outcomes obtained by the students or learners. The result of evaluation is very important for decision making of curriculum planners or implementers.

Major strengths of Tyler model are :
- In the model the clearly stated objectives are a good place to begin.
- The model emphasizes upon the active participation of the learners.
- In the model simple linear approach to the development of behaviour of objectives has been emphasized. Such emphasis is very much helpful in understanding the relevance of the behaviour for the objectives.

However, Tyler's model has been criticised on the following grounds.

- Tyler's model depends heavily on the behavioural objectives, which come from three sources (the learner, society and the specialists of the subject-matter). All three sources must agree on what objectives needs to be addressed and it is very difficult and cumbersome to arrive at such agreement.
- The behavioural objectives in the model are learner-centered, that is, these objectives or learning experiences are basically the function of interests, needs, perceptions, previous experiences of the learners. It automatically means that the learning experience is not totally within the control of the teacher. As such, teachers are deprived of any opportunity to manipulate learning experiences in such a way as to fit the desired learning outcome.
- Since the model is too dependent on behavioural objectives, it is very difficult to explain in purely behavioural units those objectives that cover skills like critical thinking, problem solving, etc. Likewise, goals of preparing children to be good citizen cannot be specified in terms of easily measurable to behaviour objectives.
- It is also said that the model is too restrictive because it covers a very limited range of student skills and knowledge.
- It does not pay enough attention to the process of evaluation. In other words, the model does not consider how the activities that lead to the attainment of objectives are carried.

Despite these limitations, Tyler's model of curriculum development has been very important and influential one and some other models have directly originated from it.

Saylor, Alexander and Lewis: Administrative Model

Administrative model of curriculum development was formulated by Saylor, Alexander and Lewsi (1981). They have described curriculum plans in terms of the relations ends (goals) to means, attention to some facts and relevant data as well as in terms of the flow of activities or procedures from beginning to the end. In this model, there are four steps involved in curriculum development. Those steps are being discussed below :

Step 1 : Educational goals, objectives and domain :

Saylor et al., (1981) have defined curriculum as a plan for providing sets of learning opportunities to achieve broad educational goals and some specific objectives for an identifyable population served by the school. Accordingly, curriculum developers begin by specifying the major educational goals and some specific objectives they tend to achieve. Each major goal represents some important domains of curriculum such as human relations, personal development, some basic skills and specialization. These educational goals and objectives are influenced by two sets of factors: *external factors* and *bases of curriculum.*

External factors include professional associations, research data, legal requirements as well as state guidelines. Bases of curriculum include learners, knowledge and society. On this point Saylor's model bears some similarities to Taylor's model.

2. Curriculum designing :

The next step in the model is to choose curriculum design. After collecting and analyzing some essential data and identified goals and objectives, the curriculum developers prepares a curriculum design, which anticipates the entire range of learning opportunities for a specified population. The curriculum design includes a subject design that utilizes specific studies in the specified curriculum area, analysis of various essential skills considered important for knowledge and competence in the area of the subject as well as a sequence plan prepared around a selection of some persistent topics or themes. Thus curriculum design, in a way, tends to anticipate the entire range of learning opportunities for the specified population.

3. Curriculum Implementation :

The next step is that of curriculum implementation where teachers now prepare instructional plans in which instructional objectives are specified and some appropriate teaching methods and various strategies are utilized for achieving the desired learning outcome for the learners. Here every effort is made to implement the curriculum plan, which guide and direct the nature and character of various learning opportunities. Therefore, curriculum planners must perceive the instruction and teaching jointly as the product of their honest efforts.

4. Curriculum Evaluation :

The last step in curriculum development in this model is curriculum evaluation, which is defined as the process of getting information for judging the worth of an educational programme, goals, educational objectives or utility of some alternative approaches designed to attain specified objectives. In this model both formative evaluation and summative evaluation are done. Formative evaluation is one in which the curriculum planners are able to get feedback regarding the effectiveness of the different facets of curriculum at every stage of the process of curriculum development such as goals and objectives, curriculum design and curriculum implementation. In summative evaluation the curriculum planners do evaluation at the end of the process and it deals with the evaluation of total plan as a whole. The feedback from such evaluation becomes the base for modifying or continuing or eliminating the curriculum. Thus both formative evaluation and summative evaluation provide feedback about the on-going process of different facets of curriculum development. The whole evaluation processes allow curriculum planners to determine whether or not the goals of the school as well as the objectives of instruction have been fulfilled. The data obtained on the basis of evaluation becomes the base for decision making in further planning.

Taba's Instructional Strategies Model

Instructional Strategies or Interactive model was formulated by Taba (1962). It is also known as *Grassroots Rationale model*, which is a modified form of Tyler's model. Taba presented her inductive model in the book entitled *'Curriculum Development: Theory and Practice'* published in 1962. She has pointed out that curriculum development should follow a sequential and logical process and she advocated for more information input in all phases of curriculum development. According to her, all curricula are made up of fundamental elements and they can be easily made successful if there is diagnosis of needs. She was the ardent supporter of the view that the nature of objectives determines what learning is to follow. In Taba's linear model there are *seven* steps that must be followed in developing a curriculum:

- Diagnosis of needs
- Formulation of objectives
- Selection of content
- Organisation of content
- Selection of learning experiences
- Organisation of learning experiences
- Determination of what to evaluate and ways and means of doing it

She strongly believed that the teachers should participate in developing a curriculum rather than the curriculum be handed down by the authority. As a grassroot approach, she begins from the bottom rather than from the top as Tyler had done in his model. She felt that teachers should begin the process by creating specific teaching-learning unit for students in schools. Such teaching-learning units would provide the basis for the curriculum design. Thus the curriculum would emerge from the instructional strategies. She had proposed an inductive approach to curriculum development. In inductive approach, the curriculum planners start with the specifics and build up to a general design as opposed to more traditional deductive approach, which starts with a general design and then, comes down to specifics.

If we pay a close attention to the seven steps of the model, it becomes clear that the model includes an organisation and relationship among five mutually major interactive elements. Those elements are : objectives, content, learning experience, teaching strategies and evaluation. With the help of these elements, a system of teaching and learning is well planned.

This model contains a number of innovative elements such as specificity in determining objectives and contents, learning experiences selected and organized in light of some specific criteria, teaching strategies specifying some methods and technology and an evaluative procedures and measures. Model also shows that some external factors lying outside the model tend to affect the internal components. Such external factors may be the nature of community in which the school is located, personal styles and characteristics of the teachers associated with the curriculum development, nature of student population, goals, resources and administrative strategies of the school, etc. Objectives help in providing a focus for the curriculum, in establishing criteria for the selection of content and learning experiences as well as for guiding and directing the learning outcomes. The process of determining goals starts with the development of overall goals, which originate from different sources like needs of students, demands of the society, etc. The objectives are further broken down into various behavioural statements, classified in terms of the kinds of student outcomes expected (such as development of thinking skills, understanding and use of important elements of knowledge, etc.) and justified on the basis of well-planned rationale.

So far as the content is concerned, it is contained within a number of teaching-learning units for each of the several grade levels. Each unit ordinarily consists of three kinds of knowledge: *knowledge regarding key concepts* (such as cultural change, social control, cooperation), *knowledge regarding main ideas* (that is, generalization derived from key concepts) and *knowledge regarding specific facts* (such as content samples chosen to illustrate and develop main ideas). The content contained in the units is incorporated into what is called learning experiences selected and organised in accordance with some clearly specified criteria like open-mindedness, transferability, justifiability, etc.

Some important teaching strategies that identify specific procedures used by the teachers are included within the curriculum. Within these strategies some innovative strategies, which promote learners' cognitive skills and encourage them for examining individual attitudes and values, are included.

The model also emphasizes upon preparing the various measures for evaluating the effectiveness of the curriculum. Several open-ended devices have been designed to assess the flexibility and variety of learners' conceptualizations as well as quality of learners' generalisation, etc.

Weinstein and Fantini : Humanistic model

Weinstein and Fantini (1970) proposed a model of curriculum development in which they have made an attempt to link sociological factors and psychological factors with cognition. Due to this reason, the authors have considered their model as a '*curriculum of a affect*'. This model is also called as *humanistic model* of curriculum development. In this model, they have shown concern with group as opposed to individuals because most learners are taught in groups. Apparently, the model looks as a part of behavioural, managerial or administrative approach but, in fact, in the model emphasis shifts from a deductive approach of curriculum to an inductive approach. The model proposes a seven-step procedures for the development of the curriculum.

- **Identification of Learners :** The first step of the humanistic model is to identify the learners. That is, here the curriculum planners, try to identify the grade level of the learners, their age as well as some cultural and ethnic characteristics. Since the learners are taught in the group, knowledge of their common characteristics and interest pattern is essential for diagnosing their individual problems.
- **Concerns of the learners :** The second step is to determine the major concerns of the learners as well as their reasons. Among the concerns, the needs, interests, self-image and self-concept of the learners are important. Since such concerns are related to some major issues, they provide some consistency to the curriculum in the long run.
- **Diagnosis :** At this stage, the teachers try to develop some strategies for instruction so that the major concerns of the learners may be successfully met. Due emphasis is given here upon the fact that the learners, as a result of diagnosis, must gain a better control over their activities and must feel ease while doing the various activities.

- **Organizing ideas :** Here the curriculum developers organize the content by selecting themes and topics around the learners' concerns rather than according to the demands of the subject-matter. The content is organized by the curriculum planners according to three major principles. These principles are called by Weinstein and Fantini as important *vehicles*. These principles are: *the experiences of the learners, their attitudes and feelings* as well as the *social context in which they reside*. In fact, it is these three types of principles, which influence the major constituents, that is, concepts, skills and values usually taught in the classroom and they also constitute the basis for what, Weinstein and Fatini have called as '*curriculum of affect*'.
- **Learning skills :** At this stage, the curriculum planners pay attention to the learning skills that the learners are likely to develop. Here learning skills include basic skills of learning how to learn which, in turn, promotes learners' coping ability and power over their social environment. It also includes self-awareness skills and personal skills. These two types of skills enable the learners to deal with their own feelings as well as how they tend to relate to other people. Various kinds of learning skills developed here also enable the learners to deal with their own feeling as well as problem solving abilities in different area.
- **Teaching Procedures :** At this stage, the curriculum planners adopt various teaching procedures, which aim at promoting learning skills, content vehicles and organizing ideas. The nature of teaching procedures should be such that they must match with the various learning styles with their common characteristics and concerns.
- **Evaluation of outcomes :** At this last step, the teacher evaluates the outcomes of the curriculum especially the cognitive and affective objectives. This evaluation component is much similar to the evaluation components of deductive models. However, here there is comparatively more emphasis upon the evaluation of affective components, that is, needs, interests and self-concept of learners.

Eisner : Systematic-Aesthetic Model

Eisner (1991) was of view that exclusively utilizing behaviouristic objectives, in fact, reflect a limitation of the curriculum through its failure to account for the context relating to the idiosyncracy of the learners. Therefore, he propounded a systematic and dimensional view of curriculum that combines the behavioural principles with aesthetic components for proposing a model of curriculum development. In his model, there are five important components. A brief discussion of these components is as under :

1. **Intentional :** This component is wider one and covers a close examination of what really matters in school. Accordingly, the curriculum planners have to address the characteristics of the curriculum, major features of teaching, forms and utility of the different methods of evaluation as well as the nature of the workplace.

2. **Structural :** This component throws light upon the structural aspects of the schools. In other words, this component relates to how schools are structured, how roles are defined and how total time is allocated. All these aspects are considered important for facilitating various educational objectives. In India today the structural organisation of schools have not much changed from the past.

- **Curriculum :** The design of the curriculum is very important. It exclusively demands creative and constructive ideas that matter, the various skills that are important for the learners and the different means through which a healthy interaction between learners and programme take place.
- **Pedagogical :** This component emphasizes that the quality of teaching must be outstanding and there should be a primary concern for the teacher irrespective of the virtues of curriculum. Schools must serve the teachers properly so that they can serve their students in a better way. A good quality of teaching enhances the effectiveness of the curriculum.
- **Evaluation :** At last, the evaluation programme is carried on. The evaluation process is not simply a way for providing a score to the learners rather it is also one means to find out how well the teachers and learners are doing better for fulfilling the already set objectives.

Summary and Review

- Curriculum basically includes all the learner experience, within or outside the school included in the programmes that have been developed to help them emotionally, developmentally, socially, morally as well as spiritually.
- There are different types of curriculum. Some of the important types are: core curriculum, null curriculum, activity-countered curriculum, learner's centered curriculum, subject-centered curriculum and problem centered curriculum.
- There are some good characteristics of any curriculum such as a curriculum must have scope for continuous monitoring and evaluation, must be flexible, must meet the needs of the learners as well as society in general, etc.
- Curriculum has some obvious objectives such as a curriculum aims at promoting good human relationship, it aims at promoting democratic sense of citizenship among the learners, it also aim at informing learners about the rubrics of the humanities, the arts, social science etc.
- There are some issues involved in the curriculum development. Important such issues are: why is curriculum formulated? Who is to face the curriculum? How is curriculum to be used? When is the curriculum to be used? Who is to formulate curriculum?
- In constructing curriculum, the planners have to take into consideration some basic principles such as principle of community centeredness, principles of child-centeredness, principle of coordination and integration, principle of balance, principle of conservation, principle of creativity, principle of utility, principle of values, etc.

- Many people consider curriculum and syllabus the same thing but the reality is that they are different from each other. The basic difference between the two is that curriculum is the some total of various subjects activities and learning experiences whereas the syllabus reflects the list of unelaborated headings.
- To study the nature of curriculum, it can be examined from the different angles such as curriculum as a process, curriculum as a plan, curriculum as an objective, curriculum as an experience, as well as curriculum as a subject-matter.
- The foundation of curriculum has been considered from three points of view: philosophical foundation, psychological foundation and sociological foundations.
- A curriculum serves many different functions such as it serves to produce some responsible citizen, it preserves and transmit cultural heritage, it promotes some basic skills and it helps in promoting a correct knowledge about the world, etc.
- There are different models of curriculum development. Those models have been loosely grouped into two categories: deductive model and inductive model. Important models covered under deductive model are Tyler's model as well as Saylor, Alexander and Lewis model. Important models covered under inductive models are Taba's model, Weinstein and Fantani's model and Eisner's model. The basic difference between deductive and inductive model is that the former proceeds from general to specific, that is, from the needs of society to specifying the instructional objects whereas the latter begins with the development of curriculum materials and subsequently, go to some generalization.

Review Questions

1. Define curriculum. Discuss the nature of curriculum.
2. Discuss the characteristics of a good curriculum.
3. Describe some of the important principles of curriculum construction or development.
4. Discuss the major issues involved in the development of curriculum.
5. Define curriculum. Discuss its functions.
6. Critically examine Tyler's model of curriculum development.
7. Describe Taba's model of curriculum development.
8. Describe Eisner's model of curriculum development.
9. Describe Saylor, Alexander and Lewis model of curriculum development.
10. Make a comparative study of deductive and inductive model.

Part-II : Major Domains of Psychometric Assessment

Chapter – 10

Planning Classroom Tests and Assessment

Learning objectives
- Steps in classroom testing and Assessments
- Purpose of classroom testing and Assessment
- Preparing specifications for Tests and Assessments
- Selecting appropriate items and Assessment tasks
- Some important considerations in preparing Relevant Test items and Assessment tasks
- Some General guidelines for writing Test Items and Assessment tasks
- How to improve learning and Instruction?
- Tests used for making decisions in the classroom.
- Summary and Review
- Review Questions

Key Terms:

Pretesting, End-of-the instruction Testing; Objective tests Items; Performance Assessments; Interpretative Exercise; Construct-irrelevant factors; Supply-type test items; Selection-type test items;

In evaluation of the learning outcomes the role of classroom tests and assessment is very vital. They provide many useful and relevant information about the performances of the students. Not only this, they also provide some indirect evidences concerning many other similar learning outcomes. However, the success of such classroom tests and assessments depends upon the careful planning of the classroom tests and assessments. In the present chapter a discussion about such planning will be primarily focused.

Steps in Classroom Testing and Assessments

When a teacher wants to know about the academic achievement of the students, he resorts to classroom testing and assessment. This requires some careful basic steps, which must be followed properly. Important such steps are as under :
- To determine what is to be measured and the purpose of such measurement
- To develop various specifications

- To select some appropriate assessment tasks
- To prepare relevant and important assessment tasks
- To assemble the assessment
- To administer the assessments
- To appraise assessment
- To use the results
- To arrive at a goal (such as improve learning and instruction, etc.)

If these steps are properly followed, it will provide the foundation for varying results that may prove to be valid one for the intended instructional uses.

Purpose of Classroom Testing and Assessment

There are varieties of instructional purposes for which the teacher uses classroom testing and assessment. These purposes can best be described in terms of their location in the instructional process. There are three important such locations where purposes of classroom testing and assessment can be exemplified : *beginning of an instructional course* (that is pretesting), *teaching and assessment during instruction* and *end-of-the-instruction testing and assessment* : A brief discussion of these three follows :

• Beginning of the instructional course (that is, pretesting) :

Classroom testing and assessment may be done at the beginning of an instructional course. Such location of testing and assessment is popularly called as *pretesting*. Such pretests aim at determining whether the students have same prerequisite skills essentially needed for the instruction or at knowing about the extent to which students have already achieved the objectives of the instruction. Pretests for determining to what extent students have already achieved the objectives of the planned instruction are in no way much different from the tests used to measure the outcomes of the instruction. In this way, the test intended to measure final achievement in a course may be given at the beginning to assess performance at entry point. If this is the case, then final test should not be the same as given at the entry point rather it should be some equivalent form of that test.

Pretests have a limited domain as well as tend to have a relatively low level of difficulty. For example, a pretest in mathematical knowledge might be continued to simple algebra and arithmetical operations. Likewise, a pretest in psychology may be continued to only the experimental psychology.

• Testing and Assessment during Instruction :

Teachers administer testing and assessment during instruction for monitoring progress, encouraging students to study more and more, and for detecting misconceptions as well as for providing feedback to teachers and students. In other words, the purpose of testing and assessment during instruction is to make formative assessment. These formative tests are variously called as *practice tests, learning tests, unit tests, quizzes*, etc. These tests and assessments usually cover some pre-defined or preannounced segment of the course such as a particular chapter or a particular portion from a given chapter or unit. When most of the students fail in such formative tests, group review is generally recommended. On the other hand, when only a few students fail or show lack of understanding of the important concept asked, alternative methods of study such as reading assignments, practice exercises, special tutoring etc., are recommended.

- **End-of-the instruction Testing and Assessment :**

At the end of the instruction, teachers do testing and assessment for the purpose of knowing the extent to which the intended learning outcomes and performance have been successfully achieved. Thus the end-of-the instruction testing and assessment aim at summative assessment. However, such end of the unit tests can also be used for encouraging students to undertake more challenging tasks, providing feedback to the student, assigning remedial work as well as for grading purposes. In a nutshell, end-of-instruction test and assessment can be used for both summative as well as formative assessment. Such tests have also been found to provide information for evaluating instructional effectiveness.

These are three basic types of classroom tests assessments. Each of these three types of classroom testing and assessment places different demands on sampling of items and tasks as well as on the type of interpretation used.

Preparing specifications for Tests and Assessments

It is essential that classroom tests and assessments must measure a representative sample of instructionally relevant tasks. For ensuring this fact, the teacher must prepare specifications that can guide the selection of test items and assessment tasks. These specifications tend to define and delimit the achievement domain to be assessed as well as describe the sample of test items and assessment tasks to be kept ready. One very popular form of specifications is a two-way chart named as a *table of specifications*. Building table of specifications generally involves :

- developing a list of instructional objectives
- preparing the instructional content
- preparing a two-way chart relating the instructional objectives to instructional content.

As a *first* step, a list of general objectives and specific learning outcomes is prepared by the teacher for a particular unit or chapter (Already discussed in chapter 8). The list of objectives is limited to those outcome, which can be measured by a classroom assessment. In fact, the nature of test items or questions or assessment may vary as a function of learning outcomes. In general, those learning outcomes that contain verbs like 'identify', 'matches' or 'distinguishes' are easily measured by objective items. Learning outcomes with verbs like 'describe', 'explain' are readily and easily measured by essay questions although they can also be measured by skillfully constructed items. Those learning outcomes starting with verbs like 'measures', 'delivers', 'constructs', etc. are effectively assessed by performance assessment tasks.

The *second* step in preparing the table of specifications is to prepare an outline of instructional content. There is *difference* between the instructional objectives and the instructional content. The instructional objectives are those which mention the type of performance the students are expected to display whereas the instructional contents are those that indicate the area in which each type of performance is to be shown. For example, the content outline for the psychological statistics (instructional objectives) may be (a) Measures of central tendency (b) Measures of variability (c) Measures of Association. The amount of the details to be included in the content outline is somewhat arbitrary but the general rule is that the outline must ensure an adequate sampling of content and proper interpretation of the results.

The *third* and final step in preparing a table of specifications is to prepare a two-way chart that relates the instructional objectives to the instructional content and specifies the nature of the desired sample of items and tasks.

Due to its broad coverage, a table of specifications is highly useful for summative tests and assessment and it is also useful in preparing for some formative tests and assessments.

Selecting appropriate Items and Assessment Tasks

Ordinarily classroom tests are of two types : *Objective test items* and *performance assessments*. Objective test items are so called because they have single best or correct answer. Such test items are highly structured and require the testtakers (a students) to select the correct answer from a number of alternative or to supply a word or two from their own side. Accordingly, they are of two types : *selection-types* and *supply types*. True-false items, multiple choice items and matching items are the examples of selection types and completion items and short answer items are the examples of supply types. (See chapter 5 for a detailed discussion). Performance assessments are those tasks that require the students to construct responses such as write an essay or perform a particular task. Performance assessment tasks do allows the student to organize and construct the answer in essay form. Some types of assessment tasks may require the students to generate hypotheses, make observations or perform before an audience, etc. For most performance assessment tasks, there is no single right or best response rather there are a variety of responses that are considered excellent or appropriate by the expert judgement.

Researchers are of view that for some instructional purposes, objective items are considered best whereas for some other instructional purposes, performance assessment are considered most effective. The appropriateness of each of these two approaches is determined by the learning outcomes to be measured as well as by the advantages and limitations of each approach.

Educators and psychologists are of view that in selecting the most appropriate types of items and assessment task, the basic fact is to select the type of item, which provides the most *direct measure of the intended learning outcome*. For example, where specific learning outcomes calls for identifying a correct answer, a selection type item will be most appropriate. If the specific learning outcome involves use of any equipment to solve a problem, performance task would be most appropriate. Where the specific learning outcomes fail to clearly specify which item type to be used, selection type items would be the most appropriate due to the greater control over the responses of students as well as due to the objectivity in scoring. A healthy practice is to include both objective test items and performance assessment tasks in any comprehensive measurement.

Some Important considerations in preparing Relevant Test Items and Assessment tasks

In general, any attempt to construct the relevant test items and assessment tasks are preceded by a number of preliminary steps such as :
- the purpose of test and assessment must be clearly spelled out
- a set of specifications should be outlined
- the most appropriate types of test items and assessment tasks should be selected
- items and tasks should be prepared in light of the specifications developed earlier.

Relevant tests items and assessment tasks, in order to be effective one, must match with specific learning outcomes. Such matching involves fitting each item or task as closely as possible to the intended learning outcomes. In fact, preparation of relevant test items and assessment tasks means analyzing the performance described in the specific learning outcomes as well as constructing the items or assessment tasks that requires the concerned performance. For example :

Specific learning outcome : Identify the function of a given body structure Relevant Tests items :

What is the function of occipital lobe?
- (a) Releases growth hormone
- (b) Eliminates waste products in the body
- (c) Maintains visual sensation
- (d) Stimulates blood pressure.

The above example clearly shows that the specific learning outcomes define the type of response the student is expected to make but it does not indicate the specific course content (that is, occipital lobe), which a student is to identify.

Sometimes, it becomes essential to prepare a general item pattern, which acts as an intermediate step between specific learning outcome and the test item. Such item pattern acts as a guide to test construction and becomes especially useful when a file of test items is to be prepared or when more than one form of the test is needed. Some examples of item patterns according to general type of learning outcomes are as follows :

Knowledge outcome
1. Who is the father of ..... ?
2. What are the characteristics of ..... ?
3. What is the name of ..... ?

Understanding outcome
1. What explanation do you provide for ..... ?
2. Which of these is an example of ..... ?
3. What is the relationship between ..... ?

Application outcomes
1. Which of these indicate a good and correct application of ..... ?
2. What are the steps in developing a ..... ?
3. What is the implication of ..... ?

Item patterns like above should derive from the specific learning outcomes, which they represent. They should never be developed haphazardly.

While constructing test items or assessment task, it is essential to obtain a representative sample of items and tasks. We know that students know thousands of facts but we can test only for a limited number of them. In each area of content and for each specific learning outcome, the researcher selects only a set of items or tasks thought to be representative of their entire knowledge and students' responses to our selected set of items and tasks are what their responses would be towards other items and tasks drawn from the same area. This automatically necessitates that our limited sample must be selected in such a way that they provide a representative sample of various areas for which test or assessment is being developed.

The number of items and tacks is a very important factor in obtaining a representative sample. The number of items and performance tasks is guided by a carefully prepared set of specifications. In fact, such number also depends upon various factors like purpose of measurement, types of test items and assessment tasks used, level of reliability of the test as well as upon the age of the students. Although there is no hard-and-fast rules for determining the number of items and tasks, from the sampling point of view, a very important consideration is the number of tasks or items to be devoted to each specific learning outcome. In general, ten objective items for each specific learning outcome is considered desirable. Where the task is extremely limited such as adding three single digit numbers etc., and where the testtakers or students are to supply rather than select the answer, the number as low as five may serve the purpose. However, for a survey test where the sample of test items covers a broad area, ten or even more objective test items for each general objective are probably sufficient.

When complex learning outcomes are being measured, some special problems regarding sampling arise because this requires more elaborate objective-type items (popularly called as *interpretative exercises*) and performance assessment tasks. Both interpretative exercise and performance assessment tasks require considerable administration time. However, any single performance task or single test item is not enough for assessing an intended learning outcome.

While constructing test items and performance assessment task, it is essential to eliminate construct-irrelevant factors that might interfere with showing good performance by the students. Example of such factors are complex sentence structure, too difficult vocabulary, unclear instructions, required nature of response being uncleared, excessive wordiness, etc. These factors tend to limit the students' responses and put a barrier in showing true levels of achievements. Therefore, these factors need to be eliminated and one important step toward such elimination is to avoid ambiguity in choosing the words and writing sentence for items.

Another important consideration especially when writing objective item items is to avoid unintended clues, which unnecessarily help students in arriving at the answer. In fact, such clues enable the students to respond correctly even though they lack necessary skills and abilities. For example :

King Charles Spaniel is an
(a) fruit
(b) animal
(c) vegetable
(d) bird

In this example the word 'an' provides unnecessary clue to the correct answer 'animal' because article 'an' is used before a word that begins with vowel. Rests of the three words begins with a consonant letter. Such clues not only occur in selection type item rather it also occur in supply type of item.

For example :

A place on land surrounded by water is known as an —

There can be two plausible answers : peninsula and island. Since 'peninsula' begins with consonant letter, it cannot be preceded by 'an'. Then, the correct answer will be naturally 'island' having vowel in the beginning. Such unintended clues in the items should be avoided.

Some General guidelines for writing Test Items and Assessment tasks

Educators and Psychologists have provided some general guidelines for writing test items and assessment tasks. They should be kept in mind while writing test items and assessment tasks. Important such guidelines are as under :

- Test and assessment specifications should be used as guide in selecting the items and assessment tasks. Since such specifications tend to describe the performance to be assessed and samples of learning outcomes to be measured, they provide guidelines to the teacher for selecting the types of items as well as tasks for writing. They also provide guidelines regarding the number of items and tasks needed for each of the achievement.
- Test items and assessment task should be prepared in advance of day of the testing. If such preparations is done well ahead, this would provide opportunity to review the items and catch up last minute ambiguity left therein.
- Each item and assessment task should be written clearly so that the task to be performed is well-understood, providing indication for the performance described in the intended learning outcome. For writing each test item and assessment task, the language used should be simple, free from unnecessary wording and unintended clues.
- More items and tasks than is needed, should be prepared. When more items and tests than needed is written, it permits a better review later, thus providing an opportunity to discard the weak and poor items and tasks.
- Whenever the test items or assessment tasks are revised, its relevance must be rechecked. When items and tasks have been revised for clarity, freedom from clues, bias, etc., its

relevance must be checked. In other words, it should be checked whether the revised items and tasks still provide a relevant measure of the intended learning outcome. This is necessary because even slight modifications sometimes may change the function of an item or task.

- Each item or task should be written in such a way that it should not provide help in responding to other items. Mostly it happens in multiple choice test where one item, if not written carefully, may provide some aid in answering the other item. For example, a date, a name or place inadvertently included in the stem of an item may be called for in a short-answer item in another portion of the test.

- The answer of the each test item should be widely agreed upon by the experts or in the case of assessment tasks, their responses judged appropriate must be agreed upon by the experts. This guideline is easy to follow when factual knowledge is being assessed but becomes more difficult to follow when complex, learning outcomes requiring essay type or other types of performance is needed.

How to Improve learning and Instruction?

A good classroom test and assessments aim at enhancing the quantity and quality of the student learning. This aim can be easily achieved if the following measures are given due attention:

- *The teacher must pay attention to the breadth and depth of content and learning outcomes measured*: When the teacher selects a representative sample instruction, it is expected that students must devote attention to all areas. By giving more weights to some areas of the course in the instruction, the students are encouraged to put their concentrative effort on those objectives. Likewise, if tests measure different types of learning outcomes, they learn so many things like how to interpret and apply facts, how to develop deep-seated conceptual understandings, how to formulate hypothesis, how to identify cause-and-effect relations and the like. All these things, in a way discourage students from depending solely on memorisation and force them to realize that mere memorization of factual information is not sufficient.

- Tests and measurements indirectly contribute to the improved teacher-students relationship, thereby having a beneficial impact upon student's learning. If the students view the tests and assessments as fair means of assessing their achievement because they perceive no bias in the items of the tests and take the test items as the most representative sample of the learning outcomes, this has a beneficial impact on student learning and it is expected that such usefulness perceived by the students will definitely improve the learning.

- Improved teaching methods also indirectly contribute to the betterment of learning and instruction.

When tests and assessments are so constructed as to measure a variety of learning outcomes, several clarifications regarding thinking skills, understandings and other complex learning outcomes are made. These clarifications help the teacher in planning the learning experience of students effectively and tend to increase basic understanding, reasoning, thinking skills in their teaching. All these have a global positive impact upon the student learning.

Tests used for making decisions in the classroom

Teachers make a variety of decisions at primary, secondary, higher secondary and college level. They must decide what information students are learning and what information they have difficulty in learning. They also have to decide whether students are ready to learn new materials and if so, they are also required to determine how much new materials already know. Teachers often use tests, which are often combined with other assessment methods, for taking these decisions. A very close observation shows that the teachers make these decisions at three levels :

- Decisions are made at the beginning of the course/instruction
- Decisions are made during instructions
- Decisions are made at the end of instruction

At the beginning of the course, the teachers often use psychological tests as *placement assessments*, which are defined as one way to determine the extent to which students possess the necessary knowledge, skills and abilities for understanding new materials as well as how much of these new materials to be taught, are already known by the students.

Decisions by teachers are also made during the instruction periodically through the academic year. In such situations, psychological tests are used as *formative assessments*, which provide help to the teachers for determining what information students are learning or are not learning during the process of instruction. Teachers don't use these test scores for assigning grades to the students. However, they are used by teachers to make immediate adjustments to their teaching methods and curriculum. In other words, the results of the formative assessments can help teachers in making adjustment to the pace of their teaching the curriculum they are covering.

If any student continues to exhibit problems write materials, teacher tends to suggest evaluating learning abilities through what it called *diagnostic assessment*, which is defined as an in-depth evaluation of a students' academic performance. For completing *diagnostic assessment*, teachers need to include a varieties of psychological tests. Often, clinical and educational psychologists conduct diagnostic assessment.

Some decisions by teachers are also made at the end of the instructional programme. Tests used for the purpose provide what is called *summative assessment*, which help teachers in determining what students know or don't know. Test scores here are used for assigning earned grades.

Sometimes teachers may use the same psychological tests as both formative and summative assessment. When psychological tests are used as a formative assessment, such tests are used to direct future instruction for providing teacher with information about what information is already learned by students. When psychological tests are used as summative assessment, test scores are used as a final evaluation, often used in determining what student should be assigned.

Summary and Review

- Through classroom tests and assessment task, the teacher is able to know about the achievement of the students. This requires at least nine carefully defined steps.
- There are varieties of instructional purposes for which the teacher uses classroom testing and assessment. There are three important locations where purposes of classroom testing and assessment are exemplified : beginning of an instructional course (that is, pretesting), testing and assessment during instruction and end-of-the-instruction testing and assessment.
- It is essential that classroom tests and assessments must measure a representative sample of instructionally relevant tasks. For ensuring this fact, the teacher must prepare some specifications which may guide the selection of tests items and assessment tasks. These specifications tend to define and delimit the achievement domain to be assessed.
- Classroom tests are of two general types : objective test items and performance assessments. For some instructional purposes, objective items are considered best whereas for some other instructional purpose, performance assessments are considered most effective.
- In preparing some relevant test items and assessment tasks, there are some important considerations that must be kept in view. For example, relevant test items and assessment tasks must match with learning outcomes. While constructing test items and assessment tasks, it is also essential that a representative sample of items or tasks be obtained. Likewise, construct-irrelevant factors must be eliminated because such factors interfere with showing good performance by the students.
- Educators and psychological have provided some general guidelines for writing test items and assessment tasks.
- Educations have also devised some important ways to improve learning outcomes and instructions.
- Tests are used by teachers for doing formative assessment, summative assessment, diagnostic assessment and placement assessment.

Review Questions

1. Discuss the important steps in planning classroom tests and assessment.
2. Discuss the important purposes of classroom testing and assessment.
3. Outline a programme for preparing specifications for tests and measurement.
4. Discuss the important considerations in preparing relevant test items and assessment tasks while preparing classroom tests.
5. Discuss the important general guidelines for writing test items and assessment tasks.
6. Discuss how psychological tests are used for making decisions in classroom by teachers.

■■■

Chapter – 11

Achievement Test

Learning objectives
- Meaning of Achievement test
- Major features of Achievement test
- Standardized achievement test vs. Teacher-made classroom achievement test.
- Major functions or uses of Achievement testing
- Types of Standardized Achievement test
- Norm-referenced achievement test vs. Criterion-reference achievement test
- Steps involved in construction and standardization of Achievement test
- Dark side of achievement testing.
- Summary and Review
- Review Questions

Key Terms:
Achievement test, Standardized achievement test, Teacher-made classroom test, Survey test batteries, Single survey test, Diagnostic test, Prognostic test.

Achievement Testing plays a central role in school programmes. Some of these tests consist of batteries of tests where as other measures individual subject or skill. Achievement tests measure achievement, which refers to the degree of learning or ability already attained as a result of exposure to some defined situations. In United Kingdom, achievement tests are known as *attainment test*. In the present chapter we shall examine different types of achievement tests most popularly used by the various school programmes, their functions or uses, the major features of standardized achievement tests, as well as some dark side of achievement testing, etc.

Meaning of Achievement Test

Achievement test is one which is designed to assess a person's current state of knowledge or skill acquired as a function of exposure to some especially designed or defined programme (Reber, Allen & Reber, 2009). Performance or attainment of a student is called achievement. The achievement of a student or learner is based on their performance rather than on their ability or potential. A learner with high ability might, under some circumstances, perform poorly as compared to a learner with lower ability. In fact, achievement tests are said to be past and present oriented and they assess the degree of learning or achievement after instruction or training. In other words, achievement tests are understood as an instrument to measure the degree of learning that has taken place as a result of exposure to a relatively defined learning experience (Cohen & Swedlik, 2005). Such learning experience may be as broad as what has been learnt in the last two years or it may be much narrower such as one fortnight, etc. Achievement test may be standardized nationally, or locally or it may not be standardized at all. In case of standardized one, it is known as *standardized achievement test*. But when it is not standardized at all, it is called *teacher-made achievement test*.

In a nutshell, achievement test has the following features :
- Achievement tests measures the degree of learning or achievement after exposure to instruction.
- Achievement tests are present and past-oriented
- Achievement rests may be standardized or teacher-made having no standardization.

Major features of Standardized Achievement Tests

The standardized achievement tests are very popular and frequently used in assessing achievement in various school programmes. They have some basic features, which must be kept in view.
- Standardized achievement tests have norms based upon the representative groups of the individuals.
- Standardized achievement tests provide directions and impose time limits and other similar controls. As a consequence, such tests make the conditions under which the tests are taken as *standard*.
- Standardized achievement tests are carefully developed and refined through the procedure of item analysis. As a consequence, in such tests every item functions appropriately and implausible distractors are deleted. Most intrinsic vagueness is also removed.
- The reliability of the standardized achievement tests is high, most frequently in between 0.80 to 0.95.
- Standardized achievement tests measure learning outcomes and contents common to many schools.
- Equivalent and comparable forms of the standardized test are usually provided along with the information concerning the extent to which forms are comparable.
- In the standardized test the manual and other related accessory materials are usually provided as important guides for administering and scoring the test, interpreting and using the results.
- The scores on the standardized achievement tests can be scientifically interpreted because such tests have manual and other guides.

- The principal validity for achievement is what is called as *content validity* or *content relevance*.

Standardized Achievement tests vs. Teacher-made classroom tests

Standardized achievement tests and teacher-made achievement tests (also called as informal achievement test) are similar in many ways. In fact, both are partners rather than competitors. Both are based upon the table of specifications and both provide clear directions to the students. They serve somewhat different purposes and provide complimentary information. Educators are of view that both kinds of tests are necessary for adequate and better evaluation of academic achievement by students in schools.

However, standardized achievement tests and teacher-made classroom achievement tests differ on several points as under :

- Standardized achievement tests put more restriction than teacher-made tests. The standardized tests have prescribed directions, time limits and other similar control than the teacher-made tests. Therefore, the former afford more meaningful bases for evaluating and comparing performance than the latter.
- Standardized achievement tests are carefully refined by item analysis so that every item is functioning appropriately, most intrinsic ambiguity is removed and implausible distractors are deleted. In case of teacher-made achievement test, such vigorous procedure is not followed due to the limited time and lack of opportunity to pretest items. As a consequence, the quality of test items of the teacher-made tests is usually lower than the standardized tests.
- The reliability of the standardized achievement tests is usually higher, commonly in between 0.80 to 0.95. But the reliability of the teacher-made test is not so high.
- On the standardized achievement tests scores can be compared with those of norm groups. The norms provided by the standardized achievement tests tend to offer a comparison with an external group. This type of comparison is considered critical for various purposes like quality control, curricular evaluation, counselling as well as for identifying exceptional students. In case of teacher-made achievement tests such external comparison is not possible. The common people feel that the average percent correct on the test directly reflects the quality of learning but they fail to realize that a very poor teacher may construct a test so easy and nondiscriminating that almost all students would tend to have perfect performance even though they have learnt little. Likewise an intelligent teacher may develop a difficult and discriminating test on which the average score may be only 50 percent or less.
- On the standardized achievement tests the interpretation of scores is easily done with the help of manual and other guides. So far as the interpretation of scores on teacher-made tests is concerned, it can be done in light of known instruction history.

Major functions or Uses of Achievement Testing

As we know, there are two general types of achievement test: group achievement test and the individual achievement test. Both these types of achievement tests permit different varieties of potential uses or functions. Some of the important ones are as under:

- Achievement tests are used in identifying the children as well as adults having known specific achievement deficits considered necessary for recognizing the cases of learning disabilities. In a nutshell, tests are helpful in making diagnostic and remedial decisions.
- Achievement tests are also useful in making placement decisions. Standardized achievement tests having a uniform score scale can prove to be very useful in identifying the performance of the student at the entry level.
- Guidance and counselling decisions are also made on the basis of scores of the achievement tests.
- Scores on the achievement tests help parents recognize the academic strengths and weaknesses of their children and thus, help in preparing some remedial programme at home.
- Scores on achievement tests help in identifying classwise or schoolwise deficiencies in achievement that forms the basis of redirection of the instructional efforts.
- Educators often group students according to similar skills in specific area on the basis of the scores of achievement tests.
- Achievement tests help in assessing the success of various educational programme by measuring the subsequent skill attainment of the students.
- Achievement tests are also used for identifying the level of instruction, which is considered most appropriate for the individual students.
- Curricular decisions between alternative programmes require a very broad base comparison where the achievement tests are very useful.

If we look attentively on the above points, it will be obvious that achievement tests serve both institutional goals as well as the individual goals. Institutional goals such as monitoring schoolwise achievement levels, are best served by group achievement test batteries whereas the individual goals such as the assessment of the individual learning difficulties is best served by individual achievement tests.

Types of Standardized Test

There are different types of standardized achievement tests. Of these types, the following four types are common:

- Survey Test batteries
- Single Survey Test
- Diagnostic test
- Prognostic tests

A discussion follow

- **Survey Test batteries**

    Survey test batteries are a group of subject-matter tests designed for a particular grade level. They are frequently used in form of standardized achievement test. A battery, as we known, consists of a series of the individual tests all standardized on the same sample of students. This makes it possible to compare test scores obtained on separate tests. As a consequences, it becomes possible to determine students' relative strength and weaknesses in the different areas covered by the tests. The major purpose of a test battery is to determine a students' general standing in a group rather than his strengths and weaknesses. Such batteries are also known as *group standardized achievement tests*. The five major such batteries are : *California Achievement Test* (CAT) published by McGrawn Hill; also know as Terra Nova), Iowa Tests of Basic skills (ITBS) published by Riverside Publishing Company; the *Comprehensive test of basic skills* (CTBS) published by Mc Graw-Hill; the *Metropolitan Achievement Test* (MAT) published by The Psychological Corporation and *Stanford Achievement Test* published by The Psychological Corporation. All these batteries cover more or less the same skills. Major general content areas covered include reading, mathematics, science, social studies, language skills, etc. These achievement batteries are used often at the elementary school level in grades 3 to 6. Such extensive usage is clearly understandable because there is a considerable uniformity in their content area. In Indian conditions some standardized achievement tests have also been developed. Achievement Test battery by Prof. R.D. Singh and General class-room Achievement Test by Prof. A.K. Singh and Prof. A. Sen Gupta are the most burning examples. Both these tests have been published by National Psychological Corporation, Agra.

    Standardized achievement batteries have some strengths and weaknesses.

    Strength
    (a) Standardized achievement batteries are the most comprehensive way of measuring achievement
    (b) Such batteries provide an overall picture of the students' standing in various subjects.
    (c) Since the different tests of the battery are standardized on the same sample of the students and scores are expressed on the same individual's performances in different subjects, they can be easily compared.

    Weakness
    (a) All parts of the test battery are usually not equally appropriate for assessing a particular school's objective.
    (b) Standardized achievement test batteries are usually time consuming.
    (c) Usually, the reliability of such test batteries is low.

- **Single Survey Test :**

    Specific tests are available to measure achievement in almost all curricular subjects. Teacher does not need to administer all tests of the batteries rather he selects some specific tests he needs.

For example, he may need to administer only a test of Mathematics to the group of students. The major strengths and weaknesses of such single survey test are as under :

Strengths
- Single survey tests usually provide the deep index of achievement because they are usually larger and more detailed than the comparable test from a test battery.
- Such test provides a thorough evaluation of the student's achievement in any particular subject.

Weakness

The major weakness of single survey tests is that they provide only one overall score, which is usually not enough to make a correct inference about the true performance of the students.

- **Diagnostic achievement tests :**

Achievement tests can also be used for diagnosis of various learning disabilities. Such tests are called as *diagnostic achievement tests*, which are commonly used individually. Such tests are used either as a supplement to measures of cognitive ability or sometimes to replace other measures. The major indicator of learning disabilities is the large scale difference between cognitive ability scores and achievement test scores. These tests provide documentation of impaired performance in some crucial academic areas such as reading, writing, calculation, etc. Not only this, such tests also provide help in identifying the particular skill deficits that may underlie learning disabilities. The popular such achievement tests are *Peabody Individual Achievement Tests* (PIAT) produced by Pearson Assessment, Kaufman Test of Education Achievement (K-TEA) published by American Guidance Science and Woodcock-Johnson III tests of achievement (WJ-III) published by Riverside Publishing Company. The Standard Diagnostic Mathematics Test is one another such test. All these test intend to diagnose various learning disabilities by obtaining scores in reading, writing, mathematics, spelling, computation, etc.

The major strength of such diagnostic achievement tests is that they provide a detailed analysis of student's disability in reading, writing, mathematics and help in determining the causes of disabilities. However, they also possess some weaknesses. They are so detailed that they become time consuming. Sometimes such tests also recommend the use of some special instruments and apparatuses. As such, they also prove t be costly.

- **Prognostic achievement tests :**

Prognostic achievement tests are those tests, which are designed to predict achievement in specific school subjects. It is like an aptitude test in its function because it intends to predict later achievement. It usually contains a wide variety of items in the concerned subject.

The major strength of prognostic achievement test is that it has proved to be very helpful in making prediction at a higher grade level. However, the weakness of prognostic achievement tests is that its uses are very much limited such as to school subjects only.

Norm-referenced achievement test vs Criterion-reference achievement Test

Another way of classifying achievement tests is to put them as norm-reference achievement test and criterion-reference achievement test. Norm-reference achievement tests are those tests, which permit interpretation with reference to a large standardization sample. Such tests permit reporting of scores in terms of standard scores, percentile rank, etc. Standardized achievement tests are examples of norm-reference achievement test. Criterion-reference achievement tests are those tests that permit interpretation with reference to the specific content mastered by the individual student. For example, a criterion-reference achievement test might provide that the student (or the examinee) must atleast correctly spell 80 words out of 100 words. The criterion-referenced test is closely tied to the specific instructional objectives of the classroom or school. Criterion-referenced achievement test looks attractive because it avoids some negative associations regarding testing by avoiding the labelling of students as failure. It is perceived as more constructive and useful because it provides specific information about the performance of the student. However, the problem with such tests is that whether it is skill such as reading comprehension or spelling or objective such as 'can add three digit numbers', a large number of items is usually required.

These two types of achievement tests are not necessarily incompatible. Reality is that most achievement test batteries provide both norm-referenced and criterion-referenced interpretation of scores.

Steps involved in Construction and Standardization of Achievement Test

There are some defined steps in the construction and standardization of achievement test. Major such steps are as under :
- Planning of the Test
- Preparation of test items
- Preparation of directions regarding items of the test
- Preparation for scoring and administration of test
- Item Analysis
- Standardization of achievement test
    - Reliability of the test
    - Validity of the test
    - Norms of the test.

A discussion of these important steps is as follows :

1. Planning of the test :

It is a very crucial preliminary stage where the tests constructor has to decide about many aspects in detail such as what he wishes to measure and what are the manifestation of the achievement test. At this stage, he is to pay especial attention to the following :

- Developing test rationale and measurable objectives
- Content of the test
- Preparation of blue-print.

A brief description follows :

Developing test rationale and measurable objectives

Here the test constructor undertakes a survey of both basic and specialized courses common to school programmes and determines objectives of the test. For example, let us assume that test constructor has decided to formulate objectives in cognitive behavioural terms focusing on knowledge, Understanding, Application, Analysis, Synthesis and Evaluation from selected units of physics commonly taught in class XI-XII of Indian schools.

Content of the test

Here the test constructor has to decide about what content is to be tested. For determining test content, the test constructor is required to examine textbooks and courses of study "in common use" and ask subject-matter specialists and teachers what topics should be included. Suppose the test constructor, after completing the necessary formalities, decides to include the following units as content in the test of achievement in physics.

(a) Matter, (b) Heat, (c) Light (d) Sound (e) Magnetism (f) Electricity (g) Motion (h) Force.

Preparation of the Blue-print

Preparation of blue-print enables the test constructor to have an objective based test showing due weightage to the different objectives, content area and number of questions or items. For taking decision about the weightage to be given to different consent areas, objectives and number of the items from a given content area objectivewise, generally experts' opinion is sought. Let us say that after seeking such express' opinion, the test constructor has prepared a blueprint for this achievement test as presented in Table 11.1.

Table 11.1 : Blue Print of the Physics Achievement test showing the Number of items in different content areas objectivewise

Objectives / Contents	Cognitive Levels of objectives						
	Knowledge	Under-standing	Applic-ation	Analysis	Synthesis	Evalua-tion	Total
Matter	2	2	3	1	1	2	11
Heat	3	3	2	2	1	1	12
Light	3	2	2	2	1	2	12
Sound	4	2	2	1	3	1	13
Magnetism	4	2	1	1	1	2	11
Electricity	4	2	3	3	3	1	16
Force	4	2	1	2	2	2	13
Motion	2	2	2	2	2	2	12
Total	26	17	16	14	14	13	100

Table 10.1 shows the number of items to be prepared in different content areas objectivewise. For example, in the content area of matter, the highest weightage in terms of number of items has been given to Application, whereas knowledge, Understanding and Evaluation have been equally weighted as 2 and minimum weight has been given to Analysis and synthesis as 1 and 1 each.

2. Preparation of Test items :

After preparing the blue print, the test constructor starts writing the items of the Physics Achievement test. Three types of items were decided to be prepared : multiple-choice items, true-false items and matching items. Total number of items framed was 100 for covering intended objectives and different content areas. For preparing these items, help of the textbooks, experts opinion, discussion with teachers and literature already available were taken into accounts. After completing the writing of these 100 items, they were examined by experts and school teachers whose comments and suggestions about structure and language of items were taken into account and accordingly, some modifications or changes in the items were introduced. Thus the first draft of the test was prepared. Table 11.2 shows the serial number of items placed in different content areas objectivewise.

Table 11.2 : Serial Number of items in the first draft of the achievement test at different cognitive levels of objective

Objectives Contents	Cognitive Levels of objectives						
	Knowledge	Under-standing	Applic-ation	Analysis	Synthesis	Evalua-tion	Total
Matter	27,29	45,47,	15,24,44	48	40	9,6	11
Heat	1,23,33	25,36,66	39,50	20,30	19	16	12
Light	28,38,57	26,17	49,69	60,70	18	8,4	12
Sound	68,3,43,65	55,76	14,11	2	5,12,56	22	13
Magnetism	86,34,13,32	67	58	21	7	54,31	11
Electricity	10,59,35,41	64,74	42,51,61	87,100,75	52,73,83	53	16
Force	99,78,63,99	77,88	80	37,62	85,95	72,98	13
Motion	82,92	81,93	91,79	84,94	90,97	96,71	12
Total						Total =	100

3. Preparation of directions regarding items of the test :

The present achievement test was prepared in three sections: A, B and C. A section contained 50 multiple choice items, B section contained 25 True-False items and Section C contained 25 matching items. For section A, the students were asked to tick the correct answer out of four answers, for section B, they were asked to put tick mark on either True or False option and for Section C, they were asked to match Column I and Column II as per direction.

4. Preparation of Directions for scoring and administration of the test :

The test constructor prepared a precise and clear direction regarding administration of the test. He also prepared a scoring key for objectively scoring the items of the three sections of the test. A score of 1 was assigned to each correct answer and zero was awarded to each wrong.

5. Item Analysis : Experimental try-out of the test :

After preparing the test items and the scoring key, the test consisting of 100 items was administered on a sample of 370 students of class XII who had opted for physics as one of the course papers. No time limit was fixed for this try out. Subsequently, the answer sheets were scored as pre-determined method of scoring.

Based on the scores of 370 students, that is, the upper 27% (N = 100) and the lower 27% (N = 100) were selected. The remaining middle 56% cases were ignored. Based on these two extreme groups, indices of item difficulty and item discrimination were calculated. The item difficulty was calculated with help of formula 11.1

$$P = \frac{R_U + R_L}{N_U + N_L} \qquad (11.1)$$

When,
P = difficulty value
R_U = number of students answered correctly in upper group
R_L = number of students answered correctly in the lower group
N_U = number of students in upper group
N_L = number of students lower group

To illustrate, the difficulty of Item No. 1 was :

$$P = \frac{90 - 40}{100 + 100} = \frac{130}{200} = 0.65$$

In this way, difficulty values of the remaining 99 items were calculated. As we know, the maximum difficulty value can be 1.00. Higher the index of difficulty value, easier is the item. Index of difficulty lower than 0.40 and above .70 were considered as too difficult and too easy respectively. Items of difficulty lower than 0.40 and above .70 were considered as too difficult and too easy respectively. Items having difficulty values in between .40 to .70 were considered as moderately difficult. Table 11.3 presents the difficulty values of 100 items by classifying them into three categories : Highly difficult, Moderately difficult and Easy items.

Table 11.3 : Difficulty values of Items

Level of difficulty	Serial Nos. of Items	Total
Highly difficult (<0.40)	15, 26, 68, 54, 41, 87, 95, 92, 64, 83	10
Moderately difficult (0.40 to 0.70)	29, 45, 47, 24, 44, 48, 40, 9, 6, 23, 33, 25, 36, 66, 39, 50, 20, 30, 19, 16, 38, 38, 97, 17, 49, 69, 60, 70, 18, 8, 3, 43, 65, 55,, 76, 11, 58, 59, 54, 86, 34, 32, 67, 89, 2, 21, 7, 10, 12, 35, 74, 42, 51, 100, 75, 52, 73, 77, 53, 99, 78, 31, 63, 77, 88, 80, 62, 85, 72, 98, 82, 81, 93, 91, 84, 94, 90, 97, 96, 46	80
Easy (>0.70)	37, 27, 1, 5, 13, 61, 71, 4, 14, 22	10

Based on the same extreme group method, discriminative index (D.I.) of the item was also calculated by formula 11.2.

$$DI = \frac{R_U}{N_U} - \frac{R_L}{N_L} \qquad (11.2)$$

When, R_U and R_L refer to the number of the examinees giving correct answers in upper group and lower group respectively and N_U and N_L are the number of students in the upper group and lower group respectively.

To illustrate the D.I for item Number 1, by formula 11.2, would be $\frac{90}{100} - \frac{40}{100} = .9 - .4 = .50$.

In this way, D.I of the remaining 99 items were calculated. As we know, higher the value of D.I, the better the item because then, it is said to discriminate better between high test scorers and low test scorers. Most researchers have provided some convenient ways of interpreting the valued of D.I. Table 11.4 summarizes those ways in brief.

Table 11.4 : Interpretation of D.I. of different range

D.I	Quality	Recommendations
>0.39	Excellent	Retain
0.30–.39	Good	Possibilities for improvement
0.20–0.29	Poor	Discard or review in depth
<0.01	Worst	Completely discard

Items having negative D.I values are also to be discarded. Table 11.5 presents the values of D.I of all 100 items of Physics Achievement Test.

Table 11.5 : D.I of items of the first try out of Physics Achievement Test

D.I	Serial No. of Items	Total
>0.39	47, 29, 44, 40, 69, 60, 70, 51, 75, 35, 23, 34, 72, 98, 90, 94, 45, 48, 9, 6, 36, 39, 50, 19, 16, 28, 38, 3, 8, 65, 55, 2, 12, 86, 33, 67, 58, 21, 7, 10, 59, 74, 42, 100, 73, 53, 78, 80	48
0.30–0.39	79, 89, 82, 99, 88, 17, 20, 11, 24, 49, 18, 30, 76, 25, 57, 66, 31, 32	18
0.20 to 0.29	26, 15, 68, 1, 5, 13, 84, 62, 91, 93, 81, 46, 97, 96, 85, 14, 43, 54, 52, 63, 56	21
0.00 to 0.20	37, 27, 41, 87, 22, 61, 71, 4, 77, 92, 64	11
<0.01	83, 95	02
	Total	100

Finally, item analysis chart was prepared in which both the difficulty value and discrimination index of items were jointly shown. Table 11.6 shows such item analysis chart.

Table 11.6 : Item Analysis chart showing index of difficulty value and discrimination value of 100 items of the test.

Difficulty level	High difficulty (<.04)	Median (0.40 to 0.70)	Easy>0.70	Total
Discrimination index >0.39	×	47, 29, 81, 40, 91, 60, 70, 51, 75, 35, 52, 34, 72, 98, 90, 94, 45, 48, 9, 6, 36, 39, 50, 19, 16, 28, 38, 3, 8, 65, 55, 2, 43, 63, 93, 67, 58, 21, 7, 10, 59, 74, 42, 100, 84, 85, 78, 80	×	48
0.30–0.39	×	46, 89, 82, 99, 88, 17, 20, 11, 24, 49, 56, 96, 76, 25, 57, 77, 31, 32	×	18
0.20–0.29	15, 68, 92, 41, 54, 64	30, 97, 62, 61, 73, 86, 18, 69	1, 5, 13, 61, 71, 4, 14	21
0–0.20	26, 83, 87	44, 23, 33, 66, 79, 53	37, 27	11
<0.01	95	.....	22	02
Total	10	80	10	100

Based on item analysis chart, a final draft of items to be retained for the test was developed. Table 11.7 presents the final draft of the test showing those items that were finally retained in the test.

Table 11.7 : Final draft of the test

Level of difficulty	Remarks	Total	Total
Discrimination Index >0.39	47, 29, 81, 40, 91, 60, 70, 51, 75, 35, 52, 34, 72, 98, 90, 94, 45, 48, 9, 6, 36, 39, 50, 19, 16, 28, 38, 3, 8, 65, 55, 2, 43, 63, 93, 67, 58, 21, 7, 10, 59, 74, 42, 100, 84, 85, 78, 80	Excellent items	48
0.30–0.39	46, 89, 82, 99, 88, 17, 20, 11, 24, 49, 56, 96, 76, 25, 57, 77, 31, 32	Good items	18
Total	66		66

6. Standardization of Physics Achievement test :

At this stage, the processes for calculating reliability, validity and norms are initiated.

- **Reliability of the test :**

Both test-retest reliability and split-half reliability of 66-item of physics achievement test were estimated. The test-retest reliability was 0.82 and the split-half reliability, after correcting for full length reliability by Spearman Brown prophecy formula, was 0.86. For estimating reliability of the test, the test was readministered on a fresh sample of 100 students of XI and XII.

- **Validity of the test :**

The content validity of the present test was established by seeking experts' judgement by using formula 5.13. Teachers who taught physics in class XI and XII acted as judges and they were asked to judge a test item by using a four-point rating scale. On the basis of match between its content and the content defined by the domain specification : very weak relevance (1), weak relevance (2), strong relevance (3) and very strong relevance (4). Using the said formula 5.13, the content validity of the physics achievement test was found to be 0.92. The construct validity of the Physics Achievement test was also established by computing the differences between two known group of students. The group consisted of those plus two students who had opted Physics as one of the papers and another group consisted of those plus two (intermediate students) who had no knowledge of Physics. Since the difference between these two groups was found to be significant (t = 7.62 P<.01), the test was said to demonstrate satisfactory level of construct validity of the test.

- **Norms of the test :**

    Grade norms of the test was developed. Norms for class XI and class XII were prepared for the purpose of meaningful interpretation of the obtained scores by the percentiles, age and grade norms quotients and standard scores. In recent years the trend has been to discourage age, grade and quotient norms in favour of percentiles and standard scores (T scores and stanines).

Dark Side of Achievement Testing

Today there is an excessive emphasis upon the scores of educational achievement tests for selection and evaluation processes. This tends to promote some inappropriate behaviour including outright fraud and cheating on the part of students as well as on officials or authorities. As a result of such fraud and cheating, there occurs inflation of achievement test scores. One popular cause is educational administrators themselves who are desperate to show the excellence of their educational institutions. Researchers have shown that cheating on achievement tests includes the following :

- Teachers coach students on test answers by proving different types of short-cuts to correct answer.
- Students are sometimes given more time than allotted
- Officials or authorities alter answer sheets.
- Teachers make copy of the tests to give to their students
- Teachers or officials provide hints or clues to the students on the test.
- Ghosts students (*Munnabhai*) answer the test items in place of real students.

In India we often heard news about I ghost students (*Munnabhai*) who appeared on behalf of the real students for enhancing the achievement test scores apparently for increasing the probability of selection or for some positive evaluation. A researcher conducted one very popular study in which he has analyzed the inappropriate testing practices being reported in achievement testing (Gregory, 2004). According to him, inappropriate testing practices included praising students who answered the test items correctly during the test, recoding a student's answer sheet because he miscoded the answer, giving students more time than allotted, giving hints or clues during the test, using last year's test questions as practice. He also reported that more than 90 percent of teachers modified their curriculum in anticipation of getting their students do better on the achievement tests and more than 70 percent teachers eliminated major topics so that they could spend more time on test-related skills.

In conclusion, it can be said that we must develop an optimistic assumption that such fraud and cheating on achievement test, in future, will be eliminated. But right now we really don't know how this will take a concrete shape.

Authentic Assessment of Achievement

There are different measures of achievement. Some are norm-referenced tests and some are criterion-reference tests. But many psychologists and educationists criticised such tests because they measure only understanding. It is also said that these tests are too structured and other contain only multiple-choice questions or true/false questions. Therefore, some experts have pointed out that the emphasis must be changed from assessing understanding to measuring application. In other words, emphasis is now being given on the abilities of the students to apply the knowledge and skills they have gained to performing some real-world tasks and solving problems of real world. This is called *authentic assessment*, which emphasizes upon the abilities of students to perform real world tasks or solve real-world problems by applying the learnt knowledge and skills. Authentic assessment depends upon more than one measure of performance as well as upon human judgement. Such assessment is also criterion-referenced. Supporters of authentic assessment are of view that students basically acquire knowledge or skills for performing a task or producing a product and any assessment is done only for evaluating their abilities to perform the task or finish a product.

Authentic assessment differs from traditional assessment (Mueller, 2011). The following are the important points of distinctions:

- In traditional assessment that uses standardized criterion-referenced tests, norm-referenced test and teacher-made tests, *curriculum drives assessment*. Assessments to measure knowledge and skills are developed and subsequently, they are administered to measure the extent to which students have acquired knowledge and skills of the curriculum. Thus curriculum is created and teachers in educational institutions deliver curriculum to teach students knowledge and skills. Teachers use traditional assessment techniques such as tests for measuring the extent to which they have gained knowledge and skills. On the other hand, in authentic assessment, *assessment drives the curriculum*.
- Traditional assessment provide students with several choice (such as a, b, c or d, True or False) with a request to select the right answer. On the other hand, authentic assessment ask students to demonstrate their understanding by performing a more complex task.
- In traditional assessment what a student can and will demonstrate is carefully structured by the persons who develops the test. The attention of the students is mainly focused on what is in the test. On the other hand, authentic assessments allows comparatively more student choice and construction in determining what is presented as evidence of efficiency and proficiency. In a nutshell, traditional assessment is teacher-structured whereas authentic assessment is student-structured.
- In traditional assessment evidence regarding students knowledge is indirect because we don't know for surety on what basis the student have selected the right answer out of given several choices. What thinking led the student to pick that answer? We really don't know. On the other hand, in authentic assessment, there is more direct evidence of application and construction of knowledge. For example, when students are asked to write a critique, it provides a more direct evidence of the skill and knowledge than asking students a series of multiple-choice questions.

Recent evidence suggests that during the early 1990s, authentic assessment in educational settings is increasing its popularity (Wiggins, 1993). Today authentic assessment plays a significant role in assessing the performance of students in educational setting.

Summary and Review

- Achievement tests measure achievement, which refers to the degree of learning or ability already attained. Achievement tests are present and past-oriented.
- Achievement tests may be standardized or teacher-made having no standardization. Standardized achievement tests have norms based upon the representative groups of individuals. Standardized achievement tests put more restriction than teacher-made tests.
- There are some obvious uses of achievement tests. Nine such uses have been recognized. One major use of achievement test is in diagnosing children as well as adults having known specific achievement deficits.
- There are four important types of standardized achievement tests : Survey test batteries, single survey test, diagnostic test and prognostic test.
- Achievement test may be norm-referenced test or criterion-referenced test. Norm-referenced achievement tests permit reporting of scores in terms of standard scores, percentile rank etc. Standardized achievement tests are examples of norm-reference test. Criterion-reference achievement tests are those that permit interpretation of scores with reference to specific content mastered by the individual student.
- Achievement test can be constructed and standardized by following some defined steps.
- There are some dark sides of achievement testing and this includes various types of fraud and cheating practiced both by the students as well as by officials.
- Today authentic assessment of achievement is getting more popularity than traditional assessment.

Review Questions

1. Define achievement test. Make distinction between standardized achievement test and teacher-made achievement test.
2. Discuss the major uses or functions of achievement tests.
3. Citing examples discuss the different types of standardized achievement test.
4. Citing example make distinction between norm-referenced achievement test and criterion-referenced achievement test.
5. Outline a plan for constructing and standardizing any specific achievement test.
6. Discuss the dark side of achievement testing.
7. Make a distinction between authentic assessment and traditional assessment of achievement.

■■■

Chapter – 12

Aptitude Tests

Learning objectives
- Meaning and Nature of Aptitude test
- Aptitude test versus Achievement test
- Types of Aptitude tests
  - Specific aptitude test
  - Global aptitude test
  - Multiple aptitude test
- Uses of Aptitude test
- Summary and Review
- Review Questions

Key Terms:

Specific aptitude test, DAT, GATB, ASVAB, MAB, Standard error of estimate, Composite score.

Everybody has some potential to achieve something. In fact, such potential not only helps but also motivates the persons to excel one in some restricted areas. Psychologists have recognized the importance of such potentials in human life. They have conducted various researches to assess these potentials and developed tools (called aptitude tests) for assessing these potentials and developed tools (called aptitude tests) for assessing those potentials. Those aptitude tests that are used in schools range from traditional scholastic aptitude tests to more comprehensive multiple aptitude tests. In the present chapter we shall discuss about the type of such tools and their significance in detail.

Meaning and Nature of Aptitude Test

Aptitude test is a test, which measures aptitude of the person. Aptitude refer to person's (or test taker) potential for learning or ability to perform in a new job or situation (Miller et al. 2013). The Oxford Dictionary of Education (2015) defined aptitude as, "the potential of a learner to assimilate knowledge, skill or understanding successfully." Aptitude is often used in specific areas of knowledge or skills. That is why, it is commonly said that a person has an aptitude for musical work or an aptitude for mechanical job etc. Aptitude tests measure the product of cumulative life experiences or what one has achieved or acquired over time. Infact, aptitude tests tend to determine what maximum can be expected from the test taker. Thus aptitude tests measure the test takers' present performance on selected tasks for providing information that may be used in estimating how the test takers will perform at some time in future or in somewhat in different situation (Thorndike and Thorndike-Christ 2015). In simple words aptitude tests are cognitive measures used to predict success in specific courses. School authorities, business and government agencies often use aptitude tests to predict how well the person will perform or tend to estimate the extent to which an individual would profit from a specified courses of training. Likewise, counsellor also use aptitude tests for help in clarifying career goals.

Major features of aptitude tests may be outlined as under :
- Aptitude tests that are cognitive measures, tend to evaluate the effect of the product of life experiences, that is, known as controlled set of experience.
- Aptitude tests evaluate the test takers' potential to profit from a course of learning or training.
- Aptitude tests heavily depend upon the procedures of predictive criterion validation. Contrary to the popular belief, aptitude tests don't measure a fixed capacity rather they provide some indication of present level of acquired abilities and can be useful in predicting the performance of future (Linn & Miller, 2011). Many tests used for prediction don't make any assumption of subsequent training or are not cognitive measures. Therefore, they are not aptitude tests. Psychologist can use any test for prediction but aptitude is measured only if the test tends to measure a cognitive skill and is used mainly for prediction, selection or placement

Aptitude Test Vs. Achievement Test

Achievement tests measure the test takers' previous learning in a specific academic area. In other words, achievement tests attempt to assess what a person has learnt following some specific course of training or instruction (Kaplan & Saccuzzo, 2002). Aptitude tests measure the test takers' performance on some selected tasks for providing information, which can be used to estimate how the test taker will perform at the sometime in future or in somewhat different situations. Thus aptitude tests evaluate the test takers' potential for learning rather than how much test taker has already learnt.

The similarity between achievement tests and aptitude tests are that both are useful for predicting future achievement and both measure learnt abilities. However, these two types of tests differ on the following points.
- Achievement tests measure those learnt abilities that are directly dependent on specific school experiences whereas aptitude tests measure abilities, which are based upon a wide range of school and out-of-school or uncontrolled set of experiences.

- Achievement test and aptitude test differ in the types of learning measured. Achievement tests measure knowledge of subject matter in various courses such as English, Mathematics, Science etc., as well as basic skills and complex learning outcomes common to many courses. On the hand, aptitude tests measure abstract reasoning abilities, verbal, numerical and general problem solving abilities as well as writing skills similar to those learnt in schools.
- Achievement test depends upon content validation procedures whereas aptitude test depends upon predictive criterion validation procedures. In other words, achievement test is considered valid if it adequately sample the domain of the construct such as geography, history, mathematics, science etc. that are being assessed. The validity of the aptitude test is mainly judged by its ability to predict the future performance.
- Achievement test assesses the product of the courses of instruction or learning whereas aptitude test tends to assess the potential of the test takers to profit from a course of learning or instruction. Therefore, aptitude tests are future-oriented whereas the achievement tests are past and present oriented.

Types of Aptitude Tests

There are different types of aptitude tests. In general, psychologists as well as educators classify aptitude tests into three categories:

- Special or Specific aptitude test
- General or Global aptitude test
- Multiple aptitude test.

Special or specific aptitude tests are those tests, which measure potential for a restricted or single capacity such as clerical aptitude, musical aptitude, mechanical aptitude etc. Such tests are considered particularly useful in making placement decision. Examples of specific aptitude tests are Clinical aptitude test, Mechanical aptitude test, Musical and Artistic aptitude test, etc.

General or Global aptitude tests are those tests that were designed to sample a wide variety of behaviours considered important in many cognitive tasks. Intelligence tests are the example of examples of general or global aptitude tests.

Multiple aptitude tests are those test batteries in which several aptitude (or factors) are assessed at a time. Such tests batteries were introduced in the early 1940s to provide better differential prediction of occasional and training potential. Two major multiple aptitude test batteries are: *Differential Aptitude Test* (DAT) primarily designed for school counselling and *General Aptitude Test Batteries* (GATB) designed primarily for use in the employment situation.

Here we shall discuss some important specific aptitude tests and multiple aptitude test batteries only. Global aptitude test (or intelligence tests) will be discussed in separate chapter later own.

Specific Aptitude Tests :

Specific aptitude tests are used to predict the probability of clinical, mechanical, artistic, musical and creative success. Since a specific aptitude test is less costly and less time consuming then multiple aptitude test batteries, it is continuing in its use even today. Some of the important such specific aptitude tests are being discussed below :

(a) Clerical aptitude tests :

There are several types of clerical aptitude tests. Of these, the most widely used clerical aptitude test is the *Minnesota clerical test*, which is a group test consisting of two parts : *number comparison* and *name comparison*. This test, in fact, is a speed test that is designed to determine the test taker's accuracy when working within a limited time period. The number comparison part consists of 200 pairs of numbers, each of which contains 3 to 12 digits. The name comparison part is similar but uses proper names in lieu of numbers. In test, the test takers are instructed to work rapidly as far as they can without errors.

Another important clerical test is the *General clerical test* which is a speed test having booklet: A (Clerical, Numerical), and B (Verbal) Booklet A has items on alphabetizing, numerical computation, error location, arithmetical reasoning and checking. Booklet B contains items on reading comprehension, vocabulary, grammar, spelling. Booklet A is suitable for assessing clerical aptitude for accounting or for payroll clerk and Booklet B is suitable for assessing clerical aptitude for secretarial jobs.

(b) Mechanical Aptitude tests :

Some experts advocating global approach maintain that mechanical aptitude is a single trait that may differ from person to person. On the other hand, there are experts who maintain that mechanical aptitude consists of a number of relatively independent abilities. Most authorities tend to support the latter view. The important such mechanical aptitude tests are as under :

(a) Mechanical Assembly tests :

In such test the test takers are required to assemble parts and their score are based on the number of parts they can assemble in a given period of time. One of the best known mechanical assembly test is the *Minnesota Assembly test*, which consists of 33 unassembled mechanical objects, which are required to be assembled within a given period of time. The number of correctly assembled parts becomes the score of the testtakers.

(b) Mechanical Reasoning tests :

Mechanical reasoning test measures the mechanical reasoning or comprehension. One of the most widely used such test is the *Bennett Mechanical Comprehension Test* (BMCT). The test consists of pictures of familiar objects and the testtakers may be asked to indicate which of the two pairs carries more weights or which of the two rooms with different amount of furniture would produce more echo and so on. (see figure 12.1)

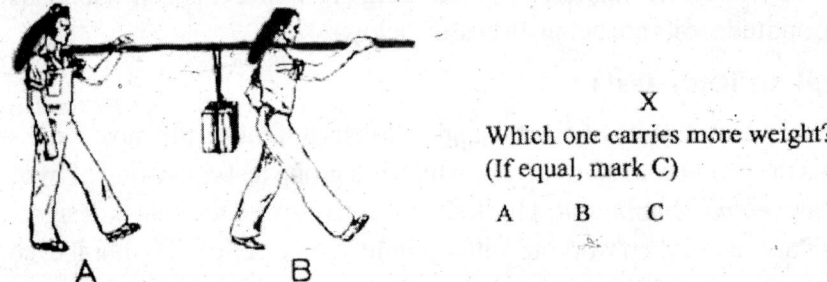

Figure 12.1 Item assessing Mechanical Comprehension Ability.

Thus the test consists of pictures with questions about mechanical principles involved in them. The test is mainly designed for 11 or 12 grade student and industrial employees.

- **Dexterity tests :**

    Some occupations require finger and hand dexterity. As we know, manual dexterity involves multiple skills. That is why one needs a particular type of dexterity test for an automobile mechanic but a different type of dexterity for a watchmaker. The *O'Connor Finger Dexterity test* and *O'Connor Tweezer Dexterity Test* are the two most popular tests and measure how fast the testtakers can insert pins into small holes both by hand and by use of tweezers. These are good and standard measure of finger dexterity and they have proven to be very useful in predicting success among sewing machine operator trainees as well as among dentistry students. Besides, there are other dexterity tests such as *Bennett Hand Tool Dexterity Test* that requires the testtakers to use the tools necessary for a given job and *Stromberg Dexterity Test* that requires testtakers to exhibit the ability to discriminate, sort and place discs as quickly as possible.

- **Spatial Relations tests :**

    Researches have indicated that these tests are very good predictors of success in many school subjects and occupations. One of the most popular space relations test is the *Revised Minnesota Paper Form Board Test*, which consists of several disarranged parts of geometric figure. The testtakers are required to select the correctly assembled figure from among different number of options. Spatial relations abilities are usually involved in engineering, drafting and art.

- **Musical and Artistic aptitude tests :**

    One of the popular musical aptitude test is *Seashore Measures of Musical Talent*. It consists of a phonograph record or tape that presents six aspects of auditory discrimination :

    In Indian loudness, pitch, time, timber, rhythm and tonal memory. All items of this test consists of pairs of stimuli. For example, in measuring pitch, the testtaker must judge whether the second stimulus sound is lower or higher than the first. Likewise, on the measure of tonal memory, the testtaker indicates whether two sequences of notes are different or same. *Wing Standardized*

Tests of Musical Intelligence is one another such aptitude test. This test attempts to measure appreciation as well as discrimination ability of the testtakers. This test has seven parts: pitch discrimination, chord analysis memory for pitch, harmony, intensity, rhythm and phrasing. The Wing test differs from seashore test in the sense that the last four tests of Wing require the testtakers to evaluate aesthetic quality of different chords.

Artistic aptitude test consist of two parts : *aesthetic judgement* and *aesthetic production*. For art critic the first one is necessary and for the artist himself, the second part is important. *Meier Art Judgement Test* is one popular aesthetic judgement test, which consists of several pairs of black and while plates. One member of each pair is masterpiece whereas the other member is the slight modification of the original one. The testtaker's task is to decide which of the two he prefers. Researchers have revealed that the Meier test is capable of discriminating among groups of testtakers that have explicit differences in art training. For assessing aesthetic production, a popular test is the *Horn Art Aptitude Inventory*, which has two subparts : *'Scribble and doodle exercise'* and *'imagery'*. The scribble and doodle section requires the testtakers to sketch several different but familiar objects such as a book, a free, a man etc. Each must be completed in a short time period such as within ten seconds. The imagery section consists of several rectangles containing several lines out of which a representational picture must be drawn.

Several of these traditional special or specific aptitudes especially clerical and mechanical aptitudes have now been incorporated in some of the multiple aptitude test batteries. In Indian conditions some specific aptitude tests have been developed. For example, *Scientific Aptitude test for College students* have been developed by Prof. A.K.P. Sinha and Prof. L.N.K. Sinha. It contains 34 items and measures scientific aptitude of college going students. Likewise, Teaching Aptitude test has been developed and standardized by Prof. S.C. Goakhar and Rajnish. It consists of 50 items and is meant for assessing teaching aptitude among B.Ed. students. Prof. S.S. Dahiya and Prof. L.C. Singh have also developed Teaching Aptitude test for B.Ed. students.

Multiple Aptitude Test Batteries :

Multiple aptitude test batteries were the practical outcomes of the technique of factor analysis. In such batteries the testtakers are tested in several, separate homogeneous aptitude areas. The development of the subtests assessing the separate areas aptitude is governed by the findings of factor analysis.

One of the first multiple aptitude test batteries was developed by Thurstone (1938) called *Primary Mental Abilities Test*, which is a set of seven tests chosen on the basis of technique of factor analysis. However, more recently, four prominent aptitude test batteries have been developed:
- Differential Aptitude Test (DAT)
- General Aptitude Test Battery (GATB)
- Armed Services Vocational Aptitude Battery (ASVAB)
- Multidimensional Aptitude Battery (MAB)

A discussion of each of these three batteries follows :

• Differential Aptitude Test (DAT)

The DAT was first published in 1947 by the Psychological Corporation, New York to act as a guidance battery for use at the secondary school level (8 to 12 grade) and later on revised in 1963, 1972, 1982 and 1992. The DAT is authored by Benett, Seashore and Wesman (1947). The battery consists of eight subtests yielding eight score. However, a ninth score, Scholastic Aptitude, may be obtained from the sum of verbal reasoning and numerical reasoning. The eight subtests are being described below :

(i) Verbal reasoning test :

Analogy type test that intends to assess the ability to reason with words or to understand and use the concepts expressed in words. Each analogy has two missing words-the first word in the first relationship and second word in second relationship. For example :

... is to night as breakfast is to ...

The test measures the ability to infer the relationship between the first pair of words and apply that relationship to the second pair of words. The verbal reasoning test has 25 minutes to answer all its item and has been considered useful in helping to predict success in academic success as well as in many other occupations having jobs involving high levels of authority and responsibility.

(ii) Numerical Reasoning test :

This test consists of numerical problems emphasizing comprehension involved in mathematical reasoning tasks rather than simple computational facility. All items are to be solved within a maximum time limit of 30 minutes. Numerical reasoning is considered important for success in such courses as physics, mathematics, engineering and chemistry.

(iii) Abstract reasoning test :

The maximum time limit for this test is 20 minutes and it is basically a non-verbal measure of reasoning ability. It attempts to measure the extent to which the testtakers can reason with geometric figures or designs. In fact, each item contains a geometric series in which the elements change according to some defined rule. The testtakers have to infer the rule(s) that are contained in the geometrical series. Abstract reasoning is useful in courses that require ability to perceive the relationship among objects in terms of size, shape, position and quantity. Examples of such courses are mathematics, pharmacy, biomedical, computer programming, music, drafting, etc.

(iv) Perceptual speed and Accuracy test :

In this test each item is made up of a number of combinations of symbols, one of which is underlined. The testtaker must mark the same combination on his or her answer sheet. The maximum time limit for answering all items of the test is 6 minutes. All test items don't call for any reasoning skill rather the emphasis is on speed. This test may predict success is certain kinds of routine clerical tasks.

(v) Space Relations test :

The space relations test assesses the ability to perceive a three-dimensional object from a two-dimensional pattern and to visualise how this object would look if rotated in space. Each item or problem shows one pattern, followed by four three-dimensional figures. The testtakers' job is to choose one figure that can be made from this pattern. The maximum time limit needed for such test is 25 minutes. The abilities measured by such test are mostly needed in drafting, architecture, art, clothing design, carpentry and dentistry.

(vi) Mechanical Reasoning test :

This test assesses the ability of the testtakers to understand basic mechanical principles of tools, motion and machinery. Each of the test consists of a pictorially presented mechanical situation followed by a simply worded question. The maximum time limit for answering all items of this test is 25 minutes. Mechanical reasoning aptitude is needed for those who are interested in learning how to repair and operate complex devices. Carpenter, engineer, electrician and machine operators do need such reasoning.

(vii) Spelling test :

This test contains a list of words, some of which are misspelled. The testtaker has to indicate whether each word is spelled correctly or incorrectly. The maximum time limit for answering all the items is 10 minutes.

(viii) Language usage test :

This test contains sentences, which are divided by marks into four subsections: A, B, C and D. The testtaker has to indicate which section contains an error or if there is no error, he must mark E. The maximum time limit for answering all items is 15 minutes.

All subtests of DTA are essentially power tests with the exception of perceptual speed and Accuracy test. Total testing time for the battery is about 3 hours (2.5 hours working time plus 30 minutes for directions and administrative works). Percentile ranks, stamines and scaled scored are available for each grade. It would be worth to clarify and mention that a *ninth score* called *scholastic aptitude* is obtained by summing the score on Verbal reasoning and Numerical reasoning.

The importance of the DAT can be easily judged from the fact that it has been translated into several languages and is widely used in Europe for vocational guidance and research applications (Gregory, 2004) DAT can also be administered in computerized version. A computerized Adaptive Testing (CAT) from DAT has been made available for such version. DAT has satisfactory degree of reliabilities and validities. DAT has also been adapted by Indian psychologists to suit local requirements.

In constructing DAT, Benett, seashore and Weisman (1984) have been guided by some explicit criteria as under :

- Each subtest of DAT should be an independent test
- DAT should yield a profile. It means that all eight separate scores can be converted to percentile ranks and could be plotted on a common profile chart.

- Tests should measure power, that is, solving difficult problems with adequate time, is of prime concern.
- Norms of DAT should be adequate and derived from larger population. That is the reason why norms have been derived from 1,00,000 students.
- Tests should be easy to administer. Each subtest has excellent 'warm up' examples and can be administered by a person with a minimum of training.
- Alternate forms should be available because such availability would tend to reduce any practice effects.
- Test materials should be practical. Thus with time limit of 6 to 30 minutes per test, DAT can be easily administered either in the morning and afternoon session of the school.

General Aptitude Test Battery (GATB)

GATB was constructed and developed by Bureau of Employment Security of the U.S.A. Department of Labour in the early 1940s from a factor analysis of 59 tests administered to thousands of male trainees in vocational courses. Analysis of those 59 different tests prepared for specific jobs revealed that there was a great deal of overlap in most of them and that only 10 different factors were measured by the complete set of tests. In fact the GATB was designed to provide measures of these different factors. In its most latest from, GATB include 12 tests and gives a scores on nine different factors. Out of 12 tests, there are eight paper-and-pencil tests and four simple apparatus measures. The entire battery takes about two and half hours and is most appropriate for high school seniors and adults. These 12 tests yield following nine factor scores :

(i) General mental learning ability (Intelligence) (G) :
The factor of general mental ability (also called intelligence) is a composite of scores resulting from three tests, namely vocabulary, Arithmetic Reasoning and Three-dimensional space.

(ii) Verbal aptitude (V) :
This aptitude score is one test, (Number 4) vocabulary test. It requires the testtakers to identify pairs of words in a set of four that are either synonyms or antonyms. For example :
a. hasten b. deprive c. expedite d. disprove

(iii) Numerical ability (N) :
This aptitude score is the composite of both the computation test (Number 2) and Arithmetic reasoning test (Number 6). For example :
Computational test :
Subtract :   6003   Multiply :   60265
              − 995   × 203

(iv) Arithmetic reasoning test :
Mohan works for Rs. 40 per hour. How much is his pay for 36 hour a week?

(v) Spatial aptitude (S) :
This aptitude score is derived from three-dimensional space test (Number 3). The testtaker

has to indicate which of four three-dimensional figures can be produced by folding a flat sheet of a specified shape, with creases at the point indicated.

(vi) Form Perception (P):

This aptitude score is derived from rapid and accurate perception of visual forms and patterns. This score is the composite of two tests: Tool Matching (Number 5) and Form Matching (Number 7). Each test requires the testtakers to find from among a set of answer choices any one that is identical with the stimulus form.

(vii) Clerical Perception (Q):

This aptitude score is the result of name comparison test (Number 1) and it involves rapid and accurate perception of the stimulus materials, which are linguistic instead of purely figural. The test presents pairs of names and requires the testtakers to indicate whether two members of the pair differ or identical. For example

        Indira Gandhi – Indra Gandhi
        Loknayak Tilak – Loknyak Tilaka

(viii) Motor coordination (K):

This aptitude score is the outcome of Mark making test (Number 8). The testtaker is, here, required to make three pencil marks as quickly as possible within each of a series of boxes on the answer sheet. The score of the testtaker is the number of boxes correctly filled with 1 minute test period.

(ix) Manual Dexterity (M):

This aptitude score is the composite of two pegboard tests: *Place test* (Number 9) and *Turn test* (Number 10). Thus this score is largely the outcome of speed and accuracy of fairly gross hand movement. In the place test, the testtaker is required to use both hands to move a series of pages from one set of holes in a pegboard to another. In Turn test, he is required to use their preferred hand to pick up a peg up from the pegboard and then, rotate it 180° and subsequently reinsert the other end of the peg in the hole. Three trials are given for each of these tests and test taker's score becomes the total number of pegs moved or turned.

(x) Finger Dexterity (F):

Finger dexterity aptitude score is the composite of two tests: Assemble test (Number 11) and Disassemble test (Number 12). Both tests use the same pieces of equipment that is a board with 50 holes in each of two sections. Each hole in one section is occupied by a small rivet. A stack of washers is piled on the spindle. In Assemble test, the test taker picks up a rivet with one hand and a washer with other hand, puts the washer on the rivet and subsequently, places the assembly in the corresponding hole in the unoccupied part of the board. Here within the period of 90 seconds, he is required to place as many rivets and washers as possible. In disassemble test, he is required to remove the assembly and return the washer to its stack and rivet to its original place. The score is the number of items assembled or disassembled.

These nine factor scores on the GATB are expressed as standard scored having mean of 100 and a standard deviation of 20. There are also three important composite measures derived from various factors. For example, G + V + N = cognitive measures, S + P + Q = perceptual measures and K + F + M = Psychomotor measures are the three measures. The alternate forms reliability coefficients for factor scores range from 0.80s to the 0.90s. The average validity coefficient of the GATB in terms of its correlation with relevant interior measures is 0.62 (Gregory, 2004).

GATB versus DAT

The basic similarities between the two most popular aptitude tests are that they are batteries of tests. Both yield several types of aptitude scores. Besides, the researchers have shown that quite substantial correlations exist between corresponding factors of DAT and GATB. However, they differ on the following points :
- DAT tests are in most cases purely power tests whereas GATB tests are quite highly speed tests.
- DAT incorporate tests of mechanical comprehension and language whereas these tests are not included by GATB. On the other hand, GATB incorporates form perception test and several types of motor tests, which are, in fact, missing in the DAT.
- GATB is more work-oriented whereas DAT is more school-oriented in its total coverage.

Armed Services Vocational Aptitude Battery (ASVAB)

ASVAB has been developed jointly for use in all USA armed services. This battery is used by the Armed Services to screen potential recruits as well as for assigning personnel to different jobs and training programmes. In other words, this battery is primarily administered to those high school students who are interested in military occupations as well as to those who have applied to enter military services. The ASVAB has ten subtests as under :
- General Science (GS) : 25-item of general knowledge in physical and biological science.
- Arithmetic Reasoning (AR) : 30-item test of arithmetical word problems based upon simple calculation.
- Word knowledge (WK) : 35-item test of vocabulary knowledge and synonyms.
- Paragraph comprehension (PC) : 15-item test of reading comprehension in short paragraphs.
- Electronic Information (EI) : 20-item test of ratio, electronics and electrical principles.
- Coding Speed (CS) : 84-item test of substitution of numeric for verbal codes within a very short period of time.
- Mathematics knowledge (MK) : 25-item test of geometry, algebra, fractions, decimals and exponents.
- Auto and shop information (AS) : 25-item test of basic knowledge of shop practices, autos and tool usage.
- Numerical operations (NO) : 50-item speed test of ability to do simple arithmetical

calculations like addition, subtraction, multiplication, division, etc.
- Mechanical Comprehension (MC) : 25-item test asking questions about mechanical and physical principles.

Out of these ten subtests, eight are power tests having adequate time limits for most subjects whereas two subtests (No and CS) are speed tests that gives emphasis upon rapid performance. A composite score consisting of AR + MK + PC + WK becomes what has been named as the Armed Forces Qualification Test (AFQT), which is used buy all armed services as a measure of general trainability in screening of potential recruits. Besides this, each of the armed services in US combines various subtests for forming composites for various classification needs and personnel selection. For example, the Army's combat composite consists of AR + CS + AS + MC.

One important feature of ASVAB is that a general factor that accounts for about 60 percent of total ASVAB variance and four group factors have been repeatedly found (Welsh et al., 1990). The four group factors are :

Verbal factor includes WK and PC
Technical factor includes AS, MC and EI
Speed factor includes No and CS
Quantitative factor includes AR and MK

The alternate-form and test-retest reliability of ASVAB are satisfactory because that ranges from 0.70 to 0.90 with one exception of PC having reliability coefficient of 0.50. The validities of ASVAB have been calculated against a variety of job and educational performance. However, validity coefficient differ substantially depending upon the type and number of criteria employed. Standard scores for ASVAB are based upon norms obtained from a representative sample of American youths.

Multidimensional Aptitude Battery (MAB)

MAB was first published in 1984 and later on revised in 1994 (Jackson, 1994). MAB is a group test designed to measure those aptitudes which are assessed by Wechsler Adult Intelligence Scale-Revised (WAIS-R). Much on the pattern of WAIS-R, MAB has five verbal scales and five performance scales as under :

Verbal Scales	Performance Scales
Information	Digit symbol
Comprehension	Picture Completion
Arithmetic	Spatial
Similarities	Picture arrangement
Vocabulary	Object assembly

These ten subtests have the same name as in the corresponding WAIS-R subtest. Only one exception is the Block Design Test of WAIS-R that has been replaced by Spatial subtest in the MAB.

MAB also yields Verbal IQ, Performance IQ and Full scale IQ. This battery is found suitable for adolescents as well as adults but it is not recommended for mentally retarded or disturbed person. The five verbal subtests have been presented in one booklet and five nonverbal or performance subtests have been presented in other booklet. Each booklet begins with some practice problems and each subtest begins with one, two or three further demonstration items. The raw scores on each of the 10 subtests are transformed to uniform scaled score equivalents, that is, standard scores with Mean = 50 and Standard deviation = 10.

Other aptitude test batteries include the Flangan Aptitude Classification Test (FACT), the Guilford-Zimmerman Aptitude Survey, and Academic Promise Tests, which are not being discussed.

Uses of Aptitude tests

Aptitude tests are used for many purposes. Some of the important uses are being discussed below :

- **Aptitude tests are used for selection purposes :**

Aptitude tests are frequently used for selection purposes. *The composite method* and the *cut-off method* are often used for selection decisions. In composite method the scores from a number of tests are combined to maximize their correlation with a criterion. The scores of the tests are combined by assigning weights that tend to maximize the contribution of each score in predicting the criterion. The regression formulas used to assign these weights maximize the correlation between the composite test scores and criterion measures. In composite method each testtaker is given a single value, which represents a weighted composite score from all measures included in the battery. Subsequently, these scores are ranked and the persons, as the need may be, selected.

In cutoff method, the scores for every testtaker is kept separate for each subtest that correlates with the criterion. The researcher, then, determines the minimum passing scores for each predictor test. Only those testtakers are selected who pass every test irrespective of what their total score is.

Which of the two methods is preferable? This decision depends on several factors. Less effort and arithmetical sophistication are required in cutoff method. But in this method compensatory behaviour is disregarded because this method assumes that the testtaker who fails one subtest will not be successful no matter how well he performs on the remaining subtests. Such compensatory behaviour may be highly justifiable in some cases such as airline pilots who must have good vision, good spatial abilities, immune to air sickness and be successful in flying a plane. A deficiency in any one of these areas would justify the elimination of that person as a pilot. But there are many such areas where it is possible to compensate for a deficiency on one subtest by increased proficiency on another. A college might retain a faculty member who is known to encourage a lively discussion with students although he has not a good relationship with most of the students or he is too aggressive in his behaviour. If a composite method is used, a person can be selected if he can compensate for his low scores in some areas by high scores in other area.

- **Aptitude tests are used for feedback :**

    Aptitude tests are also used by teachers and school counsellors for providing feedback to students regarding their potential in various fields such as mechanics, art, medicine, numerical abilities, automechanics, etc. This is again done in two ways. First, expectancy tables can be used to inform students of their probability of getting success in different areas or programmes. Second, the help of regression equation can be taken for predicting the most likely score a student is likely to receive on the criterion. While predicting using regression formula, the amount of error in prediction can be estimated by using formula for the *standard error of estimate*. If the value of standard error of estimate is larger, it means that the scores are spread out on either side of the regression line. If the standard error of estimate becomes zero, all actual scores will equal the predicted scores and the predictive validity will be perfect.

- **Aptitude tests are also used for diagnostic purpose :**

    Aptitude tests have been frequently used in the diagnosis of the problems of the students. For example, if a teacher has apprehension about difficulties in solving numerical problems by the students, he may use Numerical Ability Test for diagnosing those problems. Likewise, teachers often use Reading readiness test to know about which children will probably experience difficulties in learning specific subjects such as reading, English.

- **Aptitude tests are also used for placement decision :**

    Aptitude tests are also used for placement purposes. Placement decision assumes that individuals have been selected for the institution or organisation and that some decision must be made regarding their best and optimum deployment. Constructing good placement tests is more difficult than constructing selection tests. Therefore, the teacher or counsellor has to take special precautions when tests are used for placement. Infact, placement decision is the joint responsibility of the organisation and the individual and this is eased when done with the help of such placement tests, which have sufficient degree of predictive validity.

Summary and Review

- Aptitude tests are the cognitive measures that evaluate the testtakers' potential to profit from the course of learning or training. Such tests assess the effect of product of life experiences, that is, known or controlled set of experiences.
- There are obvious differences between achievement tests and aptitude tests. Five point of differences have been located.
- There are three types of aptitude tests : Special or Specific aptitude tests, General or Global aptitude tests and Multiple aptitude test battery.
- Specific tests are those tests, which are designed to sample a wide variety of behaviours considered important in many cognitive tasks. Intelligence tests are the best examples.
- Multiple aptitude test batteries are those tests, which are designed to tape several aptitude

(or factors) at a time. Differential Aptitude Test (DAT) and General Aptitude Test Battery (GATB) Armed services vocational Aptitude Batteries (ASVAB) and Multidimensional Aptitude Batteries (HAB) are the best examples of multiple aptitude test batteries.
- There are several uses of aptitude tests such as aptitude tests are used for selection purposes, for providing feedback, for diagnostic purposes as well as for placement decisions.

Review Questions

1. Define aptitude test. Make distinction between aptitude test and achievement test.
2. Discuss the major types of specific aptitude test.
3. Make distinction between specific aptitude test and multiple aptitude test batteries. Discuss any two types of multiple aptitude test batteries.
4. Discuss the major features of DAT.
5. What are the major features of GATB. How does it differ from DAT.
6. Write short notes on the following :
    (a) DAT
    (b) GATB
    (c) ASVAB
    (d) MAB.

Chapter – 13

Theories and Measurement of Intelligence

Learning objectives
- Definitions of Intelligence
- Theories of Intelligence
  - Biret theory
  - Spearman's g-factor theory
  - Thurstone's theory of Primary abilities
  - Guilford's structure-of-intelligence Model
  - Jensen's theory
  - Wechsler's theory
  - Cattell-Horn G_f-G_e theory
  - Carroll's Three Stratum theory
  - Cattell-Horn-Carroll (CHC) theory
  - Sternberg's Triarchic theory
  - Gardner's theory
  - DAS-Naglieri PASS Model
- Measurement of Intelligence
  - Stanford-Binet Scale
  - Wechsler Scales
  - Kaufman Scales
  - Das-Maylieri cognitive Assessment System
  - Raven's Progressive Matrices
  - Goodenangh Draw-A-Man Test
  - Test of Non-verbal Intelligence (TONI)
  - Universal Non-verbal Intelligence Test (UNIT)
  - Cognitive Abilities Test (CogAT)
  - Culture Fair Intelligence Test (CFIT)
  - Otis-Lenn-School Ability Test (OLSAT)
- Important Indian Intelligence Tests in brief historical perspectives

Key Terms:
Age equivalent, Age differentition, General mental ability, g-factor, s-factor, Primary

Mental Abilities, Group-factor theory, Structure-of-intelligence, Associative Intelligence, Abstract intelligence, Fluid intelligence, Crystallized Intelligence, Componential intelligence, Experential Intelligence, Context Intelligence, Verbal IQ, Performance IQ, Full scale IQ, General Intellectual ability, Standard Age score, Power Test, Speed Test, Cross-battery assessment.

Intelligence is one of the highly research topics in psychology and education. Thousands of research articles are published every year on this topic and new journals such as *Intelligence* and the *Journal of Psychoeducational Assessment* are wholly devoted to this field. The present chapter aims at providing a thorough discussion on definition, theories and measurement of intelligence. Such discussion will clarify many of confusions commonly found in the minds of the students. Besides, this will also clarify to what extent the psychologists and educators have been able to unfold the realities regarding the concept of intelligence.

Definition of Intelligence

Basically, there are two ways of defining any concept. Sternberg (1986) has made a distinction between 'operational' and 'real' definition. An operational definition is one that defines a concept in terms of the way it is measured. Boring (1923) has defined intelligence as "what the intelligence tests test". This is an example of operational definition of intelligence.

The operational definitions of intelligence have two major limitations. *First*, such definitions are circular in nature. Intelligence tests were invented to assess intelligence and not to define it. Test developers never intended that intelligence tests would define intelligence. *Second*, operational definitions block progress in understanding the nature of intelligence probably because they more or less close any scope of discussion of adequacy of theories of intelligence.

A real definition is the experts' viewpoints, which tell us about the nature of thing being defined. Let us now concentrate upon how some of the prominent experts in the field have defined intelligence : **Spearman** (1904, 1923) Intelligence is a general ability that involves mainly the education of relations and correlates.

Binet and Simon (1905) : Intelligence is the ability to judge well, to understand well and to reason well.

Terman (1916) : Intelligence refers to the tendency to take and maintain a definite direction, capacity to make adaptations for the purpose of attaining a desired end and the power of autocraticism.

Thurstone (1938) : Intelligence refers to the capacity to inhibit instinctive adjustments, flexibly imagine different responses and realized modified instructive adjustments into overt behaviour."

Wechsler (1939) : Intelligence refers to aggregate or global capacity of the individual to act purposefully, to think rationally and to deal effectively with environment."

Piaget (1972) : Intelligence is a generic term to indicate the superior forms of organisation or equilibrium of cognitive structuring used for adaptation to the physical and social environment."

Sternberg (1985, 1986) : Intelligence refers to mental capacity to automatize information processing and to emit contextually appropriate behaviour in response to novelty. Intelligence also includes metacomponents, performance components and knowledge-acquisition components.

Gardner (1983) : Intelligence refers to the ability or skill to solve problems or to fashion products that are valued within one or more cultural settings.

Baron (2002) : Intelligence refers to individuals' abilities to learn from experience, to engage in various forms of reasoning, to overcome obstacles by careful thought.

The above views regarding the meaning of intelligence may be representative but not exhaustive. These views are primarily Western. However, several other views of eastern experts and cross-cultural experts emphasize different types of components. For example, eastern experts emphasize much upon doing right things, benevolence, humility etc., as the primary components of intelligence. Likewise, Africans emphasize much upon maintaining harmonious and stable intergroup relations as the primary components of intelligence. Thus we find a diversity of viewpoints among experts regarding the meaning of intelligence.

Despite this diversity, a close scruity of above definitions and the viewpoints expressed by other cross-cultural experts too, we may come to the conclusion that experts broadly agree upon three things in explaining the meaning of intelligence :

(a) Intelligence is the capacity to profit or learn by past experience.
(b) Intelligence is the capacity to adapt to one's environment.
(c) Intelligence is to reason logically, plan effectively and inter perceptively.

Thus intelligence is a multifaceted capacity, which manifests in different ways across the life span.

Theories of Intelligence

In his famous book *The Republic* Plato had pointed out that the people should be given places in society based on their cognitive abilities. By the late 19th century, one of British philosophers, that is, Herbert Spencer, pointed out that general cognitive ability which he called as *intelligence*, is the most important characteristic of human being and is an excellent basis for natural selection. Towards the dawn of the 20th century, many practitioners of psychology became interested in applying this method for solving human problems of social significance.

Sir, Francis Galton, J.M. Cattell and Alfred Binet were the early most important such early practitioners. Here we shall briefly review the major theoretical models provided by such practitioners.

- **Sensory Keenness theory of Galton and Cattell :**

Sensory Keenness theory proposed by Sir Francis Galton and his disciple James McKeen Cattell was the first theory of intelligence derived in the Brass Instrument era of Psychology. In fact, Galton is remembered as being the first person to publish on heritability of intelligence, providing a ground for the contemporary nature-nurture debate. He was of view that the most intelligent persons are those who are equipped with best sensory abilities. Galton and Cattell thought that the intelligence was governed by keen sensory abilities. They had made an assumption that the information regarding external world pass through the avenues of our senses and if our senses are perceptive, our intelligence and judgement tend to cover the larger field. We find some vestiges of this viewpoint in modern chronometric analyses of intelligence such as Reaction Time-Movement Time (RT-MT) apparatus, an instrument much liked by Jenson (1980) for the cultural-reduced study of intelligence. In RT-MT studies the subject's reaction time and movement time is measured with the help of Reaction-Time-Movement Time apparatus. Jensen (1980) had reported that the indices of RT and MT correlated as high as 0.50 with the scores on the psychometric test of intelligence. Later on, Vernon (1993) had also reported a correlation as high as 0.70 between speed-of-processing RT type measures and traditional measures of intelligence. All these findings suggest that RT might be a useful addition to any standardized intelligence test batteries. However, most of the psychologists have disliked such line of research.

- **Binet's theory :**

In the beginning of 20th century there was an international movement to separate those children who could not learn due to poor ability and those children who could not learn due to any causes including poor motivation. As a result of this, Alfred Binet and Theodorm Simon together developed a test in 1905 in France for the said purpose.

Binet did not, in reality, developed any theory of cognitive ability. His work was entirely empirical and did not arise out of any theory. His viewpoint was that the cognitive ability, like physical ability, develops with age and the intelligence is expressed in terms of complex mental acts. Therefore, the intelligence scale that he developed in 1905, was basically composed of a series of mental puzzles organised in one sequence in which the children could answer them successfully. The second edition of this scale had been released in 1908. The primary objective of Binet's scale was to identify the mental level of the child or speaking in modern terminology, *age equivalent*. He was not interested in providing normative comparison but in providing what each child could actually do. In developing the scale, he was guided by two major concepts, which also underlie even today in most of the theories of intelligence. These two concepts were : *age differentiation* and *general mental ability*. These two concepts acted as the major bases of subsequent intelligence tests.

Age differentiation simply means that older children and younger children can be differentiated by their capabilities. For example, a child of eight-year old can be distinguished from a child of four-year old because the former can do simple arithmetical work and can make distinction between 50-rupee currency and 10-rupee currency whereas the latter probably cannot. In utilizing the principle of age differentiation, Binet prepared such tasks for each age level that could be completed by about 66% to 75% of the children of the given age level and also by a smaller percentage of young children as well as by a larger percentage of older children. It means that the nature of the task prepared by Binet for each age level was such that the increasing percentage of children could complete as a function of increase in age. On the basis of such task, Binet was able to estimate the mental ability of a child in terms of his completion of the tasks designed for the average child of a given age irrespective of the real or chronological age of the child. Thus a six-year old child may be able to complete the task meant for the average nine-year old child. Likewise, a six-year old child night not be able to complete that task, which is meant for the average of four-year child. Thus with the help of principle of age differentiation, one could determine the equivalent age capabilities of a child quite independent of his chronological age. Such capability was later come to be known as *mental age*.

General mental ability (GMA) is another principle emphasized by Binet. GMA is best understood as the product of various separate and distinct elements of intelligence. His decision to measure GMA was based on some practical consideration. As a consequence of this attempt, he could now search for tasks to anything related to the total or final product of intelligence. The concept of general mental ability has proved to be an important concept or principle for modern testing of human intelligence.

- **Spearman's Two factor theory :**

Charles Spearman, a British psychologist and statistician, proposed this theory in 1904. Spearman found that all of the correlations between measures of cognitive abilities and academic achievements based upon a large unbiased samples of population were *positive*. This phenomenon was called as *positive manifold*. This led him to conclude that there was a single global mental ability and he termed this global mental ability as general intelligence (or g-factor) and clearly argued that its presence in various measures was the primary cause of positive correlations between them. Later on, he revised this g-factor theory in 1927 and added one more factor called specific factor (or s-factor) to the intelligence. This is called as *two-factor theory* of intelligence. According to this theory, the person's performance on any homogeneous test or subtest of intelligence, is determined by two factors : g, the pervasive of factor and s, a factor specific to that test or subtest. Thus the reliable individual differences in a set of test scores could be explained by the presence of individual differences in g factor and s factor. The reminder was the measurement error, which could be minimized by using highly reliable instrument or test. According to this theory, some ability tests are highly loaded with g-factor whereas other tests are much representative of specific factor. Two tests that are heavily loaded with g factor could correlate highly whereas the tests which are not saturated by with g-factor, should show minimum correlation with one another. In Figure 12.1 the basis of correlation among tests according to two-factor theory has been illustrated. In this Figure, there are three tests out of which Test 2 and 3 would correlate highly with each other because each is highly saturated with g-factor as shown by the shaded areas. The white area in each test indicates specific and error factors. Test 1 would have very low correlations with Test 2 and Test 3 because it contains very little g. To understand how a single general

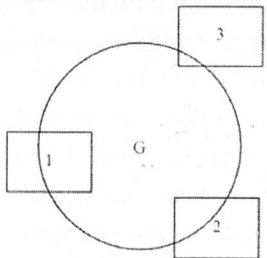

Figure 13.1 Two factor Theory based on Correlation Model.

factor underlie all intelligent behaviour, the analogy of main electric power box for the different rooms of the home may be cited. The same main electric box provides power for lights of all sizes of the different rooms. Although some lights may be brighter than others, all depend upon the power supplied from the main power box. In this way, g-factor was thought of as mental engine that emergized behaviour. The notion of g-factor was supported by Spearman through the statistical technique of factor analysis, which is basically a method of reducing a set of variables or scores to a small number of hypothetical variables called *factors*. Through this statistical technique, it is possible to show how much variance a set of test scores has in common. This common variance represents what is called g-factor.

Spearman's two-factor theory ran into trouble by the discovery of group factors. Some experts observed at that time that dissimilar tests could have correlations higher than values predicted from their respective of leadings. Such observation led to the possibility that a group of these diverse measures might be having something in common, that is, a unitary ability other than g factor. For example, several dissimilar tests might share a common unitary memorization factor that may be considered as a halfway between pure g factor and various s factors unique to each test. This indicates the existence of group factor, which is incompatible with Spearman's g factor and s factor.

- **Thurstone's theory of Mental Abilities :**

In the United State, Spearman's theory was being viewed as too simplistic. Many experts such as Thorndike and Thurstone held the view that Spearman had oversimplified the concept of intelligence. They began to view intelligence as consisting of numerous separate abilities that operate more or less independently. According to this multi-factor viewpoints, a person might be high on some components of intelligence but low on others or the vice versa. The lead of such viewpoints was taken by L.L Thurstone of the University of Chicago. In formulating theories Spearman was much impressed by the fact that the scores on different mental tasks were correlated where as Thurstone was much influenced by the fact that the correlations were far from being perfect. Therefore, Thurstone concluded that mental performance depends not on general factor but rather on several distinct group factors, which he designated as Primary Mental Abilities (PMA). On the basis of his factor analytic studies, he along with his students, proposed approximately a dozen different factors. Out of these dozen factors (or abilities) only seven have been frequently corroborated (Thurstone, 1938; Thurstone & Thurstone, 1945). These seven primary mental abilities are as under:

- *Verbal comprehension* (V) : It is measured by vocabulary tests and the principal factors in such tests are reading, comprehension, verbal reasoning, verbal analogies, proverb matching, etc.
- *Word Fluency* (W) : Measured by such tests as anagrams, rhyming, or naming words in a given category.
- *Numerical ability* (N) : Measured by tests identified with speed and accuracy of simple arithmetical calculations.
- *Associative Memory* (M) : Measured by tests requiring rote memory for paired associates.
- *Spatial Relatins (S)* : Ability to visualize and mentally rotate abstract figures in two or three dimensional space.
- *Perceptual speed* (P) : Measured by tests requiring quick and accurate grasping of visual details, differences and similarities.
- *Inducting Reasoning or General Reasoning* (I or R) : Measured by tests requiring the respondent to find a rule as we find in case of number series completion test.

Thurstone (1938) and his wife Thelma G. Thurstone had developed the Primary Mental Abilities Test consisting seven separate subtests and each test was designed to measure one primary mental ability. However, later he admitted that these seven primary mental abilities moderately correlated with each other, providing an inference about the existence of one or more second-order factors. At last, Thurstone accepted the existence of g-factor as a higher- order factor. By this time, Spearman had also accepted the existence of group factors and thus by now, the difference between Spearman and Thurstone in theoretical formulation of intelligence was largely a matter of emphasis (Gregory, 2004).

Vernon (1950), a British Psychologist, tried to provide a reapproachment between the viewpoints of Spearman and Thurstone by proposing a *hierarchical model of group factor theory*. Vernon's model has been supported by Burt (1949) and also by Humphreys (1962) in United States. According to Vernon's model, at the top of hierarchy is Spearman's g factor. At the next level, there are two major group factors such as *Verbal-Educational* (V:Ed.) and *Practical-Mechanical* (K:M). These two major group factors have been further subdivided into minor group factors. For example, V:Ed factor is subdivided into verbal and numerical subfactor. Likewise, the practical-mechanical factor is further subdivided into such sub-factors as mechanical-information, spatial and psychomotor abilities. These minor group factors are further divided into *specific factors* which, thus, occupied the bottom of hierarchy.

- **Guilford's Structure-of-Intellect model :**

    Following Thurstone's lead some other investigators have reported many specific factors. One such prominent theorist is Guilford (1967, 1988) who continued the search for factors of intelligence originally initiated by Thurstone. On the basis of about two decades of researches, Guilford (1967, 1988) proposed a boxlike model, which he called as *structure-of-intellect* (or SI) model. In this model, he has classified intellectual abilities along three dimensions :

    - *Operations* : By operations Guilford meant the kind of intellectual operation required by the test. In other words, operations are what the testtaker does. These include six types: *memory recording, cognition, memory retention, divergent production, convergentproduction* and *evaluation.*
    - *Contents* : By contents is meant the nature of materials or information upon which operations are performed by the testtakers. These includes five types of activities: visual, auditory, symbolic, semantic and behavioural.
    - *Products* : By products is meant the different kinds of mental structures that the testtaker must produce to derive a correct answer. Products are classified into six categories such as units, classes, relations, systems, transformations and implications.

    In total, then, Guilford (1988) identified six types of operations, five types of contents and six types of products, which produced $6 \times 5 \times 6 = 180$ factors of intellect. Originally, Guilford (1967) had proposal that there were 120 different intellectual abilities depending upon the combination of operations, contents and products. However, shortly before his death, that is in 1988, he expanded his theory from 120 to 180 abilities. Each combination of an operation (say divergent production), a content (say symbolic) and a product (say, unit) represents a different factor of intellect and Guilford claimed to have verified more than 100 of these factors in his research.

    This model has little impact on the development and application of the tests for the use of general purposes. According to Carroll (1993), the reanalysis of Guilford's original factor analytic data have shown that other models fitted the data better than the structure-of-Intellect model and were also more consistent with theoretical and practical considerations.

- **Jensen's theory :**

    Jenson's theory of intelligence is more or less directly derived from Spearman and Thurstone's viewpoints. His theory is based on studies of reaction time required to perform acts of varying complexity and the relationship of such measures to the general factor found in the battery of diverse tests. As we know, in 1880s, Sir Francis Galton had predicted that reaction time in performing simple acts tended to measure a person's intelligence. Beginning in the 1970s, a series of studies conducted by Jensen (1982), it was reported that reaction time measures primarily correlate with what he called Spearman's g in the intelligence tests. On the basis of such results, he has concluded that there is a biological basis for individual differences in intelligence and can be easily described as differences in the efficiency of the brain for processing information (Jensen, 1991, 1993). In fact, Jenson has proposed intelligence as a two-level process. At one level is the *associative intelligence* and at the other level is the *abstract intelligence*. Associative intelligence includes those kinds of tests, which depend upon simple verbal associations and memory. Such intelligence includes factors like verbal associations, spatial positions, stimulus-learning as well as memory for temporal sequences. Associative intelligence basically relates to the biological maturation and therefore, shows little variation in social classes and races. On the other hand, abstract intelligence includes various factors like thinking, problem-solving skills, concept learning, principle learning, multiple discrimination, etc. Abstract intelligence is based upon education and culture and therefore, is responsible for the observed differences in the abilities among different social classes or cultures.

- ## Wechsler's theory :

Wechsler (1939) who was working as a clinical psychologist at Bellevue Hospital in New York, developed an intelligence test for assessing the intelligence of adult persons. This test was named as *Wechsler-Bellevue Adult Intelligence Scale* (WBAIS). This test was really a collection of subtests, most of which had been used as group tests of Alpha and Beta batteries used in military testing programmes during World War I. Infact, Wechsler had adapted them for individual testing and administration and divided them into two scales: One scale consisted of the Form Alpha Tests, which required the use of language and the other scale was Form Beta, which was language-free. The tests from Form Alpha were named as verbal scale and its score reflected Verbal Intelligence quotient (VIQ). Likewise, the Form Beta tests produced the performance scale and its score led to what is called *Performance Intelligence Quotient* (PIQ). The sum of the two scales gave full-scale intelligence quotient (FSIQ). According to Wechsler's theory, FSIQ was equivalent to Spearman's g factor. However, Boake (2002) has traced the evolution of WBAIS from its precursors to all those tests that were in use before World War I and had clearly printed out that Wechsler's contribution was simply to combine already existing tests into a battery rather than discovering any new way of assessing cognitive ability.

Wechsler (1944) defined intelligence as, "the aggregate or global capacity of the individual to act purposefully, to think rationally and to deal effectively with the environment." Since his theory was driven by the clinical practice, the patterns in the scores were believed to have some clinical significance and could be used for diagnosis of various clinical disorders. Today Wechsler's theory which is embodied in the use of three separate tests such as *Wechsler Adult Intelligence Scale* (WAIS), *Wechsler Intelligence Scale for Children* (WISC) and *Wechsler Preschool and Primary Scale of Intelligence* (WPPSI) has given rise to thousands of research studies. However, Wechsler's theory is a clinical theory, which acts as a guide to the most informative use of test scores rather than a theory about the scientific nature of intelligence.

- ## Cattell-Horn G_f-G_c theory :

In 1943 Raymond Cattell first proposed the division of cognitive ability into two broad classes. Then the theory slept for about 20 years while Cattell focused his attention more on the study of personality. Then, in 1966, John Horn joined hand with Cattell and his study of intelligence was resumed. The theory postulated the existence of two major types of cognitive abilities : *fluid intelligence* (symbolized G_f), *Crystallized intelligence* (symbolized G_c). In fact, Cattell and Horn broke down Spearman's general intelligence into two distinct but related subtypes of g (with a correlation of about 0.50.) (see Figure 13.2)

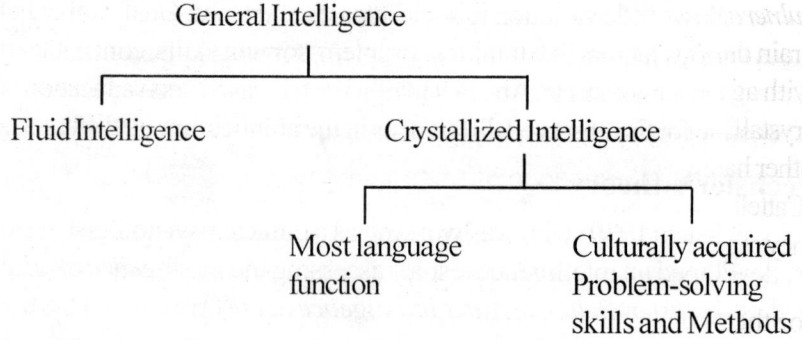

Figure : 13.2 : Fluid and Crystallized Intelligence

According to this theory, fluid intelligence is defined as the ability to deal with the new problem solving situations for which past experience does not provide a solution. The abilities that make up fluid intelligence are nonverbal relatively culture-free, and something independent of specific instruction. Fluid intelligence requires the abilities to reason abstractly, think logically and manage information in short-term memory so that new problems can be easily solved.

Crystallized intelligence may be defined as the ability to apply acquired knowledge to the solution of current problems. Such ability is dependent upon the exposure to a particular culture as well as upon formal and informal education. Long-term memory strongly contributes to crystallized intelligence through the years. Horn (1968, 1991, 1994) proposed the addition of seven factors. Most of these factors have been identified by factor analysis of batteries of tests. Those seven factors are as under :

G_q (quantitative ability) : ability to understand and manipulate numerical symbols and concepts.

G_v (visualization processing) : ability to see spatial relationship and patterns

G_s (speed of processing) : ability to reach quick decisions (reaction time) and maintain attention.

G_a (auditory processing) : ability to discriminate various sounds and detect sound patterns and relationship

G_{sm} (short-term memory) : ability to hold and use information over a short time span

G_{lr} (long-term retrieval) : ability to transfer material to permanent storage and retrieve it later at appropriate time

CDS (correct decision speed) : ability to arrive at correct judgements quickly.

Cattell and Horn have proposed that over our life span, we progress from using fluid intelligence to depending more on crystallized intelligence. In the beginning part of our life, we encounter many types of problems for the first time, so fluid intelligence is needed for their solutions. With accumulation of experiences, we become more knowledgeable and thus, we have less need to approach each situation as a new problem. Instead, we need to retrieve appropriate information and schemes from long-term memory and thus, utilize only the crystallized intelligence. Thus is said to be the essence of wisdom. According to Horn and Hofer (1992) some of the abilities such as G_v are *vulnerable abilities* because they decline with age and tend not to return to preinjury levels following brain damage. Some of other abilities such as G_q are *maintained abilities* because they don't decline with age and return to preinjury levels following brain damage. In general, performance on tests of crystallized intelligence improves during adulthood and remains stable well into the adulthood. On the other hand, performance on tests of fluid intelligence tends to decline as people enter later adulthood (Cattell, 1998). The fact that the ageing affects the two forms of intelligence differently is an evidence that they represent different classes of mental abilities (Horn & Noll, 1997; Weinert & Hany, 2003).

The latest researches have shown that the Cattell-Horn theory provides the theoretical basis for the two of the most popular tests of intelligence: the Stanford-Binet fifth edition and the Woodcock-Johnson tests of Cognitive Ability. This theory is also influencing the gradual development of Wechsler scales.

- **Carroll Three-stratum theory :**

Carroll (1993) formulated this theory which was based upon factor-analytic studies. In fact, he undertook to reanalyze the data from more than 400 factor analytic studies of abilities in a bid to find some common themes. As a consequence, he was able to formulate a hierarchical model of intelligence that contains the elements of Spearman's, Thurstone's and Cattell-Horn's models. In fact, this three-stratum model establishes three levels of mental skills–*general*, *broad* and *narrow*, all arranged in a hierarchical model. At the top or at third stratum of the model, lies Spearman's g-factor, which is thought to underlie most mental activities. Below g, lies the second stratum, which includes eight broad intellectual factors arranged from left to right in terms of the extent to which they are correlated with of factor.

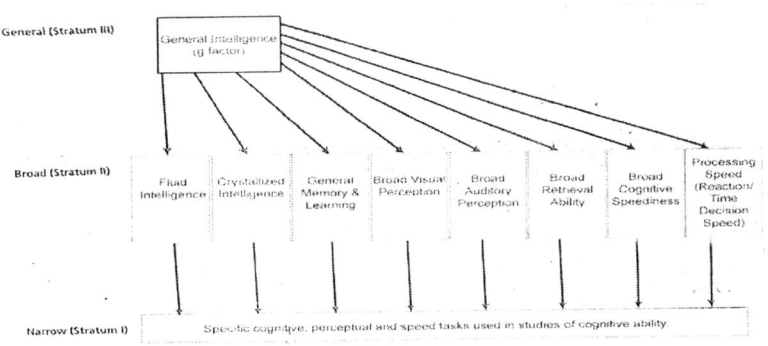

Figure 13.3 Caroll's three Stratum Model.

Figure 13.3 clearly shows that fluid intelligence is strongly related to g-factor whereas the crystallized intelligence is the next and processing speed is the least correlated with g-factor. The other *broad abilities* at the stratum II involves general memory, visual memory, auditory perception, auditory perception and the cognitive speediness. Some of these broad mental abilities resemble Thurstone's primary mental abilities. Finally, at the first stratum that represents narrow abilities, include nearly 70 highly specific cognitive abilities that feed into the broader second-stratum factors. Researchers have revealed that these specific factors lying at the first stratum correlate around 0.30 with one another, and this reflects the common g-factor at the top of the model. According to Carroll, this three-stratum model encompasses all well-known cognitive abilities and provides the most complete and detailed picture of human intelligence derived from the psychometric approach to intelligence.

One attempt to modify that has gained increasing attention blends the Cattell-Horn theory with Carroll's three-stratum theory. Although such blending was not initiated by Cattell or Horn or Carroll, it is nonetheless known as Cattell-Horn-Carroll (CHC) model of intelligence to which we will now turn.

- **Cattell-Horn-Carroll (CHC) Model of Intelligence :**

The integration of Cattell-Horn and Carroll models and therefore, CHC model was proposed by McGrew (1997). On the basis of some additional factor analytic studies, McGrew and Flanagan (1998) subsequently modified McGrew's initial CHC model. In its latest form, CHC model called McGrew-Flanagan CHC model, has ten *broad-stratum abilities* and about more than seventy narrow-stratum abilities. Each broad-stratum ability subsumes two or more narrow-stratum abilities. The names of ten broad-stratum abilities are : *Fluid intelligence* (G_f), *Crystallized intelligence* (G_c), *Quantitative knowledge* (G_q), *Short-term memory* (G_{sm}), *Reading/Writing ability* (G_{rw}), Visual processing (Gv), Auditory processing (G_a), Long-term storage and Retrieval (G_{lr}), processing speed (G_s) and Decision/Reaction Time or speed (G_t).

One of the very important features of CHC model is that it does not recognize general intellectual ability factor (or g factor). To understood the reason for such omission, it is essential to know about its background factors. In fact, the authors of CHC model intended to improve the practice of psychological assessment in education that was sometimes referred to as psychoeducational assessment by identifying tests from different batteries that could be used to provide a comprehensive assessment of the abilities of the students. This was called as cross-battery assessment of the abilities of the students or assessment that uses tests from different test batteries and entails interpretation of data from the concerned specified tests to provide a comprehensive assessment. McGrew and Flanagan (1998, p. 14) have rightly commented, "The exclusion of g does not mean that ... g does not exist. Rather, it was omitted since it has little practical relevance to cross-battery assessment and interpretations."

The major *differences* between Cattell-Horn and Carroll model are as under :

- One major difference between the two models is concerned with the existence of g-factor. In Carroll model, g is the third-stratum factor subsuming eight broad second stratum abilities. But in Cattell-Horn model, g has no place.
- In Cattell-Horn model abilities like quantitative knowledge and reading/writing ability are considered as distinct and broad abilities. But in Carroll model, all these abilities are first-stratum narrow abilities.

CHC model of intelligence as formulated by McGrew and Flanagan is considered important from the heuristic point of view. It has drew attention of many researchers and practitioners to think about exactly how many abilities really need to be assessed and how narrow or broad any approach is optimal in terms of being clinically useful.

- **Sternberg's Triarchic theory :**

Beginning in 1980s, Sternberg (1985, 1986), conducted a series of studies that led him to formulate a theory of intelligence, which was called as *triarchic theory of intelligence*. (triarchic means three). This theory of Sternberg bears some resemblance to Aristotle theory of intelligence, which also consists of three aspects, that is, theoretical, productive and practical aspects. Sternberg's theory is one of cognitive process theories, which explore the specific information-processing and cognitive processes that underlie any intellectual ability. Cognitive process theory of intelligence differs from psychometric approach to intelligence. The latter explains only *how* people differ from one another whereas the former explains *why* people differ in various mental abilities. The triarchic theory of Sternberg tries to address both the psychological processes involved in intelligent behaviour and the various diverse forms that intelligence can take.

According to Sternberg's triarchic theory of intelligence, there are three basic category or types of intelligence : *componental or analytic intelligence, experiential or creative intelligence* and *contextual or practical intelligence* (see Figure 12.4). A brief explanation of these three types are as under.

(i) Componental or analytic intelligence : Such intelligence involves abilities to think critically and analytically. This includes the kinds of academically oriented problem-solving skills assessed by the traditional intelligence tests. Persons high on this dimension usually excel on standard tests of academic potential and excellent students. Professors are usually high on this dimension of intelligence.

Componental intelligence consists of three different components : *performance components, knowledge-acquisition components* and *Metacomponents. Performance components* are those which are involved in attending to stimuli, holding information in working memory and retrieving data from long-term memory and generating responses. Thus performance components are the real mental processes used to perform the task. *Knowledge-acquisition components* are those components, which are involved in acquiring and storing new information by selectively applying the performance components to stimuli. In fact, such components allow the person to learn from experiences and combine new insights with previously acquired information. These abilities underlie the individual differences in what is called crystallized intelligence. *Metacomponents* tend to serve the managerial function of monitoring the application of performance and knowledge-acquisition components to actual solution of a problem till a satisfactory solution has been reached. Thus metacomponents are the higher-order process used to plan and regulate the task performance. Sternberg was of view that metacomponents are the fundamental source of individual differences in fluid intelligence. He was view that the traditional IQ tests measure only componential intelligence or analytical intelligence.

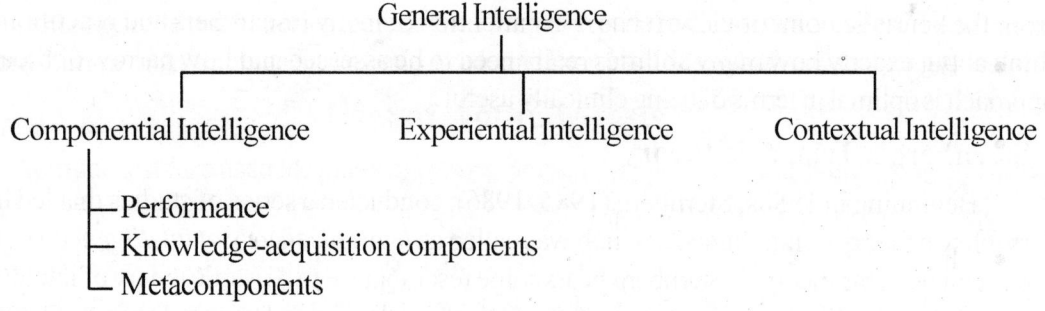

Figure 13.4 : Sternberg's triarchic theory

(ii) Experiential Intelligence : This is *creative intelligence*. People high on the measure of experiential intelligence are able to solve problems in novel way and deal with the unusual and unexpected situations in a satisfactory way. Such people are able to perform common routine tasks in a creative way almost automatically.

(iii) Contextual intelligence : It is also called as *practical intelligence*. People high on the contextual intelligence are able to cope with everyday demands and to manage themselves and other people effectively. In fact, they are '*street smarts*' so to say. Such people usually tend to capitalize on their strengths and also tend to compensate for their weaknesses. Such people either adapt to the environment well or change the environment so that they can succeed. Many people who have succeeded despite various types of hardship have a great deal of contextual intelligence. Sternberg has opined that educational programmes should teach all three classes of skills or not just analytical skills.

The triarchic theory holds promise as a model for explaining how the human mind functions. However, no practical intelligent assessment procedures have been derived from the theory. Moreover, Gottfredson (2003) has questioned many of claims done for practical implications of the triarchic theory.

- **Gardner's theory :**

Inspired by his observations of how specific human abilities are affected by brain damage, Howard Gardener (1993) developed a theory of intelligence called *theory of multiple intelligences*. Originally, he listed seven different kinds of intelligence but later in 1998, he added the eight type and then, a ninth type (Gardener, 1999a, 1999b). These nine types of intelligences are being described below :

- Linguistic intelligence : ability to use language well; found in writers and speakers.
- Logical-mathematical intelligence : ability to think logically and solve mathematical problems; found in scientists and engineers.
- Visuospatial Intelligence : ability to solve spatial problems or to understand how objects are oriented in space; found in pilot, artist, and astronaut.
- Bodily-Kinesthetic Intelligence : ability to control and manipulate body movements skillfully; found in dancers and athletes.
- Musical Intelligence : ability to perceive rhythm and pitch as well as to understand and produce music, found in musicians.
- Interpersonal Intelligence : ability to understand and relate well to others as well as ability to understand motivation of others; found in psychologist and manager.
- Intrapersonal Intelligence : ability to understand one's emotions and motivations; found in people-oriented careers.
- Naturalistic intelligence : ability to detect and understand phenomena in the natural world; found in zoologist, biologist, farmers and botanist.
- Existential Intelligence : ability to trace the questions relating to life, death and ultimate reality of the human existence; found in philosophical thinkers.

Of these nine intelligences, Gardner's first three intelligences are measured by the existing intelligence tests but the others are not assessed. Gardner was of view that these different types of intelligence requires the functioning of separate but interacting modules in the brain. His idea of multiple intelligences has great appeal especially for educators despite the fact that there is little scientific evidence that these intelligences are anything more than different abilities. Gardner's approach, though provocative no doubt, remains most controversial because it goes for beyond traditional conceptions of intelligence.

• Das-Naglieri PASS Model :

PASS model has been proposed by J.P. Das and Jack Naglieri (Das, Naglieri & Kirby, 1994). In order to explain the nature of intelligence, this theory attempts to integrate physiology and information processing and is based upon neurophysiological model of the brain proosed by A.R. Luria who had described human cognitive processes within the framework of three functional units. The function of the first unit is *cortical arousal* and *attention* whereas the second unit codes information using *simultaneous* and *successive processes*. The third unit provides for *planning*, *self-monitoring* and *structure of cognitive activities*. Thus PASS model attempts to divide intellectual performance into four basic processes : *planning* (P), *Attention arousal (P), Simultaneous processing* (S) and *successive processing* (S). The theory attempts to relate these four processes to specific neurological structures to areas which Luria calls functional systems or units (Das & Naglieri, 2001). A person gets information or input through our eyes, ears, nose, skin etc. When sensory information is analyzed, central processes become active. The above four components much as P, A, S and S make up central processing mechanisms.

Planning aspect of model is similar to Sternberg's metacomponents because it is involved in deciding where to focus attention and which type of processing the task requires. The planning process also tends to monitor the success of problem solving and modifies the approach as needed until a solution is achieved.

Another basic function is attention. A person must attend to a stimulus for processing the information it contains or solve the problem it poses. When a person is not able to pay attention, his intellectual performances decrease. Thus inability to pay attention is seen as one reason for poor intellectual performance. Once attention is directed to a stimulus, the information it contains requires either simultaneous processing or successive processing. When elements of the stimulus are surveyable because they are interrelated and accessible to inspection either directly or from being held in memory, simultaneous processing is said to occur. Simultaneous processing is involved in language comprehension and other tasks that require the perception of the stimulus as a whole. Successive processing is needed whenever the elements of stimulus of the task must be performed in a particular order. Serial recall of stimuli and performing skilled movements are examples of the task that requires application of successive processing.

The four processes are associated with different regions of the brain. Planning is broadly located in the frontal lobes of the brain. Attention or arousal is located in the frontal lobes and lower portion of the cortex. Simultaneous processing is broadly associated with occipital and parietal lobes whereas successive processing is associated with frontal-temporal lobes.

These four processes operate within an existing *knowledge* base, which is, in fact, the base of past experience, learning, emotions and motivations. These tend to provide the background for information to be processed. It is as if PASS processes are floating on sea of knowledge. Without water, they cannot operate and will definitely be meaningless. For example, if a boy does not know the letters of alphabet, he cannot process letters serially to read the word 'animal'. If the boy does not understand words, he cannot process the meaning of sentences like 'the cow is a four-footed animal', 'lion is a wild animal', etc.

The supporters of PASS model have argued that the existing tests of intelligence don't adequately assess planning. Naglieri and Das (1997) constructed the Cognitive Assessment System (CAS), a cognitive ability test that is developed to tap PASS components or factors.

Measurement of Intelligence

Intelligence is measured through various types of intelligence tests. All intelligence tests can be broadly grouped into two major general categories: *Individual intelligence tests* and *Group intelligence tests*. These two general categories may be discussed as under :

A. Individual Intelligence or General ability Tests :

Under this category, a discussion about those intelligence tests are done that are designed for individual administration by a trained examiner and in many cases in the clinical observation of the examiner. Some very widely known individual intelligence tests would be discussed here.

- **Stanford-Binet Intelligence Scale :**

Binet and Simon developed a test of intelligence published in 1905 and it consisted of 30 items presented in an increasing order of difficulty. These 30 tasks of increasing difficulty of Binet-Simon Scale provided the first important measure of human intelligence. The purpose of the test was restricted to identifying mentally disabled children in the school system of France. This scale was revised in 1908 in which two major concepts, that is, the *age scale format* and the *concept of mental age* were introduced. However, even though both concepts were eventually abandoned, they received widespread use and application in many tests as well as in subsequent revisions of the Binet scale.

Binet and Simon again revised their intelligence scale in 1911. But this third revision contained only very little improvements. By this time, the utility of this scale has been recognized throughout Europe and in the United States of America. For example, H.H. Goddard in America published translation of 1905 Binet-Simon Scale in English in 1908 and 1908 scale in 1911. However, the most significant version of the scale was published by Lewis Terman at Stanford University in 1916. It was called as Stanford-Binet Intelligence Scale or simply Stanford-Binet scale. In this revision the principle of age differentiation, general mental ability and the age scale were retained. The concept of mental age was also retained. Above all, this 1916 scale provided the first significant application of now outdated intelligence quotient (IQ) to be obtained by using the following formula.

$$IQ = \frac{Mntal\ Age}{Chronological\ Age} \times 100$$

(Multiplication of 100 is for the purpose of eliminating fraction)

The 1916 Binet-Simon scale was extensively revised in 1937 in which an alternate equivalent form was included. Thus Form L and Form M were designed and this proved to be equivalent in terms of both difficulty and content. L for Lewis and M for Merrill according to Becker (2003). This revision also included new type of tasks for use with preschool-level and adult-level testtakers. The next revision of Stanford-Binet was done in 1960, which consisted of only a single form (labelled L-M) and was composed of items considered to be the best from the two forms of the 1937 test with no new items added to the test. However, a major innovation was the use of the *deviation IQ* tables in place of the ratio IQ. The deviation IQ shows a comparison of the performance of the person with the performance of others of the same age in the standardization sample. In fact, the test performance is converted into a standard score with a mean of 100 and standard deviation of 16. If a person performs at the same level as the average person of the same age, the deviation IQ would be 100. If the performance is one standard deviation above the mean for the person's age group, the deviation IQ is 116. Still another revision of Stanford-Binet was published in 1972. In this revision a new standardization sample was used and unlike all previous norms, the 1972 revision included nonwhites. Despite this, the quality of the standardization sample was criticized. In fact, 1972 norms may also have overrepresented the west and largely urban communities (Waddell, 1980).

The fourth edition of Standard-Binet scale was very significant and it was published by Thorndike, Hagen and Sattler (1986). This edition represented a significant departure from previous versions of Stanford-Binet in important dimensions like theoretical orgnisation, test organisation, test administration, scoring and interpretation. In previous editions, the test items were grouped according to age and therefore, the test was called as *age scale*. The fourth edition of Stanford-Binet scale was point scale where test is organised into subtests by category of items not by age of the testtakers. The fourth edition was based upon G_f-G_c theory of Cattell and Horn. There were four areas or dimension: *Verbal reasoning*, *Abstract/ visual reasoning*, *Quantitative reasoning* and *Short-term memory*. The dimension of verbal reasoning is measured by four tests, namely, vocabulary test, comprehension test, absurdities test and verbal relations test. The dimension of abstract/visual reasoning is also assessed by four tests : pattern analysis test, copying test, matrices test and paper-folding-and-cutting test. Likewise, the dimension of quantitative reasoning is measured by three tests, namely, quantitative test, number-series test and equation-building test. Similarly, short-term memory is measured by four tests namely, Bead memory, memory for sentences, memory for digits and memory for objects. Thus fifteen subtests constituted the fourth edition and were organized to yield scores in the said four areas or dimensions. This fourth edition retained the use of standard scores. Raw scores on each of the fifteen subtests can be converted into the individual subtest standard age score (SAS) with a mean of 50 and standard deviation of 8. These would be combined to yield the four dimension or area score and a composite of all of these subtests. A test composite (formerly described as a deviation IQ score) may be defined as a test score or index, which is derived from the combination of and/or mathematical transformation of one or more subtest scores. The score for areas and for the total were expressed as SAS_s–normalized standardscores for that age level–with a mean of 100 and standard deviation of 16.

In February 2003 the fifth edition of Stanford-Binet intelligence Scale (SB_5) was released (Roid, 2003). The SB_5 is based upon Cattell-Horn-Carroll (CHC) theory of intelligence and it measures five CHC factors by different type of tasks and subtests at different levels. The SB_5 has been designed for administration to the examinees from age 2 years to 85 years. Table 13.1 summarizes the five CHC factors names along with their SB_5 equivalents.

Table 13.1 : CHC and Corresponding SB_5 factors

CHC factor name	SB_5 factor name
Fluid Intelligence (Gf)	Fluid Reasoning (FR)
Crystallized Knowledge (Gc)	Knowledge (KN)
Quantitative knowledge (Gq)	Quantitative Reasoning (QR)
Visual Processing (Gv)	Visual-spatial Processin (VS)
Short-term Memory (Gsm)	Working Memory (WM)

Each five factor is measured by one verbal test and one nonverbal test. Thus the test yields 10 subtest scores. The test yields a number of composite scores such as Full scale IQ, Verbal IQ, Nonverbal IQ and an abbreviated Battery IQ. All subtests scores have a mean of 10 and a standard deviation of 3. All composite scores have a mean of 100 and a standard deviation of 15. In addition, the test also yields Five Factor Index scores corresponding to each of the five factors, which the test is presumed to assess. In designing the SB_5 an attempt has been made to strike an equal balance between tasks requiring facility with language (both expressive and receptive) and tests that minimize demand for dependence upon language. In SB_5 in each of the five areas, one test is verbal and other test is nonverbal. This has been presented in Table 13.2.

Table 13.2: SB_5 factors and corresponding subtests

SB_5 Factor name	SB_5 Subtests
Fluid Intelligence (FR)	Verbal : Verbal Anologies Test
	Non Verbal : Object Series/Matrices Test
Knowledge (KN)	Verbal : Quantitative Reasoning Test
	Non Verbal : Nonverbal Quantitative Reasoning Test
Verbal-Spatial Processing (VS)	Verbal : Position and Direction Test
	Non Verbal : Form Board Test
Working Memory (WM)	Verbal : Memory for sentences
	Non Verbal : Delayed Response

The SB_5 has sound psychometric properties. The internal consistency reliability of SB_5 Full Scale IQ is high (0.97 to 0.98) and test-retest reliability coefficients are also similarly high. The content-related evidence of validity as well as criterion-related evidence for validity were determined. These were also satisfactory. Roid (2003) also presented a number of factor-analytic studies in support of construct validity of SB_5.

Wechsler scales

The Wechsler intelligence scales were developed by David Wechsler who was a clinical psychologist at Bellevue Hospital in New York city. Just two years after the Stanford-Binet Intelligence scale momental 1937 revision, he challenged its supremacy as a measure of human adult intelligence. He was of view that this test was fully inadequate for assessing adult intelligence. As a consequence, in 1939 he developed an intelligence scale for assessing intellectual performance

by adults. This scale was named as Wechsler. *Bellevue Intelligence Scale* (WBIS). David Wechsler had objected to 1937 revision of Stanford-Binet test on several grounds. Important such grounds of objection were as under :
- Wechsler did not agree with the idea of single score affected by 1937 revision of Binet test.
- The tests items were not valid for testing the intelligence of adults because items were chosen specifically to be used with children.
- Wechsler was a strong supporter of non-intellectual factors such as lack of confidence, attitude, fear of failure, etc. in determining intelligence. The 1937 Binet revision fail to incorporate these non-intellectual factors in intelligence.
- Wechsler also pointed out that mental age norms did not apply to adults.
- The 1937 Binet scale emphasized speed at the expense of accuracy.
- The 1937 Binet scale failed to consider one important fact that intellectual performance could deteriorate with advancement of age.

Since 1939 three intelligence scales have been developed and subsequently revised to assess the intellectual functioning of adults, older adolescents and children. These three scales are : *Wechsler Adult Intelligence Scale* (WAIS), *Wechsler Intelligence Scale for Children* and *Wechsler Preschool and Primary Scale of Intelligence*. A brief dsecription of these scales is being presented below :

Wechsler Adult Intelligence Scale (WAIS) :

WAIS was originally created as a revision of Wechsler-Bellevue Intelligence Scale (WBIS), which was a battery of tests published in 1939. WAIS was released in 1955. A revised form of WAIS called WAIS-R was released in 1981 and contained six verbal and five or non-verbal subtests. The verbal tests were : Information, Comprehension, Arithmetic, Digit Span, Similarities and Vocabulary. The Performance subtests were : Picture arrangement, Picture completion, Block Design, Object Assembly and Digit Symbol. A Verbal IQ, Performance IQ and Full Scale IQ were obtained. WAIS-R measures intelligence of person having age range of 16 years to 74 years. A subsequent revision of WAIS-R was released in 1997 and called as WAIS-III. It provided scores for verbal IQ, Performance IQ and Full Scale IQ along with some secondary indices such as Verbal Comprehension, Working Memory, Perceptual organisation and Processing speed. The current version of the test called WAIS-IV was released in 2008. This current version consists of 10 core subtests comprising the full scale IQ and five supplemental test. In this new WAIS-IV, the verbal performance subscales from previous versions were removed and replaced by the four index scores : *Verbal Comprehension Index* (VCI), *Perceptual Reasoning Index* (PRI), *Working Memory Index* (WMI) and *Processing Speed Index* (PSI). In addition to these four index scores, two broad scores, which can be used to summarize general intellectual abilities can also be derived. They are : Full Scale IQ (FSIQ) based on the total combined performance of the VCI, PRI, WMI and PSI as well as General Ability Index (GAI) based on only the six subtests from the VCI and PRI. A brief description of these subtests are as under :

Verbal Comprehension

This part consists of 3 core tests and one supplemental test. These test are being discussed as under :
- Similarities test : It is a core test that intends to assess abstract verbal reasoning and semantic knowledge. Here the testtakers are given two words or concepts with a request to describe how they are similar.
- Vocabulary test : It is a core test, which measures semantic knowledge and the extent to which one is able to comprehend and verbally express vocabulary. The testtakers are requested to name objects in pictures or define words or concepts presented to them.
- Information test : It is a core test, which measures the extent of general information learned in the culture. The testtakers are questioned about their general knowledge.
- Comprehension test : It is an *additional* test in verbal comprehension and measures ability to express abstract social conventions, rules and expressions.

Perceptual Reasoning

This part consists of three core tests and two supplement tests. A discussion of these four tests is as under :
- Block Design Test : It is a core test, which measures visual spatial processing, visual motor construction and problem solving.
- Matrix Reasoning Test : It is a core test, which measures nonverbal abstract problem solving and inductive reasoning.
- Visual puzzles Test : It is also a core test which measures visual spatial reasoning.
- Picture completion Test : It is a *supplemental* test, which measures ability to quickly perceive visual details.
- Figure weights Test : It is a supplemental test, which measures quantitative reasoning.

Working Memory :

This part consists of two core tests and one supplemental test. A brief description follows :
- Digit Span Test : It is a core test, which measures attention, encoding, auditory processing and working memory. Here testtakers are required to recall a series of numbers in order.
- Arithmetic Test : It measures quantitative reasoning, concentration, and mental manipulation.
- Letter-Number Sequencing Test : It is a supplemental test, which measures attention, working memory and mental control.

Perceptual Speed

It consists of two core test and one supplemental test as under :
- Symbol Search Test : It is a core test that measures processing speed.
- Coding Test : It is also a core test, which measures associative memory, processing

speed and graphomotor speed.
- Cancellation Test : It is *supplemental* test that measures processing speed.

The WAIS-IV measures intelligence of individuals aged 16-90 years Rehabilitation psychologists and neuropsychologists use WAIS-IV for assessing how the brain is functioning after injury. Specific subtest provides information on a specific cognitive function. Others employ the WAIS-RNI (*Wechsler Adult Intelligence Scale – Revised as a Neuropsychological Instrument*), which is another related measure published by Harcourt, and where each subtest is calculated with respect to neurotypical or brain injury norms.

WAIS has also been adapted and standardized for India to provide the most comprehensive and advanced measure of cognitive ability in adolescents and adults in response to demographic and clinical landscape. This is called the *Wechsler Adult Intelligence Scale-Fourth Edition India* (WAIS-IV India) published in 2013. In fact, WAIS-IV^{India} is the updated version of Wechsler Adult Intelligence Scale-Third Edition (WAIS-III). The project manager for Indian standardization is Dr. Pushpalatha Gurappa. WAIS-IV^{India} assess cognitive abilities of the persons in between 16 years to 84 years 11 months and takes about 75 minutes. It provides Full scale IQ, Index scores as well as Scaled scores by age. The major applications of WAIS-IV^{India} are as under :

- To Identity learning difficulties and giftedness.
- To obtain a comprehensive assessment of general cognitive functioning.
- To provide information for clinical and neuropsychological evaluation.
- To guide treatment planning and placement decisions.
- To identify cognitive strengths and weaknesses in a variety of neurological conditions.
- To provide reliable and valid data in academic and educational settings for research purposes.

Wechsler Intelligence Scale for Children (WISC)

Originally, WISC was constructed in 1949 as a downward extension of Wechsler. Bellevue test (Seashore, Wesman & Doppelt, 1950). Many items were taken from W-B test and esier items of the same type were added to each subtest. A revision called WISC-R was published in 1974. Its third revision called WISC-III was published in 1991 and its fourth edition called WISC-IV was published in 2003. WISC-IV yields a measure of general intellectual functioning (a full scale IQ or FSIQ) as well as four index scores : *a verbal comprehension Index*, *a perceptual Reasoning Index*, *a working Memory Index* and *a processing speed Index*. Each of these indexes is based upon scores on three to five subtests. Full scale IQ is yielded on the core subtests of each Index only. It is also possible to derive up to seven process scores, which are defined as an index designed to help understand the way the testtakers processes various kinds of information. Moreover, WISC-IV, like WIAS-IV, does not yield separate verbal and performance IQ scores. WISC-IV measures intelligence for ages 6 through 16 years 11 months. The test is based upon CHC model of intelligence.

Comparison between WISC-IV and SB$_5$

There are some similarities and differences between WISC-IV and SB$_5$. The important points of *similarities* are as under :
- WISC-IV and SB$_5$ both are individually administered tests, which take about an hour to yield Full Scale IQ based on the administration of subtests.
- The norming sample for testtakers aged 6 through 16 was 2,200 for both WISC-IV and SB$_5$.
- The test developers of both WISC-IV and SB$_5$ included exclusionary criteria in the norming sample and some separate validity studies with some of these exceptional samples were conducted in both these tests.
- Both WISC-IV and SB$_5$ were based on CHC model of intelligence.
- Both WISC-IV and SB$_5$ contain child-friendly materials.

Important differences are as under :
- The WISC-IV contains five supplemental tests whereas SB$_5$ contains no such supplemental test.
- With SB5 an *Abbreviated Battery of IQ* can be obtained from the administration of two subtests whereas WISC-IV formally have no such short-forms.
- The SB$_5$ have socio-economic status and testtaker education as stratifying variables for norming sample whereas WISC-IV did not have such stratifying variables.
- Parent education was included as one stratifying variable in WISC-IV for norming sample whereas SB$_5$ did not have such variable in norming sample.
- The major *cognitive factors* assessed by WISC-IV are working memory, processing speed, verbal comprehension and perceptual reasoning whereas the major cognitive factors assessed by SB$_5$ are working memory, visual-spatial processing, knowledge, fluid reasoning and quantitative reasoning.
- The major *nonverbal factors* assessed by WISC-IV are working memory, processing speed and perceptual reasoning whereas the major nonverbal factors assessed by SB$_5$ are working memory, visual-spatial processing, fluid reasoning, quantitative reasoning and knowledge.

Wechsler Preschool and Primary Scale of Intelligence (WPPSI)

Wechsler (1967) decided that a new intelligence scale should be developed especially for children below age 6. As a consequence, Wechsler Preschool and Primary Scale of Intelligence (WPSSI read and pronounced as Whipsy) was developed in 1967. A revision of WPSSI called WPPSI-R was published in 1989. In 2002 the new edition of WPPSI-R was released and it was named as WPPSI-III, which measured the intelligence from 3 years to 7 years 3 months. In WPPSI verbal/performance distinction was maintained. Accordingly, three composite scores are obtained here : *Verbal IQ*, *Performance IQ* and *Full Scale IQ*. There were many changes in

WPPSI-III as compared to its previous edition. Five subtests such as Arithmetic, Animal Pegs, Geometric Design, Mazes and Sentences were dropped and seven new subtests such as Matrix Reasoning, Picture concepts, Word Naming were added. On WPPSI-III, subtests are labelled as *core test, supplemental test* or *optional test*. Some tests have different age levels. For example, Receptive vocabulary is a core verbal test for children up to 3 years 11 months and an optimal Verbal test for ages 4 years and over. Picture Norming is a supplemental verbal test for ages 4 years and over. Core tests are required for the calculation of composite scores. Supplemental tests are meant for a broader sampling of intellectual functioning. Such tests are also used to derive additional scores such as a *Processing Speed Quotient*. Supplemental tests may also be substituted for a core test if a core test for some reason was not administered or was administered but somehow found unusable. Optional tests cannot be used as a substitute for core tests but may be used in derivation of optimal scores such as *General language composite*.

In WPPSI-III three of the new tests (Matrix Reasoning, Picture Concepts and Word Reasoning) were designed to tap fluid reasoning and two of the new tests, namely, coding and symbol search were designed to tap processing speed.

Thus we find Wechsler intelligence scales having all the three tests, namely, WAIS-IV, WISC-IV and WPPSI-III have many subtests. Table 13.3 presents the subtests of these three tests at a glance.

Table 13.3: The Wechsler Intelligence Scales at a glance

Subtests	WAIS-IV	WISC-IV	WPPSI-III
Information	×	×	×
Similarities	×	×	×
Comprehension	×	×	×
Vocabulary	×	×	×
Arithmetic	×	×	—
Receptive Vocabulary	—	—	×
Digit Span	×	×	—
Picture Naming	—	—	×
Letter-Number sequencing	×	×	—
Picture Arrangement	—	×	—
Picture Completion	×	×	×
Block Design	×	—	×
Object Assembly	—	×	×
Coding	×	—	×
Visual Puzzles	×	—	—
Figure weights	×	—	—
Symbol Search	×	×	×

Matrix Reasoning	×	×	×
Word Reasoning	–	×	×
Picture Concepts	–	×	×
Cancellation	×	×	–

The Wechsler intelligence scales have split-half reliabilities generally 0.80s for the individual subtests. Reliabilities of FSIQ and index scores are reported to be above 0.90. Test-retest reliabilities are somewhat lower but generally lie in 0.80s. The major evidence for validity of the index scores, comes from factor analysis of standardization data. Full scale IQs have been found to correlate substantially with other overall measures of cognitive ability.

Wechsler Abbreviated Scale of Intelligence (WASI)

The Wechsler Abbreviated Scale of Intelligence (WASI) was published in 1999 for meeting the need for a short test to screen intellectual ability of testtakers from 6 to 89 years of age. The test has been developed in two forms: one form consists of two subtests, namely, Vocabulary test and Block Design Test and it takes about 15 minutes to administer. Another form consists of four subtests, that is, Vocabulary test, Block Design test, Similarities test and Matrix Reasoning test. These four tests are WISC and WAIS type subtests that had high correlations with Full Scale IQ on the tests. The WAIS provides measure of verbal IQ. Performance IQ and Full Scale IQ. Full scale IQ is set at mean of 100 and standard deviation of 10. The normal of WAIS provides a high degree of psychometric qualities although some reviewers of the test were not fully satisfied with the way the validity research was conducted (Keith et al., 2001)

Kaufman Scales

Developed by a husband wife team of psychologists, Kaufman Scales are individually administered measures of cognitive abilities (Kaufman & Kaufman, 1983a, 1983b, 1990, 1993). There are three such Kaufman Scales : *The Kaufman Assessment Battery of Children* (K-ABC-Kaufman & Kaufman, 1983a, 1983b) the *Kaufman Adolescent and Adult Intelligence Test* (KAIT-Kaufman & Kaufman, 1993) and *Kaufman Brief Intelligence Test* (K-BIT-Kaufman & Kaufman, 1990). Of three scales, the first two represents attempt to go beyond the theoretical stance of the older intelligence scales. The test developers were interested in creating such test that may, by design, would be used in evolving theories of intelligence, would include developmentally appropriate tasks and would provide some useful information in variety of assessment situations.

Kaufman Assessment Battery for Children (K-ABC)

Kaufman Assessment Battery for children (K-ABC), a product of early 1980s, is an individual ability test meant for assessing the intelligence of children in between 2 years 6 months to 12 years 6 months. K-ABC consists of 16 subtests combined into five global scales : *SequentialProcessing*, *SimultaneousProcessing*, *mental processing composite* (a combination of sequential and simultaneous processing), *achievement* and *nonverbal*. Sequential Processing is represented by three tests; simultaneous processing is represented by seven subtests and achievement is represented by six subtests. The non-verbal scale consists of those subtests that can be administered in pantomime and responded to physically. The achievement is independent of the mental processing composite. The K-ABC measures intelligence through its mental processing scales, that is, sequential processing and simultaneous process. Offering independent and comparable scores for both intelligence and achievement in the same test is considered one of the major advantage of the K-ABC. The K-ABC non-verbal scale provides a measure of ability specifically designed for children who are linguistically deficient or handicapped.

The K-ABC is intended for clinical, psychological, minority group, preschool, neuropsychological assessment as well as research. Theoretically, the K-ABC is based upon several approaches including neurophysiological model of brain functioning of Luria (1966a; 1966b), theory of split-brain functioning of Sperry (1968) and theories of information processing especially that of Meisser (1967). In K-ABC for assessing intelligence Kaufmans have made distinction between two types of higher-brain processes, which they have called as the *sequential-simultaneous distinction*. By sequential processing is meant child's ability to solve problems by mentally arranging input in sequential or serial order. Its examples are number and alphabet order recall. On the other hand, simultaneous processing is one which takes place in parallel. Basically, it refers to a child's ability to synthesize information in order to solve the problem. Criticizing K-ABC Sternberg (1984) has said that the empirical support for the theory underlying K-ABC is questionable. In fact, he pointed out that the manual for K-ABC misrepresents the support for the theory underlying K-ABC. He noted an overemphasis on role learning as the cost of ability to learn.

Kaufman Adolescent and Adult Intelligence Test (KAIT)

This test was developed by Kaufman and Kaufman (1993) and is designed to measure intelligence for ages 11 to 85 years or older. The test is basically an attempt to integrate the theory of fluid and crystallized intelligence developed by Horn and Cattell (1966) with the theory of adult intelligence proposed by Luria (1980), Piaget (1972), etc. This test consists of a crystallized scale and a Fluid scale. The crystallized scale measures concepts acquired from schooling and acculturation. The Fluid Scale measures ability to solve new problems. The core battery consists of three subtests from each scale. An Expanded Battery is also derived by adding any of the four specified subtests and such battery is meant for use with individuals suspected of neurological damage. The test also includes a *Mental Status test* for assessing attention and orientation in persons who are cognitively too impair to be assessed by full battery.

Kaufman brief Intelligence Test (K-BIT)

K-BIT was developed by Kaufmant and Kaufman (1990) that served as a quick screening measure for assessing the level of intellectual functioning. The K-BIT has two subtests : One verbal subtest that consists of 45 expressive vocabulary items and 37 definitions for assessing verbal aspect of intelligence and one nonverbal subtest that consists of 48 matrices for assessing nonverbal ability of intelligence. The Matrices test requires solving 2×2 and 3×3 analogies using figures stimuli. It is suitable for ages 4 to 90 years and can be administered in 15 to 30 minutes. The test yields three scores : Verbal, Nonverbal and Composite and these scores are expressed in terms of deviation IQ units like other Kaufman Scales. It will be proper to mention that the K-BIT is not a shortened version of either the K-ABC or the KAIT. The K-BIT has higher degree of reliability and validity co-efficients.

Woodcock-Johnson Psycho-Educational Battery (WJ)

The Woodcock-Johnson Psycho-Educational Battery Third edition (WI-III) has been developed to measure both cognitive ability (intelligence) and achievement (Woodcock, McGrew & Mather, 2001). In fact, it is a psychoeducational test consisting of two co-normed batteries: *Tests of cognitive abilities* and *Test of Achievement*. Both of these tests are based on the Cattell-Horn-Carroll (CHC) theory of cognitive abilities and intend to assess eights of the mine factors listed in the theory. The factor of Correct Decision speed is omitted. According to Manual of test, WJ-III is designed for use with persons as young as 2 and as old as 90+. The test yields a measure of general intellectual ability (g) as well as measures of specific cognitive abilities, achievement, scholastic aptitude as well as oral language. The tests of cognitive abilities are divided into a Standard battery (10 subtests) and an Extended battery (10 additional subtests). The various subtests tapping cognitive abilities are conceptualized in terms of broad cognitive factors, primary cognitive abilities and cognitive performance clusters. Likewise, Tests of Achievement are developed in parallel forms called A and B, each of which is divided into a Standard battery (12 subtests) and an Extended battery (10 additional subtests). The various subtests tapping cognitive abilities are conceptualized in terms of broad cognitive factors, primary cognitive abilities and cognitive performance clusters. Likewise, Tests of Achievement are developed in parallel forms called A and B, each of which is divided into a Standard battery (12 subtests) and an Extended battery (10 additional subtests). The interpretation of achievement test is based upon the person's performance on clusters of tests in specific curricular areas. The WJ III is intended for diagnosing learning disabilities, planning educational programme and interventions as well as in determining discrepancies between ability and achievement. The standard battery of either cognitive abilities tests or achievement tests might be more appropriate for screenings or for brief revaluations whereas providing a more comprehensive and detailed assessment loaded with diagnostic information. The cluster scores are used for evaluating performance level, gauging educational progress as well as for identifying individual strengths and weaknesses.

The reliability of WJ-III is satisfactorily high. Individual subtest internal consistency reliabilities range from 0.76 to 0.97. Reliabilities of seven scales of the CHC model ranges from 0.81 to 0.95 and reliability for General Intellectual ability (GIA) index was 0.97 for the Standard battery and

0.98 for the Extended battery. The GIA is reported to correlate from 0.60 to 0.70 with other measures of general intelligence.

DAS-Naglieri Cognitive Assessment System

The cognitive Assessment System (CAS) was developed by Das and Naglieri (2001). It is one of the individually administered intelligence tests and it attempts to operationalize the four elements of PASS model of intelligence proposed by its authors (DAS, Naglieri & Kirby, 1994). These four elements are : *Planning, Attention, Simultaneous processing* and *Successive Processing*. Each element is measured by three subtests and thus, there are 12 subtests of CAS. First two subtests for each element are included in the *Basic Battery* and the *Standard Battery* adds one more. The tasks of the CAS are intended to assess the basic cognitive functions, which are involved in learning but presumed to be independent of schooling. The CAS is the most appropriate for use with persons between ages 5 and 17 years 11 months. It links assessment with intervention. The CAS employs verbal and non-verbal tests presented through visual and auditory sensory channels. The three subtests of the four elements of CAS are as under :

Planning (P) :
1. Matching Numbers : The testtakers underlie the two numbers in each row that are the same.
2. Planned codes : It is very much similar to the coding subtest of the WISC-IV.
3. Planned connections : The testtakers connect numbers or numbers and letters in a specified order.

Attention (A) :
1. Expressive attention : The testtakers identify words or pictures when there is interference (for example, the word red printed in yellow)
2. Number direction : The testtakers are required to circle numbers, which have a special features and not to circle numbers that don't have that feature.
3. Receptive attention : The testtakers must encircle parts of letters that share a common feature (either physical identity or lexical identity).

Simultaneous Processing (S) :
1. Nonverbal Matrices : It is similar to the Reven's Progressive Matrices.
2. Verbal-Spatial Relations : The testtakers match a verbal description with the pictures it describes.
3. Figure Memory : After seeing a geometric figure, the testtakers are required to trace the figures within a more complex geometric design.

Successive Planning (S) :
1. Word Series : It is much similar to the Digit span subtest of WISC-IV but here the testtakers use single syllable words instead of numbers.
2. Sentence Repetition : Here the testtakers read sentences in which colour words are substituted for the parts of speech and the testtakers has to repeat the sentence exactly.

3. Sentence Questions : The testtakers has to answer questions based on sentences like the ones found in sentence Repetition. When the testtakers are below the age of 8, a speech rate test is used. They must repeat a three-word sentences 10 times.

The CAS is normed for boys and girls of ages 5 and 17. Scaled scores have a mean of 10 and standard deviation of 3 for the individual subtests. Full scale scores have a mean of 100 and standard deviation of 15. The internal consistency reliability for 12 subtests range from 0.75 to 0.89 with a median of 0.82. Reliabilities of the full scale is about 0.96 and for the four scales range from 0.88 to 0.93. The test-retest reliabilities over a period of 21 days averaged 0.73 for the subtests and 0.82 for the full scale scores.

Nonverbal Measures of cognitive ability :

Nonverbal measures of intelligence were developed in order to control the growing influence of language in the measurement of intelligence. Before World War I Proteus had constructed a series of visual mazes to measure cognitive ability without the use of language and during the war, the psychologists working for US Army developed a battery of nonverbal tests designed to be used with recruits who could not read or had limited proficiency in English language. In the present section, a lucid description of four most popular nonverbal tests of intelligence would be made : *Raven's Progressive Matrices* (RPM) *Goodenough Draw-a-Man test* and *Test of Non-verbal Intelligence* (*TONI*) and *Universal Non-verbal Intelligence Test* (UNIT).

Raven's Progressive Matrices (RPM) is a nonverbal measure of Spearman's g factor or general intelligence. The test requires the *education* of relations among abstract items. The term education, here, refers to the process of figuring out relationships based upon the perceived fundamental similarities between stimuli. In other words, test is a nonverbal measure of inductive reasoning based on figural stimuli (Raven, Raven & Court 1998). The original version of this test was published by Raven in 1938 for assessing higher level abilities implied by Spearman's theory of intelligence. Since RPM is untimed test, it is a *power test* to assess the higher level of abilities. All items of the test are in *matrix format* and the test is progressive in the sense that the items get harden as one proceeds through the test. The testtakers are required to select their responses from a set of up to eight alternatives. At the simplest, items in the test require pattern recognition. The general form of a matrix item is that the stimuli are arranged in two dimensions and features of the stimuli change in two or more regular patterns. A sample item is shown with a section cut out and the testtaker's task is to select the alternative that fits. An example of sample item might look like this :

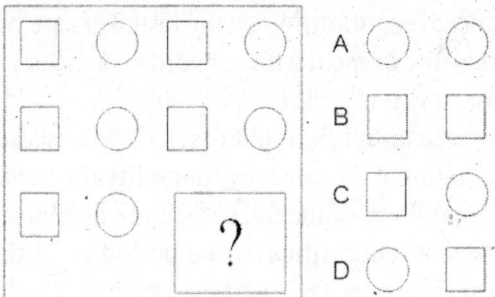

Figure 13.5 Sample Item Illustrating Raven's Progressive Matrices.
A more complex item would contain three to five features of the stimulus changing by rows, columns, diagonals, or in circular pattern.

Raven's test has three forms: the *Colour Progressive Matrices* (CPM), the *Standard Progressive Matrices* (SPM) and *the Advanced Progressive Matrices* (APM). The (CPM (1990, edition) is a 36-item test designed for children from 5 to 11 years of age. The SPM (1996 edition) is meant for the persons from 6 years and 80 years although its most of the items are so difficult that the test is most suited for adults. This test consists of 60 items grouped into 5 sets of 12 progressions. The APM (1994 edition) has a higher ceiling and it is especially meant for the above-averaged adolescents and adults. The APM consists of two sets : Set I consists of 12 problems and Set II consists of 36 problems. Reliability estimates of RPM are in the range of 0.80s but it depends on the variability of the group being tested. The test can be administered individually or in groups.

Goodenough Draw-a-Man Test

The first person to use human figure drawing (HFD) as a test of intelligence was a lady psychologist Goodenough (1926) who developed a test called Goodenough *Draw-A-Man test*. This is a nonverbal test of intelligence that can be administered individually or in group. In this test the testtaker is simply instructed to make a picture of a man as best as he or she can. An extension and revision of this test was published by Harris (1963) and named as Goodenough-Harris Drawing test. As in original test, in this version the emphasis is placed on the child's accuracy of observation as well as on the development of conceptual thinking rather than on the artistic skill. The testtaker is given credit for the inclusion of individual body parts, clothing details, perspectives, proportion as well as on similar features. A total of 73 scorable items were selected on the basis of relation to total scores, age differentiation and relation to group intelligence test scores.

In Goodenough-Harris Drawing test the testtakers are also required to draw a picture of woman and of themselves, apart from drawing a picture of man. Thus in its revised form there are three scales in the test : *Man Scale*, *Woman Scale* and *Self-Scale*. The point scores on each of three scales are transformed into standard scores with a mean of 100 and standard deviation of

15. The reliability of Goodenough-Harris Drawing test has been estimated by test-retest method, split-half method as well as by the scorer reliability. All these reliability coefficients are adequate, that is, test-retest reliabilities are in the range of 0.70s, split-half reliabilities near 0.90 and interscorer correlations are also in the range of 0.90s. The test has been validated against different criteria. Correlations between Goodenough-Harris Drawing test scores and WPPSI Full Scale IQ were found to be in the range of 0.72 to 0.80. In some other cases, correlations with individual IQ tests are over 0.50.

A new version of the Draw-a-Man test for the purpose of improving the technical qualities of the Goodenough-Harris version, has been developed by Naglieri (1988) called *Draw-A-Person : A Quantitative Scoring System* (DAP). This test (DAP) provides somewhat different administrative procedures as well as revised scoring systems, which are less ambiguous than Goodenough-Harris Drawing test. The test also includes a more detailed norms on a sample of 2,622 individuals ages 5 through 17 years of 1980 US census data. Many psychologists praise the DAP for its high reliability, clear scoring system and careful standardization.

Test of Nonverbal Intelligence-Fourth Edition (TONI-4)

The Test of Non-verbal Intelligence, now in its fourth edition (TONI-4) by Brown, Sherbenon & Johnsen (2010). The publisher is Pro-Ed, Austin. It was first published in 1982 to provide a well-normed test that did not require language, written or verbal for administration. TONI-4 is purported to assess intelligence, aptitude, abstract reasoning and problem solving. This test is designed for persons who have or suspected to have communication disorders such as dyslexia, aphasia, or other learning or speech difficulties. It is most suitable for persons having age range of 6 years to 89 years 11 months. The testtakers are requested to respond to abstract figures that present a problem along with four to six possible responses. The testtakers are asked to respond to abstract figures, which present problems along with four to six possible responses. The test consists of five practice items and 45 scored items presented in order of increasing difficulty. The test continues until three out of five items are answered incorrectly. The test is most appropriate for a wide range of ages, that is, 6 years to 89 years.

ToNI-4 has been developed with two equivalent forms, that is, Form A and Form B which enable the test to be used in situation where pretesting and posttesting are needed. Each Form consists of 60 items and all these items are abstract/figural that are obviously void of pictures or cultural symbols. As a consequence, cultural, educational and experimental backgrounds don't adversely affect the test results. All items of the tests appear in an easy-to-difficult order, with each item having one or more of eight important feature : shape, position, direction, contiguity, rotation, shading, size and movement. The test provides verbal and pantomime directions. The testing time is 15 to 20 minutes. The average internal consistency reliability using coefficient alpha is 0.96. The test-retest reliability is estimated to be between 0.88 and 0.93 for school-age students and between 0.82 to 0.84 for adults. Evidence for construct validity has been provided via correlations with other versions of ToNI and with other tests measuring similar constructs. The ToNI-4 correlated with three school achievement tests having correlation coefficients ranging from 0.55 to 0.78. The correlation of ToNI-4 with the ToNI-3 was 0.74.

Universal Nonverbal Intelligence Test (UNIT)

The Universal Nonverbal Intelligence Test (UNIT) was developed by Bruce Bracken and Stave McCallum in 1998 for assessing the general intelligence and cognitive abilities of children and adolescents from ages 5 through 17 years who may be considered as disadvantaged by traditional verbal and language-loaded measures. The test has six subtests. Of these six, three subtests utilize memory process and the remaining three as utilizing reasoning processes. These subtests are additionally classified as being symbolic or nonsymbolic in content. The test is most appropriate for a wide variety of clinical applications, including individuals with hearing impairments, and persons with limited English proficiency. The test yields five possible scores, which are called quotients by authors. They are : *a memory quotient, a nonsymbolic quotient, a reasoning quotient, a symbolic quotient and a full scale intelligence quotient*. Instructions for the test are given in pantomime and responses by the testtakers are made by pointing or manipulating objects. Practice items are provided for each of the subtest. The general item forms of the six subtests are as under :

Symbolic Memory test :

In this test the testtaker is given 10 titles each of which has the international symbol for a man, a woman, a male child, a female child or baby on it in either green or black. Subsequently, he is shown a card with one or more (upto a maximum of six) figures in a particular order for 5 seconds. The moment card is removed, the testtaker is required by hand signals to use the files for reproducing the sequence.

Spatial Memory test :

In this test the testtaker is given 16 chips of which 8 are black and 8 are green. He is also presented with a response grid, which is either 3×3 (items I-II) and 4×4 (items 12-27). A stimulus card is presented for 5 second and subsequently removed. The moment card is removed, the testtaker is instructed to place the chips of the correct colour in the correct square of the grid so that the pattern on the card may be replicated.

Cube Design test :

Here the testtaker is given upto nine cubes with two green faces, two white faces and two diagonally white and green faces. After showing a stimulus picture, he is instructed to replicate the design.

Analogic Reasoning test :

In this test there are 31 matrix items much similar to those in Raven's Progressive Matrices test.

Object Memory test :

In this test the testtaker is shown a picture containing one to seven objects for five seconds. Subsequently, the picture is removed and a second picture with the same object along with some other objects is presented. He is required to place chip on the picture of the objects, which were exposed in the first stimulus away.

Mazes test :

Here the testtaker is shown a picture of a mouse in a maze with one or several pieces of cheese outside the maze. He is required to trace a path from mouse to the correct piece of cheese.

The UNIT is most appropriate for children and adolescents between the ages of 5 years 0 month and 17 years 11 months and 30 days. Norms are provided for an Abbreviated Battery, which is composed of Symbolic Memory and Cube design subtests; a Standard Battery which adds spatial Memory and Analogic Reasoning and provides scores on all five dimensions and an Extended Battery, which is composed of all six tests. Average within-age reliabilities range from 0.64 (Mazes) to 0.91 (cube design) for the subtests, 0.91 for full scale IQ from the Abbreviated Battery, from 0.87 (Symbolic IQ) to 0.94 (full-Scale IQ) far the Standard Battery and from 0.88 (reasoning IQ) to 0.93 (full scale IQ) for the Extended battery. So far as the validity of the UNIT is concerned, the UNIT scales show appropriate correlations with different scores derived from WISC-III. The correlation between full-scale IQ on these two tests has been reported to be 0.88.

B. Group Intelligence Tests :

There are many group tests of intelligence developed by the psychologists and educationists. The origins of group intelligence test can be traced to the active efforts of European psychologists during nineteenth century. The modern group tests of intelligence owes a debt to the completion technique developed by Ebbinghaus during 1890s. His completion test was very simple and it consisted of several passage of text with words or parts of words omitted. The testtakers' job was to fill in as many blanks as possible within the time limit of five minutes. Thus the task was highly speeded. The completion was administered to all students of the class at a time. Ebbinghaus used the total number of correct completions as a basis for judging the intellectual ability of the students.

However, the first very useful and important group intelligence test was developed in United States during World War I when mass testing was necessitated. During World War I in the United States Army two such group tests were developed under the leadership of R.M. Yerkes, who was then President of American Psychological Association : *Army Alpha test* and *Army Beta Test*. The former was a verbal test and was based upon the unpublished work of Arthur S. Otis whereas the latter was a nonlanguage test. The Army Alpha test was designed for general screening and placement purpose whereas the Army Beta test was used mainly for persons who were either illiterate or could not be tested by Alpha test due to foreign-language background. Revisions of the civilian forms of both original Army tests continued in use for several years.

Later in the armed services, *Armed Forces Qualification Test* (AFQT) was developed mainly for screening purposes. This test included an equal number of arithmetic, vocabulary, spatial relations and mechanical ability items. This was followed by multiple-aptitude classification batteries for assignment to occupational specialties. Sometime later, another test called the *Armed Services Vocational Aptitude Battery* (ASVAB) was developed. The ASVAB consisted of ten subtests :

general science, arithmetic reasoning, word knowledge, paragraph comprehension, numerical operations, coding speed, auto and shop information, mechanical comprehension, mechanical operations and electronic information. These subtests are grouped into different composites, including three academic composites : *Academic ability*, *Verbal* and *Mathematics*; four occupational composites–*Mechanical and craft, Health and social, Business and clerical, Electronics and electrical*. Besides these, an overall composite reflecting general ability is also calculated. The ASVAB was used in all the armed services as a composite selection and classification battery. Since then many group tests of intelligence have been developed. In this section, some of such important tests like the Cognitive Abilities Test (CogAT), Culture Fair Intelligence Test (CFIT), Multidimensional Aptitude Battery[1] (MAB) and Raven's Progressive Matrices[*] (RPM) will be discussed.

Important Group tests of intelligence :

Important group tests of intelligence are as under :

Cognitive Abilities (CogAT) :

The cognitive Abilities test (CogAT) is a group-administered multiple-choice test intended to measure students' problem solving abilities and learned reasoning. The test is authored by D.F. Lohman, Professor emeritus at the University of Iowa. The edition of the test by Lohman and Hogen in 2001 consists of nine subtests organized into three batteries : *Verbal*, *quantitative* and *nonverbal*. Thus the test provides one score for verbal ability, one for quantitative ability and one for nonverbal ability. In addition, an overall ability score called *Standard Age Score* (SAS) is also reported. The SAS has a mean of 100 and standard deviation of 16.

These three batteries may be administered either together or separately depending upon the need. There are ten CogAT levels, all based on age, and the test is administered to students from grade 3 to grade 12. Each battery has three subtests, thus nine subtests constitute the whole test.

The verbal battery consists of Picture/Verbal analogies, sentence completion and Picture/Verbal classification. This battery is used to assess the students' vocabulary, verbal memory and efficiency, ability to understand ideas and ability to determine word relationships. The quantitative battery consists of three subtests : Quantitative relations Number series and Equation building. The purpose of this battery is to assess quantitative reasoning, problem solving abilities and level of abstract reasoning of these students. The nonverbal battery consists of three subtests : Figure analogies, Figure classification and Figure analysis. This battery is used to evaluate a student's ability to reason through the use of figural and spatial content. The CogAT Form 7 is available for those students who don't speak English natively as its content is primarily nonverbal.

---

1. MAB has already been discussed in chapter 11 and RPM has also been discussed earlier in this chapter.

The CogAT contains between 118 and 176 questions. The students are provided between 30-45 minutes for each battery and it takes between two to three hours to complete all three batteries. Although the CogAT provides three separate scores (Verbal, Quantitative and Non-verbal), they are not unrelated. In fact, correlations among these three scores are quite high, averaging about 0.70. This commonality reflects a general cognitive ability factor. In addition, each of the nine subtests involves some elements of abiity that are unique to the subtest (specific factor), the nature of which is not clear by the factor analysis.

Culture Fair Intelligence Test (CFIT)

Since beginning psychologists were interested in developing such an intelligence test which could measure intelligence completely devoid of cultural climate, educational level and verbal fluency. Such test was termed as *culture-free test of intelligence*. An important assumption of culture-free test of intelligence was that the effect of culture can be controlled through elimination of verbal items and exclusive reliance on nonverbal and performance items. But soon it was realized that no test could legitimately called as 'culture-free' test because it not possible to wholly eliminate the cultural influences. Therefore, a number of such tests were now termed as *culture fair test of intelligence*.

The culture Fair Intelligence Test (CFIT) was developed by Cattell (1949). It is nonverbal measure of intelligence and was developed in an attempt to measure cognitive abilities, which were devoid of sociocultural and environmental influences. As we know, Cattell had proposed that general intelligence consists of both fluid intelligence (Gf) and crystallized intelligence (Gc). Gf is biologically and constitutionally based whereas Gc is person's cognitive ability based on augmentation of Gf through schooling, sociocultural and experimental learning. The CFIT is the measure of fluid intelligence, which interacts with environmental experience for producing what is called crystallized intelligence.

The CFIT consists of three scales : Scale I, Scale II and Scale III. Scale I is used for children ages four to eight and is most appropriate for mentally defective adults and children. Scale II is meant for adults of average intelligence and children aged eight to thirteen Scale III is meant for adults of high intelligence and for high schools and college students. Scale I consists of eight subtests of mazes, copying symbols, recognizing similar drawings and some other non-verbal tasks. Both scales II and III consists of four subtests. Scale II and Scale III are truly group tests whereas Scale I is partially individual test and partially group test because four of its subteests must be administered individually. The CFIT has gone several revisions, emerging in its current form in 1961. For scales (Scale II and III), two equivalent forms called Form A and Form B are available. Both forms must be administered to each testtaker for obtaining a full test. Each form consists of four subtests : Series, Classification, Matrices and Conditions (topology). In series subtest the testtaker is presented with an incomplete, progressive matrices and he is instructed to select from among given choices, the answer which best continues the series. In classification subtest the testtaker is presented with five figures. In Scale II, he is required to select one, which is different from the remaining four. In scale III, he is required to identify two figures, which are in some way different from the remaining three. Likewise, in Matrices subtest the testtaker is required to correctly complete the design or matrix presented at the left of the row. In conditions (topology) subtest the testtaker is required to select from among given five choices, the one which duplicates the conditions given in the far left box. The CFIT is a speed test and each form of Scale III and Scale III takes about 30 minutes to administer but only 12.5 minutes devoted to actual testtaking.

The reliability of Scale I is 0.9, Scale II 0.87 and Scale III is 0.85. The reliabilities of the full test are generally in the mid 0.80s. The validity of CFIT is satisfactorily high. The CFIT shows consistently robust correlation with other measures of general intelligence such as WAIS, WISC, Stanford-Binet and Raven's Progressive Matrices. These correlations largely in the range of 0.70s to 0.80, provide sufficient evidence for the validity of the test.

Otis-Lennon School Ability Test (OLSAT) :

The name Otis-Lennon shows the surnames of two people–the developer of the test Arthur Sinten Otis (who died before OLSAT was published) and the test editor and publishing executive Roger Thomas Lennon who marketed Otis' concepts as a *school ability test*. OLSAT was originally published in 1979 by Harcourt and the test is at present running in its eighth edition popularly called $OLSAT_8$ published by Pearson.

$OLSAT_8$ measures abstract thinking and reasoning ability of children from Pre-K to 12. In particular, it measures those cognitive abilities that relate to student's academic success in schools. It is a grouped-administered (except pre-school) multiple choice test that measures verbal, quantitative and spatial reasoning ability. The test yields verbal and nonverbal scores and from this, a total score called School Ability Index (SAI) is derived. The SAI is a normalized standard score with a mean of 100 and standard deviation of 16.

The test has been organized into five areas : Verbal comprehension, Verbal reasoning, Pictorial reasoning, Figural reasoning and Quantitative reasoning. The verbal comprehension and verbal reasoning together constitute verbal section. The remaining three together constitutes non-verbal section. The verbal comprehension subsection has four types of questions : Antonyms, Following Directions, Sentence completion and Sentence Arrangement. Verbal Reasoning subsection has seven types of verbal reasoning questions: Arithmetic reasoning, Aural reasoning, Logical selection, Word/Letter Matrix, Verbal analogies, Verbal classification, and Inference. The purpose of the verbal comprehension is to evaluate the children's ability to observe and comprehend relationships among words, to comprehend different definitions of words according to variations in the context, as well as to build sentences. On the other hand, the purpose of verbal reasoning subsection is to evaluate the children's ability to determine relationship among words, to study similarities and differences as well as to apply conclusions in different situations.

The Non-verbal section consists of three subsections : *Pictorial Reasoning, Figural Reasoning* and *Quantitative Reasoning*. In Pictorial Reasoning subsection there are three types of questions : Picture classification, Picture analogies and Picture series. Its purpose is to assess the children's ability to reason by using different illustrations and images, to find out similarities and differences as well as to comprehend and continue some progressions. In figural reasoning subsection there are four question types : Figural classification, Figural analogies, Pattern matrix and Figural series. The major aim of this section is to assess the children's ability to use geometrical shapes and figures for determining relationships, compare and contrast different figures as well as comprehend and continue progression. Quantitative reasoning subsection contain three different types of questions, namely, Number series, Numeric inference and Number Matrix. This section aims at assessing children's ability to determine relationships along with numbers as well as figure out and utilize different computation rules to solve numerical problems.

The completion time varies by level and the maximum time is 75 minutes. OLSAT-8 can be administered with Stanford Achievement Test Series (10th edition) for relating a student's actual achievement with his or her total school ability. When these tests are administered in combination, an Achievement/Ability Comparison (AAC) score is obtained. AAC describes the achievement of a student in relation to the achievement of students with the same measured school ability. AAC scores provide educators a better understanding of the student's potential for getting success in school.

Individual test vs. Group test of Intelligence

Individual test of intelligence differs from group test of intelligence in many ways. Some of the important such distinction are as under :

- Group tests of intelligence generally employ multiple-choice items although in early group tests, open-ended items had been used. But such open-ended items were later on dropped because of excessive amount of time needed for its scoring. As a result of multiple-choice formats, group tests are easily and quickly scored objectively by a computer. Individual tests of intelligence lack these facilities.
- Group tests and individual tests also differ in control of item difficulty. In the individual test, the test administrator has to follow entry rules as well as basal and ceiling rules for ensuring that each testtaker may be examined with items most suited to his or her level of ability. On the other hand, in group tests the items of more or less similar content are arranged in increasing order of difficulty within *independently timed subtests*. Such item organisation provides help in trying each type of item such as arithmetic, vocabulary, etc. and in completing easier items of each type before trying more difficult ones.
- Group tests and individual tests also differ in the mode of administration. In group tests the role of examiner or test administrator, from the pointview of test administration, is limited to reading instructions and enforcing the given time limits. Here there is little opportunity for one-to-one interaction between testtakers and test administrators. Lack of such interaction may produce disastrous results especially from those testtakers who are shy and confused. In individual tests of intelligence a close interaction between testtakers and test administrator occurs and he (test administrator) plays an active role in test administration.

- Group tests of intelligence are generally standardized on a larger sample, that is, on several thousands of carefully selected individuals. The standardization sample, in case of individual tests of intelligence, is comparatively smaller.
- Individual tests of intelligence have proved a great aid in the diagnosis and remediation of various types of learning difficulties whereas the group tests are commonly used for mass screening for strengthening the institutional decision making process. They are used in schools at every level.

Advantages and Disadvantages of Group testing

Group testing has some advantages and disadvantages. Important advantages are as under:
- Group tests have the advantage of economy of time, because such tests can be administered simultaneously to as many persons as can be fitted and seated comfortably within the available space and easily reached through microphone. Thus the group testing has eliminated the need for one-to-one relations between testtaker and test giver.
- Group testing has made the test giver's role very simple and easy. For group testing, the test administrator need not be highly trained or specialized as it is needed in Standford-Binet or Wechsler scales. Here the test administrator only needs ability to read simple instructions to the testtakers during the conduct of the test. Now-a-days, the use of tape-recorded instructions as well as computer administration has further simplified group testing.
- In group testing the scoring is more objective and more reliable and can be done even by outsider such as a clerk. Most group tests, today, are being scored by computers.
- Group tests provide better established norms than the individual tests. This happens because in group testing there is an ease of gathering data on the larger and representative standardization sample.

However, group testing has some disadvantages as under :
- In group testing the test administrator has poor opportunity to establish rapport and obtain cooperation of the testtakers. Moreover, any temporary conditions of the testtaker such as headache, illness, fatigue, anxiety, etc., that is likely to affect the performance, is less likely to be detected in group testing. It has also been observed that the persons unfamiliar with testing, are generally more handicapped in group than in individual.
- In group testing multiple-choice items or standard item types such as analogies, similarities, classification, etc. are used and such items may penalise implications in the answers. Thus group testing has been criticised because they put restrictions on the testtakers' responses.
- Group testing lacks flexibility in as far as every testtaker is tested on all the items of the test. Such testing does not permit the testtaker in concentrating on those items, which are appropriate to his or her ability level. On the other hand, individual tests provide opportunity for the test givers to select items on the basis of the testtakers' own prior record or ability.

Important Indian Intelligence Tests in Brief Historical Perspectives

In India intelligence testing has remained for a long time one of the most popular academic pursuits for both psychologists and educators. As we know, scientific psychology in India started with the establishment of the Department of Psychology in 1916 at the University of Calcutta under the chairmanship of Prof. N.N. Sen Gupta, who was a student of Hugo Munsterberg. The next Chaimran of this department was Prof. G.S. Bose who pioneered Psychoanalysis in India after proper correspondence with Sigmund Freud. The second department of Psychology was established in 1924 at Mysore under the Chairmanship of Prof. M.V. Gopalaswami who was the student of Charles Spearman at the University of London. These early Indian psychologists frequently laid emphasis upon the utility of psychological tests especially the intelligence test.

In the first few decades of the 20th century there was a growing academic interest in India that was characterized by selection and guidance. Some of the important Western tests like Binet-Simon test, Terman-Merrill test, the Stanford-Binet test, Wechsler-Bellevue Scale and some separate performance tests like Koh's Block Degin test, Alexander's Pass Along test were used in the first few decades of the 20th century with translation in local language with minimal changes in items.

If we trace the history, it will become clear that in India up to 1922, the research on construction and validation of psychological tests was conducted by Christian Missionaries engaged in educational work. Among these, the names of Miss Gordon, Spence, West, Wyatt and Rice are important (ICSSR, 1972). Work on Binet tests was meticulously done by Miss Gordan at Madras, West at Dacca, Spence at Jabalpur and Wyatt and rice at Lahore. Dr. C.H. Rice at Lahore (a city in preparatition India) around 1922, first attempted a standardization of Binet Simon test in Urdu and Punjabi. In 1927 Dr. J. Munry of Christian College constructed verbal group tests of intelligence in Urdu, Hindi and English. Likewise, group tests based on the work of Otis were devised by J.M. Sen in collaboration with G. Das Gupta of David Hare Training College at Calcutta and a research paper to this effect was also presented at the Indian Science Congress in 1925 and 1926. In 1934, Mahalanobis in the Indian Statistical Institute at Calcutta started group intelligence testing in Bengali (Mukerjee, 1993). Standardization of some indigenous intelligence tests were also undertaken in 1941. Shohan Lal in Allahabad published a norm for intelligence test in Hindi. Gopeshwar Pal, at Calcutta, encouraged standardization of intelligence tests in Bengali. S.M. Mohsin developed Bihar Intelligence test in the 1940s. One of the most significant aspect of his test was that he preferred an Index of brightness rather than IQ. Later on, several tests were standardized by Krishnadas Pande for Hindi speaking high schools and college students. According to Kuppuswami (1964) standardization of tests was done in different languages such as Marathi, Gujrati, Kannad, Malayalm, Nepalese, etc. During the fourth and fifth decades of the last century, some Bureaus established in different states also were instrumental in developing many psychological tests including intelligence tests. Some of the first such Bureaus were established at Patna (Bihar) under the leadership of S.M. Mohsin, at Kolkata (Calcutta) in West Bengal under the leadership of R. Das, in Lucknow (Uttar Pradesh) under the leadership of Sohan Lal, and later under C.M. Bhatia and S.N. Mehrotra.

In Bombay (Mumbai) a Bureau named as the Parsi Panchayat Vocational Guidance Bureau was established. The Bureau also started publication of a journal named *Journal of Vocational and Educational Guidance* (Dalal, 2011). The *first Mental measurement handbook for India* by Long and Mehta (1966) included 103 tests of intelligence in different Indian languages. This handbook is considered very important for most Indian tests on grounds of reliability, validity and norms.

In the first two Surveys of research released by ICSSR and edited by Mitra (1972) and Pareek (1980, 1981) there was a negligible inclusion of intelligence test. The National library of Educational and Psychological Tests (NLEPT) at National Council of Educational Research and Training (NCERT) aims at collecting Indian and Foreign tests and periodically brought out catalogues and reviews of available tests. It (NCERT) published the *Indian Mental Measurement Handbook : Intelligence and Aptitude Tests* (1991), which provides descriptions of 43 published intelligence tests in India and five unpublished tests (Srivastava, Tripathi & Misra, 1996). Out of these tests, the maximum number of tests (51%) was available in Hindi, 19% were nonverbal, 14% were in English, 5% in Gujarati and 2% each in Bengali, Marathi and Tamil. Majority of these tests (81%) were administrable in group.

M.V. Gopalswami who headed the Department of Psychology at Mysore and was trained at London University with Spearman in mental testing, developed Indian adaptations of several Western Intelligence tests. S.K. Kulashrestha (1971) adapted the Stanford-Binet test in Hindi. In this adaptation, most of the features of L-M form remained unchanged. Changes were made in those items that were not fitting the Indian culture. V.V. Kamat (1934) prepared the adaptation of the Stanford-Binet in Kannada and Marathi. The scales are nonverbal in nature and popularly used with children having some intellectual disabilities. C.M. Bhatia (1955) adapted the Performance Scale of WAIS. It has five subtests, namely, Koh's Block Design Test, Pass-Along test, Pattern drawing test, Tests for immediate memory and Picture construction Test. This test was standardized on 11 to 16 year old students of Uttar Pradesh. Murthy (1966) advocated the use of a short Battery of Bhatia Performance Scale consisting of Koh's Block Design Test and Alexander Pass Along Test because it was considered to be as sensitive as the full battery. P. Ramalingaswami (1974) adapted Performance Scale of WAIS called WAPIS. It has five subtests from WAIS Scale : Picture completion test, Picture arrangement test, Digit symbol test, Block design Test and Object assembly test. The test was standardized on a sample of 604 persons between 15 and 45 years. Pershad and Verma (1988) adapted Verbal part of WAIS called Verbal Adult Intelligence Scale (VAIS). It contains four subscales, namely, Information test, Digit Span test, Arithmetic test, and Comprehension test. The test provides separate quotients for the subscales and verbal IQ is derived from average of the separate subscales. The VAIS is meant for adult subjects in the age range of 20 and above. Malin (1969) has adapted WISC meant for assessing the intelligence of the children. Recently, WISC-IV has also been adapted in India at the initiative of Psychocorp (Pearson India) in 2013.

Pramila Pathak (1987) has adapted the Draw-A-Man Test for children aged above 6 years. T.R. Sharma (1972) has adapted the Draw-a-Bicycle Test for children between 11 and 16 years.

Likewise, Kohli, Kaur and Malhotra (2006) has developed a Draw-a-Pearson test for children in between 6 and 12 years.

Some group tests of intelligence that have been popularly used are the CIE verbal group test of intelligence by Uday Shanker, the CIE Non-verbal group Test of intelligence as well as the Group Test of General Mental Ability by Jalota (1972), the Group Test of Intelligence by Bureau of Psychology, Allahabad as well as the Group Test of Intelligence by Prayag Mehta (1962; Kochhar, 2006). Some other group tests are the Group Test of Intelligence by G.C. Ahiya (1976) and Group Test of Intelligence by P. Ahuja (1974). Another important group intelligence test has been developed by P.M. Mehrotra that includes both verbal and Nonverbal components. The verbal test consists of Analogy Test, Classification Test, Number Series Test, Vocabulary Test and Reasoning Test. The non-verbal test consists of Analogy Test, Arrangement Test, Classification Test, Digit Symbol Test and Part-fitting test (Kochhar, 2006).

These are some of the important intelligence tests developed in Indian condition. These tests have been used frequently in India. However, it is difficult to judge accurately the relative frequency of test usage. Shrivastava Tripathi & Misra (1996) conducted a survey of school counsellors who mentioned that in daily practice, they use mostly four tests such as *Standard Progressive Matrices* (96%), *Test of General mental Ability by Jalota* (27%), *Bhatia Performance Battery* (22%) and *Stanford-Binet test* adapted by Kulashrestha (4%). However, counsellors reported that these tests were not enough for fulfilling the need. They have to depend upon other sources of information such as students' interview, parents' interview, teachers' rating, etc.

Summary and Review

- Intelligence has been explained and understood in different ways because different experts have described it in their own way. Despite these variations, there are three key things that are reflected in various viewpoints. *First*, intelligence is the capacity to profit or learn by past experiences. *Second*, Intelligence is to reason logically, plan effectively and infer perceptively. Thus intelligence is a multifaceted capacity, which manifests in different ways across the life span.
- There are different theories of intelligence. Important such theories are : Binet's theory, Spearman's two factor theory, Thurstone's theory of Primary Mental abilities, Guilford's Structure-of-intellect theory, Wechsler theory, Cattell-Horn Gf-Gc. Theory, Carroll's Three Stratum theory, Cattell-Horn-Carroll (CHC) theory, Sternberg' Triarchic theory, Das-Naglieri PASI Model, etc.
- There are two popular ways of measuring intelligence : individual intelligence test and group intelligence test. Individual tests are those, which are administered to one person at a time. Group tests are those, which can be administered to more than one person at a time. Important individual intelligence tests are Stanford-Binet intelligence test, Wechsler scales, Kaufman Scales, Woodcock-Johnson Psycho-Educational Battery, Das Naglieri-Cognitive Assessment system. Important nonverbal measures of intelligence are Raven's

Progressive Matrices (RPM), Goodenough Draw-A-Man Test, Test of Nonverbal Intelligence (TONI) and Universal Non-verbal intelligence test (UNIT). Important group tests of intelligence are : Army Alpha test, Army Beta test, Armed Forces Qualification Test (AFQT), Armed Services Vocational Aptitude Test (AIVAB). Cognitive Abilities Test (CogAT), Culture Free Test of Intelligence (CFIT), Otis-Lennon School Ability Test (OLSAT).

- Individual tests of intelligence and group tests of intelligence have their own unique features as well as advantages and disadvantages.
- In India also several intelligence tests have been developed and standardized. Among them, performance scale of intelligence developed by C.M. Bhatia, Group Test of General Mental ability by Jalota, Group Test of Intelligence by Prayag Mehta, Bihar Test of intelligence by Mohsin and Group Test of intelligence by G.C. Ahuja are important. Likewise, several foreign made intelligence tests have been adapted by Indian psychologists to suit local condition. Important such adaptations are those of adaptations of Stanford-Binet test by S.P. Kulashrestha, Performance Scale of WAIS (called Wechsler Adult Performance Intelligence Scale) by P. Ramalinga-swami, Verbal part of WAIS (called Verbal Adult Intelligence Scale) by Pershad and Verma, Draw-A-Man Test by Pramila Pathak, and Draw-A-Bicycle Test by T.R. Sharma.

Review Questions

1. Define intelligence. How is it measured.
2. Discuss Spearman's theory of intelligence. Also point out, by citing some examples, how this theory is considered as the parents of several later theories of intelligence.
3. Examine critically Cattell-Horn G_f-G_c theory of intelligence.
4. Describe Carroll's three-stratum theory of intelligence.
5. Examine Sternberg's Triarchic theory of intelligence.
6. Describe the major features of Das-Naglieri PASS Model of intelligence.
7. Discuss the important individual intelligence tests that are frequently used in assessing intelligence of the persons.
8. Describe the important group intelligence tests that are commonly used in assessing intelligence of the persons.
9. Outline a historical development of intelligence tests in India.
10. Write Short notes on the following :
    (a) g-factor
    (b) s-factor
    (c) Advantages of group tests
    (d) Individual intelligence test vs. Group intelligence test.

Chapter – 14

Personality Testing

Learning Objectives

- Personality assessment
- Traits, States and Types
- Nature of Projective Tests
- Classification of Projective tests
  - Association Techniques
  - Pictorial techniques
  - Verbal techniques
  - Expressive techniques
- Criticism of Projective techniques
- Structured Personality tests
  - Logical content strategy
  - Criterion-group strategy
  - Factor-analysis strategy
  - Theoretical strategy
- Behavioural Assessment methods
  - Behavioural Observations and Rating scales
  - Situational Performance measures
  - Psychophysiological Methods
- Summary and Review
- Review Questions
- Review Questions
- Key Terms

Key Terms:

Personality assessment, Traits, States, Types, Projective hypothesis, Inkblot test, Location, Determinant, contents, Therapeutic Apperception test, Words Association test, Sentence Completion test, Draw-A-Person test, MMPT, MBTI, NEO Personality Inventory, Self-monitoring, Rating scale, Role Playing, Biofeedback.

Much of our time is spent in trying to understand others and in wishing others to let us know in better than they do. We often hear that no two people are the exactly alike and that each of us is unique. When people track about uniqueness, obviously they are referring to the personality which is popularly understood as unique and stable pattern of characteristics and behaviours. For assessing personality, personality tests are used. There are different types of personality tests such as projective tests and self-report personality inventories. Projective tests are mostly used for individual assessment whereas self-report personality inventories are more suitable for group administration although many of them have been successfully employed in individual assessment. Besides, behavioural assessment methods are also employed for personality assessment. In the present chapter a focus on above types of personality measurements would be given. In formulating personality inventories several theoretical approaches have also been followed. The present chapter also intends to give focus on those approaches briefly.

Personality Assessment

Personality assessment refers to the process of measurement and evaluation of psychological traits, states, values, interest, attitudes, etc. and/or related individual characteristics. There are two popular methods or techniques of personality assessment frequently used in psychological testing. These two techniques are : *Projective techniques* and *Self-report personality inventories*. Before discussing these two techniques of personality assessment, it is essential to have a clear picture of some background terms like *traits*, *types* and *states* for a better understanding.

Traits, States and Types

Psychological traits :

If we go into the depth, it shall be obvious that there is little consensus regarding the meaning of trait. Theorists such as Allport (1937) has considered trait as "physical entities which are bonafide mental structures in each personality." For him, a trait is a "generalized and formalized neuropsychic system... that guide consistent forms of adaptative and expressive behaviour." For Cattell (1950) traits are defined as mental structures but he made it clear that structure did not necessarily imply actual physical status. Likewise, Holt (1971) considered traits as some real structures inside people that tend to determine the behaviour of the individual in lawful ways.

There are a few examples, which reflect that there exists no consensus regarding the meaning of trait. One intelligence way, in such situation, to clarify the meaning of trait is to shy away from these definitions and simply concentrate upon the fact that traits refer to consistency in behaviours. (Colman, 2015) Put it in a scientific terms, traits are relatively stable cognitive, emotional and behaviour characteristics of the persons that distinguish them from one another by establishing their individual identities. In this context, a definition of personality trait proposal by Guilford (1959) seems to be most relevant. He said that traits refer to "any distinguishable, relatively enduring way in which one individual varies from another". In this definition the term 'distinguishable' indicates that the behaviours labelled with different trait terms are really different from one another. For example, a behaviour termed 'reserved' should be distinguishable from a behaviour labelled as 'outgoing'. The situation or context in which behaviour is displayed is important in applying trait names to behaviours. For example, a behaviour done in one context may be labelled with one trait term but the same behaviour shown in another context may be better described using another trait term. For example, a person talking with friend on dining table may be demonstrating friendliness whereas that the same person talking to the same friend during wedding ceremony of his sister may be considered rude. Obviously, it means that the trait term selected by the observer is dependent both on the context as well as upon the behaviour itself.

Another important point in Guilford's definition is the use of the term *relatively enduring*. He did not emphasize upon the fact that traits represent enduring ways in which persons vary from one another. Rather, he emphasized upon the word relatively enduring. The inclusion of the word relatively indicates that exactly how a trait is manifested depends, at least to some extent, upon the situation or context. For example, a student may behave aggressively towards his classfellow in a rather subdued way before the teacher but he may be much more aggressive in the presence of his family members.

Personality states :

In personality assessment the term 'state' has been used in at least two different ways. *First*, the word personality state is understood as an inferred psychodynamic disposition, which conveys the dynamic quality of id, ego and super ego in perceptual conflict (Cohen & Swerdlik, 2005). Such psychodynamic dispositions are usually assessed by the popular psychoanalytic techniques such as words association, free association, dream analysis and analysis of psychopathologies of everyday life. *Second*, the word 'state' is used for the *transitory exhibition* of a trait of personality. So, state is a *mini-manifestation* of traits. Putting it in other words, state is an indicative of temporary predisposition whereas trait is the indicative of relatively enduring behavioural disposition (Chaplin et al., 1988). For example, Ramesh may be described as being in 'anxious state' before appearing for UPSC examination though one who knows Ramesh well may describe him as 'an anxious person'. Very few existing personality tests seek to make distinction between traits and states. Seminal work in this field was done by Spielberger et al. (1985). These researchers develop some personality inventories that seek to make distinction between states and traits. State-Trait Anxiety Inventory (STAI) developed by Spielberger and Sydeman (1994) is one example where a distinction between state anxiety and trait anxiety has been shown. State anxiety refers to temporary experience of tension due to a particular situation whereas trait anxiety refers to a relatively enduring or stable personality characteristic. The items of STAI consists of short descriptive statements and subjects are instructed to indicate either how they feel right now (indicative of state anxiety) or how they generally feel (indicative of trait anxiety). The trait-state differentiation was also applied by Spielberger et al., (1985) in another developed inventory called *State-Trait Anger Expression Inventory* (STAXI).

Personality Types :

Personality type is a constellation of traits and states that are similar in pattern to one category of personality as proposed by taxonomy of personalities. Thus type is simply a category of individuals said to share a common collection of characteristics. Thus type is different from trait. Traits are *characteristics* possessed by the persons whereas types are *descriptions* of people by placing them into discrete categories types. Since Greek physician Hippocretes' classification of people into four types (Sanguine, Choleric, Melancholic and Phlegmatic) there have been many attempts to provide personality typology. In 1940s Sheldon and Stevens (1942) proposed a type theory that was based upon temperament and body build. This typology later on faded away. A personality typology devised by Jung (1923) became also popular and this typology was the basis for the personality test Myers-Briggs Type Indicator (MBTI) (Myers & Briggs, 1943). Holland (1973, 1985, 1997) also put forward one typology. According to this personality typology, there are six personality types (Enterprising, Artistic, Social, Realistic, Investigative and Conventional). Based on this typology, Holland et al., (1994) developed a personality test called Self-Directed Search (SDS), which is a self-administered, self-scored and self-interpretative in nature. Another personality typology proposed by Freidman and Rosenman (1974) provides only two types–Type A personality and Type B personality. Type A personality is characterized by restlessness, competitiveness, impatience, feeling of time-pressured, strong needs for achievement and dominance. Type B personality is characterized by the opposites of Type A's traits. The poplar inventory developed by Jenkins et al., (1979) called *Jenkins Activity Survey* (JAS) is used to categorize respondents as Type A or Type B personalities.

Nature of Projective Test

Projective tests of personality assessment were borne in the spirit of rebel against normative data as well as against the attempt made by psychologists to break down the study of personality in terms of specific traits of varying strengths. This orientation is much emphasized by Frank (1939) who introduced the term *projective method* for describing a category of tests for studying personality with the help of some unstructured stimuli. In projective test the respondent (or testtaker) is given some vague and ambiguous stimuli (called unstructured stimuli) and he is required to respond with his or her own constructions or responses. Such stimuli permit an almost unlimited variety of possible responses. One of the basic assumptions of the projective test is that the personal constructions or response toward unstructured stimuli tend to reflect the unconscious needs, motive and conflicts of the respondent. This is known as *projective hypothesis*.

Major features of projective tests are as under :
- Projective tests are characterized by a *global approach* to the assessment of personality because here attention is focused on a composite picture of the whole personality rather than upon the measurement of separate traits.
- Projective tests present unstructured stimuli to the resttakers.

- Projective tests represent a disguised testing procedures where the testtaker has no idea of the interpretation to be done from their responses.
- Projective tests have flourished within the clinical setting and have been a predominant tool in the hands of clinical psychologist.

The major *assumptions* of projective tests are as under :
- There is an unconscious mind within the individual.
- The more unstructured the stimulus materials, the more the testtakers reveal about the personality.
- Every response done by the testtakers provides meaning for analysis of personality.
- Projection is usually greater to the stimulus materials that are very similar to the testtakers in physical appearances, gender, occupation, etc.
- The testtakers are not aware of the meaning of what they respond.
- The response on test has no right or wrong answers
- There is a parallel between behaviour obtained on a projective test and behaviour displayed in social situation.

If we trace the origin of projective tests, it dates back to the nineteenth century. Francis Galton in 1879 developed the first projective technique called *word association test*. Later, Hermann Ebbinghans used *sentence completion test* as a measure of intelligence but it was realized later that it is a good measure of personality. (Gregory, 2004) The word association tests proved to be very attractive to the researchers and it was adapted soon by Kent and Rosanoff (1910) and Carl Jung popularized this test in therapy. Hermann Rorschach who was much influenced by psychoanalytic theory of personality, published his inkblot test in 1921. In 1905, Alfred Binet used a precursor to story-telling or thematic apperception test when he used several verbal responses to pictures as important measure of intelligence. Encouraged by the success of such story-telling technique, a test called *Thematic* Apperception Test (TAT) was first published by Murry in 1935 but later on Murray and Magan (1938) working at Harvard Psychological Clinic published a book entitled *Exploration in Personality* in which TAT was presented in detail. Subsequently, some projective tests were developed in form of drawings. Draw-A-Person Test by a lady psychologist Machover (1949) and House-Tree-Person Test by Duck (1948, 1992) are its good examples.

Classification of Projective Tests

There are different types of classification of projective tests. The most popular classification of projective tests are : *Association techniques*, *Pictorial techniques*, *Verbal Techniques* and *Expressive techniques*. A discussion of various projective tests coming under each of these four categories is presented below :

Association techniques :

Association techniques are those techniques where the testatkers are provided with some unstructured stimuli in terms of some vague pictures and they are requested to tell what they see in or associate with it. Under this category, three tests are very popular. Rorschach Inkblot Test, Holtzman Inkblot Test, Somatic Inkblot Series (SIS).

Rorschach Inkblot Test:

Before discussing Rorschach Inkblot test, let us have a brief look on the *history of inkblot techniques*.

The notion of using inkblots in studying human functioning did not start all on a sudden rather it has a brief history. Zubin, Eron and Schumer (1965) reported that the concept of formless stimuli as used in inkblot technique, can be traced back to Leonardo De Vinci and Botticelli in the fifteenth century. Exner (1969) pointed out that during the latter half of the twentieth century in Europe, there was a public interest in inkblots not as a test but as a game. In the popular game of that time called 'Blotto', the inkblot designs were used. In this game, the challenge before players was to associate an image to the design taken from many inkblots. In 1857 J. Kerner reported that persons frequently report some unique personal feelings when they see inkblot stimuli. Such unique personal responses open way for studying persons in details because some meanings lie hidden in such responses. At Harvard University Dearborn published an article in 1897 in which the potentials of utilizing inkblot techniques in experimental psychology was discussed. In 1848, as Tulchin had had pointed out, Dearborn published the results of applying an inkblot technique to a group of sixteen subjects. In this study he had used twelve sets of inkblots each having ten blots very similar in nature. Pre-Rorschach work of Sharp, Krikpatrick, Whippe, Pyle, Bartlette and Parsons, all of whom published materials concerning inkblot methodology between 1900 and 1917 in United States and England, is also worthmentioning. (Exner, 1974)

A review of literature shows that, there are *three* important periods in the history of inkblot use in projective testing *Pre-experimental period*, *Psychological experimental period* and *Period of Rorschach Innovative research*.

Preexperimental period covers the period of Nineteenth century during which artists use to point indeterminate forms to stimulate creative imagination. Psychological experimental period started with Alfred Binet and V. Henri in 1895 with the assessment techniques, which assessed imagination as an index of cognitive ability and personality. At that time H. Rorschach was only 10 years old child. This idea of Binet and Henri. got support from several quarters. This support led to the publication of first set of standardized inkblots by Whipple (1910). The third historical period started in 1911 when Rorshach's innovative research on the 'interpretation of accidental forms started. As a schoolboy, he exhibited interest in art forms and perception. He was very affectionately nicknamed 'inkblots' or 'klex' by his class fellows. In fact, Rorschach's major purpose for developing an inkblot was to investigate the subjects' reflex hallucination through viewing inkblots. Cessell (1965) extended the use of inkblot technique to study the subjects' body perceptions and somatic symptoms where the clinicians could better recognize the sufferings of a person. Rorschach test is one of the most popular inkblot test developed by Hermann Rorschach in 1921. This test was developed for the diagnostic investigation of personality as a whole. In this test there are ten inkblot cards. Of these ten cards, Card No. I, IV, V, VI and VII (five cards) are in black and white, two cards, that is, Card No. II and III contain only some patches of colours and the remaining three cards, that is, Card No. VIII, IX and X are fully coloured cards. An inkblot of Rorschach Test type is presented in Figure 14.1

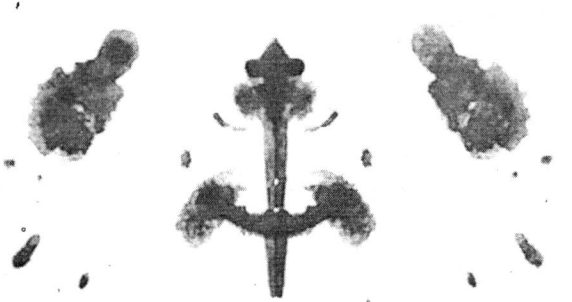

Figure 14.1 An Inkblot similar to Rorschach Inkblot.

The administration of Rorschach test consists of two phases : *free association phase* and *inquiry phases*. Sometimes a third phase called *testing-of-the-limit* is also needed. Each card is administered twice—first in free association phase and then, in inquiry phase. The first phase is free association phase where the testgiver (or the examiner) presents the testtaker (the subject) Card I and asks, *"What might be this?"* If the testtaker asks for any clarification, the testgiver always reply in a non-directive way, such as 'It's up to you'. The testtaker may give more than one response per card. All these responses are recorded by the testgiver in verbatim. However, this is not required and the testtaker may reject the card as a whole without giving any response. But it happens rarely. The remaining nine cards are presented in a similar way one after the another in a numbered order. The testgiver also records how long for the first time it takes testtaker to respond to a card (reaction time) and the position of the card such as *upside down*, *sideways* etc., while responding. Thus the test-giver records all relevant information including the testtaker's verbatim response, non-verbal gestures, length of time before the first response to each card, the position of the card and soon.

Next, the inquiry phase begins where the testgiver shows the cards again and scores the responses of the testtaker. Now, responses are scored on the basis of at least following dimensions : *location, determinant, content, popularity vs. originality.*

In *location* the testgiver tends to determine where the testtaker's perception is located. In using for location the testgiver codes the responses. When the testtaker used the whole blot, this response is coded as W; when a well-defined part of the blot was used, it is coded as D; when unusual or poorly defined part of the inkblot was used, the response was coded as Dd. When the percept was seen on white space, it is coded as S. Location may be scored for other responses such as confabulatory responses like DW where the testtaker generalizes from a part to whole.

In *determinant*, the testgiver tries to find out what *determined* the response. In other words, here attempt is made by the testgiver to find out what it was in the inkblot that led to the testtaker to see that particular percept. In general, at least four properties of an inkblot may determine or produce the response: its *form or shape* its *perceived movement* its perceive *colour* and it *shading*. When the shape or form of the blot determined the response, the response is coded as F. In scoring *form quality*, it is established that to what extent the percept match the stimulus properties of the inkblot. Here responses are commonly coded as F, F+ and F−. When the percept matched stimulus properties of the blot, it is coded as F; when the percept matched the stimulus properties of the blot in an exceptionally better way it is coded as F+ and when the percept matched the stimulus properties of inkblot very poorly, it is coded as F−. When response occurred because of movement, it is coded as M, FM and m. M is coded for human movement such as two people hugging, FM is coded for animal movement such as two dogs playing and m is coded for inanimate movement such as 'car is running'. When response is determined by colour (such as pink clouds), it is coded as C. The colour response may be coded as CF (Colour/Form) or FC (Form/Colour) depending upon whether form is primary or secondary to colour as determined. When the response is determined by shading or texture features such as a furry bear (because of the shading) it is coded as T.

Identifying the determinant is considered as the most difficult aspect of Rorschach test. This is because of two reasons. *First*, there are some difficulties involved in conducting inquiry and *second*, there is no standardized administration procedures. As a consequence, the testgivers vary widely in conducting their inquiries and such difference among testgivers tend to influence the scoring and administration. As a consequence, the reliable experimental investigations of the Rorschach test are rare (Lewandowski & Saccuzzo, 1976).

In scoring *content*, an attempt is made to find out what was the percept? Although scoring content varies from one Rorschach expert to another, certain major categories are commonly employed by most authorities. Most popular among them are human figures (H), human details (or parts of human figures, Hd), Animal (A), animal details (or parts of animal details or Ad), plants, maps, blood, X-rays, clothing, art objects, landscapes and sexual percepts. An inquiry is not necessary for determining the content.

So far as the *popular-original responses* are concerned, most experts agree that popular are those responses that are frequently given for each card. Exner (1993) Comprehensive system has defined popular responses as those responses that occur once in three protocols on the average and this has been accepted as standard method of scoring popular response. Likewise, original responses are those responses that include rare, unusual or deviant verbalizations done by the testtaker during administration of Rorschach test.

Sometimes a third component in administration called *testing-of-the limits* is also included. This procedure helps the testgiver to restructure the situation by asking some specific questions that provide some additional information regarding personality. For example, suppose the testtaker has given whole response throughout the entire inkblot when forming percept. The testgiver might want to determine if the details within the inkbolts could be elaborated. Accordingly, the testgiver might ask, "Sometimes individuals, use a part of the blot to see something" or the testgiver might point to a specific area of the card and ask." What does this look like?" One additional purpose of the *testing-of-the-limits* may be to help the testgiver in determining whether the testtaker is able to refocus percepts if given a new frame of reference.

Table 14.1 summarized the whole discussion of Rorschach scoring. As we know that Rorschach died before he could complete his scoring methods, the very systematization of Rorschach scoring was left his followers. In fact, five American psychologists produced independent but overlapping approaches to the scoring and interpretation of the test: Samuel Beck, Bruno Klopfer, Marguerite Hertz, Zygmunt Piotrowski and David Rappaport. The scoring methods of Rorschach responses vary from one system to other. Fortunately, Exner and his colleagues have tried to synthesize these earlier approaches into one system called comprehensive scoring system (Exner, 1993; Exner & Weiner, 1995). At present, Exner's this Comprehensive scoring system has supplanted all other approaches and is regarded as the best system of scoring.

After the entire protocol has been coded, a number of summary scores is computed for hypothesizing about the personality of testtaker. Hypotheses regarding various aspects of personality are based both on the number of responses of a particular category as well as upon the interrelationships among the categories. For example, the number of W response is associated with conceptual thought process. Form responses are related to the reality testing.

Table-14.1: A Summary of scoring codes of Rorschach test responses

I.	*Location*:		Where on the blot was the percept seen or located?
	W:	Whole	Entire inkblot used
	D:	Common detail	Well-defined part of inkblot used
	Dd:	Unusual detail	Poorly defined unusual part used
	S:	Space	Percept defined by white space
II.	*Determinant*:		What feature of the blot determined the response?
	F:	Form	Shape or Outline of the blot used
	F+:	Form$^+$	Clear and excellent match of percept and inkblot
	F−:	Form$^-$	Very poor match of percept and inkblot.
	M:	Human Movement	Human Movement seen in inkblot
	FM:	Animal movement	Animal movement seen in inkblot
	m:	Inanimate movement	Inanimate movement seen in inkblot
	C:	Colour	Colour determined the response
	FC:	Form colour	Form being primary and colour being secondary in determining the response
	CF:	Colour Form	Colour being primary and form being secondary in determining the response

		T :	Texture Shading involved in the response
III.	*Content* :	What was the percept?	
		H : Human	Percept of whole human form
		Hd : Human details	Human form incomplete in any way
		A : Animal	Percept of whole animal form
		Ad : Animal details	Animal form incomplete in any way
		Xy : X-ray	X-ray of any human part; shading involved
IV.	*Popular vs. Original*		
		P : Popular	Responses given by many normal individuals
		O : Original	Rare and usual verbalizations

The above lists are incomplete and is designed to illustrate only quantitative scoring.

When F+ responses fall below 70 percent, it is hypothesized that the testtaker exhibits severe pathology, brain impairment or some kinds of intellectual deficit (Exner, 1993). Thus a psychotic person is expected to score very poor on form level. F+ percent is also taken to be indicative of ego strength and higher scores of F+ indicates a greater capacity to deal effectively with stress. Human movement is thought to be indicative of creative imagination. Likewise, colour responses are indicative of emotional reactivity.

Rorschach protocols are also evaluated not only for its quantitative data but also for qualitative features such as specific content, pattern of response, recurrent themes and their relationship among different categories. (Moreland et al., 1995, Exner, 1999) One very important aspect of qualitative data is an evaluation of content frequently reported by emotionally disturbed or brain-damaged individuals but infrequently reported by the normal population. Such responses are used to discriminate between normal and disordered persons. (Moreland et al., 1995) Sometimes the testtakers give *confabulatory responses* where they overgeneralize from a part to a whole. Such responses also illustrate the value of qualitative interpretation. When the testtaker says, "It looked like my father because of beard. My father has also a long beard just like this." Here the testtaker sees the detail—the large beard and overgeneralizes so that the blot looks like his father. When the testtaker gives so many such confabulatory responses, it becomes indicative of the fact that he or she is in some disordered state like being brain-damaged, mentally retarded or emotionally disturbed individuals. Normal people rarely give such confabulatory responses.

Evaluation of Rorschach test on psychometric properties like standardization, reliability, validity etc., is very difficult. It has been reported that the procedure of administration and scoring of Rorschach test is not standardized. For example, one system scores for human movement whenever a human in action is perceived by the testtaker whereas another system has developed some elaborate and strict rule for scoring such human movement. In absence of standardized scoring, it is very difficult to determine the frequency, consistency and meaning of a response on the Rorschach test. In India Banerjee and Kundu (1983) have developed a computer programme for scoring Rorschach protocol.

So far as the reliability of investigations are concerned, they have produced varied and inconsistent results. In some studies internal consistency coefficient were found to be in the range of 0.80 and 0.90s whereas for some other studies such co-efficients range is that of 0.10 or 0.01. However, Exner (1993) have provided some evidences that acceptable levels of inter-scorer reliability can be attained with the Rorschach tests. So far as the validity is concerned, the literature reflects contradictory findings. In meta-analytic research the indices of convergent validity for the Rorschach test has been found to be comparable to those obtained with MMPI (Atkinson et al., 1986; Parker, Hanson & Hunsley, 1988). Exner's own extensive work with Rorschach has also extended support to many constructs measured through his system. However, when Rorschach responses are examined in the context of diagnosing complex current conditions or predicting future behaviour, mixed results have been obtained. (Wood, Nezworski & Stejskal, 1996a, 1996b)

Frequently cited arguments that go against the Rorschach test can be enlisted as under:
- The test lacks universally accepted standard of administration, scoring and interpretation.
- The results of the test are unstable over time
- The evaluation of data obtained on the test is subjective
- Most of the modern experts consider the test as unscientific
- Test is considered inadequarte on all important traditional standards.
- Test lacks predictive or prognostic validity with respect to outcome treatment or later behaviour.

Moreover, Fiske and Baughman (1953) found that practitioners often feel that the variation in the total number of response is too much and as a consequence, it is difficult to interpret the ratio scores of Rorschach test adequately. Dubey (1982) conducted a study involving 300 subjects and reported that most of Rorschach indices are dependent upon the number of responses. In this study when groups of normal, neurotics and schizophrenics were divided on the basis of number of responses into high and low productivity, many of Rorschach indices which were found to be significant, lost their significance. This finding substantiated the findings of Fiske and Baughman (1953).

Hotlzman Inkblot Test (HIT)

As we know, the major problems of the Rorschach test were its variable number of responses from one testtaker to another, lack of an alternative form and lack of standard procedures. The Holtzman Inkblot Test was developed to remove these problems while maintaining the advantages of Rorschach test (Holtzman, 1961; 1986; Holtzman et al., 1961). In HIT there are 45 cards. Both coloured and achromatic cards are included here and few inkblots are markedly asymmetrical also. The testtaker is asked to give only one response per card. Each response is followed by simple *two-fold* questions: where was the percept represented in the blot, and what about the blot suggested the percept? The HIT has been developed in parallel forms. (Form A and Form B, each having 45 cards) The 45 responses to the HIT are scored for 22 different variables, which were derived from early Rorschach scoring system. Important such variables are reaction time, location, space, Form responses, colour, shading, movement, content scores, anxiety, hostility, etc. For each variable percentile scores are available for normal adult and children and for a number of deviant groups. (Holtzman, 1975)

The scoring system of HIT is highly reliable and the standardization of the test appears to be adequate. Interscorer reliability for the different categories was found to range between 0.95 to 1.00 for most categories. Test-retest reliabilities and split-half reliabilities are also satisfactory, the former being in the range of 0.70 to 0.80.

The validity of HIT has incorporated a number of approaches including cross-cultural comparisons, correlation with other tests and with behavioural indicators of personality characteristics as well as contrasted group comparisons with both normal persons and psychiatric patients. In general, the relationships are the modest but somehow supportive of HIT validity. (Holtzman, 1988; Sacchi & Minzi, 1989)

Thus while making comparison to Rorschach test, HIT appears to be psychometrically more sound. The availability of parallel forms permits not only the measurement of test-retest reliability but also adequate follow-up studies. Since in HIT there is restriction of responses to one per card, it holds response productivity constant for each respondent, thus avoiding many of the limitations of Rorschach scoring. Despite these advantages, there is relative paucity of information on HIT as compared to Rorschach Test and still more data are required to establish diagnostic significance of various scores as well as the construct validity of the test (Dush, 1985)

There has came up a recent variant of HIT, which requires 25 responses to each of a carefully selected subset of 25 cards from Form A. This new variant has been named as HIT-25 to distinguish it from HIT. The new test has been considered as a very useful instrument for making diagnosis of schizophrenia. Using completely objective criteria and simple decision rules, HIT-25 has been correctly able to classify 26 of 30 schizophrenics and 28 of 30 normal college students (Holtzman, 1988). The total results on HIT are summed algebraically, yielding a normalcy score, which becomes the basis for simple diagnostic decisions. Scores obtained above zero, suggest normalcy whereas scores obtained below zero indicates schizophrenia. A score of zero is considered indeterminate.

Somatic Inkblot Series (SIS)

Somatic inkblot series (SIS) is the recent and latest development in the inkblot techniques (Cassell, 1965, 1980a, 1980b, 1984). The SIS is a structured but projective technique. It is structured by a sequential presentation of intentionally designed inkblot-like images, which demonstrate typical and atypical response potentials. The SIS is projective in the sense that it is based on spontaneous, individually generated responses to semi-ambiguous figures that tend to produce intrapsychic associations specific to persons. The SIS is adjunct to counselling and psychotherapy.

The SIS was first conceived in 1959 and named as SIS.I (20 inkblot images). After several years of research on SIS-I, SIS-II with 62 inkblot images in series A and B in the Booklet form was developed in 1980. This booklet form SIS-II was a tremendous success. Encouraged by this success, SIS-Video (Video version of 62 images Form) was developed in 1984. The test is designed for self-administration. It can also be administered as group administered test. There are five forms of SIS test:
- SIS-20 card set
- SIS-II 62 images booklet
- SIS-Video (Video tape/CD of SIS-II 62 images)
- SIS-I (Video tape/CD of SIS-I 20 cards)
- SIS-living images (40-image video exclusively meant for female)

The administration of SIS is designed in such a way that it can maximize the pulling power of inkblot projection. Most of the ambiguous structure relates to the specific life situations and post-traumatic dream content, which have found to be of clinical importance. Researchers have established that viewing the SIS has both diagnostic and therapeutic applications. The viewer writes responses on an answer sheet. By writing responses rather than verbalizing them, the subject (or testtaker) is likely to release more painful material. In addition to this, the subject may gain some insight and sense of mastery of traumatic material.

It has been shown that the applications of the procedure of SIS particularly SIS-II (video/CD) have proven to be far-reaching and of international significance. In fact, the use of graphic images, rather than words, have allowed SIS-video to cross both cultural and language boundaries. Its number of uses in psychological assessment and psychotherapy is rapidly growing.

Pictorial Techniques :

This technique asks the testtakers to construct a story or provide a complex and meaningfully organized verbal responses on the basis of test stimuli appearing in some form of pictures. One of the best example of this technique is Thematic Apperception Test a TAT. The history of using picture as projective stimuli dates back long before TAT. One of the earliest uses of picture as projective stimuli came at the beginning of the twentieth century (Brittain, 1907) in which nine pictures, one by one, shown to children who told a story on the basis of these pictures. A clear sex difference was reported. Girls showed more inclination towards religious and moral themes than boys. Another study involving storytelling technique on the basis of pictures incorporated the children's imagination (Libby, 1908). Schwartz (1932) who was a psychiatrist at clinic for Juvenile Research in Detroit developed Social Situation Picture working at Harvard Psychological clinic Christina D. Morgan and Henry Murray developed *Thematic Apperception Test* in 1935. Due to its unparallel importance, TAT is being described in detail.

The first articles describing Thematic Apperception Test (TAT) were written by Christiana D. Morgan (1897-1967), a lady psychologist (Morgan, 1938) or C.D. Morgan and H.A. Murray with Morgan listed as senior author at Harvard psychological clinic. (Morgan and Murray, 1935,

1938) In a mimeographed manuscript of Harvard University archives, the early version of the test was even titled as Morgan-Murray Thematic Apperception Test (White et al., 1941). However, W.G. Morgan (1995) pointed out that the name of Christiana D. Morgan was dropped as an author of the 1943 version of the test.

TAT was originally developed to reveal the person's basic personality characteristics through interpretation of their imaginative productions in response to several pictures. Thus Test was basically developed as a vehicle for the study of personality from a psychoanalytic point of view. Although the test is designed to reveal some conflicts, attitudes, goals, and repressed materials, it actually produces material that is a mixture of these plus situational influences, cultural influences and so on. TAT was originally a method of stimulating fantasy. Later on, Murray revised various aspects into a theory of *needs* and *presses* (environmental forced). The combination of needs and presses was called as *thema*.

TAT consists of 31 cards including a blank card. These cards depict people in different situations. Some cards contain only objects. Some of the cards are said to be useful for boys and men, some for girls and women and some for both sexes. Murray had suggested that 20 out of 31 cards be selected for administrating to a single person. These 20 cards also include one blank card. These 20 cards are usually administered in two sessions of one hour each. 10 cards are administered in each session. The cards reserved for the second session were deliberately chosen to be more unusual, bizarre and dramatic. But in real clinical practice, the clinicians typically select somewhere between 6 to 12 cards for administration to a given person. (Phares, 1984). The testtaker is introduced with a cover story that it is a testing imagination in which their task is to make up a story in which he is required to tell who the people are, what they are doing, what they are thinking or feeling, what led upto the scene and how it will turn out. Regarding the blank card, he is instructed to imagine some picture on the card, describe it and then, construct a story about it. The person's productions are written or transcribed verbatim by the testgiver or clinician. In some cases, person may be asked to write out their stories.

In original method of interpretating TAT stories, the testgiver determines first the 'hero' of the story. Hero is the character of either sex with whom the respondent has presumably identified himself or herself. Then the content of the stories is analyzed in terms of needs, press, theme and outcome. The needs and press are analyzed with reference to Murray's list of needs and press. Murray (1943) favoured rating of both needs and press. Important needs assessed by TAT are achievement need, affiliation need, dependency need, power need, sex need, etc. As we know, press refers to the environmental forces that may facilitate or interfere with the satisfaction of needs. Exposure to physical danger, being comforted, being attacked or criticized, receiving affection are the examples of press. The continuation of need and press is called themas. When the clinician is assessing the importance or strength of a particular need, press themas and outcome for the persons, special attention is paid to the intensity, duration and frequency of its occurrence in several stories as well as to the uniqueness of its association with a given picture. For example, the *frequency* of theme such as depression and *outcomes* such as failure indicate their importance. Usually, the frequency and intensity of each need and press are rated on 1 to 5 scale and themes and outcomes of each story are noted as well.

The most unfortunate part of TAT is the wide diversity of administration and scoring. This diversity has made it very difficult to investigate the psychometric properties of TAT as a distinct psychological test. Moreover, some research data are available to show that conditions like sleep deprivation, hunger, and social frustration tend to affect TAT responses. (Atkinson, 1958) Besides, some situational factors like who the examiner is, how the test is administered, and the testtaker's prior experiences of the similar test also affect the TAT scores. It is very difficult to evaluate the reliability and validity of TAT in any formal sense. Many variations in instructions, methods of administration, number of cards used and type of scoring system are employed. As a consequence, hard conclusions are virtually impossible. In view of the above facts, split-half, test-retest and alternate-form reliability measures are inappropriate for TAT. To some extent, it is possible to investigate theme reliability but since we can't expect word-for-word similarity from one situation to the next, we cannot depend upon this. However, we can study the reliability of judges' interpretations. When there is an explicit derived set of scoring instructions as some quantitative ratings are involved, it is possible to arrive at interjudge agreement. Some quantitative scoring schemes and rating scales have been developed that yield good scorer reliability. But their application is time-consuming. Therefore, this has not been popular in clinical setting. Several attempts have been made to establish validity of the TAT. These include (a) comparison of TAT interpretations with case data (b) comparison between psychiatric diagnoses derived from TAT and pyschiatrists' judgement (c) establishment of the validity of certain general principles of interpretation, and (d) matching technique and blind analyses. Since TAT is basically a subjective test, its validity cannot be assessed apart from the skills of clinician who administrators it.

TAT has been adapted by Prof. Uma Choudhary (1960) in Indian condition. It consists of 14 cards which have human figures and environmental details in relation to Indian set-up. Out of these 14 cards, some are reserved for males, some are reserved for females and remainings are for both sexes.

Variants of TAT

There are several alternatives or variants of Thematic Apperception Test. Hence a brief introduction to some of the important and popular such variants have been made.

One popular such test is the *Children's Apperception Test* (CAT) developed by Bellak and Bellak (1952) to meet the special needs of children aged 03 through 10. The CAT consists of animals rather than human figures, presumably because children find it easier to identify with animal characters. These pictures intend to evoke fantasies relating to the problems of feeding, sibling rivalry, aggression, toilet training and other childhood experiences. Subsequently, a human modification of CAT called CAT-H was developed for use with older children especially for those having mental age beyond 10 years (Bellak & Hurvich, 1966)

Another such test is *Tell-Me-A-Story* (TEMAS), which has been prepared for children aged from 5 to 18 years (Constantino, Malgady & Rogler, 1988). TEMAS intends to assess the cognitive, affective and personality characteristics of children. The test has two parallel sets of fully coloured cards—one for ethnic minority children and one for white children. The stimulus of TEMAS intend to facilitate verbal production and stimulate stories dealing with conflicting-goods such as immediate versus delayed gratification. The minority set of cards depict features most typical of Black or Hispanic origin. Although TEMAS is considered as improvement over TAT in so far as the suitability for African American and Hispanic American children is concerned, its reliability especially internal consistency reliability and test-retest reliability have been questioned. (Dana, 1993)

Another popular apperception test is *Roberts Apperception Test* for children (RATC) developed by McArthur and Roberts (1982). The test consists of two overlapping sets of 16 stimulus cards—one for boys and one for girls. The pictures on the cards show children interacting with adults and other children. Explicit guidelines allow fairly objective scoring of responses. This test represents a serious effort to combine the flexibility of projective tests with administration and scoring of a standardized objective test. The validity studies of RATC continued to yield some favourable result.

The *Appercetive Personality Test* (APT) is another projective test, which tries to introduce objectivity into the scoring system as well as try to address some long-standing criticisms of TAT. (Karp et al., 1990) APT consists of eight stimulus cards, which depict recognizable people in everyday life settings (Holmstrom et al., 1990). The pictures depict males and females of different ages as well as of minority group members. After telling a story about each card of APT orally or in writing, the testtakers are required to respond to a series of multiple-choice questions, which saves two purposes. First, such questionnaire supplys quantitative information and second, it fills in information gaps from stories that are too brief to score. In this way, responses are subjected to both clinical and actuarial interpretation.

The *Gerontological Apperception Test* (GAT) has been developed for the aged. (Wolk & Wolk, 1971) The stimuli of GAT make use of one or more elderly individuals involved in a scene with a theme relevant to the elderly such as loneliness, family conflicts, helplessness, etc. The *Senior Apperception Test* (SAT) is an alternative to Gerontological Apperception Test and is parallel in content, that is, it is also meant for elderly people and its stimulus cards also depict elderly people in a scene relevant to the elderly. (Bellak, 1975; Bellak & Bellak, 1973) Researches, however, have shown that neither GAT nor SAT has any advantages over TAT in testing the personality of adults. (Foote & Khan, 1979)

Rosenweig Picture-Frustration study is another test using pictures as projective stimuli (Rosenweig, 1945, 1975). This test utilizes cartoons depicting frustrating situations. The test is derived from Rosenswig's theory of frustration and aggression. It represents a series of cartoons in which one person frustrates another or calls attention to the frustrating condition. In the blankspace provided the testtaker writes what the frustrated person would prefer to reply. (see figure 14.2)

Figure 14.2 Rosenweig Picture Frustration Study

The test is available in separate forms for adults, aged 14 and over (Rosenzweig, 1978a 1978b) for adolescents aged 12 to 18 (Rosenzweig, 1970, 1981) and or children (Rosenweig, 1960, 1977, 1988).

Responses to P.F study are scored in terms of the *type* of reaction elicited and *direction* of the aggression expressed. Type of aggression may be obstacle-dominance (ephasizing the frustrating object), ego-defense (focusing attention on the protection of frustrated person) and need-persistence (focusing upon the constructive solution of the frustration problem). The *direction* of the aggression may be extrapunitive or extragressive (turned inward upon the self) and inaggressive or inpunitive (aggression is evaded so as to avoid or gloss over the situation). For each scoring category, the percentage of responses is determined and compared with normative data. Subsequently, a group conformity rating (GCR) is derived and this tells about the extent to which one's responses conform to or are typical of those of the standardization group.

Another test utilizing picture-story method is that of *Hand test* (Wagner, 1983), which consists of nine cards with pictures of hand on them. There is also tenth card which is a blank card. The testtaker is asked to tell what the hands on each card might be doing. In case of blank card, the testtaker is instructed to imagine a pair of hands on the card and then tell what they might be doing. Testtakers' responses to each card are recorded. Responses are interpreted according to 24 categories of needs such as affection, dependence and aggression, etc.

The *Blacky Pictures Test* is another interesting projective technique (Blum, 1950, 1968) based on story-telling technique. The test consists of a series of cartoon pictures that depict the adventures of dog named Blacky and his family. Each of these cards presumably relates to a particular stage of psychosexual development. The test is principally used with children for eliciting information that is especially useful from a psychoanalytic point of view. Problems with the test are absence of norms and poor stability scores.

Verbal Techniques :

Verbal techniques or verbal projective techniques are those techniques, which are wholly verbal and utilize word or open-ended phrases and sentences in both stimulus materials and responses. Therefore, they are called as *Semi-structured techniques*. Such techniques, of course, presuppose a minimum reading level and thorough familiarity with language in which the test had been developed. Therefore, these techniques cannot be administered to children or illiterates. Perhaps, the two best examples of verbal projective techniques are : *Words Association tests* and *Sentence Completion tests*.

Words Association tests, in general, is defined as a semi-structured, individually administered projective test in which a list of stimulus words is presented to the testtaker who responds verbally in writing with whatever comes to mind first upon hearing the word. Subsequently, responses are analysed in terms of content and other variables. Originally known as *free association test*, this technique was first introduced by Galton in 1879 (Anastasi & Urbina, 1997). With lapse of time, interest in the phenomenon of word association test increased and resulted in additional studies. Some concrete methods were developed for recording the responses given by the testtakers as well as for the length of time elapsed before obtaining a response (Cattell, 1887; Trantscholdt, 1883). Kraeplin (1896) had studied word association and examined the impact of hunger, fatigue as well as practice on the phenomenon of word association. All these studies and similar others at least revealed one important thing: the associations made by the individual to words were not simply chance associations but were the outcome of the interplay between the person's attitudes, experiences and his unique personality characteristics.

The first use of words association test on clinical basis was done by Jung (1910) as well as by Kent and Rosanoff (1910). Jung was of view that by selecting certain key words that represented the possible areas of conflict, words association test could be used as a method for uncovering the emotional complexes hidden in the personality. He prepared a list of 100 words and presented it to the patient. He recorded the responses to each word and through analysis of the kinds of responses, he was able to get clue to the nature of the hidden conflicts. Jung's list of 100 words became very popular in clinics in Europe. In United States of America Kent and Rosanoff (1910) developed another list of 100 words, which were standardized on normal 1000 adults. These words were all commonly used words and believed to be neutral with respect to emotional impact. A frequency table based on these 1000 cases were developed. These tables were used to evaluate examinees' responses according to the clinical judgement. In general, psychiatric patients were found to exhibit a lower frequency of popular responses than the normal subjects in the standardization group.

Jung's as well as Kent and Rosanoff's experiments served as an inspiration to Rappaport, Gill and Schafer (1946) who developed another Words association test at the Menninger clinic. This test was considered superior to other earlier tests. This test consisted of 60 words, some words were neutral (book, water etc.) and some words were traumatic such as breast, suicide, etc.). The administration of the test is done in three parts. In the first part each word of the list was administered to the testtaker who had been instructed to respond quickly with first word that come to his mind. The testgiver recorded the time taken by the testtaker in responding to each item. In the second part of the test, each word was again presented to the testtaker who is instructed to reproduce the original responses. Both deviations between original and their second response as well as the length of time before reacting were recorded. The third part was that of inquiry where the testgiver puts questions to clarify the relationship that existed between the stimulus word of the list and response. For example, he may put questions like: what were you thinking about? What was going in your mind at the time of responding?

Responses of the testtakers to the words association test are analysed at two levels: the *formal level* and the *content level*. The formal analysis includes the study of reaction time and the various forms taken by the response. The content analysis is done to study the symbolic meaning of the words. A longer reaction time (longer than a few seconds to respond to stimulus words), indicates that the word touches off painful associations, is uncomfortably close to repressed experience or threatens to bring into consciousness materials that are anxiety-provoking. For understanding the dynamic nature of the conflict, content analysis is done. Here the meaning of the response word is examined for whatever light it might throw on the unconscious conflict of motivation of the person.

Some experts object to the word association test because they feel that the test yields very limited information regarding the personality as a whole. As a result, the test is not used widely in modern psychodiagnosis.

Sentence completion tests

The sentence completion test is test in which the testtaker is presented with a series of partial sentences and he is required to complete the meaning. For example :

I with my parents_____
My sex life is_____
I like to_____
I am here because_____
I think that my mind is_____

Originally, sentences completion test was used in Germany by famous psychologist Hermann Ebbinghaus as a nonprojective test of intellectual level. The first serious attempt to use the test as one of the methods of exploring personality began in 1930s and one of the earliest tests of this type was constructed by Amanda Rohde and Gertrude Hildreth in 1940. During World War II several sentence completion tests were developed for various assessment programmes. One such test was used as a projective technique in the Office of Strategic Services and another test was used for evaluating the adjustment of patients in Army hospital.

Several standardized sentence completion tests are available to the clinicians. One such widely used test is *Rotter Incomplete Sentences Blank* (Rotter & Rafferty, 1950). There are 40 incomplete sentence items and the testtaker is instructed to respond to each item in a way that expresses their real feelings. The test is available for three levels: *high school* (grades 9 through 12), *college* (grades 13 through 16) and *adult*. Responses on this test are interpreted according to several categories: family attitudes, social and sexual attitudes, general attitudes and character traits. Each response on the test is evaluated on a 7-point scale that ranges from *need for therapy* to *extremely good adjustment*. Inter-scorer reliability was reported to be in the 0.90s. Apart from several original validity studies, sociometric techniques have been used to demonstrate the validity of this test as a measure of adjustment (Lah, 1989).

Although sentence completion tests are useful for obtaining diverse information about the person's interests, future goals, fears, conflicts needs and so on such tests are vulnerable, to faking on the part of the testtaker's intention to make a good or bad impression.

Expressive Techniques

Expressive projective techniques are those techniques that ask the testtakers to express themselves in some way, most commonly by drawing a picture. Two best examples of expressive techniques are *Draw-A-Person test* and *House-Tree-Persons test*.

Encouraged by Goodenough's (1926) Draw-A-Man test of assessing intelligence, Karen Machover (1949), developed a test of assessing personality on similar line. His test is known as *Draw-A-Person (DAP)* test, which is a projective assessment of personality. In this test the testtaker (usually a child) is given pencil and a blank sheet of white paper and told to draw a person. Immediately after the first drawing is completed, he or she is given a second sheet of paper with the instruction to draw a picture of a person of the sex opposite of the person just drawn on the first sheet. Subsequently, many experts ask questions about the drawing such as 'Tell me a story about the figure', "Tell me about that boy a girl", "What is nice or not nice about the person", or "What is the persons doing?" etc.

Drawings generated by DAP have been formally evaluated through analysis of the different characteristics of the drawing. Special attention is paid to such factors as length of time required to complete the drawings, size of the figures, placement of figures, pressure of the pencil used, symmetry, shading, line quality, facial expressions, overall appearance, posture, clothing, etc. Based on these factors, various hypotheses have been generated. For example, unusually light pressure suggests that the testtaker has some character disturbances (Exner, 1962). Likewise, the testtaker who draws a tiny figure at the bottom of the paper might be depressed or insecure having a poor self-concept. The testtaker who draws a picture that goes beyond one sheet of paper and goes off the paper, is considered to be impulsive Buck (1948, 1950) have derived various meanings from the *placement* of drawings on the page. Placement of drawings on the right of the paper conveys orientation towards the past. Placement on lower left side of the paper suggests depression with a strong desire to go to past and placement at the upper right suggests a strong desire to suppress unpleasant experience of the past followed by an optimistic future.

The characteristics of the person drawn have also been paid attention in analyzing the figure. For example, unusually large breasts drawn by a male is usually interpreted to reflect unresolved oedipal problems with dependence upon mother (Joltes, 1952). Unusual large ears or eyes suggests some paranoid thinking (Shneidman, 1958). Long and conspicuous ties tend to suggest sexual aggressiveness (Machover, 1949). Emphasis on button in figure suggests dependent and inadequate personality (Halpern, 1958).

A latest and more appropriate application of this test has been in screening children suspected of behaviour disorder and emotional disturbance. For this purpose, Naglieri et al. (1991) developed *The Draw-A-Person: Screening Procedure for Emotional Disturbance (DAP: SPED)*. This test is based upon the assumption that the rendering of unusual features in figure drawings signals emotional problems.

Another projective drawing test is *House-Tree-Person test* developed by Buck (1948) in which the testtaker draws a picture of house, tree and person. Originally, H-T-P test was concerned as a measure of intelligence (Buck, 1948) but clinician soon abandoned the use of H-T-P as a measure of intelligence and it is now almost exclusively used as a projective measure of personality.

In H-T-P test the testtaker is given complete freedom in sketching the three asked objects : He is requested to produce pencil drawing and crayon drawing separately. According to Buck (1992) the testtaker is given four sheets of paper. Identification information is put on the first sheet and pages two, three and four are separately titled as House, Tree and Person respectively. For each testtaker, two drawings form are needed-one for pencil drawings and another for crayon drawings. Buck (1992) has also recommended a separate four-page form for a postdrawing interrogation that consists of questions to be asked from the testtaker regarding his opinion about the elements of drawings. However, many experts have questioned the value of separate crayon drawings as well as postinterrogation questionnaire.

So far as the interpretation of H-T-P test is concerned, it rests upon three general assumptions. House drawings indicate home life and intrafamilial relationship of the testtaker. Tree drawings indicate the manner in which the testtaker perceives and deals with the environment and Person drawings indicate the testtaker's interpersonal relationship. Several interpretative hypotheses for both quantitative and qualitative aspects of three drawing have also been provided (Buck, 1992). The reliability and validity of the H-T-P has not been satisfactorily established due to various impediments to such researches. Among such impediments one important impediment is that no H-T-P sign has but one meaning (Gregory, 2004). In a *Kinetic Family Drawing test*, which is to some extent similar, the testtaker is required to draw a picture of his or her family (Burns & Kaufman, 1970, 1972).

Like other projective techniques, figure drawing tests are vulnerable to the basic assumption that drawings are essentially self-representations and represent something far more than drawing ability (Swensen, 1968). For example, in Draw-A-Person test one system suggests that the large head is taken as indicative of intellectual functioning whereas other system suggests that it means brain damage (Machover, 1949). Although different systems have been devised for scoring of figure drawings, solid support for the validity of such systems have been questioned (Watson et al., 1967).

Criticisms of Projective tests

Projective tests have been attacked on several grounds. These grounds can be, in general, categorized as related to *assumptions*, *situational variables* and *psychometric considerations*.

- **Assumptions**

Many academicians and clinicians have questioned the assumptions of projective test. One assumption of the projective test is that something called unconscious exists. Although the term unconscious has been widely used in psychology as if its existence were given, conclusions of many studies about the existence of unconscious based on experimental testings of predictions derived from hypnotic phenomena, from specific personality theories as well as from signal detection theory have been inconclusive (Brody, 1972). Another important assumption of the projective test is that more ambiguous the stiumuli, the more testtakers reveal about the personality. Murstein (1961) has said that the stimulus material is only one aspect of the total situation. Reality is that various factors such as environmental variables, response sets and reactions to the examiner all contribute to the response patterns. Besides in situations where stimulus materials were designed to be unclear or hazy or presented with incomplete lives, projection on the part of the testtakers was not found to increase. Likewise, the assumption that the projection is greater onto stimulus materials that are similar to the testtakers with respect to physical appearance, occupation, gender, etc. has also been questioned.

- **Situational variables.**

According to Frank (1939) the projective tests are fully capable of tapping personality patterns without disturbing the pattern being tapped. If this is really true, then variables related to the test situation should have no effect on the data obtained. But results of the several studies have shown that the situational variables affect the data obtained on the projective tests. Bernstein (1956) has reported that TAT stories written in lone situation are likely to be less optimistic, less guarded and more affectionately involved than those TAT stories written in the presence of the testgivers. Likewise, variables like specific instructions (Henry & Rotter, 1956), subtle reinforcement cues given by the testgiver (Wickes, 1956) and the age of the testgiver (Mussen & Scodel, 1955) do affect the protocols of the projective tests.

Moreover, in any given situation several variables may be placed in a mixed situation. Researches have shown that the interaction of those variables tend to influence the judgements regarding the meaning of the responses of the test. Variables like the testtaker' social class, clinician's training, and motivation to manage a desired impression are all capable of influencing the clinical judgement (Langer & Abelson, 1974). Masling (1960) has carefully reviewed the literatures regarding the influence of situational and interpersonal variables in projective testing and has provided the evidence that three factors influences the projection. Masling (1965) pointed out that through gestural, postural and facial cues, the testtakers on Rorschach test are capable of eliciting the responses they expect. He further argued that the testgivers on the projective tests interpret the projective data with regard to their own needs and expectations as well as to that of their own subjective feelings about the testtakers. Not only this, even the testtakers utilize every possible cues in testing situation, including cues related to the action or appearance of the testgivers or examiners.

- **Situational variables.**

The psychometric soundness of many projective tests has not been well demonstrated. Researchers have shown that various variables like inappropriate sample, inadequate control groups, poor external criteria and uncontrolled variations in protocol length tend to contribute spuriously to the ratings of validity (Cohen & Swerdlik, 2005). Moreover, there are some methodological obstacles in determining psychometric properties as many test-rest or split-half methods are inappropriate. Projective tests have no standard grading scales and have both low reliability and validity (Lilienfeld, 1999; Cicarelle & White, 2017). It is also very challenging to execute a validity study that effectively take into account various situational variables that influence the administration of the tests.

Eysenck (1959) has also criticised the projective tests on various grounds and most emphatically he pointed out that there is no empirical evidences for most of the postulated relationships between projective test indicators and personality traits. Moreover, no meaningful and testtable theories underlie the projective tests.

Structured Personality Tests

Structured Personality tests, unlike Projective tests, present a structured and clearly defined stimuli to the respondents who understand the meaning of their responses. One of the major assumptions of structured personality tests is that human beings possess characteristics or traits that are stable, vary from individual to individual and can be measured. There are so many structured personality tests, which have been developed following one or more of several strategies. The major strategies to develop structured personality test are :

- Logical content strategy
- Criterion-group strategy
- Factor-analytic strategy
- Theoretical strategy

A discussion of the various important structured personality tests under each of the above strategy is presented have

Logical-content strategy :

As its name implies, the logical-content strategy utilizes various reasons and deductive logic in the development of personality test. Therefore, it is a deductive strategy. Here the test developers tries to logically deduce the types of contents or items that should measure the characteristics intended to be assessed. Suppose one wants to measure thirst behaviour, then statement like 'I frequently drink after three to four hours, will be considered logically relevant content but item like, 'I try to arrive at meeting in time, would be the example of logically irrelevant content. Thus logical content approaches depends upon a literal or veridical interpretation of the items of the test. Initial efforts to measure personality used this logical-content strategy. Using this strategy, the important personality tests developed were as under :

Woodworth Personal Data sheet was the first personality inventory that was developed using logical content strategy in 1917. This test was developed for screening World War I recruits for personality and adjustment. Later on, this inventory was known as *Woodworth Psychometric Inventory,* which sustained items to assess fears, sleep disorders, and other problems deemed symptomatic of psychoneuroticism. The Woodworth Personal data, sheet contained 116 questions to which the persons respond as either 'Yes' or 'No'. This test yielded a single score, providing a global measure of functioning.

Besides Woodworth Personal data sheet, some other personality tests using logical content strategy were also developed. Important such tests were the *Bell Adjustment Inventory, Bernreuter Personality Inventory, Mooney Problem Checklist* and *Symptom Checklist–90 Revised*. The Bell Adjustment Inventory and the Bernreuter Personality Inventory were developed in early 1930s. The former tends to assess adjustment in different areas such as home, health, social and emotional whereas the latter tends to assess six personality traits like sociability, confidence and introversion. Both those tests yielded not one but more than one overall score. In fact, these tests laid the foundation of many such modern tests of personality that yielded multiple scores.

Mooney Problem checklist using this logical-content strategy was published in 1950. This test contains a list of problems that frequently recurred in clinical case history and in the written statements of problems often reported by high school students. The testtakers who checked an excessiving number of items are considered to exhibit difficulties. Suppose if the testtaker checks items related to finances, it means that he is exhibiting financial problems.

Symptom checklist-90 Revised (SCl-90-R) is the modern example of logical-content strategy (Derogatis, 1994). This test consists of 90 symptoms such as faintness, poor appetite, poor sleep. The test intends to screen for psychological problems and symptoms of psychopathology. The testtakers are requested to indicate, using a five-point scale, to what extent and how much they have been distressed by each of the problems during the last seven days. Separate norms for male and females adult and adolescents have been provided in the SCL-90-R.

The chief advantage of the logical-content strategy is its simplicity and directness. These features made this strategy very popular in developing a test. However, the strategy has been criticised. In structured personality test, developed on the logic of face validity, the testtakers had ample opportunity for conscious attempts for manipulation of the results. As a consequence, tests based upon this strategy don't have features designed to detect response biases (Ellis, 1946, McNemar & Landis, 1935). Due to this exclusive reliance on such tests is not appreciable. These criticisms, to a greater extent, were removed by a new conceptualization in the personality testing to be called as the criterion-group strategy.

Criterion-group strategy

As its make name implies, in criterion-group strategy the test developer selects a criterion group or a group of individuals who share a characteristic such as schizophrenia. Subsequently, he selects a normal group or control group of normal people. The test is administered to both groups, that is, criterion group and control group and those items are located that make a distinction or contrast between these groups. In a nutshell, in this approach the responses are treated as diagnostic or symptomatic of the criterion behaviour with which they are found to be associated. The next step is to cross-validate the test. For cross-validating the test or scale, attempt is made to determine how well it distinguishes an independent criterion sample consisting of the persons also known to possess the characteristics to be measured from that of a control group. If the test significantly distinguishes the two groups, then it is said to have cross-validated. The next step in this strategy is to conduct some additional researches to determine the meaning of the situation when the testtakers endorse a large number of items on a particular scale. For example, an independents group of individuals securing high score on aggressiveness may be intensively studied and determined how they are described by others, how they themselves describe as well as the features of their family backgrounds, etc.

There are several personality tests that have been developed following this approach. However, here a discussion of only four such most popular tests will be done :

Minnesota Multiphasic Personality Inventory (MMPI), *California Psychological Inventory, Personality Inventory for Children* (PIC) and *Million clinical multiaxial Inventories* (MCMI).

Minnesota Multiphasic Personality Inventory (MMPI) is one outstanding example of criterion-group strategy. It is a true-false self-report questionnaire measure of personality developed in early 1940s by Hathaway (a clinical psychologist) and a neuropsychiatrist McKinely (1943). Originally, this inventory has 550 items and each item to be answered in terms of one of three options : 'True', 'False' and 'Cannot say'. There were ten clinical scales and four validity scales. The names of 10 clinical scales were : *Hypochondriasis* (HS), *Depression* (D), *Hysteria* (Hy) *Psychopathic deviate* (Pd), *Masculinity-feminity* (Mf), *Paranoia* (Pa), *Psychasthenia* (Pt), *Schizophrenia* (SC), *Mania* (Ma) and *Social Interoversion* (SI). The four validity scales were : *lie scale* (L), *Infrequency scale* (F), *Correction scale* (K) and *cannotsay* (or ?). The purpose of the clinical scales was to assess the concerned pathological traits and the purpose of validity scales was to check against carelessness, misunderstanding, malingering and other different kinds of responses sets and testtaking aptitudes of the testtakers. These validity scales have nothing to do with the validity of the test in the technical sense of the term.

MMPI has been revised and reconstructed into two separate versions: *MMPI-2* (Butcher, Dahlstrom, Graham, Tellegen & Kaemmer, 1989) and *MMPI-Adolescent* (MMPI-A) (Butcher, Williams, Graham, Archer, Tellegen, Ben-Porath & Kaemmer, 1992). MMPI-2 consists of 567 affirmative statements to which the testtakers respond in terms of either True or False. The first 370 items are identical to those in MMPI except for editorial changes and tend to assess 10 clinical scales and validity scales. The remaining 197 items constitute the new, revised and retained validity, content, supplementary and subscales of the complete inventory. Besides 10 clinical scales and nine validity scales, MMPI-2 has 15 content scales, 27 component scales 21 supplementary scales and 28 Harris-Lingoes subscales (Anastasi & Urbna, 1997). The nine validity scales are : *Cannot say score (?), Infrequency scale (F), Infrequency scale (FB), Infrequency scae (Fp), Lie scale (L), Defensiveness scale (K), Superlative self-presentation scale (S), Response Inconsistency scale (VRIN) and Response Inconsistency scale (TRIN)* (Butcher et al., 2017). Of 197 items, 107 are new items. The range of these items is very wide covering areas like general health, sexual, political and social attitudes, affective, neurological and motor symptoms, educational, occupational, family and marital questions, and some neurotic or psychotic behaviour like hallucination, delusion, phobias, ideas of reference, obsessive and compulsive states, sadistic and masochistic trends, etc. Of the 21 supplementary scales of MMPI-2, there are three new validity indicators for assessing care and veracity with which the testtakers respond to items. They are: Back F (Fb) scale, Variable Response Inconsistency scale (VRIN) and True Response Inconsistency scale (TRIN). Whereas Fb scale is the extension of the original F scale, VRIN and TRIN scales aim at detecting inconsistent or contradictory responses. Back F (Fb) assesses testtakers' falling bad (in the last half of the test). VRIN assess testtakers' answering similar/opposite question pairs inconsistently and TRIN assesses testtakers' answering questions all true/all false.

So far as the reliability and validity of MMPI-2 are concerned, the internal consistency of the inventory is lower. However, the test-retest validity is higher indicating that the scores remain somewhat consistent over time (Rajdev et al., 1994). Many researches regarding the validity of MMPI-2 done in various situations are available. Many of these researches was critical having charges regarding low validity.

Initial interpretation of MMPI-2 was done with the help of McCall's T, with a mean of 50 and a standard deviation of 10. Individual with a T scale of 70 was considered two standard deviations above the mean and was considered elevated for the MMPI. However, with the new norms of MMPI-2, T scores of 65 are now considered significant (Kaplan & Saccuzzu, 2002). One of the major shortcomings of the MMPI-2 is that the inter-correlations among clinical scales are very high, which suggests that items overlap between MMPI scales. An alternative version of the test called the *MMPI-2-RF* (Restructured Form) was published in 2008. It retains some aspects of the traditional MMPI assessment but adopts a different 338 of the original 567 items of MMPIL.

MMPI-2-RF retains only Restructured clinical (RC) scales developed in 2003 and was subsequently, subjected to extensive research. Validity scales of MMPI-2-were retained and includes an over reporting scale of somatic symptoms (F_5) as well as revision versions of the validity scales of MMPI-2 such as VRIN-r, TRIN-F-r, FP-r, Fs-r, L-v and Kv. There are also new scales that tend to assess somatic complaints. The MMP2-RF scales in contrast to most of MMPI and MMPI-2 clinical scales that are heterogeneous, are fairly homogeneous and tend to assess distinct symptom constellations or disorders. From a theoretical perspective, MMPI-2-RF scales assures that psychopathology is a homogeneous condition that is additive.

MMPI-Adolescent is somewhat a new form designed especially for the assessment of personality of adolescents aged 14 to 18 years. MMPI-A consists of 478 items and it incorporates most features of the MMPI and MMPI-2 including all the 10 clinical scales and three validity scales. Besides, it also incorporates items covering areas specifically relevant to adolescents such as school and family problems. In a nutshell, MMPI-A has original 10 clinical scales (D, Hs, Hy, Pd, Pt, Mf, Sc, Pa Ma and Si), six validity scales (L, F, F1, F2, ?, K, VRI-N and TRIN), 31 Harris-Lingoes subscales, 15 content component scales, the Personality Psychopathology Five[1] (Psy-5) scales (PSYC, AGG-R, DISC, NEGE and INTR), three social introversion scales (Social avoidance, Alienation and Shyness/self-consciousness) and six supplementary scales. Besides, there is also a shorten form of 350 items covering only the basic scales and validity scales.

Major criticism of MMPI-A are overlaps in what clinical scales measure, irrelevance of Mf scale as well as lengthy nature of the test. A restructured form of the MMPI-A called MMPI-A-RF is set to release shortly.

California Psychological Inventory :

California Psychological Inventory (CPI) was originally published in 1957 and its recent revision was published in 1987 by H.G. Gough. This revised third edition of CPI consists of 434 items. The inventory, though has been shortened from its original length of 480 items down to 462 in 1987revision and most recently to 434 items. According to the ETS (Educational Testing Services) test collection catalogue, the CPI contains 20 scales of which three are validity scales. For 13 of these 17 non-validity scales, items were selected on the basis of contrasted group responses against such criteria as social class membership, ratings, participation in extracurricular activities, etc. For the remaining 4 non-validity scales, items were originally grouped subjectively and subsequently, checked for internal consistency. The names of these 20 scales along with their abbreviations are presented in Table 14.2

The PSY-5 is a set of scales that assess dimensional traits of personality disorders originally developed from factor analysis of personality disorder content of DSM (Harkness et al., 1995). These scales are named as Aggressiveness, Psychoticism, constraint, Negative Emotionality Neuroticism, Positive Emotionality/extraversion.

Table 13.2: 20 scales of Californial Psychological Inventory

Scales	Scales
Dominance (Do)	Tolerance (To)
Capacity for Status (Cs)	Good Impression (GI)
Sociability (Sy)	Communality (Cm)
Social Presence (Sp)	Achievement via conformance (Ac)
Social-Acceptance (Sa)	Achievement via Independence (Ai)
Independence (In)	Intellectual efficiency (Ie)
Empathy (Em)	Psychological Mindedness (Py)
Responsibility (Re)	Flexibility (Fx)
Socialization (So)	Feminity (Fe)
Self-cantors (Sc)	Sense of well-being (Wb)

Of these 20 scales three scales, namely, Good Impression (GI), communality (Cm) and sense of well-being (Wb) are validity scales. GI is designed to detect subjects who fake good, Wb is designed to detect subjects who fake bad and Cm is designed to detect subjects who respond randomly.

Nearly half of items of CPI has been drawn form original MMPI and it intends to assess personality of normally adjusted adult individuals like MMPI, CPI shows considerable intercorrelation among its scales.

Another important component of this test is that the scales are based on 'folk' concepts of personality. These scales are called 'folk' because they attempt to assess personality themes that are cross-cultural and easily understood by laypersons and psychologists alike. Thus the inventory attempts to tap all these personality factors that arise without exception to some varying degree in human beings of all cultural groups.

Like MMPI-2, the scores on all CPI scales are converted in terms of standard score scale with a mean of 50 and SD of 10. The advantage of CPI is that it can be used with normal individual. MMPI and MMPI-2 do not apply to the normal individuals. Moreover, since CPI focuses on assessment of common interpersonal behaviour (that is, dominance, self-control, etc.) among normal adults, extreme scores on the some of the scales provide information on specific maladjustments the persons might be experiencing. In addition, the folk concepts used in CPI makes the inventory more adaptable to many cultures.

The major criticism of CPI is that the inventory fails to provide a parsimonious and theory-oriented description of normal personality. In addition, there is also lack of theoretical justification of the various criteria used in developing folk concepts. It is also said that the normative samples of CPI does not represent the general population. Approximately, 50% of the sample consists of high school students and 16.7% of the sample are undergraduate students. Persons working in professional occupations are underpresented in the normative sample (Atkinson, 2003). This affects the real interpretation of CPI scale scores.

Personality Inventory for children (PIC):

The PIC was developed through same 20 years of research by a group of several investigations at the University of Minnesota (Wirt & Lachar, 1981; Wirt et al., 1991). According to these investigator, PIC has not utilized MMPI items but it has been constructed on the same principles or methodology. The inventory is suitable for children and adolescents between ages 3 and 16 years.

The original PIC consists of a total of 600 items. It consists of 12 clinical scales, three validity scales and a general screening scale. Thus a total 16 scales have been incorporated in PIC. The 12 clinical scales intend to assess the children's cognitive development and academic achievement, psychological climate of the family as well as several emotional and interpersonal problems. The three validity scales are the *Lie scale*, the *Frequency scale* and the *Defensiveness scale*. The Lie scale has items that make the child appear in an unrealistically favourable light, the Frequency scale consists of rarely endorsed items and the Defensiveness scale intends to assess mental defensiveness about the behaviour of the child. The screening scale (such as Adjustment) is used to recognize children in need of psychological evaluation. One of the features that distinguish PIC from MMPI is that the true-false items (or questions) of PIC are not answered by the child rather they are answered by knowledgeable adult such as mother or father.

PIC has been revised and this revised version is called as PIC-R (Revised) in which items have been recordered and also reduced from 600 to 420. These 420 items have been grouped into three progressively longer parts : Part I (items 01 to 131), Part II (items 132 to 280 items) and Part III (items 281 to 420 items). It is to be noted that PIC-R is not a self-report inventory rather an inventory of observed behaviour. The reliability of PIC-R is satisfactory with test-retest reliability in the range of 0.81 to 0.92 and internal consistency coefficients in the range of 0.81 to 0.92. Several criterion-related validity studies have provided satisfactory degree of correlations with independent ratings from clinician.

PIC-R has again been substantially revised (Lachar & Gruber, 2001) and renamed as PIC-2. PIC-2 consists of 275 true-false items that are completed by parent or parental surrogate. It is most suitable for children having age of 5 to 19 years. It provides the most comprehensive, multiview perspective on child's emotional and behavioural adjustment in the school, home and community. There are two *complimentary* tests called the *personality Inventory for Youth* (PIY) and *Student Behaviour Survey* (SBS). PIY is filled out by the child and SBS is filled by the teacher.

Scale raw scores of PIC-2 are converted to T-scores with a mean of 50 and standard direction of 10. Higher the T scores, the greater is the probability of psychopathology or deficit. One of the common limitations of PIC is that the responses on scales of PIC may reflect, as least partly, the motivation, attitude and personal or cultural standards of the parent. As a consequence, some inconsistencies among the reports of different observers such as the two parents or between the parental report and self-reports of the children may occur. A very good way to deal with such inconsistencies is to gather and compare reports from more than one observer.

Millon Clinical Multiaxial Inventories (MCMI)

The Millon Clinical Multiaxial inventories consists of four versions of the same test, that is, MCMI, MCMI-II MCMI-III and MCMI-IV. The MCMI was published in 1977 and contains 11 personality scales and 9 clinical syndrome scales (Millon, 2008). The publication of MCMI corresponds with DSM-III. This test intends to provide information regarding personality traits and psychological disorders. MCMI was standardized on clinical populations and is meant for adults (18 and over). MCMI-II, a new version of MCMI, was published in 1987 and this version corresponds with the publication of DSM-III-R. This version consists of 13 personality scales and a clinical syndrome scale. Again, this inventory was revised in 1994 and it was called MMI-III to incorporate changes made in DSM-IV. MSMI-III consists of 175 brief, self-description statements to be marked 'true' or 'false' by the testtakers. It is highly grounded in Millon's biopsychosocial views of personality functioning and psychopathology. (Millon, 1981, 1990). This third version consists of 24 clinical scales grouped into four major categories, that is, *Clinical Personality Patterns*, *Severe Personality Pathology*, *Clinical syndromes* and *Severe syndromes*. The first two categories contain scales intended to assess enduring Axis II of personality disorders of DSM. Whereas the remaining two categories encompass some of the Axis I syndromes of DSM. The 11 scales included in clinical Personality Patterns are : *Schizoid, Avoidant, Depressive, Defendant, Histrionic, Narcissistic, Antisocial, Aggressive* (*Sadistic*), *Compulsive, Passive-Aggressive* (negativistic) and *Self-defeating*. There are three scales included in severe personality disorder: *Schizotypal, Boderline* and *Paranoid*. The seven clinical syndromes scale are : Anxiety, Somatoform, Bipolar: *Manic, Dysthymia, Alcohol dependence, Drug dependence* and *Post-Traumatic stress disorder*. Each of the 24 scales is based upon 12 to 24 overlapping items that often appear in as many as three different scales. Besides these subscales, there are three modifying indices (*Disclosure, Desirability* and *Debasement*) and a validity check intended to detect test-taking biases and atypical responses.

MCMI-IV, revision of MCMI-III, has been released in 2015 and this version consists of 195 true-false items. (Million, Grossman & Million, 2015). The changes in MCMI-IV have been done to accommodate changes in DSM-V. MCMI-IV consists of 15 personality scales, 10 clinical syndrome scales and 5 validity scales. The current scale composition of MCMI-IV consists of a total of 30 scales divided into 25 clinical scales and 5 validity scales. The 25 clinical scales are further divided into 15 personality scales and 10 clinical syndrome scale. The 15 personality scales are further divided into 12 clinical personality patterns and 3 severe Personality Pathology scales. The names of 12 clinical personality patterns are: *Schizoid, Avoidant, Melancholic, Dependent, Histrionic, Turbulent, Narcissistic, Antisocial, Sadistic, Compulsive, Negativistic* and *Masochist*ic. The three personality pathology are : *Schizotypal, Boderline* and *Paranoid*. The 10 clinical syndromes scales correspond with clinical disorders mentioned in DSM-V. These ten clinical syndrome scales are broken down into three severe clinical syndrome scales and seven clinical syndrome scales. The three severe clinical syndrome scales are : *Thought disorder, Major* *depression* and *Delusional disorder*. The seven clinical syndrome scales are: *Generalized Anxiety, Somatic symptom, Bipolar disorder, Persistent depression, Alcohol use, Drug use,* and *Post-traumatic stress disorder*.

The psychometric properties especially reliability and validity of different scales of MCMI-IV have been satisfactory. The test-retest reliability of MCMI-IV ranged from 0.73 to 0.93 with many values above 0.80 (Million, Grossman & Million, 2015). The support for the validity of MCMI-IV, is mixed one.

Factor-analytic strategy :

Factor analytic strategy of development of personality test is based upon factor analysis, which is a statistical technique for reducing the redundancy in a set of intercorrelated scores. A factor analyst starts with a database consisting of the intercorrelation of a large number of items of a tests. Subsequently, he factor analyzes these intercorrelations for finding out the minimum number of factors that account for as much of variability in the data as possible. Then, he attempts to label these factors by ascertaining what the items related to a particular factor, have in common. For Example, a test may have form scales: suspicion, defensiveness, hostility and aggression. These scales may correlate highly, which means that these four scales overlap in what they measure, that is, they share common variance. All the four scales may be related to paramid personality. If the factor analyst establishes that a substantial proportion of the variability in all the four scales is related to some common factor (paranoid personality, he possesses the sufficient ground to argue that the test actually has only one scale measuring paranoid personality. Some of the popular personality tests developed out of factor analytic strategy are : *Guilford-Zimmerman Temperament Survey, Cattell's Sixteen Personality Factor Questionnaire, Eysenck Personality Questionnaire,* and *NEO-Personality Inventory-Revised*. A discussion of these personality tests begins :

Guilford-Zimmerman Temperament Survey :

Guilford and his associates made a pioneer efforts in which they determined the intercorrelation of several tests and subsequently, factor analyzed the results in order to locate the main dimensions underlying all personality tests. The result of this initial attempt was a series of inventories, which was published in 1940s and which ultimately capsuled into a single inventory called Guilford-Zimmerman Temperament Survey (Guilford & Zimmerman, 1956).

Guilford-Zimmerman Temperament Survey (GZTS) was the first major factor analytic structured personality test, which had ten dimensions each measured by a set of 30 different items. The names of ten dimensions are : *general activity, restraint, ascendance, sociability, emotional stability, objectivity, friendliness, thoughtfulness, masculinity* and *Personal relations*. The test presents a list of statements, most of which are self-statements. The testtaker has to respond to each statement in terms of 'Yes' or 'No'. Like MMPI and MMPI-2, there are three verification keys, which intend to detect falsification as well as evaluate the validity of the test profile.

Today GZTS is only historical importance and it failed to gain prominence partly because it was overshadowed by MMPI and partly because of its subjective and arbitrary way of naming the factors.

Cattell's Sixteen Personality Factor Questionnaire :

The Sixteen Personality Factor (16 PF) questionnaire was developed not all on a sudden. Rather, it was developed gradually by R.B. Cattell who was the main author. In fact, Cattell had a strong background in physical sciences especially in physics and chemistry. Cattell had moved from physical sciences into the field of Psychology in the 1920s. As a consequence of such training, he had a belief that all fields can best be understood by first seeking to find the fundamental elements underlying within the concerned domain and then, develop a valid way to assess those various elements (Cattell, 1965). The major reasoning of Cattell was that human personality has basic underlying universal dimensions just as the physical world has the basic building blocks like oxygen and hydrogen. He was of view that if one is able to discover the basic building blocks of personality one can easily predict and understand various human behaviour. The primary goal of Cattell in developing 16 PF questionnaire was to discover the number of fundamental points of human personality and to assess those traits. Cattell applied the technique of factor analysis for attaining this goal. His interest in factor analytic technique was much developed due to famous psychologist/mathematician Charles Spearman under whom he served as a research assistant.

Over several decades of factor-analytic researches, Cattell and his co-workers developed the 16 Personality Factor questionnaire, currently in its fifth edition (Cattell et al. 1993). The 16 PF aims at assessing normal personality for ages 16 and over and yields 16 scores on 16 source traits. This 16 PF was first published in 1949, the second edition in 1956, third edition in 1962, and fourth edition having five alternative forms, were published between 1967 and 1969 (Cattell, Eber & Tatsuka, 1970). The fifth edition of the 16 PF released in 1993 contains 185 multiple-choice items, which are non-threatening and ask simple questions about daily behaviour and interests. The US version of 16 PF questionnaire was also restandardized in 2002 along with development of forms for children and adolescent. The 16 PF provides scores on 16 source traits and 5 global personality scales, all of which are bipolar (that is, high and low and each scale has a distinct and meaningful definition). The 16 PF test also includes three validity scales: *Impression Management (IM) scale, Acquiescence* (ACQ) *scale* and Infrequency (INF) scale. The fifth addition of the 16 PF containing 185 items is available in only one form. It takes about 35-50 minutes for paper-and-pencil version and 30 minutes by computer.

The 16 PF questionnaire provides scores on 16 primary (source) personality scales and five global personality scales. Table 14.2 presents 16 primary scales and Table 14.4 provides five global personality scales.

Table 14.3: 16 Primary traits measured by the 16 PF Questionnaire

Description of low scores	Primary Factor	Meaning of High Scores Description
Reserved, Cool, aloof impersonal, formal	Warmth (A)	Outgoing, warm, kindly, easy-going, participating
Less intelligent, concrete thinking, lower scholastic mental capacity, slow learner	Reasoning (B)	More intelligent, abstract thinking, fast learner, higher general mental capacity
Emotionally less stable, easily annoyed, changeable, easily upset	Emotion stability (C)	Emotionally stable, adaptive, faces reality calmly, mature
Humble, mild, easily led, docile, accommodating, cooperative	Dominance (E)	Assertive, aggressive, competitive, Dominant, bossy, stubborn.
Silent introspective, prudent, serious, restrained	Liveliness (F)	Expressive, spontaneous, cheerful, impulsive, happy-go-lucky, enthusiasm
Self-indulgent, disobeys rules, nonconforming, expedient	Rule-consci-ousness (G)	Conforming, moralistic, dutiful conscientious, rule bound
Timid, hesitant, shy, intimidated, expedient	Social Boldness (H)	Bold, uninhibited, spontaneous, Venturesome,
Tough minded, self-reliant, rough, utilitarian, objective	Sensitivity (I)	Sentimental, tender-minded sensitive, intuitive, refined
Unsuspecting, unconditional, adoptable, accepting, easy	Vigilance (L)	Suspicious, skeptical, vigilant, distrustful, oppositional, hard to fool
Practical, conventional, prosaic, solution-oriented	Abstractedness (M)	Absent-minded, impractical, imaginative, abstract, absorbed in ideas
Open, naive, genuine, forthright, unpretentious	Privateness (N)	Shrewd, polished, nondisclosing, private, discreet, diplomatic, socially alert, calculating
Free of guilt, confident, self-satisfied, unworried, self-assured	Apprehension (O)	Self-blaming, worrying, guilt-prone, apprehensive, insecure
Conservative, traditional, respecting traditional ideas	Openness to change (Q_1)	Liberal, analytical, flexibility, free thinking, open to change

Affiliative, group-oriented, sociable	*Self-reliance* (Q_2)	Resourceful, self-sufficient, self-reliant, solitary
Flexible, undisciplined, self-conflict, impulsive, loves disorder	*Perfectionism* (Q_3)	Self-disciplined, will power, self-sentimental, organized, perfectionistic
Tranquil, placid, relaxed, low drive	*Tension* (Q_4)	Frustrated, tense, driven, high energy, impatient

Table 14.4: Five global factors and corresponding primary factors.

Global factors	Primary factors
1. Extraversion/Introversion	A, F, H, N and Q_2
2. High Anxiety/Low Anxiety	C, I, O and Q_4
3. Receptivity/Tough Mindeness	A, I, M and Q_1
4. Independence/Accommodation	E, H, L and Q_1
5. Lack of Restraint/Self-control	F, G, M and Q_3

Factors B has not been included because it is not a part of hierarchical structure of personality in the manner other factors are :

Some other important features of the test are its provision of parallel *High School Personality Questionnaire* (HSPQ) now the *Adolescent Personality Questionnaire* (APQ) (Schuergh, 2001). Another parallel extension called *Children's Personality Questionnaire* (CPQ) are for ages 8 to 12 (Porter & Cattell, 1985). Early School Personality Questionnaire (ESPQ) is still another downward extension (Goan & Cattell, 1959). Lichtenstein, Dreger & Cattell (1986) also developed a *Pre-school Personality Questionnaire* (PSPQ). To extend the use of the test to clinical populations, items related to psychological disorders have been factor analyzed, resulting in 12 new factors (source traits) in addition to the 16 already discovered to assess normal personalities. These new 12 factors constituted a clinical instrument called *Clinical Analysis Questionnaire* (CAQ) (Delhees & Cattell, 1971). Table 14.5 presents 12 abnormal source traits measured by CAQ.

Table 14.5: Major Abnormal source traits measured by CPQ.

Normal source Trait	Factor	Abnormal source Trait
Low Hypochondriasis	D_1	High hypochondriasis
Zestfulness	D_2	Suicidal disgust
Low brooding discontent	D_3	High brooding discontent
Low anxious depression	D_4	High anxious depression

High energy euphoria	D_5	Low energy depression
Low guilt and resentment	D_6	High guilt and resentment
Low bored depression	D_7	High bored depression
Low paranoia	Pa	High Paranoia
Low psychopathic deviation	Pp	High psychopathic deviation
Low schizophrenia	Sc	High schizophrenia
Low Psychasthenia	As	High psychashenia
Low general psychosis	Ps	High general psychosis

With discovery of these 12 abnormal source traits and 16 normal source traits, Cattell believes that he has now identified the major source traits of both normal and abnormal personality. Of these various tests developed by Cattell and his coworkers, the 16 PF questionnaire has been very popular and has been translated into more than 30 languages with proper shapes of adaptation. These adaptations have been done with local standardization samples plus reliability and validity information.

Eysenck Personality Questionnaire (EPQ)

Eysenck Personality Questionnaire (EPQ) has been developed by H.J. Eysenck and his second wife S.B.G. Eysenck (1975). H.J. Eysenck, like Cattell, is considered as a psychometrist of personality. EPQ is the latest in the series of test developed by Eysenck to assess the major dimensions of normal and abnormal personality. Other tests to measure personality were the *Maudsley Personality Inventory* (MPI) and the *Eysenck Personality Inventory* (EPI).

Eysenck identified three major dimensions of personality: Extraversion/Introversion (E), Neuroticism/stability (N) and Psychoticism/socialisation (P). EPQ consists of scales to assess these dimensions. Later, it also incorporated a Lie (L) scale to assess the validity of the testtakers' responses. (Lajunena & Scherler, 1999) The EPQ consists of 90 statements to be answered as 'Yes' or 'No' and is considered most suitable for persons aged 16 and older. In 1985 a revised version of EPQ called EPQ-R was published (Lajunena & Scherler, 1999). EPQ-R has 100 Yes/No questions in its full version and 48 Yes/No questions in its shorter version.

All three dimensions E, N and P are biologically-based independent dimensions and are measured on a continuum. A brief description of these three dimensions is presented here :

1. Extraversion/Introversion (E) :

Extraversion (or extrovert) is characterized by being lively, sociable, active, assertive, sensation-seeking, carefree, dominant, surgent, venture and such persons like excitement and are much oriented towards external reality. On the other hand, introversion or introvert is quiet, introspective individuals who are more oriented towards inner reality and prefer a well-ordered life. Introverts are fond of books rather than people. Such person tends to plan ahead and 'looks before he leaps'. Eysenck has proposed the arousal theory to explain the differences in behaviours of extraversion and introversion. The theory maintains that there is an optimal level of cortical arousal and the performance of the person deteriorates as the arousal level exceeds or becomes lower than the optimal level. Arousal can be assessed through brain waves, or sweating or skin conductance. Extraverts, according to Eysenck, are underaroused and are, therefore, in need of external stimulation so that their arousal level may be at or above the optimal level of arousal. On the other hand, introverts are assumed to have higher levels of arousal in comparison to the optimal level. As a result, they are more sensitive to the stimulation. Thus any given level of incoming stimulation is amplified by high cortical arousability of introverts. Such persons are in need of peace and calm to bring down their arousal level to an optimal level. It has been reported that about 16 percent of people fall in extraversion, another 16 percent of people fall in introversion and remaining 68 percent lie in the midrange of Extraversion/introversion continuum, an area called ambiversion (Bartol and Bartol, 2008).

A review of literature reveals some interesting differences: between extraverts and introverts. (Pervin & John, 1997) Some of the major such differences are as under:

- Introverts are more sensitive to pain than extraverts
- Introverts become fatigued easily than do extraverts
- In case of introverts excitement interferes with their performance whereas it enhances performance in case of extraverts
- Extraverts are more active sexually in terms of frequency and different partners as compared to introverts.
- Introverts do better at signal detection tasks than do extraverts
- Extraverts prefer jobs involving interactions with other people whereas introverts tend to prefer more solitary jobs.
- Introverts prefer more intellectual forms of humour whereas extraverts like explicit sexual and aggressive humour.
- Extraverts are more suggestible than introverts.

2. Neuroticism/Stability (N)

Neuroticism (or emotionality) is characterized by negative affect such as being anxious, depressed, tense, irrational, shy, moody, emotional and showing guilt-feelings and low self-esteem. In a nutshell, neurotics are emotionally unstable individuals and exhibit anxiety level disproportionate to the realities of the situation (Eysenck, 1965). In his arousal theory Eysenck locates the seat of neuroticism in the visceral brain limbic system, which contains amygdala, hippocampus, cingulum, septum and hypothalamus. All these organs of the brain are involved in generating emotionality. People high in neuroticism have lower thresholds for activity in visceral brain and high responsibility of the sympathetic nervous system. (Eysenck and Eysenck, 1985) As a consequence, neurotics overreact to even mild form of stimulation. Since neurotics have low activation thresholds, they are unable to inhibit or control their emotional reactions. Emotionally stable persons, on the other hand, have high activation threshold and better emotional control. Therefore, they remain calm even under pressure. Such persons experience negative affect only in the face of very major stressors.

The relationship of these two dimensions, that is, extraversion-introversion (E) and neurotocism-stability (N) to four temperamental types distinguished by Greek physician Hippocrates and Galen define four quadrants. These consists of *stable extraverts, unstable extraverts, stable introverts* and *unstable introverts*. The major characteristics of these four types derived from factor analytic technique are as under :

- Stable introverts are more sensitive to pain than extraverts (Sanguine qualities) and are sociable, outgoing, talkative, easygoing, carefree, responsive, lively, preference towards leadership.
- Unstable extraverts: (choleric qualities): Touchy, restless, impulsive, aggressive, active, changeable, excitable and optimistic.
- Stable introverts: (Phlegmatic qualities): Passive, careful, Peaceful, controlled, reliable, even-tempered, calm and thoughtful.
- Unstable introverts: (Melancholic qualities): Moody, rigid, anxious, sober, reserved, unsociable, quiet, anxious and pessimistic.

3. Psychoticism/Socialisation (P)

Psychoticism (P) is not equivalent to psychosis such as schizophrenia although a schizophrenic would be expected to score high on the dimension of P. The dimension of psychoticism is characterised by poor concentration, poor memory, cruelty, disregard for danger and convention, considered peculiar by others, tough-minded, aggressive, unsympathetic, dogmatic, masculine, showing recklessness, impulsiveness, etc. Eysenck assumed that biologically psychoticism is related to maleness, which is very closely linked to secretion of androgens. (Eysenck, 1982) In addition, those who score high on P dimension show a relative lack of serotonin and the presence of certain types of antigens in their bodies. However, this speculation of Eysenck regarding the relationship between psychoticism and hormones remains to be confirmed by the researchers.

In general, the psychometric qualities of EPQ are satisfactory and even stronger than those found in most self-report inventories. The practical utility of the test has been supported by many researches.

NEO–Personality Inventory–Revised

NEO–Personality Inventory–Revised is the outcome of decades of factor-analytic researches done on clinical and normal adult populations. This inventory measures five broad dimensions and is based upon a common language descriptors of personality (lexical hypothetic) grouped together using factor analysis. These five factors are *extraversion* (E) *agreeableness* (A), *openness to experience* (O), *conscientiousness* (C) and *neuroticism* (N), often listed under acronyms OCEAN or CANOE. In fact, this inventory is based upon both factor analysis and a theory in item development and test construction. (Kaplan & Saccuzzo, 2002)

In three stages of test development and revision, Costa and McCrae (1992) developed the NEO-Personality Inventory Revised (NEO-PI-R) for measuring the above Big Five personality factors. Originally, they had concentrated on only three factors of Neuroticism. Extraversion and

Openness and therefore, the title was NEO-Personality Inventory. Later on, they added two more factors, that is, agreeableness and conscientiousness to conform to five-factor model. They have differentiated each of these Big Five factors into six more specific *facets*, which are more specific traits that together constitute the given domain. (*Cf* Table 14.6) Each facet it measured by 8 items. Therefore, the NEO-PI-R consists of a total 240 items. (5 factors × 6 facets × 8 items) For each item, the testtakers indicate the extent to which they agree or disagree on five-point rating scale; *strongly agree, agree, neutral, disagree and strongly disagree.*

Another important feature of NEO-PI-R is that it is available in two forms: Form S and Form R. Form S is for self-report and Form R is for outside observers such as spouse of a client. The internal consistency reliability of the inventory is highly satisfactory, that is, 0.86 to 0.95 for the domain scales and 0.56 to 0.90 for the facet scales. Temporal stability coefficient range from 0.51 to 0.83 in three-to-seven-year longitudinal studies. The evidence for validity coefficient of the inventory is substantial and is based upon correlations with other personality questionnaire such as Cattel's 16 PF and Eysenck's inventories, correspondence of ratings between self and spouse, etc.

Table 14.6: NEO-PI-R Big Five factors and associated facets

Extraversion (E)	: gregariousness, warmth, activity, excitement-seeking, positive emotions, assertiveness.
Neuroticism (N)	: angry hostility, Anxiety, depression, self-consciousness, impulsiveness and vulnerability.
Openness to Experience (O)	: actions, ideas, feelings, fantasy, values, aesthetics.
Agreeableness (A)	: altruism, compliance, modesty, trust, straightforwardness, tendermindedness.
Conscientiousness (C)	: dutifulness, competence, order, achievement-striving, self-discipline, deliberation.

A brief description of the five trait scales along with characteristics of high scorers and low scorers is presented below :

Extraversion (E). This scale assesses the quality and intensity of interpersonal interaction, activity, need for stimulation and capacity for joy. High scorers on E scale are characterized by being sociable, talkative, person-oriented, optimistic, affectionate and fun-loving. Low scorers on this scale are characterized by being reserved, sober, aloof, task-oriented, retiring and quiet.

Neuroticism (N) : This scale measures adjustment vs. emotional instability. It tries to identify persons prone to psychological distress, excessive cravings, unrealistic ideas, and maladaptive coping reactions. High scorers on this scale are characterized by being nerves, emotional, insecure, hypochondrical, inadequate, worrying whereas the low scorers are characterized by being relaxed, calm, hardy, unemotional, self-satisfied and secure.

Openness to Experience (O) : This scale assesses proactive seeking and appreciation of experience for its own sake as well as tolerance for exploration of the unfamiliar. Openness reflects degree of intellectual curiosity, creativity and depicts a preference for a variety of activities over a strict routine. High scorers on their scale are characterized by being curious, creative, original, imaginative, untraditional and showing broad interests whereas the low scorers are characterized by being unartistic, unanalytical, conventional, down-to-earth and showing narrow interest.

Agreeableness (A) : This scale assesses the quality of an individual's interpersonal orientation ranging from compassion to antagonism in feelings, thoughts and actions. In other words, this scale measures individual differences for social harmony. High scorers on this scale are characterized by being soft-hearted, trusting, helpful, forgiving, gullible, straightforward and of better-matured. On the other hand, those who are low scorers, are characterized by rude, suspicious, cynical, uncooperative, vengeful, ruthless, manipulative and irritable.

Conscientiousness (C) : This scale measures the individual's degree of organisation, persistence and motivation in goal-directed behaviour. In a nutshell, this scale measures conscientiousness, which is basically a tendency to show self-discipline and act dutifully. It is related to the way the people control, regulate and direct their impulses. High scorers on this scale are characterized by being reliable, hardworking, punctual, self-disciplined ambitious, perseverating and organised. Low scorers on this scale are characterized by being unreliable, lazy, careless, lax, aimless, negligent and weak-willed.

Theoretical strategy

As its name implies, the theoretical strategy begins with a formal or informal theory of personality and the construction of personality inventories is based upon this theory. The test constructor develops the inventory around a preexisting theory. Items of the inventory are made consistent with the theory. For example, if the theory hypothesizes that personality can be betters understood in terms of five major areas, theoretical strategy demands that every item in the scale be related to the characteristic being assessed. Prominent examples of theory-guided inventories are *Edwards Personal Preference Schedule (EPPS)* based upon Murray's Need-press theory of personality, *Myers-Briggs Type Indicator* (MBTI) based upon Carl Jung's theory of personality types and *Million Clinical Multiaxial Inventories*[1] based upon Million's highly influential theory of personality disorders. (Craig & Olson, 1995) Theoretical strategy differs from factor-analytic approach in the sense that here items are constructed according to demands of the theory whereas the factor-analytic approach often produces a retrospective theory based upon some initial test findings. Here we shall discuss those popular tests, which are based upon theoretical strategy.
Edwards Personal Preference Schedule (EPPS) :

---

1. Million Clinical Multiaxial Inventories have already been discussed under criterion-group strategy. It qualifies for both criterion-group strategy and theoretical strategy.

EPPS was the first attempt to assess Murray's psychogenic needs with structured personality inventory. In fact, Murray (1938) had proposed a theory of psychogenic needs and also a projective test called Thematic Apperception Test (TAT) to assess these needs. Much influenced by this theory, Edwards (1954, 1959) a Washington Professor, proceeded to develop a structured personality inventory for assessing these needs. Therefore, he selected 15 needs from Murray's list and constructed forced-choiced items with content validity for each need. This inventory was named as *Edwards Personal Preference Schedule (EPPS)*, which was a objective and non-projective personality inventory and attempts to assess fifteen normal needs or motives. The list of these needs with a brief description has been presented in Table 14.7.

The EPPS consists of 210 pairs of statement in which items from each of 15 scales are paired with items from the other 14. The items, thus, are in forced-choice format in which the testtaker must choose the one statement from each pair which, he thinks, the most representative one. The inventory also incorporates a consistency check by repeating 15 items in identical format.

Table 14.7: Description of 15 needs of EPPS

Need		*Description*
Achievement	:	A need to accomplish tasks in excellent way.
Abasement	:	A need to accept blame, criticisms and punishment
Affiliation	:	A need to form friendship and associations as well as to cooperative and converse sociably with others
Aggression	:	A need to fight and revenge an injury
Autonomy	:	A need to resist coercion as well as restriction and independent
Change	:	A need to seek new experiences and avoid routine
Deference	:	A need to conform to custom and to yield easily to the influence of an allied other.
Dominance	:	A need to influence and control others as well as to lead and direct
Exhibition	:	A need to make an impression or be centre of attention
Endurance	:	A need to follow through on tests and complete assignments
Intraception	:	A need to analyze the behaviour and feelings of other person
Heterosexuality	:	A need to be associated with members of opposite sex
Nurturance	:	A need to assist, protect and comfort someone who is helpless
Order	:	A need to plan well and put things in order
Succorance	:	A need to stay close to protector and to have one's need satisfied by a sympathetic person

Edwards preferred forced-choice format in his test for detecting faking and social desirability. The feature of EPPS is that pairs of statements in each item are matched for social desirability (Edwards, 1957). Since each statement in an item pair is of equal social desirability, the content of each statement will excert more pressure upon the testtaker in determining his choice. The result of the test is considered valid if the consistency checks for more than 9 out of 15 paired items. In EPPS, it is possible to make all possible pairwise comparisons between statements embodying the 15 needs. Therefore, the test produces a measure of relative strength of each of 15 needs. When the testtaker makes a choice, he selects between one of two needs because in EPPS, Edwards listed items for each of the 15 scale and then, paired them with items from the other 14 scales. It means that in each choice the testtaker selects one need at the cost of other. It automatically means that with this procedure, the testtaker selects the items on one scale relative to the selection of items on another and thus, produces an *ipsative score*, which compare the person against himself or herself and produce data that shows relative strength of each need for the person. In fact, ipsative scores present result in relative terms rather than in an absolute total. Two persons with identical ipsative scores may differ significantly in the absolute strength of a particular need.

The psychometric properties of the EPPS are satisfactory. The split-half reliability ranges from 0.60 to 0.87. The test-retest reliability with one week interval ranges from 0.74 to 0.88. The test has been validated against correlating the scores with 16 PF, California Psychological Inventory, Thematic Apperception Test, etc. In these studies, there are often statistically significant correlations among the scales of these scales and EPPS, but these correlations are usually low-to-moderate.

Despite its impressive use in college counselling, EPPS has been criticised on several grounds. Several studies have shown that like other personality inventories the EPPS can be faked despite the use of force-choice technique. Some data available raised doubt regarding EPPS' ability to control social desirability effects (Steward, Gimenaz & Jackson, 1995).

Myers-Briggs Type Indicator (MBTI)

MBTI has been developed in 1943 by mother-daughter duo, that is, by K.C. Briggs and her daughter I.B. Myers for assessing personality preferences. The test is based upon psychological-type framework of personality proposed by Carl Jung (Quenk, 2000). These two women had no formal training in psychology or measurement but were highly inspired by the research work of Carl Jung about different psychological types. (Cohen & Swerdlik, 2005). This inspiration was instrumental in developing MBTI. MBTI is scored on four theoretically independent or orthogonal measures: *Extraversion-Introversion (E-I). Sensing-intuition (S-I). Thinking-feeling (T-F) and Judging-perceiving (J-P)*. Although the testtaker's scores on each dimension are continuous, his scores are generally summarized in a typological manner. Since there are two poles in each dimension, the number of possible personality types is $2^4 = 16$. For example, a testtaker may score more towards Introversion, sensing, thinking and judging and therefore, the resulting type would be ISTJ. The other fifteen personality types are : ISFJ, ISTP, ISFP, ESTP, ESFP, ESTJ, ESFJ, INFJ, INTJ, INFP, INTP, ENFP, ENJP, ENFJ and ENTJ.

MBTI is most suitable for adults and adolescents aged 14 years or over. The test is often used for conflict management, leadership and coaching, communication styles career exploration and team development. The MBTI has also been successfully used to study relationship between personality and financial success. (Mabon 1998) The test is available in 30 languages and it takes about 15-25 minutes in its completion (Miller et al. 2013)

Various versions of MBTI are available. Form M consisting of 93 items, Form Q consisting of 144 items (93 items from Form M plus 51 additional items), Form G consisting of 126 items, Form F consisting of 166 items, Form J consists of 290 items (94 items from Form F plus 196 additional items) and From K consists of 131 items (94 items from Form G plus 37 additional items). Each form is used for a specialized purpose. Of these, Form M is referred to as standard form since 1998. Apart from these forms, there is another form designed for children ages 8 to 18 years. This form has 70 items and is named as Murphy-Meisgeier Type Indicator (MMTIC).

The MBTI has satisfactory psychometric properties. Test-retest coefficients range between 0.73 and 0.83 (E-I), 0.69 and 0.87 (S-I), 0.56 and 0.82 (T-F)) and 0.60 and 0.87 (J-P). The internal consistency coefficient range between 0.55 to .065 (E-I), 0.64 and 0.73 (S-I), 0.43 to 0.75 (T-F) and 0.58 and 0.84 (J-P). So far as the validity coefficients are concerned, numerous studies have been reported and test is correlated against several other personality inventories like 16 PF, California Psychological inventory, NEO-Personality Inventory, Strong Interest inventory, Adjective checklist etc. The validity coefficients of the four scales range between 0.66 and 0.76 (E-I), 0.37 and 0.71 (S-N), 0.23 and 0.78 (T-F) and 0.39 and 0.73 (J-P).

In literature no evidence exists that MBTI correlates with various measures of job performance. Therefore, MBTI is not considered appropriate for job screening, placement and selection.

Besides EPPS and MBTI, there are some other personality tests, which are based upon some theories of personality. For example, Jackson Personality (JPI), Personality Research Form (PRF), Jenkeins Activity Survey (JAS) and State-Trait Anxiety Inventory (STAI) are theory-grounded inventories. JPI and PRF are based upon Murray's (1938) theory of needs (Kaplan & Saccuzzo, 2002). JPI is used on normal persons for assessing various aspects of personality including cognitive, interpersonal and value orientations. PRF measures the various needs proposed by Murray. PRF is available in several forces. The longer forms tend to provide 22 scale scores each having 20 items including two validity scales—Infrequency and Desirability. The shorter forms have 15 scales only, each scale having 20 items. JPS is designed to assess Type-A coronary-prone behaviour pattern STAI is designed to assess state anxiety and trait anxiety. STAI consists of 40 items, of which 20 items assess state anxiety and 20 items assess trait anxiety. State anxiety is defined as the transitory feeling of being worry or fear whereas trait anxiety is relatively stable tendency of an individual to respond anxiously to the threatening situation.

Behavioural Assessment Methods

Personality is also measured through various behavioural assessment methods. In behavioural assessment the focus of assessment is *behaviour* and not traits or states. In other words, in behavioural assessment the focus is on what an individual does in given situation rather than on inferences about what traits or attributes he possesses more globally. (Mischel, 1968) The behavioural assessment is conducted through various means such as behavioural observation and rating scales, interview analogue studies, situational performance methods and self-monitoring. A brief discussion follows :

Behavioural observation and Rating scales :

In behavioural observations, as its name implies, the researchers, clinicians or counsellors watch the activities of the targeted persons and maintain some kind of records of those activities. Sometimes researchers, clinicials or counsellors don't themselves serve as observers rather they assign this work to some trained assistants. The examples of behaviour observation are observing the children in a play room by a child psychologist through a one-way glass and observing the employees doing their work at their workplace by the industrial psychologists.

Behavioural observation may take many forms. One form may be that in which the observer may record the running events using tools such as paper and pencil, a video-film, a still camera or a cassette recorder. Here the observer acts as a naturalist. Mehl and Pennebaker (2003) used such behavioural approach in the study of social life of students. In fact, they recorded the conversations of 52 undergraduates across 2 two-day periods by means of a tape recorder. Another form of behavioural observation is to use what is called *behaviour ratingscale*, which is defined as a pre-printed sheet containing boxes having coded terms on which the observer notes the presence or intensity of the target behaviours. Rating scales are considered useful because they provide some standardized format. Such scales help in focusing rater's attention on all relevant behaviours to be assessed so that some are not overlooked or some others are weighed too heavily.

However, there are some problems with rating scales. Often, there is poor agreement among raters in their evaluation of the same behaviours of the same individuals. Another problem involved in rating scales is the tendency of the raters to be unduely influenced in the overall evaluation of the individual by one or a few favourable or unfavourable traits. This is called *halo effect*.

Interview

Personality is also assessed through interview. A clinical psychologist often use interview in diagnosis and treatment of the patients. Employers use the technique of interview in evaluating the job applicants and employees for job promotions. In personality assessments both structured and unstructured interviews are used. In structured interview the content of the questions and even the manner in which they are to be asked are carefully preplanned and the interviewer tries not to deviate in any way from the structured format so that meaningful comparisons can be made between different individuals. In the unstructured interview the questions to be asked and direction that will be taken are not planned before hand and therefore, the interview becomes highly personalized. Sometimes, unstructured interview becomes so loose that little objective information is derived from it.

Self-monitoring

Self-monitoring may be defined as the act of systematically observing and recording one's own behaviour and/or events related to that behaviour. In other words, self-monitoring refers to the persons observing and recoding their own behaviours, thoughts and emotions. Thus self-monitoring, different from self-report, depends upon the observations of the behaviour of interest at the time and place of its actual occurrence. (Cone, 1999) In a nutshell, the person is asked to maintain behavioural logs over some pre-determined time. Such a log easily provide a running record of the intensity, frequency and duration of certain target behaviour along with the stimulus conditions that accompanied such conditions and item consequences that followed. Self-monitoring has been successfully used for recording specific thoughts, feelings and behaviour. The ultimate success and utility of self-monitoring depends upon competence and motivation of the assessee. However, some scientific methods have been devised to ensure compliance in the process. (Wilson & Vitousek, 1999) For example, a handheld computer has been devised to provide aid in such process by providing beep as a cue to observe and record behaviour. (Schiffman et al., 1997) One common problem in self-monitoring is that it incorporates the potential problem of reactivity which refers to the possible changes in assessee's behaviours and thinking and therefore, in performance. For example, if a person is on weight-loose programme and self-monitoring himself, he may not show any inclination towards cheese or cheese-made foods. Such effect may be counteracted with proper training and education to the assessee. Another problem is that sometimes persons are inaccurate or may deliberately distort their observations or recordings for various reasons. There are also persons who are found to resist the entire procedure of self-monitoring.

Despite these problems, self-monitoring is an useful and efficient technique of behavioural assessment of personality variables.

Situational Performance Measures :

Situational performance measures may be defined as the procedure in which the behaviour of the individual is observed and evaluated under some standard set of situation or circumstances. In such measures, performance of some specific task under real or simulated conditions is assessed. A person interesting in getting driving licence for road transport undergoes a road test. This is an example of situational performance measure. Another good example of situational performance measure is leaderless group technique. In this technique a group of individuals are given a task to complete and the observer records information related to the individual's initiative, cooperative, leadership and other related variables. An individuals well know that their behaviours are being observed. Intentionally vague instructions are provided to the individuals and there is no one is position of leadership. Individuals themselves determine how they will accomplish the task and who will be doing what. Such leaderless group situation provides an opportunity to observe the degree of cooperation, initiativeness, leadership, etc, by each individual and the extent to which each individual is able to function with team spirit. Besides these, situational tests have also been used as technique of assessing.

Psychophysiological methods :

Behaviourally oriented clinicians and researchers tend to use psychophysiological methods for understanding and predicting behaviour. These experts tend to study the various physiological indices such as heart rate and blood pressure, which are readily influenced by psychological factors. Hence, the name psychophysiological methods. The important psychophysiological methods used to study behaviour are *biofeedback, plethysmograph, lie detector or polygraph.*

Biofeedback may be defined as a class of psychophysiological assessment techniques designed to display and record a continuous monitoring of some selected psychological processes such as heart rate, blood pressure, alpha wave, EEG activity etc. In fact, in this technique the person learns to control autonomic responses such as blood pressure, heart rate, EEG activity, etc by attending to the output of a device, which monitors the response continuously. Thus, some aspects of the person's physiological functioning is monitored by the apparatus and then, information is fed back to the person in form of some auditory or visual signal. The person is required then to modify that signal by changing physiological function. Early experimentation using biofeedback or human beings have demonstrated a capacity to produce certain types of brain waves or command (Kamiya, 1968). Since then, biofeedback has been successfully used in several therapeutic and assessment-related applications. (Zhang et al., 1997)

Plethysmograph is another device that records changes in the volume of a part of the body occurring due to variations in blood supply. Researches have shown significant difference in the blood supply of normal and psychoneurotic groups by using plethysmograph for assessing blood supply in the forearm. A special type of Plethysmograph designed to assess changes in blood flow to penis, called penile Plethysmograph has been successfully used in the assessment of male sex offenders.

Lie detector or polygraph (literally more than one graph) is another technique in which the interviewer or polygrapher asks a series of questions and the person has to give reply. Whenever the person speaks lie, the detectable physical changes occur and these changes are recorded by the lie detector variously called as a graph, chart or polygram. However, the reliability of the judgement, made by polygraphers is a debatable matter. Moreover, the polygraphic instrument is also not much standardized (Abrams, 1977).

Role playing has also been successfully used in behavioural assessment following the suggestion of Rotter and Wickens (1948). In this technique the person enacts a certain role in a simulated situation and through a careful observation and recording of the behaviour done, an inference is drawn about the personality variables of the role player. The technique of role playing has been successfully used in the assessment of social skills and assertiveness Twentyman and McFall (1975) have studied social skills in shy males by using a role playing technique in which the subjects faced six social behaviour situations that required them to play a role. They are instructed to respond as aloud as possible assuming that they were in real situation. Once a subject started speaking, a trained female assistant responded to the efforts of the subject. The conversation continued until the subject terminated the conversation. Likewise, McFall and Lillesand (1971) utilized the technique of role playing in assessing assertiveness. Such technique provides a simple and efficient means of assessing the person's behavioural skills and deficits.

As a concluding comment regarding behavioural assessment techniques, it can be said that their similarity resides less in formal structure than in goals. Their goals that bind them together, are essentially the accumulation of information about the behaviour in question, the conditions or stimuli that produce that behaviour and maintain that behaviour and the consequences that follow that behaviour.

Summary and Review

- Personality assessment refers to the process of measurement and evaluation of psychological traits, states, values, interests, attitudes, etc. and/or related individual characteristics.
- Traits refer to any distinguishable relatively enduring way in which one individual differs from the other. They are more or less consistent ways of behaving. 'State' is a mini-manifestation of trait. It means that state is indicative of temporary exhibition of a trait of personality. Personality type is a constellation of traits and states having similar pattern.
- Projective tests are the measures of personality based on projective hypothesis. Some unstructured stimuli are shown to the testtakers who respond to them. It is assumed that through these responses the testtaker projects his motives, conflicts and interests.
- Projective tests have been classified into different categories such as Association techniques, Pictorial techniques Verbal techniques and Expressive techniques.
- Important projective tests placed under the category of Association techniques are Rorschach Inkblot test Heltzman Inkblot test and Somatic Inkblot test.
- Major Projective tests covered by Pictorial techniques are : Thematic Apperception test, and variants of TAT like CAT, TEMAS, APT, GAT and Rosenboverz Picture-Frustration study.
- Verbal techniques of projective tests include Words, Association Test and Sentence completion Test.
- Important expressive techniques include Draw-A-Person Test and House-Tree-Person Test.
- Projective tests have been criticised on many grounds. Major criticisms are related to their assumptions, situational variables, and various other psychometric considerations.
- There are also some structured Personality tests, which assesses the personality in some standardized ways. The structured personality tests have been developed using various strategies such as logical-content strategy, criterion group strategy, factor-analytic strategy and theoretical strategy.
- Important structured personality tests developed under logical-content strategy are Woodworth Personal Data sheet, Mooney Problem checklist, Bell Adjustment Inventory, Bernreuter Personality Inventory, etc. The chief disadvantage of personality tests developed using this strategy is that such tests fail to detect response bias because the

content of the test provide ample opportunity for testtakers to manipulate the true responses.
- Important tests developed under criterion-group strategy are MMPI, CPI, PIC and MCMI. Of these, MMPI is the most popular test, which measures pathological traits of human personality.
- Major personality tests developed under factor-analytic strategy are Guilford-Zimmerman Temperament Survey, Cattell's Sixteen Personality Factor questionnaire, Eysenck Personality Questionnaire (EPQ), and NEO-Personality Inventory-Revised.
- Some important personality tests utilizing theoretical strategy are Edwards Personal Preference Schedule (EPPS) and Myers-Briggs Type Indicator (MBTI).
- Personality has also been assessed by various behavioural assessment methods such as by behavioural observation, rating scales, situational performance measures, psychophysical methods, playing, etc.

Review Questions

1. Define projective test. Discuss its important assumptions.
2. Giving examples outline a plan for classifying projective test.
3. Discuss the main features of Rorschach test.
4. Present the major characteristics of TAT. Discuss some of the important variants of TAT.
5. Discuss any two important projective techniques belonging to verbal techniques.
6. Point out the major differences between verbal projective techniques and expressive projective techniques.
7. Discuss the major features of MMPI. Discuss its important variants.
8. Discuss the important features of NEO Personality Inventory-Revised.
9. Make a comparative study of MMPI and MCMI
10. Outline the important behavioural assessment methods used for assessing personality.
11. Write Short notes on the following :
    (a) Traits, States and Type
    (b) MBTI
    (c) Cattell's Sixteen PF questionnaire
    (d) EPQ.

Chapter – 15

Measurement of Interest, Values and Attitudes

Learning Objectives
- Interest Measurement
- Strong Interest Inventory
- Jackson Vocational Interest Survey
- Self-Directed Search
- Minnesota Vocational Interest Inventory
- Career Interest Inventory
- Value Scales
- Study of Values
- Rokeach value
- Measurement of attitudes
- Thurstone Scale
- Likert Scale
- Guttman Scale
- Semantic Differential Scale
- Summary and Review
- Review questions
- Key Terms

Key Terms:

Attitude, Interest, Value, Realistic, Investigative, Artistic, Social, Enterprising, Conventional, Theoretical values, Artistic value, Social value, Political value, Religious value, Instrumental value, Terminal value, Lost-letter technique, Equal-appearing interval method, Method of summating rating, Cumulative scaling, Semantic differential scaling.

The measures of interests and attitudes have been considered very vital because they reflect some basic information about personality. When one is able to know about the interest and attitudes of a person, he is, to a greater extent, is able to predict about the academic and occupational achievement of that person. Not only this, it also enables to draw some important conclusions about his interpersonal relations as well as about the likely enjoyment the person my derive from different phases of his life. Psychologists have made encouraging efforts to assess both interests and attitudes. As a consequence, several tests are available for both these constructs. One important feature of the tests of both interests and attitudes is that they *cannot* be rigidly classified to such categories like interests and attitudes. In fact, they tend to overlap. For example, a test designed to measure the strength of interest in investigative occupation, can also be said to assess the person's attitude towards pure science. In the present chapter a lucid discussion of various important measures of interests and attitudes will be made. Such discussion will be of immense importance for researcher.

Values are also related to the life choices and are often discussed with interests and attitudes. Many researchers have been conducted in this field and as a consequence, some interesting and wide-ranging investigations of value studies have occurred across cultures. In the present chapter, attempt has been made to focus on some important measures of value. However, most of the formal assessment of values is being incorporated within instruments designed to facilitate correct decision making as well as within the instruments designed to assess motives and work-related attitude.

Interest of Measurement

The interest of the person is measured by interest inventory, which is defined as an instrument to provide information about general patterns of likes and dislikes of the persons. The common observation shows that people vary widely in their interest patterns. For example, if the persons are given choices about how to spend leisure time, one might choose to read books while other may choose to engage in gossip and still other may choose to do some physical activities.

Many interest inventories have been developed for assessing the likes and dislikes of the persons. Here a discussion of only some major interest inventories would be undertaken. These are: *Strong Interest Inventory*, *Kuder Occupational Interest Survey*, *Jackson vocational Interest Inventory*, *Self-Directed Search*, *Minnesota Vocational Interest Inventory* and *Career Assessment Inventory*. A discussion follows :

Strong Interest Inventory (SII)

Strong Interest Inventory (SII) is the current version of the test originally developed by E.K. Strong who first publishes it in 1927 and then, named as Strong Vocational Interest Blank (SVIB). A very important revision of SVIB was published in 1974 and then, it was called as *Strong-Campbell Interest Inventory* (SCII). The greatest boost to the credibility of the inventory came with subsequent introduction of third revision in 1981 and fourth revision in 1985. In fact, the inventory has been revised six times over the years reflecting significant and continued development in the field. (Donnoy, 1997) This was all done with the co-author Professor Jo-Ida Hansen. The sixth and the latest revision of the inventory was done in 2004. The latest revision incorporates all of the best researches over the decades of Strong, Campbell and Hansen as well as the popular typology of John L. Holland.

The latest version of the inventory consists of 291 questions asking the testtakers to indicate his preference for wide range of occupations, school subjects, activities as well as about the types of people. The inventory *never* measures abilities and skills of the testtakers rather it measures the interest of the testtakers. This latest version measures the testtaker's interest in six areas: *Occupations* (107 items), *Subject areas* (46 items), *Activities* (85 items), *Leisure Activities* (28 items), *People* (16 items) and *Your characteristics* (9 items). The first 282 items of the inventory covering the first five areas are answered by the testtakers, by choosing one of the, following options: 'strongly like', 'like', 'indifferent', 'dislike' and 'strongly dislike'. The nine items of 'your characteristics' are answered by the testtakers in the same way but with different options such as 'strongly like me', 'like me', 'don't know', 'unlike me' and 'strongly unlike me'. This newly revised version of the inventory typically takes about 30-45 minutes in its completion.

The results of the inventory include the scores derived on the following scales/codes.

- Scores on the level of interest on each of the six Holland codes or General occupational Themes (GOT), which includes Realistic (R), Investigative (I), Artistic (A), Social (S), Enterprising (E) and Conventional (C) called R-I-A-S-E-C. Holland (1985) had concluded that there are six themes of vocational interest that represent six different types of individuals and six types of work environments. He further expressed the view that the salience of one or more of these themes in a person could be identified by the individuals expressed likes and dislikes for occupations.
- Scores on 30 basic interest scales (such as art, science, teaching, social service, etc.) These scales can best be understood as subdivisions of GOT.
- Scores on 244 occupational scales which basically reflect the similarity between the testtakers' interest and those of people working in each of 122 occupations.
- Scores on 5 Personal Style Scale each of which reflects preference for broad styles of living and working.
- Scores 3 Administrative Scales that tend to provide information about the testtaker's responses including the total number of responses, percentages of 'likes', 'dislike', 'indifferent' responses as well as about total number of infrequent responses. In a nutshell, these scales tend to provide information about test errors or unusual profiles.

The strong Interest inventory has higher degree of test-retest reliability over various time intervals (Thorndike & Thorndike and Christ, 2015). The inventory is high in both concurrent validity and predictive validity (Donnay, 1997).

Jackson Vocational Interest Survey (JVIS)

The JVIS was developed by Jackson (1977). It is a Career interest assessment considered appropriate for use with students of high school, college and university as well as for career planning with adults. It consists of 289 pairs of statement describing job-related activities. The survey incorporates forced-choice format that asks testtakers to indicate preference between two equally popular interests minimizing the susceptibility to response bias and providing help to the testtaker in discriminating between career interest. The JVIS contains 34 basic interest scales, which cover 26 work role and 8 work styles. Work role indicates what a person does on the job. Some of these jobs are very closely associated with a particular occupation or type of occupation. Work styles reflect the preferences for working environments or situations in which a particular type of behaviour is expected. Such styles are directly or indirectly related to the values of the individual. The survey is equally applicable to both sexes although separate percentile norms for male and female population are available A high score on any of 34 basic interest scale indicates the person interest in things that he does in specific field of work and the way he is expected to act in that work context.

The scores derived from 34 basic interest scales are modelled after Holland's six themes, and include 10 General Occupational themes such as Expressive, Practical, Assertive, Logical, Inquiring, Socialised, Helping, Communicative, Conventional and Enterprising. Here the test developer has also provided a measure of Academic satisfaction, indices of Response consistency and Infrequency as well as the measure of unscorable responses.

Kuder Occupational Interest Survey (KOIS)

Frederic Kuder developed interest inventories working over several decades. The first of these inventories was *Kuder Preference Record-vocational* published in 1939. This inventory has some unique features not found in Strong's inventories. One such feature was the use of forced choice triad items where the testtaker has to decide which of the three activities they would like most and which least. Another distinguishing feature of this inventory was that scores were found not for specific occupations but for 10 broad interest areas, namely, *mechanical, scientific, outdoor, computational, persuasive, artistic musical, social science, literary and chemical*. The Kuder Preference Record underwent several revision. As a result of revision and downward extension of Kuder Preference Record, new inventory called the Kuder General Interest Survey (KGIS) was developed. The KGIS was designed for grade 6 to 12 and uses very simpler language and easier vocabulary A still later revision resulted in what is called as the Kuder Occupational Interest Survey (KOIS).

KOIS is a self-report vocational interest test mostly used for vocational guidance and counselling (Kuder & Dianmel, 1979; Kuder & Zytowski, 1991). Making a simple comparison of Kuder's Vocational Interest test to Strong Interest Inventory, it can be said that whereas Strong employs a general reference group, that is, he compares the interest of the person to those of certain groups of people holding certain occupations, Kuder pays attention on measuring the person's broad areas of interest. And the testtaker's score on each occupational scale is expressed as a correlation between his interest pattern and the interest pattern of the particular occupational group. The KOIS yields the testtaker's scores along ten vocational interest scales such as mechanical, outdoor, chemical, scientific, literary, social science, pervasive, artistic, musical and computational. It is a paper-and-pencil test, which consists of 100 forced-choice triads of activities. It takes about 30 minutes to complete. The results of KOIS are presented as Percentile scores separately for men and women. It then compares the testtaker's scores on these 10 scales to scores obtained by individuals holding certain professions.

The internal consistency reliability of vocational interest scales range from 0.47 to 0.85 with a median of 0.66. The temporal stability coefficient over fortnight period ranged from 0.86 to 0.90. The validity of KOIS is satisfactory and is based upon factor analysis and scale scores matching the actual occupations of the participants. It has also a dependability scale, which indicates caution in interpreting the results.

Self-Directed Search:

Holland (1985,1997) developed the self-Directed Search (SDS), which is self-directed, self-scared, and self-interfered vocational counselling instrument. The most important unique feature of SDS is that it was designed to be used for career exploration without taking any help of a career counsellor. In other words, since SDS is self-explanatory, an individual can complete score and interpret the instrument on his own. The SDS is usually completed individually, but it can also be used in small groups. Recently revised in 1994, the SDS consists of two booklets: the *assessment booklet* and the *occupations Finder*. This interest test can be completed within 40 to 50 minutes (Holland et al, 1997).

The SDS is based upon Holland's model, which specifics that people and working environments can be classified according to six basic types: *Realistic, Investigative, Artistic, Social, Enterprising* and *Conventional*. These six occupational themes or personality types are together known as RIASEC. The assessment booklet of SDS provides the individual an opportunity to consider preferred activities and competencies and to obtain scores on six occupational themes. The scores on three highest themes are used to develop a three letter summary code. For example, an individual who scores highest on Artistic theme, followed by Investigative and Social, would receive three letter summary code of AIS. Subsequently, the individual refers to the occupations Finder booklet for a list of occupations associated with that particular letter code. The occupations Finder contains more than one thousand occupations, each of which has been classified according to the Holland codes. The individual is encouraged to learn more and more about occupations associated with three letter code.

Researches have shown that the reliability coefficients of SDS are generally satisfactory for the summary codes. The construct validity and the predictive validity of the SDS tend to fluctuate according to the sex, age, educational level and distribution of types (Holland *et al*, 1994). Major criticisms of SDS canter around some of its scoring and interpretive procedures outlined (Manuele-Adkins, 1989).

Minnesota Vocational Interest Inventory (MVII)

The Minnesota vocational Interest Inventory (MVII) is designed for those high school students (grade 9-12) who don't plan to attend college but emphasises upon skilled and semi-skilled trades. It is modelled after SVIB scales. The MVI has nine basic interest scales as well as 21 specific occupational scales. The nine interest scales are: *mechanical, health service, office work, electronics, food service, carpentry, sales office clear hands and outdoor.* Some of the important 21 non-professional scales are plumber, carpenter, truck driver, mild wagon driver, machine operator, mechanist electrician, radio-TV repairman, hospital attendant, ware houseman etc. The MVII is, thus designed for the purpose of measuring the individual's similarity of interests with various professional occupational groups. It is used extensively by the military and by guidance programmes for individuals not inclined to go to college for higher studies and specialisation. It is available in both the expendable and reusable booklet format. In the expendable booklet format, the direction provided make the inventory practically self-administering. The inventory takes about 45 minutes—total working time for its completion. A separate answer sheet and a reusable booklet would, although time consuming, tend to indicate that it is possible to attempt what is called central administration procedures.

Career Assessment Inventory (CAI)

The career assessment Inventory (CAI) was developed by Johansson (1991) in order to fill the gap created by the strong interest inventory, which focused only on professional occupations, making it less useful for those who don't opt for college degree but seek immediate job entry. There are two visions of the inventory; *vocational version* and *extended version* modelled on the pattern of Strong Interest Inventory. The vocational version of the CAI provides scores on Holland's six general occupational themes, 22 basic interest scales and 91 occupational scales. The 91 occupational scales vary widely in type and level (such as truck driver, nurse, school teacher, etc). The enhanced version consists of 370-items using a 5-point scale for responses. It takes about 35 to 40 minutes in its completion. This enhanced version provides information for 111 occupations having wider distribution of educational requirements. A very important feature of CAI is that the scores are not reported separately for males and females.

The psychometric properties of CAI are highly satisfactory. Researchers have shown that the test-retest reliability ranged from 0.91 to 0.96 for the general theme scales, 0.88 to 0.95 for the basic intension scale and 0.81 to 0.96 for the occupational scales (Zarella & Schuerger, 1990). The internal consistency reliability ranged from 0.89 to 0.92. Of course, it is a matter of concern in assessing the validity of the CAI especially when judging it against the Strong Interest Inventory as well as its larger sample sizes.

In India some interest inventories have been developed. Important ones are *Career Maturity Inventory* by Nirmala Gupta, *Multiphasic Interest Inventory* by S.K. Bawa, *Comprehensive Interest Schedule* by Sanjay Vohra, Educational *Interest Record* by S.P. Kulshrestha, *Vocational Interest Record* by S.P. Kulshrestha and *Interest Inventory* by J.S. Sodhi and H. Bhatnagar. All these Indian interest inventories have been published by National psychological corporation, Agra.

Value Scales

A value in life of a person is considered very important energizer. It may be defined as a shared and enduring belief about the ideal mode of behaviour or some end state of existence (Gregory, 2004). The values held by the person shape his interest in life, which, in turn, shape his life choices. Moreover, values not only link the individual to the world of work but they are also interwined in moral, spiritual and religious matters. The assessment of values is considered important because values instill action, shape attitudes and guide how to influence others in a better way. In a nutshell, values are considered important because they tend to provide a pervasive framework for personal actions and judgements.

There are two important measures of personal values. *Study of Values* and *Rockeach Value Survey*.

Study of Values (SOV)

A very early instrument to assess personal values by psychometrically sound self-report measure was the Study of Values (SOV). The initial publication of SOV was done by Vernon and Allport (1931) and its subsequent revision was done by Allport, Vernon and Lindzey (1960) and its revised third edition by Allport, Vernon and Lindzey (1970). Allport was a student of Eduard Spranger, an American philosopher and pyschologist. Based upon the Spranger's view (1928) that an understanding about the person's value philosophy best captures the essence of the person, Allport and his colleagues Vernon and Lindzey constructed this SOV, which was the third most cited non-projective measure of personality after MMPI and EPPS. The SOV measures six types of values: Theoretical (T), Economic (E) Aesthetic (A), Social (S), Political (P) and Religious (R). These six values were directly outlined after Spranger's (1928) *Type of Men*. A brief description of these six types of values is as under:

- Theoretical (T): A person dominated by theoretical value is mainly interested in the discovery of truth.
- Aesthetic (A): Such person perceives highest value or interest in form and harmony.
- Economic (E): A person dominated by economic value is mainly interested in what is useful.
- Political (P): A person dominated by political value is interested mainly in power in all realms.
- Religious (R): Such person places highest value upon the mystical unity with the cosmos.
- Social (S): Such person places highest value for the love of people.

The Study of Value consists of two Parts: Part I and Part II. Part I consists of 30 statements or questions with two possible alternative answers provided for each. The testtaker's task is to select the alternative in each instance that is relatively more acceptable to him or her. In so doing the testtaker has three points to distribute between two alternatives for each item. For example, if he strongly agrees with alternative 'a' and does not agree with 'b' alternative, he assigns three points to the former and zero point to the latter. If he slightly likes 'a' over 'b', he assigns two points to 'a' and one point to 'b'. These two situations would stand reversed if his preference was alternative 'b' over 'a'. Thus the testtaker is required not only to indicate his preference but also xhibit the relative strength of these preferences. Part II of the SOV consists of 15 questions, each having four alternatives. The testtaker is asked to order these four alternatives in terms of relative preferences, assigning a rumencal value of 4 to the first choice, 3 to the second, 2 to third and 1 to the fourth. Thus in Part II, like Part I, the test formats were designed to elicit the relative strengths of the person's preferences.

Since Part I has two alternative answers to each of thirty question and four alternatives to each of the remaining 15 questions, there are altogether 120 possible responses for Part I and Part II taken together. Out of these 120 responses, 20 responses refer to each of the six values ($6 \times 20 = 120$). When the testtaker's scores on each value dimension are summed, they are plotted on a profile so that the testtaker's standing on the six values can simultaneously be observed.

The study of value has provoked considerable discussion among students and teachers of psychology but has not been an influential test. A major problem with SOV is that six values are vaguely defined and are too general from the pointview of general use.

By the early 1980s, SOV has fallen into disuse due to its archaic content and out dated language. However, Kopelman et al., (2003) have recently updated the SOV so that it may prove more relevant today. In this revision 15 items of the original 45 items were modified-7 items from part I and 8 items from part II. This revision by Kopelman and his colleagues was called updated version of SOV (or SOV-U). The reliability coefficient of SOV-U across six values domains were similar to the reliability coefficient of original SOV (SOV-O). Inter-form coefficients for the six value domains produced a median coefficient of 0.74.

Rokeach Value Survey (RVS)

The Rokeach Value Survey (RVS) was initially developed by Rokeach (1973). It is one of the most extensively used ipsative measures of human values and is frequency used by career counsellor for assessing client's value. Rokeach recognized two types of values–*instrumental value* and *terminal value*. Instrumental values are defined as desirable modes of conduct whereas terminal values are desirable end states of existence. For example, broadmindedness is an instrumental value whereas self-respect is a terminal value. The RVS consists of 36 items assessing 18 instrumental values and 18 terminal values Table 15.1 shows 36 kinds of values proposed by Rokeach. The testtakers are asked to rank separately the 18 terminal values and 18 instrumental values based upon their importance in their life. The rank for each item becomes the score for that value. Since ties are not allowed, the value scores will range from 01 to 18, with lower scores indicating greater importance.

The median test-retest reliability coefficient for terminal values ranged from 0.76 to 0.80 and 0.65 to 0.72 for instrumental values. Rokeach also calculated the test-retest reliability of individual value scores separately across all the testtakers. Using this approach, the reliability of the individual scales is lower, that is, about 0.65 for the terminal values and 0.56 for the instrumental values. The validity of Rokeach value survey against different tests was found to be satisfactory.

Table 15.1: 36 kinds of values identified by Rokeach

Instrumental values	
Ambitions	Imaginative
Broadminded	Intellectual
Cheerful	Logical
Capable	Loving
Clean	Obedient
Courageous	Polite
Forgiving	Responsible
Helpful Self-controlled	
Honest Independent	

Terminal values	
A comfortable life	Inner Harmony
A exciting life	Mature love
A Sense of Accomplishment	National Security
A World at Peace	Pleasure
A World of Beauty	Self-respect
Equality	Salvation
Family Security	Social Recognition
Freedom	True friendship
Happiness	Wisdom

There are limitations of the RVS. The individual scales of the RVS possess marginal reliability. It means, then, that the survey should not be used for individual guidance. Another limitation stems from its ipsative nature of the survey. Due to ipsative nature, it is difficult to assess absolute measure of the value for the testtaker. Moreover, the survey omits several important values such as individual rights, physical well-being, carefreeness, etc.

Measurement of Attitude

Attitude is one of the most important aspect of human behaviour. Allport (1935) has considered it as 'distinctive' and 'indispensible' to social psychology. An attitude is best defined as a certain regularity of a person's feeling, thoughts and predisposition to act towards an object (Secord & Backman, 1964). Thus attitude has three components: feeling (affect component), predisposition to act (behavioural component) and thoughts (cognitive component). This is called as *ABC component* of attitude—a *tripartitle* model of attitude.

Attitude are very closely related to the *values*, *interest opinion* and *beliefs*. Hence, distinction among these terms is required. As discussed earlier, *interest* refers to a pattern of likes and dislikes of the person. So it emphasizes more upon affective component of attitude. *Value* is shared and enduring idea about what is ideal. So, it refers to what is ultimate or best. *Opinion* can be regarded as overt and verbal demonstration of an attitude. Opinion is more changeable, more factual, more specific but less central than attitude. (Gregory, 2004) Another point of distinction is that opinions can be expressed in words but attitudes cannot. Belief refer to a conviction that something is true even though it cannot be proved. The belief that God exists, certainly belongs to this category. Thus belief lies somewhere between attitude and knowledge.

Attitude is an undesirable and hypothetical construct. Therefore, it is inferred from measurable responses indicating either positive or negative evaluations of the object of the attitude. Attitude can be easily inferred from cognitive responses (that is from knowledge about the object of attitude), from affective responses (that is, from the expressed feeling towards the object of attitude) and from behavioural responses (that is, from intentions or actions with respect to object). Of these three measures, affective measure are regarded as the most important ones. (Gregory, 2004)

There are three broad approaches to the measurement of attitudes :
- Behavioural approach
- Covert or Implicit approach
- Overt or direct approach (questionnaire)

A brief discussion of these approaches is as under :

Behavioural approach

This approach involves measurement of intentions or behaviours done with respect to the object of the attitude. For example, of a person asks for donation to the 'social upliftment fund' by making door to door canvassing, it means that the person has a positive attitude towards such fund. Some other examples of behavioural approach to assessment of attitude would be asking people to put their signature for construction of a road bridge over a river. Those who sign the petition reflect positive attitude towards the cause and those who decline to sign the petition reflect negative attitude towards the cause.

Convert or Implicit approach

Convent or implicit approach to the measurement of attitude involves unobtrusive procedures, response latency measures and physiological measures. Such convert or implicit measures have been developed for overcoming the problems with intentional and unintentional distortion of attitude measurement.

Unobtrusive measures are measures that disguise what is being measured and/or conceal the measurement itself. One such popular measure is what is called as the *lost-letter technique*. In Milgram's Classic lost letter technique the ostensibly lost letter are placed in public places (Milgram *et al*., 1965). The address on the envelopes are mentioned. The assumption here is that the individuals with more positive attitude towards the addressees are likely to pick up the envelopes and put it in the mailbox. The rate and speed of return for these letters is taken as indicator of attitude towards the addressee.

In response latency measure an attempt is made to determine attitude activation from the impact that an attitude object has on the speed with which an individual can make a judgement. *Implicit Association Test* (IAT) is one such measure that was proposed by Greenward, Mc Ghee and Schwartz (1998). The implicit association test is also known as *implicit attitude test*. In this test the implicit attitude is measured by pairing the object of the attitude with pleasant and unpleasant words. Such test is thought to assess the associative strength between each target concept and a particular attribute. For example, for assessing implicit attitude towards the dwarf, the respondent may be presented with the images of the dwarf person with images of the normal person. Interspersed with pleasant words such as *happy*, *attractive*, *pleasure*, *intelligent* and some unpleasant words such as *ugly*, *rotten*, *dull*, etc. In one phase of the test, the testtaker's task would be to react as quickly as possible by pressing a computer key whenever a face of dwarf or unpleasant word is presented and a different key whenever the image of a normal person or the pleasant word is presented. In the second phase, the pairing may be reversed so that the testtaker is required to press one key for a dwarf person or pleasant word and a different key for a normal person or an unpleasant word. Thus the first phase would be congruent phase and the second phase (reversed phase) would be an incongruent phase. The testtaker with negative or unfavourable implicit attitude towards the dwarf would find the first phase easier than the second phase and therefore, would be responding more quickly in the first phase as compared to the second phase. Thus the difference between the average latencies between the two phase provide a measure of the implicit attitude. Since IAT is an unobtrusive and non-reactive measure and responses to it are difficult to be faked, it is a very popular method of assessing socially stigmatized attitudes. It has also been successfully used to test theories in social psychology (Gregory, 2004).

The supporter of IAT claim that it gets at unconscious attitudes, which are inaccessible through the various conventional methods. Critics of IAT claim that its reliability and validity are comparatively low. A methodologically strong study by Cunnigham et al. (2001), however, lends support for the validity and reliability of the method but it is considered very cumbersome for the large scale uses.

Physiological measures of attitudes have also been frequently used. Such measures intend to capture the physiological correlates of the evaluative responses. Early attempts to use physiological responses for attitude measurement focused on Galvanic skin response (GSR) and pupillary responses. Rankin and Campbell (1955) used GSR for the first time as a measure of the ability of skin to conduct electricity in attitude measurement. In their experiment white subjects demonstrated an elevated GSR during interactions with an African American experimenter as compared to interactions with white experimenter. Likewise, Hess (1965) used pupillary responses for attitude measurement. The results of Hess research revealed that positive evaluations of the picture done by the participants led to the dilation of purpil whereas the negative evaluations of the picture done by the participants led to the constriction of pupil. Another physiological measure of attitude tries to investigate subtle muscle activity in specific areas of face, commonly over the brow (frowning) and the cheek (smiling). Cacioppo et al. (1986) showed that electromyographic (EMG) in these areas displayed distinct patterns following exposure to positive or negative stimulus. Another attitude measure based on facial EMG attempt to assess the modulation of eyeblink reflexes during exposure to an object. Lang et al. (1990) reported that the exposure to the positively evaluated images or stimuli was associated with *eye-blink inhibition* whereas negatively evaluated images or stimuli was associated with amplification of eye-blink reflexes. Physiological attitude measures are also based upon assessment of brain activities through the *Positron Emission Tomography* (PET) and *functional Magnetic resonance imagery* (FMRI). Researches have shown that activity in amygdala, is linked to the processing of negatively evaluated stimuli (LeDoux, 1996). Recently, Phelps et al., (2000) recorded the activities of amygdala for white participants while they were observing the images of African American and white faces. They reported a difference in activities of amygdala in these two instances, with enhanced activities in observation of the faces of African American. Likewise, Hart et al. (2000) reported enhanced amygdata activities in response to outgroup faces for both African American and white participants. Still another technique of attitude measurement based upon brain activities called *event-related brain potentials* (ERP) has been proposed by Cacioppo et al., (1993, 1994). In ERP the neural activity is recorded via electrodes placed on scalp and changes in it are recorded following the presentation of an event such as presentation of an attitude object. This procedure is based upon a particular component of the ERP wavefrom called the P_{300}. This component is very sensitive to the meaning of event performed during an ERP. Researches have shown that when the subjects are asked to classify stimuli according to certain dimension such as high tones vs. low tones, low tone followed by a series of high tones etc., evoke a larger P_{300} in some specific location of scalp. (Fabiani et al., 1987)

Overt or Direct approach (Questionnaire)

The majority of attitude measures are questionnaire based upon established scaling methods. The important attitude scales are: *Thurstone Scale*, *Likert Scale*, *Guttman Scale*, *Cumulative Scale*, *Scale discrimination technique* and *Semantic differential scale*. A discussion of these scales follows :

Thurstone scale :

Thurstone (1931) developed several scaling techniques. The most widely used scaling technique is the *equal-appearing interval method*, which was used in developing an attitude scale. In this method about 70 to 75 items or statements regarding the object of attitude are prepared. These statements are, then, submitted to a group of judges who are asked to sort these statements on a scale having eleven categories that appear to cover equal portions of the scale. One end of the eleven point category indicates favourableness towards the object of the attitude and the other end of the scale is designated to mean unfavourableness towards the object of the attitude. The middle point of the scale (F) is designated as neutral (see Figure 14.1)

1	2	3	4	5	6	7	8	9	10	11
A	B	C	D	E	F	G	H	I	J	K

Unfavourable   *Favourable*

Figure 14.1: Thurstone Equal appearing interval scale

These 11 categories varying from A to K are commonly designated the values from 1 to 11 respectively. Each judge rates the statement in any one of the eleven categories, which he considers appropriate to the extremity and direction of the statement. One of the major assumptions here is that the judges don't express their attitudes in sorting the statement rather they simply decide the degree to which the given items are favourable, unfavourable or neutral. There are two major purposes of the judges rating :

- To find out where on the favourableness-unfavourableness dimension each statement falls.
- How well a group of judges agree regarding the meaning of the statement.

The position of any item on the scale of favourableness-unfavourableness is indicated by median, which is calculated on the basis of all judgements regarding a statement. The median becomes the scale value of the statement. Another measure called Q (Semi-interquartile range) is calculated for each statement on the basis of the distribution of judgements obtained for that statement. The position of any item on the scale of favourableness-unfavourableness is indicated by median and its ambiguity by Q-value. Higher the value of median, higher degree of favourableness and higher the value of Q, higher the ambiguity is indicated due to the spread of ratings. Let us consider the examples of the following three statements which was rated by judges on 11-point scale.

	Items	Median	Q
1.	Abortion should be legal whenever the pregnancy is due to rape or incest	9.2	1.2
2.	Abortion should be easily available to any unmarried woman who wishes it	5.8	6.7
3.	Abortion should be legal when the life of mother is endangered	10.1	1.6

The above three items intend to reflect altitude towards abortion. The first statement seems to be promising for inclusion in the attitude sale because it indicates a high degree of favourableness but low degree of ambiguity (Q=1.2). The second statement is not fit for being included in the final inclusion because it comes near to the neutrality and it contains higher degree of ambiguity (Q=6.7). Item number 3, on the pattern of item no. 1, is fit for final inclusion in the attitude scale.

When the scale has been constructed, it is used to assess the attitude of the person who simply selects those items or statements with which he agrees. His attitude score becomes the average of the scale values of the statements with which he agrees. For example, if he agrees with four statements having scale values of 6.2, 7.3, 9.6 and 8.5, his attitude score would be 31.6/4 = 7.9, which meant that the person has an overall favourable attitude towards the object of the attitude.

Likert Scale

Shortly after publication of Thurstone scale, Likert (1932) developed an attitude scale based upon the *method of summated ratings*. Likert scale is the most popular method of attitude measurement. Here the researcher begins constructing large number of statements about the attitude object. These statements are submitted to a group of judges who indicate their own attitudes by responding to one of the five response options like *strongly agree, agree, neutral*, disagree and *strongly disagree*. Weights of 1, 2, 3, 4 and 5 are given respectively to each of these five options for favourable items and these weights are reversed respectively for the unfavourable items. An example showing an attempt to measure attitude towards family planning follows :

Favourable item :

Family planning is a healthy step towards improving economy of the country in the long run.

Strongly Agree	*Agree*	*Neutral*	*Disagree*	*Strongly disagree*
5	4	3	2	1

Unfavourable item:

Family planning retards the smooth growth of economy of the country in the long run :

Strongly Agree	*Agree*	*Neutral*	*Disagree*	*Strongly disagree*
1	2	3	4	5

In the above example the first statement measures a favourable attitude towards Family planning and the second statement measures an unfavourable attitude towards Family planning. Accordingly, the first statement is weighted as 5, 4, 3, 2 and 1 for strongly agree, agree, neutral, disagree and strongly disagree respectively and the second statement has been weighted in reversed order. After giving weight to the individual items, the total score for each subject or judge is determined by adding the individual scores. Finally, an internal consistency analytics or item analysis is carried out. This is done by computing correlation between scores on an item and the total scores on all items. Then, those items yielding higher correlation are retained for the final scale. It is the use of this item analysis that makes Likert's scale distinct from Thurstone scale. Usually, a set of 20 to 22 items are finally selected for the scale. On Likert's scale, higher score, indicates a favourable attitude and lower scores indicates an unfavourable attitude towards the attitude object.

Guttman scale

Guttman (1950) developed a scaling technique called *cumulative scaling* or *scalogram analysis*, which intends to develop a set of items for attitude measurement that will be unidimensional in nature. Guttman, in fact, developed such a scale in which response to any single item could be determined by a total score on the whole set of items. In such situation, a person with higher score than another person on the same set of items will get a ranking just as higher on every statement in the set as the other individual. Thus on Guttman's scale, an attitude scale is considered unidimensional if on every item, a person with a more favourable attitude, gives a response, which is more favourable than what is obtained from a person with less favourable attitude. In such situation, his response to every item will be perfectly consistent with his overall position on the dimension of attitude. The nature of unidimensional scale can be easily illustrated with a hypothetical example. Suppose an attitude scale consists of four items and each item can yield five possible scores 4, 3, 2, 1 and 0 representing an agreement with all four items on one end and disagreement with all four items on the other hand. If five subjects are to take this scale (considering it as a unidimensional scale), the scores and the pattern of responses would be something like those presented in Table 15.2.

Table 15.2: Hypothetical example of an unidimensional scale

Subject	Score	Agree with item			
		1	2	3	4
1	4	×	×	×	×
2	3	×	×	×	
3	2	×	×		
4	1	×			
5	0				

Table 15.2 reveals that items are perfectly consistent. A score of 4 is made by agreeing with items 1,2,3 and 4. Likewise, a score of 3 in made by agreeing with items 1,2 and 3 and disagreeing with item 4. Likewise, a score of 2 is made by agreeing with item 1 and 2 and disagreeing with item 3 and 4. Similarly, a score of 1 is made by agreeing with items 1 and disagreeing with items 1, 2 and 3. All these happen due to perfect consistency. Still another sign of consistency is picked up by the very observation that the subject who agrees with items 4, also agrees with item 3, 2 and 1 and the subject who agrees with item 3 also agrees with item 2 and 1 and so on.

One basic characteristic of uni-dimensional scale is that here pattern of responses is reproducible from the knowledge of the scale score. In the above example, the subject who has made a score of 4 agrees with 1, 2, 3 and 4 and the subject who has made a core of 3, agrees with item 1, 2 and 3 but disagrees with item 4. In this way for every other score, pattern of responses can be reproduced. In practice, however, such perfect consistency or reproducibility or unidimensionality is rarely achieved. Therefore, Guttman (1950) has recommended that errors are permitted to a certain extent and therefore, he suggested that the pattern of response must be at least 90% reproducible in order for a scale to be treated as uni-dimensional one. Clearly then, 10% of responses may fall outside the range of reproducibility.

Scale Discrimination Technique

This technique, developed by Edwards and Kilpatrick (1948), attempts to synthesize the method of attitude scale construction developed by Thurstone, Likert and Guttman. The major steps involved in this technique are as under :

A large number of statements regarding attitude object is prepared by the researcher. As in Thurstone's scaling technique, these statements are submitted to the judge who sort these statements into categories according to the degree of favourableness. Items that are not sorted consistency are termed as vague or ambiguous and therefore, they are rejected. The remaining statements are constructed in multiple-choice formats with various response categories like *strongly agree, agree, mildly agree, mildly disagree, disagree and strongly disagree*. These statements are now given to a group of new subjects who respond to each item by choosing the response option that best describes his own viewpoint. Each statement is then scored and scores are added to derive a total score for the subject. Subsequently, like Likert scale, each item is subjected to item analysis and the items that are found to be non-discriminating ones, after item analysis, are rejected. The remaining items are, then, dichotomized and subjected to cumulative scaling like that of Guttman's scale.

In this way, the scale discrimination technique represents a mixture of Thurstone's scale, Likert's scale and Guttman's scale. However, this approach needs to be verified and tested by the subsequent researches.

Semantic Differential Scale :

This scale is a tool of measuring attitude developed by Osgood, Suci and Tannenbaum (1957) as a part of their attempt to study the measurement of meaning. The semantic differential scale is a technique for assessing the *connotative meaning* of the objects. The connotative meaning is considered important because it generates emotional or emotive response. According to Osgood, Suci and Tannenbaum, the connotative meanings are multidimensional. They identified three dimension–*evaluative*, *potency* and *activity*. Evaluative dimension is characterized by adjective pairs like fair-unfair good-bad, potency is characterized by adjective pair like dominant-submissive, weak-strong and activity is characterized by pair like fast-slow, busy-idle.

In using the semantic differential scale, the researcher presents the participants with a series of bipolar adjective scales. Each scale has two ends, with two adjectives having opposite meanings (see Table 15.3). In the semantic differential scale, the concept is rated on 5-point scale to 9-point scale. However, 7-point scale is more common. A respondent is asked to indicate how the target concept or person appears to him by making a mark at the appropriate point between the two adjectives. Thus a respondent who thinks that the police

Table 15.3: A Semantic Differential scale to measure connotative meaning of the concept of 'Police Officer'.

Good	|_|_|_|_|_|_|_|	Bad
Strong	|_|_|_|_|_|_|_|	Weak
Energetic	|_|_|_|_|_|_|_|	Lazy
Nice	|_|_|_|_|_|_|_|	Nasty
Capable	|_|_|_|_|_|_|_|	Helpless
Busy	|_|_|_|_|_|_|_|	Idle
Fair	|_|_|_|_|_|_|_|	Unfair
Dominant	|_|_|_|_|_|_|_|	Submissive
Active	|_|_|_|_|_|_|_|	Passive

officer is very good, will put a mark in the space nearest to 'good' and another respondent who perceives police officer as neither good or bad, would put a mark halfway between the two extremes and so forth. After collecting the responses, the researcher analyzes them by using various statistical analyses.

In a nutshell, it can be said that the researchers have tried to assess attitude through indirect and direct measures. Of these, direct measures of attitude through various questionnaires based upon scaling methods are comparatively very popular. Likert scale of attitude measurement is the most popular direct measure.

Summary and Review

- Interest measurement is done through various interest inventories. Interest is understood as the pattern of likes and dislikes. Important interest inventories developed for the purpose are Strong Interest Inventory, Jackson Vocational Interest Survey, Kuder's Occupational Interest Survey, Self-Directed Search, Minnesota Vocational Interest Inventory, and Career Assessment Inventory.
- Value scales have been developed to assess values of the persons. Values are defined as shared and enduring belief about the ideal mode of behaviour or some end state of existence. Important value scales are Study of Value (SOV) and Rokeach Value Survey (RVS).
- Attitudes are understood as a certain regularity of a person's feelings, thoughts and predisposition to act towards an object. For assessing attitude various measures such as behavioural measures, covert or implicit measures and overt or direct measures are available. In behavioural measure attempt is made to assess attitude through intentions or

behaviours done with respect to the object of the attitude. Covert or implicit measures include measures like lost-letter technique, physiological measures and implicit association test. Overt or direct measures include questionnaire based upon established scaling methods. Important attitude scales are Thurstone Scale, Likert Scale, Guttman Scale, Scale discrimination technique and Semantic differential scale.

Review Questions

1. Define interest. How is interest measured?
2. Discuss the major features of Strong Interest Inventory and Kuder Occupational Interest Survey.
3. Make distinction between value and interest. Outline a plan for assessing value of the person.
4. Outline a plan for assessing attitude through some direct or overt measures.
5. Discuss the main features of Likert Scale of attitude measurement.
6. Outline the major characteristics of Thurstone Scale of attitude measurement.
7. What is cumulative scaling? How does it differ from semantic differential scale.

Chapter – 16
Measurement of Creativity

Learning Objectives
- Meaning of creativity
- Different approaches to creativity
- Cognitive approach
- Psychosocial approach
- Confluence approach
- Measurement of creativity
- Ways of enhancing creativity
- Summary and Review
- Review questions
- Key Terms

Key Terms:

Originality, Flexibility, Elaboration, Fluency Mini-c, Little-c, Big-c, Confluence approach, Divergent thinking, Torrence Test, Remote Associates Test, Social-Personality approach, Psychometric approach.

We are often placed in amaze when we imagine how writers, composers, directors or painters think in such a way to produce something original and novel pieces of work. These novelty and originality become the foundation stone of creativity. Trustfully, creativity is an area of problem solving. If we pay attention to what a creative person does, it would be obvious that it involves moving from an original state to the goal state. The creative person here, tries to solve ill-defined problem. Certainly, then, creativity is very important because it provides us with new knowledge and new inventions that can improve the quality of human life. In the present chapter a discussion regarding the various aspects of creativity such as its definition, types, ways of measurement, its relationship with intelligence will be discussed.

Meaning of Creativity

If one goes to define creativity, a very simple way to do this is to state that the creativity involves coming up with a new and unusual answer to a problem or question. But, in reality, this is not enough. The answer or idea must not only be new but it must also be practical and useful. For example, suppose Mohan asks Sohan to come up with a creative answer to the question. "How can you roast a goat?" If Sohan's reply is to put the goat into a small cage type room and burn the room. The answer is no doubt would be credited with novelty but it lacks the usefulness requirement. Thus creativity, in general, refers to finding a solution that is both unusual and useful.

There are many definitions of creativity. According to Wood and Wood (1996), creativity is best understood as the ability to produce original, appropriate and valuable ideas or solution to the problems. Likewise, Baron (2002) has opined that creativity may be defined as ability to produce work that is both novel (or original) and appropriate or useful. Ciccarelli & White (2017) have opined that creativity involves coming up with entirely new ways of looking at the problem or unusual, inventive solutions. In a similar tune, Sternberg (2011) has pointed out that the production of something original and worthwhile is creativity. According to Colman (2015), creativity is production of ideas or objects that are both novel or original and worthwhile or appropriate.

Analyzing these definitions, we come to the following basic points regarding the meaning of creativity:

- Creativity involves novelty or originality.
- Creativity involves usefulness or appropriateness or meaningfulness.
- Creativity relates to both ideas and/or work.

Sometimes people think that creative people are different from other people. But the reality is that creative people are actually normal people. Csikszentmihalyi (1997) has point out that

- Creative people are not afraid to be different rather they are more open to new experiences than many others. They have more vivid dreams and daydreams than others.
- Creative people value their independence.
- Creative people are often unconventional in their work but not in other respect.
- Creative people usually have a sufficient and broad range of knowledge about different subjects and use their mental imagery very broadly.

Different approaches to creativity

Psychologists have taken interest in answering one basic question about creativity: "What factors produce it?" Several approaches have been developed to answer this question.

• Cognitive approach :

The cognitive approach to creativity has been dominant and most popular in psychology. The cognitive psychologists have focused on the basic processes that underlie cognitive thought. Important such basic processes are retrieval from memory, association, synthesis, transformation and so on. Some cognitive psychologists have pointed out that creativity is the part of our everyday processes of lives. When we speak something new or do something new, creativity is said to occur. Such kind of everyday creativity has been called as *mundane creativity* to be contrasted with *exceptional creativity* which involves the emergence of something really new and practical.

- **Psychosocial approach :**

    This approach has been chiefly emphasized by social psychologists who are of view that there are certain personality traits that make people creative. Besides, there are environmental conditions that either encourage or discourage creativity. (Simonton, 1994) Among the traits needed for creativity are intellectual habits, levels of ideation, openness, autonomy, expertise, exploratory behaviour and so on.

- **'Four C' model :**

    This model was introduced by Kaufman and Beghetto (2009). According to this model, there are four c : *mini-c*, *little-c*, *Pro-c* and *Big-c*. Mini-c involves transformative learning including some meaningful interpretations of experiences and actions. *Little c* involves everyday creative expression and problem solving. *Pro-c* is expressed by people who are professionally creative though not necessarily eminent in the concerned field. Big-c involves creativity honoured and considered great in the given field. Of these, Big c and little c have been widely used. Kozbelt et al., (2010) have used the contrast of big c and little c in reviewing major theories of creativity. Boden (2004) has distinguished between what is called h-creativity (historical) and P-creativity (personal).

- **Confluence approach :**

    This approach suggests that for creativity to occur, several components must converge. (Amabile, 1983) Lubart (1994) has pointed out that the creativity requires a confluence of at least six distinct components.
    - *Intellectual abilities* : Several mental abilities like ability to recognize the worth and useful idea, ability to see problem in new way, ability to convince others about the new ideas.
    - *Knowledge* : Proper knowledge about the field in which problem is located.
    - *Personality traits* : Traits like willingness to take risk and to tolerate ambiguity.
    - *Typical styles of thinking* : Showing preference for thinking in novel ways as well as ability to think globally and locally.
    - *Supportive environment* : The environment in which creative ideas are to be occurred, must be congenial and supportive
    - *Intrinsic, task-focused motivation* : Creative people show preference and love for what they are doing and find intrinsic reward in their work.

    When these six components are present, a high level of creativity is likely to occur.

Measurement of Creativity

Measuring creativity is a difficult task. It is not a hyperbole nor an oxymoron. It is a science. There are two approaches to the measurement of creativity: *Psychometric approach* and *Social-Personality approach*. Before we go into the depth of these two approaches, it would be proper to discuss the four terms most common to many measures of creativity. They are: *originality*, *fluency, flexibility* and *elaboration*. The originality refers to the ability to produce something new, innovative or nonobvious. This may be something abstract like an idea or tangible like a poem or some outwork. Fluency refers to the ease with which responses are reproduced and is usually assessed by the total number of response produced. For example, if the testtaker is asked to respond within next one minute as many words as possible that begin with letter 'e', it would constitute an example of fluency. Flexibility refers to the varieties of ideas produced and the ability to shift from one approach to the other. Elaboration refers to richness in detail in any verbal explanation or pictorial display. Now let us discuss the above two approaches in detail.

Psychometric approach

The formal psychometric approach to the measurement of creativity is usually considered to have begun with J.P. Guilford's 1950 address to the American Psychological Association (Sternberg & Lubart, 1999). This address not only popularized the topic but also paid attention to the scientific approach to conceptualizing creativity. Guilford (1950, 1967) constructed several tests to measure creativity. One of his popular such test is Guilford Test of divergent thinking. In divergent thinking there is a reasoning process in which the person's thinking moves in many directions making several possible solutions. Such thinking deeply needs flexibility of thought, originality and imagination. Divergent thinking is different from convergent thinking, which is a process of deductive reasoning that incorporates recall of facts as well as a series of logical judgement to narrow down solutions and finally, arrive at one solution. In Guilford's Divergent Production tests, different tasks for assessing creativity were used. Some of a few examples are below:

- How many pictures of real objects you can make using a circle within a few minute?
- Many words begin with E and end with N. List as many words as possible in one minute period?
- What would be consequences if people reach at final height of the body at the age of 3. List the several possible consequences within a one-minute period?
- Name as many unusual uses of a pen as you can think of in one minute time.
- Classify the given words in as many ways as possible.

Responses to these varied tasks are scored for four components: originality, fluency, flexibility and elaboration. Originality is based on each response compared to the total amount of responses from a defined group of testtakers. Responses given by 5% of the group are awarded one point and responses given by only 1% of the group are awarded two point. The score for flexibility is based upon the difference of categories. The score for fluency is based upon relevant answers and the score for elaboration is given on the amount of detail given in the response.

Torrance (1974), building upon Guilford research work, developed one important test of creativity called *Torrance tests of creative thinking* (TTCT). The test originally measured creativity through four factors developed by Guilford - originally, fluency, flexibility and elaboration but in its third edition in 1984 flexibility scale was eliminated and two new scales were added. The names of new scales are: *Resistance to Premature closure* and *Abstractness of Titles*. TTCT has two parts: *TTCT-Verbal* and *TTCT-Figural*. TTCT-verbal consists of five tasks: ask-and-guess task, product-improvement task, unusual uses task, unusual questions task and just suppose task. TTCT-Figural consists of picture construction test picture completion task and repeated figures of lines and circles. A brief description of tasks of is TTCT-Verbal is as under:

- *Ask-and-guess task* : Such task requires the person first to ask such questions about a picture that cannot be answered by just looking at the picture. Then, he is asked to make a guess about the possible causes of event depicted in the picture and subsequently, their both immediate and remote consequences.
- *Product improvement task* : In this task common toys are used and children are directed to think of as many improvements as they can so that the toy may appear more fun to play with. They are then asked to think of unusual uses of these toys.
- *Unusual uses task* : Here the children are directed to think of most interesting and most unusual uses of the given toy, other than playing.
- *Unusual uses task* : Unusual uses task involves the modifications of Guilford's Brick uses test. In such task Terrance substituted tin cans and books for bricks on the assumption that the children would be able to handle tin cans and books easily than bricks.
- *Just suppose task* : Such task is similar to Guilford's consequence task. Here children is confronted with an improbable situation and asked to predict the possible outcome from the introduction of a new or unknown factor.

The three tasks of TTCT-Figural are as under:

- *Picture construction task* : In this task the children are given a shape of triangle and a sheet of white paper. The children are requested to think of a picture in which the given shape is an integral part. They may paste it wherever they like to paste on the white paper and lines with pencil to make any novel picture. They are also requested to think of a name for the picture and write it at the bottom.
- *Picture completion task* : In this task on white paper an area of 54 square inches is divided into ten squares each containing a different stimulus figure. The children are requested to sketch some novel objects or design by adding as many lines as they can to same ten figures.
- *Circle and Squares task* : In this task two forms are used. In one form the children are exposed to a page of forty two circles and asked to sketch such pictures, which have circles as a major part. In an alternate form, squares as used instead of circles.

On the basis of answers provided by the testtakers, an inference about the level of creativity is made.

Another creativity test called *Remote Associate Test (RAT)* was developed by Mednick and Mednick (1967). In this test the testtakers are provided three words and their task is to find a fourth word associated with the other three. For example, the testtaker may be presented three words such as Type, Ghost, Story. He is required to tell the fourth such word that connect all these three. The answer may be 'writer'. In fact, Mednick tried to assess creativity in terms of the ability to link remote and combine them into new associations that meet certain criteria.

These are some of the important tests of creativity. Besides these, there are some comparatively less frequently used tests of creativity like *self-assessment tests*, *Artistic assessment test*, *Wallach and Kogen creative Battery*, etc. In self-assessment test the testtakers respond to the questions about how creative they feel or whether they possess certain creative traits or attribute. Torrance along with his colleagues Joe Khatena developed a self-assessment test, which is named as *Khatena-Torrance Creative Perception Inventory* (KTCPI) that consists of two subtests: *Something About Myself* (SAM) and *What kind of Person Are You* (WKOPAY). The former measures artistic inclination, individuality, sensitivity, intelligence and self-strength whereas the latter measures self-confidence, inquisitiveness, awareness of others, appeal to authority and imagination. Another self-assessment test is '*openness to experience*' scale of NEO-PR developed by Costa and McCrae. Psychologists prefer to assess creativity through this scale. Artistic assessment tests also measure creativity. In this test the assessment of an artistic product such as a drawing, a short story, a dance, a sculpture, and musical composition, etc. is done. Experts rate and judge these products for determining creativity. Such type of creativity is considered domain-specific, which meant that creativity is not considered a general skill but is something specific to a discipline or domain. Amabile, a researcher at Harward University, developed one such technique called the *Consensual Assessment Technique* (CAT) for assessing such domain-specific creativity. In Wallach and Kogan Creative Battery, attempt has been made to focus on requiring the testtaker to form associative elements into new combinations, which meet specific requirements. For example, children may be asked to name all the round things they can think of and to find similarities between them.

In India some psychologists have developed creativity tests. For example, B.K. Passi has developed a test named as *Passi Test of Creativity*, which measures creativity through si sub tests, namely, Seeing Problem test, Unusual test, Consequence test, Inquisitiveness test, Square puzzle test and Block consequences test. The first three tests are verbal and the last three tests are nonverbal in nature. This test measures three components of creativity: *fluency*, *originality* and *flexibility*. The test is most suitable for grades X to XI. Baquer Mehdi has also developed both verbal test and nonverbal test of creative thinking. The verbal test provides scores for fluency, flexibility and originality for class VII and VIII. The non-verbal test of creativity provides scores for elaboration and originality for class VII and class VIII. K.N. Sharma has developed a test of creativity called *Divergent Production Abilities*, which has six subtests: Word production, uses of things, similarities, sentence construction, making titles, and elaboration. It provides scores for four components of creativity: fluency, flexibility, originality, elaboration as well as total creativity. One of the major features of this test is that it can be successfully used on children, adolescent and adults.

Social-Personality approach :

In this approach to the measurement of creativity some personality traits such as self-confidence, independence of judgement, attraction to complexity, risk taking are taken as the measures of creativity. After conducting metaanalysis, Fiest (1998) arrived at a conclusion that the creative people tend to be more confident, self-accepting, ambitious, dominant, less conscientious, less conventional and more open to experiences. Batey et al. (2010) have, likewise, reported that openness to experience was consistently related to the different ways of assessing creativity. In order to understand the relatedness of creativity and personality traits, Fiest (1998) made a comparative study of artists and nonartists as well as scientists and non scientists. He reported that the artists tend to have higher levels of openness to experience and lower levels of conscientiousness. Likewise, scientists were more open to experience and higher in confidence and dominance facets of extraversion as compared to nonscientists.

Ways of Enhancing Creativity

Creativity has been identified as one of the basic modern skills and as one of the Four Cs (*critical thinking, creativity, collaboration and communication*) of the 21st century. There are several ways of encouraging creativity. Some of the important such approach are as under:

- Osborn (1957) has proposed one common group approach for enhancing creativity. This is called *brainstorming principle*. A we known, brainstorming is usually conducted in group setting and it involves guidelines and directions as under:
- Evaluation is withheld until later. Therefore, criticism is ruled out.
- The wilder idea is most preferred. It is easier to frame an idea later than to think one immediately.
- The greater number of ideas is preferred.
- Persons can combine two or more ideas proposed by others.

In the technique of brainstorming, persons, thus encourage themselves as well as encourage each other. For success of brainstorming, the complete friendliness and a relaxed frame of mind are very important.

- Synectics is another approach developed for enhancing creativity. This approach has been developed by Gordon (1961). This approach utilizes the use of various types of analogies in creative thinking where the person makes strange familiar and familiar strange through the use of various kinds of analogies. In synectics, the following four kinds of analogy are the most popular ones.
- *Direct analogy* : In this analogy the person is encouraged to find something that solves the problem he is trying to solve. Graham bell used direct analogy when he considered how the relatively huge bones of human ear could be moved by some dedicate membrane and he invented telephone where a piece of steal is moved by a membrane.
- *Personal analogy* : In this analogy the person is encouraged to place himself in the situation. For example, if a person wants to know that how a machine work efficiently, he must imagine that he is himself a machine.
- *Symbolic analogy* : In this analogy the person uses an objective and impersonal image for describing a problem.

- *Fantasy analogy*: In this analogy the person is freed from the boundaries of normal restriction. Here the person may imagine himself sprouting wings and flying over several kilometers for solving a problem of specific nature.

Nickerson (1999) has provided some special techniques for enhancing creativity. Some of these techniques are as under:
- Provoking opportunity for discovery and choice
- Developing metacognitive skills
- Acquisition of domain-specific knowledge
- Stimulating and rewarding curiosity and exploration
- Building internal motivation
- Providing supportable beliefs about creativity
- Building basic skills
- Focusing on mastery and self-completion
- Encouraging confidence and willingness to take risks
- Teaching strategies for facilitating creative performance.

Summary and Review

- Creativity refers to the ability to produce original, appropriate and valuable ideas or solutions to the problem.
- Creativity involves the components of originality, flexibility, fluency and elaboration. Among others, it primarily involve divergent thinking.
- There are different approaches to creativity. Important such approaches over cognitive approach, and confluence approach.
- There are various techniques of assessing creativity Psychometric approach to the measurement involves use of various creativity tests like Guilford Divergent Production test, Torrance Test of creativity, Remote Associate test and many other such psychometric tests. Creativity has also been measured by assessing some personality traits like openness to experience, self-confidence, etc.
- Various ways of enhancing creativity have been proposed. Osborn, Gordon and Nickerson have proposed some convincing and satisfactory ways of enhancing the creativity.

Review Questions

1. Define creativity. Discuss the various psychometric tests commonly used for assessing creativity.
2. Citing examples discuss the various components of creativity.
3. Discuss the different approaches to understanding the nature of creativity.
4. Discuss the important ways for enhancing the creativity.

Chapter – 17

Neuropsychological Assessment

Learning Objectives

- Nature and goals of neuropsychological assessment
- Tests of neuropsychological assessment
- Characteristics of useful neuropsychological batteries
- Halstead-Reitan Neuropsychological test battery
- Luria-Nebraska Neuropsychological test battery
- Some Indian Neuropsychological test battery
- Comparison of Halstead-Reitan test battery and Luria-Nebraska test battery
- Standard Psychological Testing vs. Neuropsychological logical testing
- Summary and Review
- Review Questions
- Key Terms

Key Terms:

Neuropsychological test, Halstead-Reitan Neuropsychological test, Luria-Nebraska Neuropsychological test, PGI battery of Brain dysfunction, AIIMS Comprehensive Neuropsychological battery, NIMHANS Neuropsychological battery.

There are many sources of individual differences. One important source of individual difference is neurological source. Understanding the neurological sources of the differences among individual helps the researchers in identifying some brain-based disorders in areas like memory, attention, learning, intelligence, personality, cognition, emotional expression and so on. Understanding these differences in learning can define current and future expectations of the lifestyle of the individuals. A central part of neurological assessment is the administration of neuropsychological tests for the assessment of cognitive function. Neuropsychologists also realize that neuropsychological assessment must also include an evaluation of the person's mental status. In the present chapter various tools of neuropsychological assessment including ways of evaluating the person's mental status will be discuss in detail so that a better understanding of brain-behaviours relationship could be easily-understood. Neuropsychologists try to locate the factors that tend to influence how the brain is working for understanding and explaining various types of abnormalities.

Nature and Goals of Neuropsychological Assessment

Neuropsychological assessment is a performance-based technique to assess the cognitive functioning of the person. In fact, such assessment intends to examine the cognitive consequences of brain damage, brain disease and other mental illnesses. The neuropsychological assessment is done due to some obvious reasons. *First*, it is carried out for clinical evaluation for understanding the pattern of cognitive strength as well as difficulties a person tend to experience. *Second*, it is also done for scientifically investigating a hypothesis relating to brain-behaviour functioning.

There are three important *goals* of neuropsychological assessment :
- One goal of neuropsychological assessment is to make diagnosis, that is, to determine the nature of the problem. In other words such assessment aims at differential diagnosis that is to confirm or classify a diagnosis.
- Neuropsychological assessment also aims to measure changes in functioning overtime so that the impact of any rehabilitation programme may be assessed. On the basis of neuropsychological assessment recommendations for cognitive disorders and psychological adjustment for guiding rehabilitation, educational, vocational or other services, are made.
- Neuropsychological assessment aims at understanding the nature of brain injury or its impact upon the individual's behaviour.

There are several neuropsychological tests that tend to evaluate one or more of the following categories

- Attention and concentration
- Language and Reading skills
- Verbal and visual memory
- Auditory and visual processing
- Auditory and visual processing
- Visual spatial functioning
- Gross and fine motor development
- Sensory development and sensory integration
  - Executive functioning
  - Logical analysis
  - Concept formation
  - Problem solving
  - Reasoning
  - Inhibitory control
  - Planning
  - Flexibility of thinking
- Social skill development
- Emotional and Personality development

Tests of Neuropsychological Assessment

All neuropsychological tests can be placed under any of the two categories. *Individual neuropsychological test* and *batteries of neuropsychological assessment*. For assessing attention impairment, individual neuropsychological tests like *Test of Everyday Attention (TEA) , Continuous Performance Test (CPT), Paced Auditory Serial Addition Task (PASAT)* and

Subtracting Serial Sevens have been developed. For assessing learning and memory tests like *Wechsler Memory Scale, Rey Auditory Verbal Learning test* and *Fuld object-Memory Evaluation* are the most common ones. For assessing language and functions like aphasia[1], tests such as *Multilingual Aphasia Examination, Boston Diagnostic Aphasia Examination, Porch Index of Communicative ability* and *Token test* are commonly used. Besides these, a nonstandardized clinical examination is also conducted for diagnosing aphasia. For testing spatial and manipulatory ability, drawing tests such as *Bender Visual Motor Gestalt Test* has been most popular. Likewise, for assessing executive functions, tests like *Wisconsin Card Sorting test, Proteus Mazes, Tinkertoy test, Category test* from Halstead-Reitan Battery are important. For assessment of gross and fine motor development various tests like *Pegboard performance tests, Finger typing test* and *Line tracing test are* very important.

    Besides these tests, there have been developed some test in batteries in neuropsychological assessment that incorporates a series of different tests. These neuropsychological test batteries are given routinely as a group. If we pay attention to various test batteries, we will find that there are *two* types of tests batteries: those test batteries that constitute a formal battery of commercially available tests and those that are informally composite batteries assembled for use with particular populations. Of these two types of batteries, the formal batteries are preferred. We shall here discuss two important such formal batteries: *Halstead-Reitan Neuropsychological test battery* and *Luria-Nebraska Neuropsychological test battery.* Before discussing these test batteries, it would be wise to discuss some important *characteristics* of useful neuropsychological batteries.

Characteristics of Some useful Neuropsychological Batteries

    There are some defined criteria or characteristics of neuropsychological batteries. They are being discussed as under:

- *Thoroughness* : To be useful a neuropsychological tests battery must be thorough enough to assess a wide variety of functions. For example, such tests must measure general intelligence and memory; they must measure sensory, perceptual and motor functions, they must identify the hemisphere containing speech; they must assess language functions and they must also examine left and right frontal lobe function.
- *Time*: Neuropsychological test batteries should not take far long time in their administration. This is so because the subjects are often weak and they get easily tired. Therefore, test batteries must be such that they should be completed within short period of time without compromising thoroughness.
- *Adaptability*: Test batteries must be adaptable and portable to match the varying constraints determined by the health of the subjects. Often the subjects are wheel chaired

---

[1]. Aphasia refers to any deviation in language performance caused by brain damage.

or even bedridden. Therefore, tests batteries must be adaptable to the demands of the health of the subjects who are wheelcharged or even bedridden. Therefore, test batteries must be adaptable to the demands of the health of the subjects.

- *Flexibility*: Test batteries should be such that they could be improved in the light of new data. They could be improved with new innovation and research. Such batteries should not employ formulas a cut off scares because they tend to lower the flexibility.
- *Ease and cost*: Test batteries should be easy to administer and score. Tests must be prepared in standard format and scored objectively so that consistent results from the test batteries must be obtained. Test batteries should also be inexpensive. Costly test bakeries may force the neuropsychologists for obtaining a partial assessment protocol.

Halstead-Reitan Neuropsychological Test Battery or HRNB

This is a very popular neuropsychological test battery by Halstead who was chairman of the department of psychology at the University of Chicago and his student Reitan in the early 1940s. In fact, the test battery was developed by Reitan from the work of Halstead (1947). (Anastasi & Urbina, 997) It began as a battery of seven tests that intended to discriminate between patients of fornted lobe lesions and other patient groups or normal persons. Test battery is used to assess the condition and functioning of damage in individuals aged 15 years and older. Its complete administration requires about 5 to 6 hours. The battery also provides useful information regarding the cause of damage as well as regarding whether the damage occurred during childhood and whether the damage is getting worse or staying the same.

The current version of battery consists of eight core tests and rests are ancillary tests, commonly used in combination with the basic battery. A brief description of these tests is as under :

- **Category test :** This is a nonverbal test, which measures person's ability to formulate abstract principles. The tests consist of 208 geometric pictures and is presented in seven subtests. For each picture, the persons are asked to decide whether they are reminded of the number 1, 2, 3 and 4. They have to depress a key (on the instrument) that corresponds to their number of choice. If they depressed the correct number, a door chime will ring and if they depress one of the incorrect numbers, a buzzer will sound. Based on this feedback of chime and buzzer, the person must determine the underlying principle in each of the seven test category. The last subtest contains two underlying principles. Usually, the test takes about *one hour* to complete but the persons with severe brain damage may take as long as two hours. The scoring of test is done on the basis of the number of errors. Reitan has suggested a cut off of 50 or 51 errors. Scores above that are taken to be indicative of brain impairment for ages 15 to 45 and scores above 46 indicate impairment for ages 46 and older.

The children's version of this Category test consists of 80 items and five subtests for young children. For older children, it consists of 168 items and six subtests.

- **Trail Making test :** This test consists of two parts–Part A and Part B. Part A comprises 25 numbered circles all randomly arranged. The persons are directed to draw lines between the circles in sequential order until they arrive the particular circled labelled 'End'. Part B consists of circles having letter A through L and 13 numbered circles, which are intermixed and randomly arranged. The persons are directed to connect the circles by sequential order until they arrive at a circle labelled 'End'. The testtakers take about 5 to 10 minutes to complete. The test can be administered orally if the person is incapable of writing. Scoring is the time to complete each part. When errors occur, it naturally increase the total time for adults. Scores above 40 seconds for Part A and 91 seconds for Part B are taken as indicative of brain impairment. For children and persons of different culture the colour Trail test has been developed. The colour Trail test uses colours instead of numbers and letters.
- **Tactual performance test :** This test assesses sensory ability, motor functions, memory for shapes and spatial location as well as the brain's ability to transfer information between its two hemisphere. The test also determines on which side of the brain, damage may have occurred. In this test before a blindfolded person a form board having ten cut-out shapes and ten wooden blocks matching those shapes are placed. The person, then, is directed to use only their dominant hand to place the blocks in their appropriate space on the form boards. This procedure is repeated with nondominant hand and then, using both hands. After this, both formboard and blocks are removed followed by the blindfold. Now, the person, from memory, is instructed to draw the formboard and shapes in their proper locations. A time twist of 15 minutes exists for each performance segment or trial and the total time for completion of the test is somewhere from 15 to 50 minutes.

In scoring the time to complete each of the three blindfolded trials is recorded. Thus, total time for all trials combined (called time score) the number of shapes recalled (called memory score) and the number of shapes drawn in the correct locations (called localization score) are determined. Halstead and Reitan throughout believed that this is a frontal lobe test. However many other are of view that the frontal lobe patients are not impaired at this task but right parietal patients are

- **Finger Tapping test :** Also known as finger oscillation test, it measures the person's finger-tapping speed on a key resembling a Morse code key. The person here is instructed to rest the index finger of the dominant hand on a key that is attached to a counting device. He taps this index finger as quickly as possible for ten seconds. In this way, 5 to 10 times, the trial is repeated. He is also given rest between each 10 second trial. The entire procedure is repeated with the non-dominant hand. This test measures motor speed and helps determine particular areas of the brain that may be damaged. Scoring involves using the five accepted trials to calculate an average number of taps per trial for both hands. In general, the dominant hand performs ten percent better than the nondominant hand. This test is generally interpreted in combination with other tests of the battery. Some children's version uses an electronic tapper in lieu of a manual one.

- **Rhythm test :** Also known as the Seashore. Rhythm tests, it requires the person to discriminate between like and unlike pairs of musical beats. In fact, in this test 30 pairs of tape-recorded, non-verbal sounds are presented to the subject. He is required to decided if the two sounds are similar or different. The test assesses auditory attention, concentration as well as ability to discriminate non-verbal sounds. It has been found that one person with right temporal lobe removals are impaired at this task. The children's version of this test does not include rhythm test.

- **Speech Sounds Perception test :** It is an auditory acquity test in which the person listens to a tape recording of 60 spoken non sense words, all of which have an 'ee' vowel sound in the middle with different beginning and ending sounds. Such as 'weem', 'Meer' etc. After each word, the person underlines from a set of four written words, the spelling that represents the word he heard. The test assesses auditory attention and concentration as well as the ability to discriminate between verbal sounds. Basically, this is a test of left hemisphere function especially the left posterior temporal-parietal cortex around Wernicke's areas or passibly the left face area. The children's version includes fewer word choices.

- **Reitan-Indiana Aphasia Screening test :** This test is a modification of Halstead-Wepman Aphasia screening test. It assesses aphasia, which is the loss of ability to understand and use written or spoken language due to brain damage. In this test the person is presented with different questions and tasks that would be easy for a person without impairment. Some examples of such task would be writing the name of shown pictures without saying the name aloud, verbally naming pictures, reading printed materials, solving simple arithmetic problems, drawing shapes lifting the pencil, etc. The test, on the whole, detects the possible signs of aphasia, which may require further evaluation.

- **Reitan-Kolve Sensory-Perceptual Examination :** This test tends to evaluate the extent to which the person is not able to perceive stimulation on one side of the body when both sides are stimulated simultaneously. The test contains various types of components such as tactile, auditory and visual ones. This test measures several abilities like ability to specify whether touch, sound or visible movement is occurring on right, left or both sides of the body, ability to recall numbers assigned to particular figures, ability to identify the shape of a wood block placed in one hand by pointing to its shape on a formboard in the opposite hand as well as to identify numbers on fingerprints when eyes are closed.

Besides these core tests, there are some ancillary tests, which are tests commonly used in combination with Halstead-Reitan battery. Important such ancillary tests are *Minnesota Multiphasic Personality Inventory, Wechsler Adult Intelligence scale or Wechsler Intelligence Scales for* children, *Rey Auditory Verbal Memory test, Rey Complex Figure test, Test of Memory and Learning, California Verbal learning test, Wechsler Memory scale, Grip Strength test, Grooved Pegboard test, Reitan-Klove Lateral Dominance Examination, Buschke Selective Reminding test, Wide-Range Achievement test* etc.

Scores on Halstead-Reitan test are generally analysed as under :

- Overall performance on the battery is determined. For this, two types of parameter are established : *the Halstead Impairment Index (HIT) and the General Neuropsychological Deficit Scale* (GNDS). The HII is calculated by counting the total number of test in impaired range and dividing it by the total tests administered. The HII in between 0.0 to 0.2 indicates normal functioning, 0.3 to 0.4 indicates mild impairment, 0.5 to 0.7 indicates moderate impairment and 0.8 to 1.0 indicates severe impairment. The GMDS is determined by assigning a value between zero and four to forty two variables involved in the tests. The GNDS coming in between 0 to 25 indicates normal functioning, 26 to 40 indicates mild impairment, 41 to 67 indicates moderate impairment and 68 plus indicates severe impairment.
- **Reflection regarding laterization and localization :** Performance on sensory and motor tasks provides such reflections. In such reflection the particular region of the brain stands damaged.
- **Performance on individual tests :** Here each test is interpreted in the context of other tests in the battery. If the pattern of poor performance is found on three or more tests, brain impairment is most likely to occur. But poor performance on any one test may be due to factors other than impairment.

There are also children version of battery. For older children ages 9 to 14, there is *Halstead Neuropsychological Test Battery* (Reitan & Davison, 1974) and for ages 5 to 8, there is *Reitan Indiana Neuropsychological Test Battery.* (Reitan, 1969) These two children's batteries are based upon adult version of Healstead-Reitan. (Reitan & Wolfsun, 1993)

Luria-Nebraska Neuropsychological Test Battery or LNNB

Luria-Nebraska Neuropsychological test battery (LNNB) is one of the fixed batteries used for measuring neuropsychological deficiencies. Among the neuropsychological deficiencies, it primarily assesses learning, experience and cognitive skills. The work of the Russia neuropsychologist, Alexander Luria, has been the basis for the theory behind the development of LNNB. This test was developed by Charles Golden in 1981. This test is one of the most widely used fixed batteries. (Golden et al., 1985)

The LNNB has 269 items grouped into eleven clinical scales, that is, C_1 through C_{11} and there are three *Summary scales*, that is, S_1, S_2 and S_3. Similar items are grouped together into eleven clinical scales which are presented in Table 17.1. Raw scores on each clinical scales are converted into T scores with a mean of 50 and Standard deviation of 10. Higher scores reflects brain damage and scores above 70 are especially suggestive of brain damage.

Table 17.1: Eleven Clinical Scales of LNNB with tasks included by each Scale

Scales	Major tasks included
C1 Motor	Coordination, drawing, speed and other motor abilities
C2 rhythm	Discriminate and produce verbal and nonverbal rhythmic stimuli
C3 Tactile	Identify subtle tectile stimuli
C4 Visual	Solve progressive matrices, identify drawings and unfocused objects
C5 Receptive speech	Comprehends words, sentences and discriminate phonemes
C6 Expressive speech	Articulate words and sentences fluently
C7 Writing	Use motor writing abilities in general; write from dictation
C8 Reading	Read letter, words and sentences; synthesize letters into words
C9 Arithmetic	Comprehend arithmetical signs and number structure
C10 Memory	Remember verbal and Nonverbal stimuli in different conditions
C11 Intelligence	Reasoning, concept formation and complex mathematical problem solving

Besides these clinical scales, there are three summary scales, which are derived from test performance: S_1 (Pathognomonic), S_2 (Left hemisphere) and S_3 (Right hemisphere). The pathognomonic scale indicates the degree of compensation gained such as actual physical recovery as well as functional reorganization of brain. Higher scores indicate poor compensation. The left hemisphere and the right hemisphere scales are used to determine whether the injury is diffused or lateralized.

LNNB is used with persons aged 15 and above items are scored 0.1 and 2 according to criteria mentioned in the manual. There is also a children version of LNNB called *Luria-Nebraska Neuropsychological Battery for children* (LNNB-C), which is used for children with ages 8 to 12. This tests has 149 different items, which are scored from 0, 1 and 2. Both LNNB and LNNB-C takes roughly about 2 to 3 hours to administer.

The reliability of LNNB has been estimated by the usual methods like split-half, internal consistency and test-retest and the reliability coefficients were found to be satisfactory. The average test-retest reliability for the clinical scales was near 0.90. So far as the validity is concerned, many validity studies of classification of brain-damage persons versus other criterion groups have clearly shown that LNNB has hit rates of 80 percent or still better.

LNMB has been successfully applied to detect brain functions as well as to pin-point what mental disorder is present. The battery has also been able to detect disorders like brain trauma, boarderline personality, schizophrenia, metabolic problems and degenerative disorders.

Besides these two test batteries, that is, Halstead-Reitan test battery and Luria-Nebreska test battery, there are also some other test batteries. Important such batteries are *Dean-Woodcock Neuropsychology Assessment System* (DWNAS), *Cambridge Neuropsychological Test Automated Battery* (CANTAB), *Barcelona Neuropsychological Test* (BNT), *Mini Mental Status Examination* (MMSE), *Cognitive Assessment Screening Instrument* (CASI), *Cognitive Function Scanner* (CFS), etc.

Some Indian Neuropsychological Tests

In India also some neuropsychological tests have been developed. Among these, three are worthmentioning : *PGI Battery of Brain Dysfunction* (PGIBBD), *AIIMS Comprehensive Neuropsychological Battery* and *NINHANS Neuropsychological Battery*. A brief discussion of these tests follows :

- **PGI Battery of Brain Dysfunction (PGIBBD) :**

PGIBBD was developed by D. Prasad and S.K. Verma in 1990 at the Department of Psychiatry, PGI, Chandigarh. This test is based on unitary approach, following factorial sampling design. The test yields a psychometric profile, which can differentiate persons with organic brain dysfunction from those having no organic pathology. The test was standardized with subjects having the age range of 20-45 years. This test battery includes five tests: *PGI memory scales* (PGIMS), *Revised Bhatia Short Scale of Performance test of Intelligence*, *Verbal Adult Intelligence Scale* (VAIS), *Nahor-Benson Test* (NBT) of *Perceptual Acuity and Bender Visual Motor Gestalt Test* (BVMG). The complete administration of the battery takes about 2 hours. The reliability of the different subtests ranges from 0.69 to 0.85. Its validity is also fully established and widely acclaimed. Norms have been developed for adult groups aged 20 to 50 years with respect to age, sex and education.

- **AIIMS Comprehensive Neuropsychological Battery**

This battery was developed by Gupta et al., (2000). It is largely based upon Luria-Nebraska Neuropsychological Battery and is considered useful for both diagnosis and rehabilitation. It is considered useful for clinical and research settings facilitating identification, lateralization and localization of brain lesions in age group of 15 to 80 years. It consists of 160 items in Hindi and has 10 basic scales such as *Tactile scale, Visual scale, Receptive speech scale, Expressive speech scale, Reading scale, Writing scale, Arithmetic scale, Memory scale, Intellectual processes scale, Pathognomonic scale, Left hemisphere scale* and *Righ hemisphere scale*. Norms are available in forms of T-Scores. The reliability of the battery has been established by methods like test-retest method, internal consistency method and inter-rater reliability. The test-retest reliability ranges from 0.79 to 0.98; internal consistency reliability ranges from 0.79 to 0.99 and inter-rater reliability ranges from 0.98 to 1.0. The validity studies show that the battery was effective in discriminating brain damaged patients from normal and schizophrenics.

- **NIMHANS Neuropsychological Battery**

This battery was developed at National Institute of Mental Health and Neuroscience (NIMHANS) by Rao et al., (2004). The battery includes 18 subtests measuring neuropsychological deficits in the various cognitive domains like Speed, Attention, Memory, Executive function and Comprehension. The speed is assessed by Finer Tapping test and Digit Symbol substitution Test. The attention is assessed by Colour Traits test, Digit vigilance test and Triads test. The memory is assessed by four tests like Auditory Verbal hearing test, Logical memory test, Complex Figure test and Design learning test. The executive function is measured by eight different tests such as controlled word Association test, Animal Names test, Design Fluency test, N Back test (Verbal & Visual), Self-ordered Pointing test, Tower of London test, Wisconsin Card Sorting test and Stroop test. The comprehension is assessed by Token test. Normative data are available for adults in the age range of 16 to 65 years following a factorial sampling design. The reliability and validity of the test are well-established. The battery takes about 4 to 5 hours in its complete administration. The battery is available separately for elders and children.

Comparison between Halstead-Reitan Test battery and Luria-Nebraska Test Battery

There are some similarities and differences between Halstead-Reitan test battery and Lura-Nebraska Test battery. The major *similarities* are :
- Both batteries measure cognitive deficits.
- Both batteries have version for children and share a common purpose.

The major *differences* are :
- Halstead-Reitan test battery is older than Luria-Nebraska test battery.
- Halstead-Reitan test battery offers some flexibility in the number and selection of tests to be administered.
- Luria-Nebraska test battery is based upon some of Luria's theories and diagnostic procedures.
- Luria-Nebraska test battery is comparatively more standardized in content, materials, administration and scoring than Halstead-Reitan test battery.
- Luria-Nebraska test battery takes less time in complete administration than Halstead-Reitan test battery.

Today clinical neuropsychological assessment has been much influenced by recent developments in the direct assessment of brain impairment through electroencephalography and neuroimaging techniques such as Position emission tomography (PET) and Magnetic resonance imaging (MRI). These techniques have modernized and improved the neuropsychological assessment no doubt, no technique alone is cent-percent perfect. In fact, the integration of neuropsychological and neuroimaging methodologies offers the greatest promise in developing scientific knowledge about brain-behaviour relationship.

Standard Psychological Testing vs. Neuropsychological testing

The Standard psychological assessment basically evaluates general cognitive and personality patterns/traits. Such assessment intends to diagnose psychiatric syndrome such as ADHD based on behavioural patterns. However, such assessment lacks the capacity to understand the underlying neurological process that is generating symptoms. As a consequence, such psychological assessment can't provide recommendation for treatment.

Neuropsychological assessment, on the other hand, makes a comprehensive assessment of cognitive processes. Neuropsychology integrates the genetic, developmental and environmental history with psychological testing for better understanding of the functioning of the brain. With such comprehensive assessment of the cognitive and personality functioning, the neuropsychiatrist can easily specify the origin and development of psychological disorder and also can make some specific recommendations for treatment.

Summary and Review

- Neuropsychological assessment is a performance based technique to assess the cognitive functioning of the person. The major goals of neuropsychological assessment is to make diagnosis of Neuropsychological disorder, the impact of brain injury upon the individual's behaviour as well as to assess changes in functioning overtime.
- There are two categories of tests of neuropsychological assessment: individual neuropsychological test and batteries of neuropsychological assessment.
- There are five important characteristics of neuropsychological batteries, thoroughness, time, adaptability, flexibility and ease as well as the cost.
- There are two popular neuropsychological test batteries: Halstead-Reitan neuropsychological test battery and Luria-Nebraska neuropsychological test battery.
- In India several neuropsychological test batteries have also been developed. Of these, PGI Battery of brain dysfunction, AIIMS. Comprehensive neuropsychological battery and NIMHANS neuropsychological battery are popular ones.
- Neuropsychological assessment differs from Standard psychological assessment.
- There are some similarities and differences between Halstead-Reitan test battery and Luria-Nebraska Test battery.

Review Questions

1. Define neuropsychological assessment. Discuss the major features of neuropsychological test batteries.
2. Discuss the primary features of Halstead-Reitan Neuropsychological test batteries.
3. Explain the major characteristics of Luria-Nebraska Neuropsychological test battery.
4. Make a comparative study of Halstead-Reitan test battery and Luria-Nebraska Test battery.

■■■

Chapter – 18

Grading and Reporting

Learning Objectives

- Meaning of Grading and Reporting
- Functions of Grading and Reporting system
- Type of Grading and Reporting system
- Multiple Grading and Reporting system
- Guidelines for developing Multiple grading
- Guidelines for developing Multiple grading and Reporting system
- How to assign letter grades?
- Summary and Review
- Review questions
- Key Terms

Key Terms:

Absolute grading, Reporting, Letter grade, Pass-fail system, Checklist of objectives, Portfolio of student work, Multiple grading and reporting system, Relative grading.

Grading and reporting are the most important foundational elements in every educational institution. In fact, grading and reporting progress in one of the complex and frustrating aspects of teaching because these two elements involves so many factors to be considered and so many decisions to be made. In the present chapter a very critical attempt has been made to remove various types of complexities involved therein. In this attempt various types of grading and reporting system will be discussed and some guidelines for their effective uses will also be discussed.

Meaning of Grading and Reporting

Grading means teachers' evaluation formative or summative—of the performance of students. Thus grading is the process summarizing evaluation information and assigning a letter or grade on the report card. Reporting means how those evaluations are communicated to the students, parents or others in understandable form. In other words, reporting is the process of communicating the achievement of students formally or informally in understandable language. Because of their importance, teachers and educators must ensure that grading and reporting must meet the criteria for reliability and validity. Besides, they must also ensure that grading and reporting are *meaningful, accurate* and *fair.* Commonly it has been observed that in determining grades, teachers merge scores for major examination, projects along with evidence from homework, classroom participation, work habits, punctuality in the class, etc. Computerized grading programmes help teachers apply different weight to each of these various categories. Subsequently they are grouped in idiosyncrative ways (McMillian, 2001; McMillian et al., 2002). The resulting grade becomes impossible to interpret accurably or meaningfully. (Brookhart & Nitko, 2008) Therefore, it is essential that the grades must be made meaningful and so also the reporting.

Functions of Grading and Reporting System

Grading and reporting in educational system are designed to serve a variety of functions. There are three such important functions: *instructional uses, reports to the parents/ guardians* and *administrative and guidancesuses.* A discussion follows :

1. Instructional uses :

The major focus of grading and reporting should be improvement of student learning and development. This can be achieved when the report (a) clarifies instructional objective for the students (b) shows students' strength and weakness in learning (c) provides information regarding the student's personal and social development (d) indicates where teaching might be modified and (e) evaluates the student's motivation. In general, the student learning is best improved by the day-today assessment of learning as well as from the feedback from tests and other assessment procedures. Besides this, a well-designed report form, together with a portfolio of students work which contains carefully selected examples, can provide the systematic summary of the learning progress. Apart from these, periodic progress reports can also contribute to the motivation of students by providing knowledge of results and short-term goals.

2. Reports to parents/guardians :

One of the important functions of grading and reporting is to inform parents regarding the progress of students in the class. Such reports help parents in knowing how well their children are achieving the intended learning outcomes of the programme of the school. Educators and teachers are of view that such information is considered important from several points of view:

- Information regarding success and failures of the students enable the parents to provide emotional support and encouragement urgently needed to the children.
- When parents/guardians come to know what the school is attempting to do, they are better able to cooperate with the management of school in promoting children's learning and development.
- When parents/guardians come to know about strength and weakness of their children in learning, its helps them in making a more effective educational and vocational plans.

At the elementary level of education, the reporting is frequently supplemented by parent-teacher conferences.

3. Administrative and guidance uses :

Grading and reporting serve many administrative and guidance functions. Some of the administrative functions are :
- Used in determining promotion, graduation and awarding honor to the students.
- Used in determining about athletic eligibility
- Used in reporting to other schools and prospective employers.

For most of these administrative used, usually a single letter grade is commonly adopted. Some of guidance uses of grading and reporting are as under :
- Such reports help counsellor in providing input for realistic educational, vocational and personal counselling.
- Reports that contain ratings on personal and social characteristics are very much needed in helping students with adjustment problems.

These above guidance functions of grading and reporting are best served when they are both comprehensive and detailed.

Thus it is clear that grading and reporting system serve many functions. For serving these functions properly, it is essential that a more elaborate reports be presented than simple traditional single letter grade.

Types of Grading and Reporting Systems

There are different types of grading and reporting systems in educational institutions. While developing different types of grading and reporting, attempt is made to maintain a proper balance between the need for detailed information and need for simplicity and conciseness. The important types of grading and reporting are: *Traditional letter-grade system*, *Pass-fail system*, *Checklists of objectives*, *Letters to parents/guardians*, *Portfolios of student work* and *parent-teacher conference*. A discussion follows :

• Traditional letter-grade system :

A look at past would clearly show that letter grades have been the most important and popular method of grading and reporting student progress in school throughout the history of education. In this system, single letter grade such as A (Excellent), B (Good), C (Average), D (Needs improvement) and F (Failure) for each subject is assigned. In some cases, a single number such as 5, 4, 3, 2, 1, etc. is used in lieu of letter. This letter-grading system is considered very simple, precise and convenient. Grades are also easily averaged and they are useful in predicting future achievement. However, such system has some obvious limitations when used as the sole method of reporting.
- Such letter grades are a combination of many things such as achievement, work habits, effort and good behaviour
- Such grades don't indicate the specific strength and weakness of student in learning

- The proportion of students assigned to each letter-grade varies from one teacher to another and sometimes from situation to situation.

As a consequences of these limitations, letter grade system is very difficult to interpret and use. For example, a grade of C may be interpreted a good achievement but reflecting poor work habit followed by some undesirable behaviour. The same C grade may also be interpreted as showing poor achievement but showing strong effort to achieve more and more followed by some good behaviour. In order to meet the challenges posed by these criticisms, some educators have recommended to reduce the number of grades to two such as S (for satisfactory) and U (for unsatisfactory). Sometimes a third grade such as H (for honors) is also added. But these small number of categories have not been widely applauded because they provide even less information than the traditional grade system.

- **Pass-fail system :**

This is a simple two-category system that is commonly used in elementary school, high school and colleges. In this system the student has simply to pass the course and if he is not able to do so, he is placed under 'fail' category. It encourages, students to take some courses, usually elective courses, which are not included in their grade-point system. Such system also removes fear of getting a lower grade-point average and this allows students greater freedom to select their learning experiences. Educators and teachers are of view that it also permits students to focus on, those aspects of the course that relate directly to his major area of interest and neglect those areas, which are of little interest.

Although pass-fail system is very easy to adapt, it has its own limitations. *First*, it offers less information than the traditional letter-grade system. *Second*, students tend to work at the minimum just to pass the course. *Third*, it provides no indication of learning and therefore, its value for describing present performance or predicting future performance stands lost. Despite these limitations, the pass-fail system serves the purpose for which it was formulated if its use is kept limited to some minimum number of course.

- **Checklists of objectives :**

In this grading and reporting system a list of objectives to be checked or rated is provided before the teachers. Such system provides rating of progress toward the major objectives in each subject-matter area. This system of grading and reporting is most common at elementary school level. For example, the following statements for writing illustrate the nature of the report.

Writing
- Writes with good and clear handwriting
- Writes correctly
- Grammatical structure of the written language is good
- Writes at a satisfactory speed

Various symbols are used to students on each of these major objectives. The commonly used symbols are O (outstanding), S (satisfactory), P (proficient), PP (Partially proficient) and N (Needs improvement).

There are some obvious advantages of checklist form of reporting. It provides a detailed analysis of student's strengths and weaknesses so that a planned and constructive action can be taken to help improve learning. Such grading and reporting system also provides students, parents and others with a frequent reminder of the objectives of the school (Linn & Miller, 2011). However, there are some difficulties associated with checklists of objectives. *First*, such reporting system is very time consuming. *Second*, it becomes a difficult job to keep the list of objectives or statements down to a workable number. *Third*, another difficulty often encountered is how to state those statements in simple and concise terms so that they could be readily understood by all users.

- **Letters to parents/Guardians :**

Some educational institutions prefer to write letters to parents/guardians regarding the student's progress in the institutions. Through such letters, it becomes possible to report about the strengths, weaknesses and learning needs of each student. Not only this, letters also suggest specific plan for improvement. Such letters also make it the student's progress in all relevant areas of development.

Despite the fact that the letters contain a lot of flexibility and provide a good supplement to other types of reports, their usefulness as a sole method of reporting progress has its own limitations. Some of these limitations are as under :
- Preparing a comprehensive letter requires a lot of time and skill on the part of the teachers.
- Letters fail to provide a systematic and cumulative record of student progress toward the objectives of the school.
- Accounts of weaknesses written in letter are often misinterpreted by parents and guardians.

- **Portfolios of Student work :**

The portfolio of a student may be defined as a purposeful collection of pieces of student work. Portfolios are sometimes described as *portraits* of person's accomplishment. The portfolio of the student work often contain commentary on entries by both students and teachers. The entries in the portfolio are selected deliberately in such a way as to illustrate the range of student work such as results of laboratory, experiments, types of writing, types of mathematical problems, etc. Entries are also selected in a way that may reflect progress during the year and in some cases progress from early to later stages of completing a work or any project, if any. Such portfolio also teaches students about objectives and standards they are expected to meet.

- **Parent-teacher Conferences :**

Some schools regularly arrange open house where parent-teacher conferences are held where a two-way communication between home and school occurs. In such conference, besides receiving a report from teacher, parents (or guardians) get an opportunity to provide information regarding the student's out-of-school activities. Such conference is very much flexible and permits teachers and parents to ask questions, discuss their common concerns and cooperately plan a program for improving the student's learning and development. This give-and-take in the conference makes it possible to overcome any doubts and misunderstandings concerning the progress of the students.

The parent-teacher conference as a method of reporting about the progress of the student is widely applauded. However, it has some limitations. *First*, such conference fails to provide a systematic record of student progress. *Second*, the conference requires substantial amount of time and skill. *Third*, some parents don't turn up to such conference. In view of these limitations, it can be said that it can be useful as only one of the supplementary methods of reporting.

Multiple Grading and Reporting System

In multiple grading and reporting system the traditional letter (or number) grades are not removed rather attempt is made to improve letter grade system and supplement it with a meaningful and more detailed reports about student learning progress. In a typical multiple grading and reporting system the letter grades or numbers are retained and they are supplemented by grades with checklists of objectives. In some cases two grades are assigned to each subject : one for academic achievement and the other for making effort for improvement and growth.

Teachers who report multiple grades using different criteria also become instrumental in increasing validity, reliability and fairness of the grading process. Apart from this, the extent to which the classroom achievements of student learning are linked with student learning outcomes addressed in large scale assessment, the relationship between achievement grades and the accountability of assessment results becomes stronger. (Guskey & Bailey, 2010)

Guidelines for Developing Multiple Grading and Reporting System

As we know, no one grading and reporting system is found satisfactory in all educational institutions. Therefore, each educational institution has to develop methods that fit well with its requirements. There are some general principles for devising multiple grading and reporting system. These principles act as guidelines for the said purpose. Some of those principles are as under:

- The grading and reporting system should be developed cooperatively by students, parents and school authorities. For developing a healthy multiple grading and reporting system a committee consisting of representatives of parent groups, student organisations, elementary and secondary school teachers, counsellors and administrators should be formed. Various ideas and suggestions are given to the committee through the representatives and the members go to their respective groups for modifications and approval of the committee's tentative plans.
- The grading and reporting system should be based on a clear statement of various educational objectives. The grading and reporting system should be based upon some clear and defined objectives. Some of these objectives will be general school objectives and some will be unique to a particular course or areas of study. Although final report

from is modified by a number of practical considerations, the major focus should be on the objectives of school and course of the study.
- The development of grading and reporting should primarily be guided by the functions to be served. Any grading and reporting system must contain the information most needed by the report users. Although it is seldom possible to meet all the users' need, a satisfactory compromise is most likely if they are known. In fact, it would be most useful if the letter grades in each subject are supplemented with a separate reports on effort, course objective, work habits, personal and social characteristics. Such reporting system is needed by most report users.
- The reporting and grading system must match with school standards. The grading system needs to be aligned with various performance standard. In this direction, one of the important steps is the use of the same categories for grades and performance standard. There is also urgent need to align, the performance, which the students are expected to achieve in a course with the verbal description of performance standards.
- The grading and reporting system should not be only diagnostic rather they should also be practical. In general, a comprehensive picture of strength and weakness of the students' learning and development must be reflected in the reporting system. While preparing a comprehensive report, it is also important to maintain a balance with respect to its practical demand such as preparation of report should not take too much time, the report must be understandable to all concern and the report must be easily summarized by authorities for maintaining school records. Most of the educators and teachers are of view that a compromise between comprehensiveness and practicality can best be shown by supplementing letter-grade system with more detailed reports on different aspects of development and learning by students.

In a nutshell, it can be said that the grading and reporting system considers different needs of students, teachers, parents and school personnels. When the common method of grading such as letter-grade system is supplemented by other methods of reporting, the grades become more meaningful.

How to Assign Letter Grades?

Since the letter grading system such as A, B, C, D and F is most common among school teachers, it is essential to know how to assign such grading. The problem of assigning letter grades involves addressing the following four things : *determining what to include in a grade, combining datain assigning grades, selecting the appropriate frame of reference for the grading* and *determining the distribution of grades*. A discussion follows :

• What should be included in letter grade?

As we know, letter grades are most likely to be meaningful and useful if they represent only achievement of the students. If they are mixed with other factor such as effort, amount of work completed, personal conduct, etc. the interpretation of the letter grade becomes a challenge as well as very much confusing. Most teachers feel that the letter grade should be strictly based upon the achievement of the students but they also consider that effort should be considered along with achievement of student whom they judge to be somehow less able. Although such attempt seems reasonable at the outset, it incorporates some drawbacks. First drawback is that for a teacher it becomes very difficult to assess the effort of the student in true sense of the term. The second drawback is that using different bases for grading for different students signals a mixed message and may prove to be unfair to students who are judged more able than others. Still another drawback is that ever with most sophisticated instrument, it is very difficult to distinguish between aptitude and achievement because both depend upon the student's learning.

Educators are of view that if letter grades are to be valid indicator of achievement, they must be based solely on valid measures of achievement. For assessing achievement various measures like tests, ratings of performances, written reports o other measures are generally incorporated. How much emphasis should be given to these various measures in letter grades, is determined by the nature and objectives of the course. For example, a grade in music might be determined by tests and rating on various musical performance skills, a grade in science might be determined by tests and assessments of laboratory performance and so on. In determining the grade, the role of instructional objectives in particular, is considered very important. Other things being equal, the more important the instructional objective is, the greater the weight it should receive in determining course grade. In a nutshell, letter grades should be such that they must reflect degree to which the students have successfully achieved the learning outcomes as specified in the course objectives.

- **How should data relating to achievement be combined in assigning letter grades?**

The next step in assigning letter grade is to combine various elements or aspects of achievement such as tests, written reports or performance ratings so that each element receives its intended weight. A very common practice is to combine the elements into a composite score by assigning appropriate weights to each element and subsequently, use these composite scores as a basis for grading. Let us assume that one wants to combine scores on a final examination and on a test and one wants them to be given equal weight. Our range of scores on these two measures are as under:

	Range of scores
Final examination	80 to 100
Test	10 to 50

One apparently simple method of equating the scores is to add together the final examination score and test score for each student. The effectiveness of this simple method can be checked by comparing a composite score of a student who has obtained highest score on the one final examination and lowest score on the test scores (100+10=110) with a student who has obtained lowest score on the final examination and highest on the test score (80+50=130). Obviously, such comparison where two scores are added is not giving them equal representation. Another method of equalising scores is to make maximum possible scores the same for both sets of scores. In the above example, if one multiply the test score by 2, one would get equal top scores, that is, 100 on both measures. Let us apply this method to the same two extreme cases. In such case our first student would get a score of 120, that is, (20+100) whereas the second student would get a score of 180, that is, (80+100). It is obvious that this second method also does not equate the scores. This method produces even the larger difference between the two composite scores. This is due to the fact that the influence of each component on the composite score depends upon variability or spread of scores and such variability is not taken into account by the method. Therefore, it is essential to evolve such method, which may take variability of scores into account. Here in this example the final examination score has a range of 100–80=20 and the test scores have the range of 50–10=40. It means that if one multiply each final examination score by 2, one would be able to get the desired equal weight. Let us check the effectiveness of this third method. Now a student getting highest mark on the final examination and the lowest mark on the test would get a score of 200+10=210 and the student getting lowest on the final examination and the highest mark on the test would get 160+50=210. This check makes it clear that this method gives two sets of scores equal weight in the composite score. Still a more refined method of weighting system is to use standard deviation as the measure of variabilities. The scores on each test or examination can be converting to T score, which can then be multiplied by whatever weight is desired and summed to yield a composite score on the basis of which grades are assigned.

- **What frame of reference should be used in letter grading?**

Letter grades are assigned on the basis of some frame of reference. They are assigned on the basis of one of the following frames of reference :
- Relative grading, that is, performance in relation to other group members
- Absolute grading, that is, performance in relation to some specified standards
- Performance in relation to learning ability or amount of improvement.

A brief discussion is presented below :

When letter grades are assigned on relative basis, it means that the students performance is compared with the performance of a reference group, particularly his class fellows. In this method, grade to be given to a student is determined by his relative ranking in his total group rather than by some absolute standard of achievement. In other words, here grading is dependent upon relative performance. Since grading is dependent upon relative performance, it is influenced by both the performance of student as well as the performance of the group. As a consequence, if a student has been placed in low-achievement group, he is expected to be much better gradewise than when placed in high-achieving group. This is both unfair and misleading.

Assigning grade on the basis of absolute performance involves comparing the performance of a student to some standard set up by the teachers. These standards may be concerned with the mastery to be achieved by students and may be specified as a task to be done such as typing 40 words per minute without making any error or as the percentage of correct answers to be obtained on a certain test or examination. Since this is a standard-based system, letter grades are assigned on the basis of the absolute standard of performance and not on relative basis. As a consequence, if in a group or class all students show a high level of mastery consistent with established standard of performance, all will receive high grades. In addition, percentage-correct scores are also used in setting absolute standards. Various schools use absolute grading based on the percentage of correct scores like this:

Percentage of scores	Grades
95-100	A
84-94	B
75-83	C
65-73	D
Below 65	F

Researchers have revealed that when grading is done on the basis of absolute level of achievement, some requirements must be fulfilled:
a) The domain of learning task must be clearly defined
b) The standards of performance must be clearly spelled out and justified.
c) The measures of the achievement of students should be criterion-referenced.

Letter grades are also assigned in relation to learning ability or amount of improvement. This is a common practice adapted by teachers at the elementary school level. This is a straightforward system found inconsistent with standard-based system of grading and reporting. Moreover, this system has many other difficulties, too. Making a reliable measure of any learning ability, with or without test, is a very difficult task because achievement on such test is contaminated by some unknown and uncontrolled factors. Likewise, improvement such as growth in achievement over a specified time period is very difficult to be reliably assessed with classroom measures of achievement. As a consequence, the dependability on the resulting grades becomes low. If such grading is to be used at all, it should be used as a supplementary grade. For example, one letter grade may indicate student's level of achievement (relative or absolute) and the second letter grade may indicate achievement in relation to ability or degree of improvement exhibited since last grading time. This is called *dual letter grading*.

• How should the distribution of letter grades be determined?

There are two important methods of determining letter grade for measuring level of achievement of student. Relative grading and absolute grading. A discussion follows:

Relative grading :

Relative grading is based upon the relative level of achievement of student. Such grading is upon the ranking of students in order of overall achievement and assigning letter grades on the basis of those rank. Such ranking may be done on the basis of single classroom group or it might be based upon the combined distributions of several classroom groups taking the same course. In any case, before, assigning letter grades, the proportion of A's, B's, C's, D's and F's must be determined. Today some schools assign grades on the basis of normal curve and this is especially done when the course or combined courses have a relatively large and unselected group of students. However, assigning grades for the classroom group is not defensible because such group is usually too small to yield a normal distribution. If grading is to be done on the basis of normal curve, separate distributions for gifted and remedial classes and for introductory and advance course should be done so that there must be flexibility in the distribution to the extent that variations in the performance of students from one cause to another and from one time to another in the same course may be easily allowed. Such flexibility is easily demonstrated when ranges rather than fixed percentages of students should received each letter grade. For example letter A indicating 10% to 20% of students, B indicating 20% to 30% of the students C indicating 30% to 50% of the students, D indicating 10% to 20% of the students and F indicating 0% to 10% of students. This is just a suggested distribution for a course.

- **Absolute grading :**

Absolute grading is most appropriate when the instructional objectives have been clearly specified and standard for mastery has been appropriately set. Letter grade in such system reflects degree to which the objectives have been attained and the student has reached the standard for mastery. In such situation the following letter grades are very common :

A = Outstanding (The student has mastered all types (both major or minor) of instructional objectives.)
B = Very good (The student has mastered all the course's major instructional objectives as well as most of the minor ones.)
C = Satisfactory (The student has mastered all major instructional objectives and just a few of minor ones.)
D = Very weak (The student has mastered just a few of the major and minor instructional objectives, lacks essentials needed for the next highest level of instructional objectives Remedial work is desirable.)
F = Unsatisfactory (The student has not mastered any of the course's major instructional objectives and lacks the essentials needed for the next level of instruction. Remedial work is needed.)

If the measures of achievement have been designed to yield scores in terms of percentage of correct answers, absolute grading might be, then, defined as follows :

A = 95% to 100%
B = 84% to 94%
C = 75% to 83%
D = 65% to 74%
F = below 65%

Letter-grading based on percentage of correct scores is considered most appropriate only when the necessary conditions for an absolute grading system has been fulfilled. When the measuring instruments are based upon some undefined hodgepodge of learning tasks, use of percentage-correct score tends to produce letter grades, which are difficult to be interpreted.

In reality, absolute grading system for reporting on student progress seldom uses only letter grade. A comprehensive report is utilized that includes a checklist of objectives to inform both student and parent that which objectives have been successfully mastered and which have not been mustered even by the end of each grading period. In some standard-based programmes letter grade or numbers are assigned to each objective for reflecting the level of proficiency achieved. This may be done as under:

 4 = Good proficiency, has developed skill
 3 = Proficiency could be improved, skill developed satisfactory
 2 = Low proficiency, skill developed, needs additional work
 1 = Basic skill not acquired.

Summary and Review

- Grading means teachers' evaluation either formative or summative of the performance of the student. Reporting means how these evaluations are communicated to students, teachers or parents in some understandable form.
- Grading and reporting have three important functions: instructional uses, reports to the parents/gradient and administrative and guidance uses.
- There are different types of grading and reporting system. Important ones are: traditional letter-grade system, pass-fail system, checklist objectives, letters to parents/guardian, portfolios of student work and parent-teacher conference.
- In multiple grading and reporting system, the traditional letter (or number) grades are improved (rather than removed) and supplemented it with a meaningful and more detailed reports about students learning progress.
- There are some obvious guidelines proposed by educators for developing multiple grading and reporting system. Six such guidelines have been emphasized.
- In addressing the problem of how to assign letter grade four questions are important: what should be included in letter grade? How should data relating to achievement be combined in assigning letter grade? What frame of reference should be used in letter grading? and How should the distribution of letter grade be determined?

Review Questions

1. Give the meaning of grading and reporting. Discuss their main functions or uses.
2. Discuss the different types of grading and reporting system frequently used in schools.
3. Discuss the important guidelines provided by educators for developing multiple grading and reporting system.
4. Write are essay or how to assign letter grades.

Chapter – 19

Applications of Psychological Testing in different fields

Learning Objectives
- Application of Psychological testing in clinical setting
- Application of Psychological testing in counselling and guidance
- Application of Psychological testing in educational setting
- Application of Psychological testing in industry, organisation and business
- Application of Psychological testing in sports
- Application of Psychological testing in military
- Summary and Review
- Review questions

Key Terms:
Intelligence test, clinical Interview, Aptitude test, Interest test, Achievement test, Scholastic Assessment test, Career planning, Educational planning, Projective tests, Psychometric tests, Ranking methods, Athletic Motivation Inventory, Profile of Mood states, Mental toughness.

Psychological tests are only one tool in the process of assessment. Psychological testing deals with the instrumentation part of the assessment where psychometric instruments are used for gathering information. As we know, the roots of psychological testing can be traced backed to over 3,000 years ago, when Chinese had used written tests in filling the civil service posts and systematic testing of both physical and mental progress was used by the ancient Greeks. At the turn of the 20th century, the experimental psychologists like Cattell and biologists such as Galton and Person contributed to the development and construction of psychological tests. However, the very strong impetus for the development of psychological tests have been practical rather than theoretical. As a consequence, psychological tests and testing have been successfully applied in many different fields such as in clinical setting, in industrial organization and business, in educational setting, in counselling and guidance, in military, in sports, in politics and law as well as in transport and traffic. In the present chapter an attempt has been made to examine only the major applications of psychological testing in some select fields.

Application of Psychological Testing in Clinical Setting

As we know, clinical psychology is the branch of psychology that focuses on the prevention, diagnosis and treatment of abnormal behaviour. As such, clinical psychologists conduct psychological assessments aimed at diagnostic, prognostic and thereauptic decision in various mental health settings (Butcher, 1995; Maruish, 1994). Clinical psychologists use a wide variety of tests in assessing individual clients. The common popular stereotype of a clinical psychologist suggests that they use diagnostic tests extensively. As we know, diagnosis refers to the process of determining when an individual's problem fulfils the criteria for a psychological disorder. The process of arriving at a diagnosis is called as *screening*. Indeed, there are a number of objective and projective personality tests and diagnostic tests that are popularly used by clinical psychologists. The popular measures of personality used by clinical psychologists are MMPI-2, Sixteen personality factors test, Edwards Personal Preference Schedule, NEO Personality Inventors, California Psychological Inventory Million Clinical Multiaxial Inventory-III, etc. Among projective tests of personality popularly applied in clinical settings are Rorschach Inkblot test, Thematic Apperception Test, Rotter's Incomplete Sentences Blank, and Draw-a-Person test. Among diagnostic testing MMPI (Minnesota Multiphasic Personality Inventory) and Bender- Gestalt are very popular.

However, clinical application of psychological tests is by no means limited to personality tests. Clinical psychologists also employ neuropsychological test batteries. Neuropsychological assessment requires specialized training and is usually conducted by trained psychologists or psychiatrists. The most widely used fixed neuropsychological test battery is the *Halstead-Reitan Neuropsychological Battery*. Another neuropsychological test battery is the *Luria-Nebraska Neuropsychological Battery* (LNNB) which takes about a third of the time it takes to administer Halstead-Reitan battery. Besides these two batteries, there are also other test batteries that are used to probe deeply into any one area of neuropsychological functioning. A few examples of such battery are *Neurosensory center comprehensive Examination of Aphasia (NCCEA), Montreal Neurological Institute Battery and Southern California Integration test*. NCCEA focuses on communication deficit; the Montreal battery has been useful in locating specific kind of lesions and the Southern California test aims at assessing sensory-integrative and motor functioning of children aged 4 to 9 years.

Perhaps the biggest advancement in the field of neuropsychological assessment have come form of high technology and mutually beneficial relationship that has gradually developed between psychologists and medical personnels. As a consequence, electrophysiological techniques are being used in neuropsychological assessment. Some examples of these electrophysiological techniques are *electroencephalography* and neuroimaging methods like *positron emission tomography* (PET) and *magnetic resonance imaging* (MRI). Two important electrophysiological methods frequently applied for investigating brain functions are *electroencephalogram* (EEG), and event-related potential (ERP). EEG shows continuous written record of brainwave activity and ERP shows a record of the brain's electrical response to the occurrence of a specific event.

Many clinical psychologists have specialized training in administering intelligence tests. These tests not only provide data regarding intellectual performance of the testees, but they also provide a valuable opportunity to observe their behaviour in response to a challenging situation. Important individual intelligence tests commonly used by clinical psychologists in clinical setting are the *Stanford-Binet test*, the *Wechsler Adult Intelligence Scale* (WAIS), the *Wechsler Intelligence Scale for Children* (WISC), the *Wechsler Preschool and Primary Scale of Intelligence* (WPPSI), the *Kaufman Assessment Battery for Children* (K-ABC), etc. Among group intelligence tests, in clinical setting the use of *Raven's Progressive Matrices* (RPM) and *Culture Fair Intelligence test* are very popular. In clinical setting the clinical psychologists also obtain information derived from observation, interviewing and the case history. They combine such information with test scores to provide an integrated picture of the individual. (Beutler & Berren, 1995)

Such procedure helps the clinical psychologists in making informed decisions pertaining to differential diagnosis. Such decision-making usually proceeds by means of the collection, analysis integration and thoughtful reporting about relevant behavioural data. (Anastasi & Urbina, 2002) In psychological assessment using psychological test two things, namely hypothesis generation and hypothesis testing about the individual case take place regularly. Each item of information whether it is generated in clinical case history or by a test score, tends to suggest a hypothesis about the individual client that will be either refused or confirmed as other relevant facts are gathered.

Besides using intelligence tests for assessing the individual's level of intellectual functioning, clinical psychologists also use *pattern* or *profile* of test scores. Such profile analysis provides data that are of help in the diagnosis of brain damage and various forms of pathology that may affect intellectual functioning. The Wechsler scales make use of such profile analysis. Since Wechsler scales, profile analysis has been used for other tests also. Most profile analysis uses variants of three important procedures. The first procedures involves knowing about the extent of variation among test scores such as variations between verbal subtests and performance subtests. The second procedure comprises analyzing the salient features of the profile in light of *base rate data* about the rarity or frequency of such features within the normative group. The third procedure is linked to score patterns associated with particular pathology such as anxiety state, learning disabilities Korskoff syndrome, amnesia, etc.

Whatever the psychological tests are applied by psychologists in clinical settings, they use three models of clinical assessment, which allows the therapists to obtain a complete picture of a client and to understand and reduce human suffering. (Barlow & Durand, 2009) Those models are : *information-gathering model*, *therapeutic model* and *differential treatment model*. When clinical psychologists use assessment mainly as a way to gather information for making a diagnosis and facilitating communication, he is said to use information-gathering model. In this model the used psychological tests provide a standardized comparison with similar people and allow the practitioner to make predictions about the client's behaviour outside the assessment settings. Besides this, test results also provide a baseline that the clinical psychologists often use to identify disorders and design an individualized treatment programme. In *therapeutic model*, new experiences and new information are provided to the client who often uses these information in making changes in their lives. This model evolved during 1950s and 1960s when humanistic movement was at its peak. In fact, the clinical psychologist often uses the assessment information for encouraging self-discovery and growth. Such assessment when used properly, provides valuable tool for increasing insight and growth. In differential treatment model, clinical psychologists often use psychological tests for conducting research and evaluating programme outcome. Here used psychological tests provide definitive answers as to whether or not clients as a group has responded positively to a particular therapy or intervention. Of these three models, clinical psychologists are usually predisposed to use information-gathering model where he is considered as the primary user of the test results, which provides the basis for diagnosis and designing an intervention.

In clinical setting most clinical psychologists also use *clinical interview* as a supplement to data obtained on the basis of administering psychological tests. The clinical interview that may be *structured*, *semi-structured* or *nondirective* basically involves a discussion between the client and the clinician-often a clinical psychologist during which the clinician makes a serious attempt to collect information about the client's attitude, emotions, life history as well as about his past behaviour. (Durand & Barlow, 2006)

In a nutshell, it can be said that the psychologists in the clinical setting use various types of the tests for the purpose of diagnosing the psychopathology of the client and making an appropriate therapeutic recommendations for the welfare of the client.

Application of Psychological Testing in Counselling and Guidance

Like a clinical psychology, counselling psychology is a branch of psychology that is concerned with the prevention, diagnosis and treatment of abnormal behaviour. The basic difference between a clinical psychologists and a counselling psychologist is that the clinical psychologist focuses their research and treatment efforts on *severe* forms of abnormality whereas a counselling psychologist tends to focus more on every day ordinary problems like difficulties in concentration, poor study habits, career decisions, marital and family communications, etc. However, the commonality between the two branches is that both strive to foster personal growth in their clients and the psychological tests employed by both professionals overlap considerably. Both counselling psychologists and clinical psychologist use psychological testing for obtaining information considered necessary for providing services to clients. On the whole, the counselling psychologists are found to show readiness to use the therapeutic model-one of three models of psychological assessment. In counselling psychology, the major focus of test use is the test taker who is perceived as the primary user of the test results (Campbell, 1990). A counselling psychologist originally is specialized exclusively in vocational assessment as well as in guidance. (Drummond, 1996)

Counselling is a kind of interaction, which occurs between two individuals called *counsellor* and *client*, takes place in a professional setting and is mainly initiated and maintained to facilitate some changes in the behaviour of the client. Here the counsellor is trained and focuses on some aspects of a client's adjustment, developmental or decision-making needs. Counselling is basically concerned with bringing voluntary changes in the client. Counselling is *not* advice, nor making suggestions or recommendations.

As we know, psychological tests are used for different purposes. Of the many uses of psychological tests, one important use is to assist the counsellor in the process of counselling. In counselling different types of psychological tests are used. The common psychological tests used in counselling are intelligence test, aptitude test, interest test, personality test, etc. Computerized adapting testing (CAT) is also employed by the counsellors. In other words in counselling all those tests are used that are commonly used in clinical setting by the clinical psychologists.

The first intelligence test was designed by a Frenchman, Alfred Binet, to be administered individually. Several American versions of this test were developed. The most popular of these is the *Stanford-Binet,* based on the work of Lewis Terman at Stanford University, and was published in 1916. This test has remained a very popular choice of the counsellor and the most recent revision being the fifth edition. The Wechsler intelligence scales are also popularly used in counselling process. Wechsler intelligence scales are also popularly used in counselling process. Wechsler intelligence scales are based upon the assumption that intelligence is the sum total of the individual's abilities to think in a rational manner, to act purposefully and to deal in an effective way with him or her environment. Wechsler intelligence scales were developed for adults, as well as for both preschool and primary school children. Still another popular individually administered measure of intelligence for children having age range from 2 years to 12 years is *Kaufman Assessment Battery for children.* This test consists of various subtests of mental and processing skills, which yield five major scores in *Sequential processing*, *Simultaneous processing*, *Mental processing composite*, *Achievement* and *Nonverbal*. Besides this, there are also Kaufman Adolescent and Adult Intelligence test that measure ability to solve problems and make important decisions based on knowledge, verbal conceptualization, life experiences acculturation etc. A composite score provides a measure of overall intellectual functioning.

Another popular intelligence test used in counselling situation is *Slossan Intelligence test* that assess the cognitive abilities of children and adults. One of the most popular group intelligence test is *Otis-Lennon School Ability test* (OLSAT). In India also some intelligence tests have been developed and they have been frequently used in counselling and clinical setting. Draw-a-Person test developed by Adarsh Kohli, Manjeet Kaur and Ramma Malhotra for 6 to 12 years of children and Bhatia Battery of Performance Intelligence test for 11 to 16 years of children are the popular choice of counsellor.

Aptitude tests are another classes of test used in counselling. Aptitude tests are those tests that seek to measure a person's potential ability to perform or to acquire proficiency in some occupation or other type of activities. Such aptitude tests are used by counsellors because they may a) recognize potential abilities about which the person is not aware b) promote the development of potential abilities of a given person c) provide valuable information in making educational and career decisions or other choices between different competing alternatives d) help in anticipating the academic and vocational success a person might attain and e) help in persons group with similar aptitudes for developmental and other academic purposes.

The most popular used multiple aptitude batteries commonly used by counsellors are *Differential Aptitude Battery* (DAT), *General Aptitude Test Battery* (GATB), *Armed Services Vocational Aptitude Battery* (ASVAB) and *Multidimensional Aptitude Battery* (MAB). Besides these, counsellors sometimes use Special aptitude test such as *Clinical aptitude test*, *Mechanical aptitude test*, etc.

One precaution that the counsellor must take is that he must expect a person to demonstrate the considerable differences across a range of aptitude, he must be alert to the possibility that a person may not demonstrate the same level for given aptitude every time. In other words, aptitude measures are actuarial rather than absolute.

Interest tests are also used by counsellor. Such tests provide information about an individual's general pattern of likes and dislikes. Various interest inventories have been designed to assess an individual's interest and relate them to those of various occupational areas. The development of interest tests evolved from studies which indicated that the person in a given occupation seemed to be characterized by a cluster of common interests that made him distinct from people in other occupations. In counselling process the popular interest inventories frequently preferred by counsellors are *Kuder Preference Record, VocationalStrong Interest Inventory*, *Ohio Vocational Interest Survey II* (OVIS-II), *Self-Directed Search*, *Jackson Vocational Interest Survey*, etc. In India also some interest inventories have been developed and they are used by counsellors. Important ones are *Comprehensive Interest Schedule* by Sanjay Vohra, *Interest Inventory* by T.S. Sodhi and H. Bhatnagar and *Multifactor Interest Questionnaire* by S.D. Kapoor and R.M. Singh, *Educational Interest Record* and *Vocational Interest Record* both by S.P. Kulshrestha.

The Kuder Preference Record is the most popular of the various Kuder Interest inventories. The results are analyzed and profiled for occupational areas of outdoor, mechanical, occupational, scientific, persuasive, artistic, literary, music, social service and clinical. The Kuder General Interest Survey is the revision of original preference record and extends the use downwards to the sixth grade. The Kuder occupational Interest Survey is still another version, which provides scores showing similarities with occupational and college-level areas.

The Strong Interest Inventory is a revision of the earlier forms of Strong vocational Interest Blank. Scores are reported in five main forms: *Occupational Scales*, *General Occupational Themes*, *Basic Interest scales*, *Personal Style scales* and *Administrative Indexes*. Items of the inventory include occupations, school subjects, activities, leisure activities and types of people. The score report is organized to demonstrate to the clients the overall trends in personal interest as well as to note how these trends are related to the different works.

The self-directed search (SDS) is a self-scoring tool that person could use to classify themselves according to the six occupational themes (*realistic, artistic, investigative, conventional, social* and *enterprising*). The SDS is not only self-scored but it is self-administered and self-interpreted.

The *Ohio Vocational Interest Survey* (OVIS-II) has been developed for use with high school students and adults. This inventory is separated into three parts: questionnaire, a local information survey and interest inventory. It yields data on several job activities that are derived from the interest scales of survey.

The *Jackson Vocational Interest Survey* has been designed for high school, College and adult age groups. It consists of several pairs of statements that describe job-related activities. After scoring, this scale is profiled along 34 basic interest scales. These scales tend to cover career role dimensions considered relevant to a variety of careers and work-style scales indicative of work environment preferences.

The *Comprehensive Interest Schedule* was developed by Sanjay Vohra and it measures interest in eight vocational areas–*influential-administrative-enterprising, venturous-defence-sports, artistic-creative-performing, scientific-medical-technical, analytical-expressive-computational, social-humanitarian-education* and *nature and clerical*. Interest Inventory developed by T.S. Sodhi and H. Bhatnagar measures interest of adolescent girls in eleven areas- *literary, outdoor, mechanical, scientific, persuasive, social service, artistic, clerical, teaching, home management* and *administrative*. Multi-Factor Interest Questionnaire developed by S.D. Kapoor and R.M. Singh measures interest in eight areas-*business, chemical, agricultural, mechanical, scientific, outdooractivities, aesthetics* and *social of adolescents*. Educational Interest record developed by S.P. Kulshrestha measures interest in seven areas-*agriculture, commerce, fine arts, home science, humanities, science and technology* Vocational Interest Record developed by SP Kulshrestha measures interest in ten areas-*literary, scientificexecutive, commercial, constructive, artistic, agriculture, persuasive, social* and *household*.

Some benefits results from the use of standardised interest inventories with which a good counsellor who assists youth and adults in career and related decision-making must be aware. Some of these benefits are as under:

- A verification of a person's claimed interest is easily done
- A comparative and contesting inventory of the interests of the individual is prepared
- Identification of possible level of interests for various activities is done
- Contrasting picture of interest with abilities and achievements is obtained
- Probability to indentify previously unrecognized interests occurs.
- Possible levels of interests for various career activities are identified
- Problems associated with career decision making are identified
- A stimulus for career exploration is found.

While using various interest tests, the counsellor must remember that interest inventories measure only interest. They must also be aware of the fact that interests in case of adolescents and young's change very rapidly. Counsellors should also keep in mind that interest inventories measure only broad general areas of interests.

Counsellor has to interpret the obtained test scores in a scientific way. Goldman (1961) has identified four dimensions of interest test interpretation as under :

- **Descriptive :** Such interpretation provides descriptive information, that is, how the client compares his likes and dislikes with others, whether he is sociable or likes to be left alone, whether his approach to others is aggressive or submissive.
- **Genetic :** Here test data are interpreted to answer 'how' and 'why' of client's behaviour. In such type of interpretation counsellor commonly uses *ex post facto* approach. In other words, here the counsellor provides indications as to what happened in the past and that are inferred from the present ones.
- **Predictive :** In such type of interpretation the counsellor tries to explain what is likely to happen to client in future on the basis of present test scores. In such interpretation, the predictive validity is involved. When a counsellor predicts on the basis of low academic aptitude of a student that he would not be able to do better at college level, this illustrates predictive interpretation.
- **Evaluation :** Unlike the previous three interpretation, here the counsellor goes beyond the test data. On the basis of test data, the counsellor here tends to make specific suggestion, recommendation or some professional advice. Such interpretation is based upon the terms on the basis of which the counsellor makes recommendations.

So far as the guidance is concerned, it is closed related to counselling, though guidance is not counselling. In fact, guidance is essentially the assistance given to persons in making some intelligent choices and adjustments. Thus the objective of guidance is to provide help to other through giving some information. The person seeking guidance is expected to make his own choices. The person who is giving guidance does not choose but helps in making guidance seeking person in making desirable or intelligent choice. In providing guidance the practitioner giving the guidance has to know about the broad potentialities of the person receiving the guidance. As a consequence, the practitioner has to take help of some psychological tests for knowing about the potentialities of the person. Those tests are basically the same that are ordinarily used by the counsellors.

Application of Psychological Testing in Educational setting

Psychological tests have been widely used in educational setting. All important types of psychological tests such as intelligence test, aptitude test, achievement test interest test and personality tests are used by the practitioners in educational settings. In this section, a discussion of application of important psychological tests in educational setting will be done. The discussion would be, for convenience, in three major parts : *Types of education tests*, *Psychological testing in institutional settings* and *Psychological testing in individual decision-making*.

1. Types of Educational tests :

There are many such psychological tests, which have been frequently used in educational settings. Although it is a herculion task to review all those tests, some of the popular educational tests will be included in discussion.

a) General Achievement Batteries :

Several test batteries are available for assessing general education achievement in those areas, which are frequently covered by institutional curriculum. The achievement test battery measures largely what the person has learnt in basic skills school or college courses. A batteries typically include an assessment of basic skills in language, reading, mathematics, in combination with varying amount of content knowledge in science and social studies.

Achievement tests may be *group tests* or *individual tests*. Group achievement tests are mainly used in classroom given simultaneously to dozen of students at the same time. Such tests are also called as *educational achievement test*. Individual achievement tests are administered on one student as a time. Such individual achievement test is considered very important in diagnosing learning disability (LD). Institutional goals such as monitoring school wide achievement levels are best served by group achievement test batteries whereas individual achievement test is commonly used for assessing individual learning difficulties.

A practical application of group achievement test batteries and individual achievement test commonly serve the following functions

- Children and adults with some specific achievement deficits needing more attention for assessment of learning disabilities are identified.
- The success of any educational programmes is assessed by measuring the subsequent skills developed by students.
- Students can be easily grouped according to similar skill level in specific academic domains.
- The classwide or schoolwide achievement deficiency is identified on the basis of achievement test scores.
- Scores on achievement tests help parents in identifying the academic strength and weaknesses of their children and thus, foster individual remedial programmes at home.
- Achievement tests also help in identifying the standard of medium of instruction that is considered appropriate for individual students.
- Important decisions such as grading decisions, placement decisions, guidance and counselling decision, selection decision, curricular decision, etc. are done on the basis achievement test scores.

There are different types of group standardized achievement tests of which five such tests have become very popular. They are: *Iowa Tests of Basic Skills* (ITBS) published by Riverside Publishing Company, *Metropolitan Achievement Tests* (MAT) published by the Psychological Corporation, *Comprehensive Tests of Basic Skills* (CTBS) published by Mc-Graw Hills, *California Achievement Test* (CAT) published by McGraw-Hill and *Sanford Achievement Test* (SAT) published by Psychological Corporation. All these tests assess more or less the same skills but each break the test data in different slices and some begin the coverage of a given skill at earlier ages than others. The common content areas covered by these tests are language skills, reading, science, mathematics, social study/research skills etc.

So far as the individually administered achievement tests are concerned, they have been mainly directed to identify learning disabilities and learning difficulties. One of the first individual achievement test was *Peabody Individual Achievement* originally published in 1970 and revised in 1989. Scores are obtained for general information, reading recognition, reading comprehension, total reading, mathematics, spelling test total, written expression, and written languages. The test is used with students from kindergarten through age 22-11. This test was published by American Guidance Service. Another popular individual achievement test is *Kaufman Test of Educational Achievement (K-TEA)* published by American Guidance Service. Its revised version called K-TEA II was published in 2004. It provides scores for Reading, Mathematics, Spelling and Battery Composite. The third important individual achievement test is *Woodcock-Johnson-III Tests of Achievement (WJ-III)* published by Riverside Publishing Company. The test provides six general academic cluster scores : Broad Reading, Oral language, Broad Math, Math Calculation, Broad written language and written expression.

b) Tests of Minimum Competency in Basic Skills :

Such tests are basically designed for adults for their special uses in adult education class, educational programmes conducted in penal institutions as well as for job training programmes. (Anastasi & Urbina, 2002) One example of such test is *Tests of Adult Basic Education* (TABE Forms 7 & 8). This battery covers five different content areas, which include reading, language and applied mathematics. The results of this battery are reported as norm-referenced scores.

c) Teacher-Made Classroom Test :

When no standardized test is available, the teachers are forced to prepare their own tests to be used in classroom. The teacher makes a plan about the test, write items and conduct simple item analysis. Such test is used for assessing the designated course content. But use of such test is not preferred much because it is commonly observed that while writing items teachers lay overemphasis on rote memory and trivial details and give emphasis only on some areas ignoring others, etc.

d) College level tests :

The Scholastic Assessment Test (SAT) and the American College test (ACT) are the two most popular college level tests used by educators. The SAT comprises SAT-I Reasoning tests and SAT II Subject tests. The SAT I assesses vocabulary in context, reading comprehension and critical reasoning. The SAT II emphases the application of mathematical concepts, the interpretation of data and actual construction of a response.

The American College Test ACT) designed for college-bound students, yields four scores, that is, score in English, Mathematics, Reading and Science reasoning. One common criticism of Act is that in this test heavy emphasis is given upon reading comprehension that saturated all the four tests.

Graduate Record Form (GRE) is another college-level test widely used by graduate programmes in different fields as one component in the selection of candidates for training. The GRE offers subject examination in different fields but the core of the test is the general test designed to asses verbal, quantitative and analytical writing aptitudes. (Gregory, 2004)

In Indian situation some achievement tests have been developed and popularly used by educators for assessing academic achievement. Important among them are : *General Classroom Achievement Test* by A.K. Singh and A. Sen Gupta, *Mathematics Achievement Test* by Ali Imam and Tahira Khatoon, Hindi Achievement Test by L.M. Dubey, Physics Achievement Test by S.N.L. Bhargava, *Science Achievement Test* by R.D. Singh.

2. Psychological Testing in Institutional settings :

In institutional settings psychological testing is done for various purposes. Some of these are as under :

• Admission and Selection decisions :

School systems, in most of the countries, have heavily relied on some form of aptitude and intelligence testing to determine at what level the students should be placed for optimal learning on their part. The goal has been to determine appropriate instructional levels for students. In pursuing the goal honestly, the educational institutions depend heavily on the tests of general ability, scholastic aptitude and educational achievement.

In early childhood education Readiness test is a very popular test. Such tests are usually administered upon school entrance. School readiness is defined as the attainment of prerequisite skills, knowledge, attitudes, motivations and other related appropriate behavioural traits that enable the student to take maximum advantage from school instruction. (Anastasi & Urbina, 2002) Such tests contain much in common with intelligence tests for primary grades, they also give emphasis upon learning to read and write. Specific functions frequently covered by such tests are motor control, aural comprehension, vocabulary, quantitative concepts, visual and auditory discrimination and general information. One good example of Readiness test is the *Metropolitan Readiness Tests*. Another similar tests are *Boehm Test of Basic Concept* and *Bracken Basic Concept Scale*.

• Identification of exceptionality among students :

Educational institutions frequently need to identify the exceptional students who are at either end of ability continuum. There are two categories of such students. One category consists of those students who are at the lower end of continuum because of their developmental delays in the maturation and therefore, need early detection and remedial help. The mentally retarded children and the children suffering from learning disabilities are placed in this category. Usually some aptitude tests are applied to identify this group of children.

One the other hand, there lies those children who are placed at the upper end of the continuum because they are intellectually gifted students. Institutions need to identify such talented students also who will benefit most from an enriched educational curricula. A more challenging learning environments are to be created for such students. Failure to provide adequate stimulating environment often leads to poor motivation and performance on the part of such talented students. Here ability tests play a very critical role in the identification of such talented students.

- **Application to the problem of tracking in the classroom :**

Often aptitude tests in educational setting are used to solve the problem of tracking in the classroom. In such situation students are grouped according to their ability so that adequate instruction according to the level of their ability may be given. In fact, such application of aptitude testing aims as maximizing the motivation of students and minimizing their frustration by placing them in classes where instructions as well as materials presented are neither too difficult nor too easy.

While tests are far from being perfect for this purpose, they usually provide a very valuable aid in this direction. The danger in their use is that the school authority or the administrator may rely on them too heavily, without taking into consideration the essential data considered relevant in decision-making process.

3. Psychological Testing in Individual decision-making :

In the previous section a discussion was made on how tests are used by the educational institutional in taking some administrative decisions. Here the discussion will shift to know how tests are employed in the educational and career decisions made by the individual. When students progress through the educational system, they face many decision points where various types of questions do arise in their mind. Important such questions are whether to take vocationally relevant course or follow the traditional course, which subject should be selected as honours subject and which ones should be taken as electives ones. How are psychological tests applied to these individual decision-making tasks? To what extent they are useful?

These issues can be discussed under two major heads: *Career planning* and *Educational planning*. A discussion follows :

- **Career planning :**

One very popular application in individual testing in educational setting involves career decision-making. In fact, throughout life the individuals are faced with the decisions about what type of work they should pursue. In the early period, there is need for general areas to be clearly identified so that they may be explored further Persons of middle-age career reveals another problem. In fact, persons of this age tend to seek support and encouragement in the change process and they are always in the need of such data that may either reinforce or contraindicate their plans. At all these stages, psychological testing play a very significant role in decision-making process. This may well be illustrated from one case from early developmental period.

Mohini was a college student who had often been wondering what she should do after finishing her college education. She has recently been thinking to take some coaching classes for preparing the Banking examination after graduation. As a part of career counselling process in college, she was given an interest test. The test scores revealed that she had a strong preference for science. When further ability testing was done, she learned that she could be really a scientist if she honestly tried. Such information provided by psychological testing opened a new way and she planned her career in such a way that could easily facilitate in being her scientist.

Thus at individual level the psychological testing help the person in taking fruitful career decisions.

- **Educational planning :**

Every student wants to select an appropriate educational programmes but at the time, most of the students are reluctant to seek help from the counsellor. They want to take some easy alternative without taking help from anybody. This, infact, creates frustration in the long run. Psychological testing here plays a very important role in such educational planning. This can be illustrated from the following case.

Mohan was a college student who was graduating from arts subjects. He had joined college somewhat undecided about what he really wanted to pursue. In school days, he had not heard much about the importance of any college subject. He was, thus, confused. Mohan sought help from the counselling center in the campus and he was administered some interest tests as a part of counselling process. Tests results showed that Mohan had a strong interest in the those business occupations that were person-oriented. This suggestion by counsellor brought a change in his mind and he started thinking to explore sales and marketing. Thus his educational planning changed from simply going through the arts subjects to some business occupations.

This case clearly shows how psychological testing, when applied appropriately, tends to shape the educational planning of the person. Thus, psychological testing played a key role in helping individuals make important decisions about their education plans.

In a nutshell, psychological testing plays a central role both in making institutional decisions as well as in making decisions by individuals in making their own educational goals.

Application of Psychological Testing in Industry, Organization and Business

In work settings ranging from small business to big industrial organizations, decisions are made regarding selection of personnels, their hiring, training, promotion, and the like. For most part, these personnel decisions involve either predictive or evaluations of job performance. Psychological testing is of great help in both prediction issues as well as evaluation issues. On the basis of tests scores, the workers are hired, promoted or placed in new jobs for achieving maximal level of maturity. Likewise, organizational rewards given to

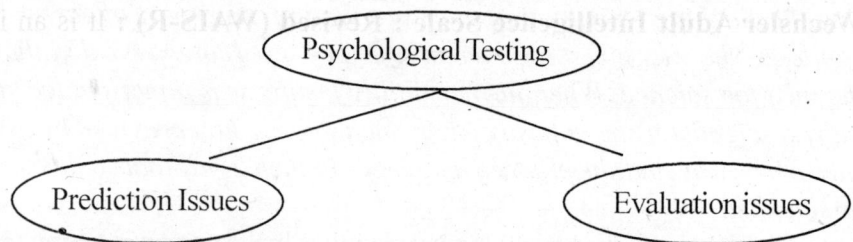

Figure 19.1: Psychological testing in Prediction and evaluation of workers' performance

the workers, at least in part, are dependent upon the evaluation of past performance. The promotion, raises, bonuses and other rewards are based upon evaluation of past as well as present performance. In such evaluation decisions again, psychological testing is of immense importance. Figure 19.1 shows the two most important areas of psychological testing in industrial organization and business. A discussion of psychological testing in these two separate but related fields of business and industrial organization follows :

1. Psychological Testing : Prediction issues

Psychological tests are very useful instruments that provide information about those abilities, skills, experiences and characteristics of the workers that can be used to predict job performance. There are different measurement techniques and this diversity clearly reflects the apparent diversity of factors which affect job performance. Keeping in view the nature of job, the performance may be a function of some cognitive or non-cognitive abilities, previous experience, social skills as well as some specific task-oriented skills. Researchers have shown that some testing techniques may be more appropriate for specific attributes than others. For example, paper-and-pencil test may be more appropriate for assessing cognitive abilities and aptitudes whereas the interview might prove to be more appropriate for assessing motivation and social skills. Likewise, for assessing background experience application blank, background checks and references might be more appropriate. The current performance of the worker may be suitably assessed by work sample test and holistic evaluation can best be done through assessment centers. Some of the important tests which are frequently used in making predictions about workers' performance are being discussed here.

a) Cognitive Ability test :

Most of the cognitive ability tests used in business and industry are basically similar to those used in educational and clinical settings. Those cognitive abilities tests are better known as intelligence tests. The following intelligence tests are mostly commonly used in business and industrial organization.

- **Otis self-administration tests of Mental ability :** This test has proven very useful for screening applicants for a wide variety of jobs including office clerks, lower level supervisors, assembly-line workers as well as for all those jobs which don't require high level of intelligence. Since this is a group test, it takes little time to complete.

- **Wechsler Adult Intelligence Scale : Revised (WAIS-R) :** It is an individually administered test used in industry mainly for selection of senior management officers. The test consists of verbal section and performance section and it yields three IQ score : *Verbal IQ Score*, *Performance IQ score* and *Full Scale IQ score*. Computer-assisted interpretation is also available for the test.
- **Wonderlic Personnel Test :** This test is one version of Otis Series of tests and is popularly used in industrial selection because it takes a mere 12 minutes to complete. This is also a group test and includes verbal, numerical and spatial content items. It is very useful in predicting success in some lower level clerical jobs.

b) Aptitude tests :

In industry and business organization several types of aptitude tests such as clerical aptitude tests and mechanical aptitude tests are frequently used for predicting success in the concerned jobs. Among clerical aptitude tests, the *Minnesota clerical test* and the *General clerical test* are very common. The Minnesota clerical test is a speed test, which determines the individual accuracy when the employee is working under limited period of time. The test consists of two parts—*number comparison* and *name comparison*. The examinees are instructed to work as fast as they can without any errors. The General Clerical test is a group speed test, which has two booklets: A and B. The booklet A contains various items on alphabetizing, numerical computation, checking, error location, and arithmetic reasoning. It is most suitable for predicting success in job of accounting or comprehension, spelling, vocabulary, grammar and is most suitable for predicting success in secretarial jobs.

Among the mechanical aptitude tests, the *Revised Minnesota Paper Form Board Test* is comparatively very popular. The Revised Minnesota Paper Form Board Test basically assesses spatial relations or visualization and the manipulation of objects in space. This test has proved to be most beneficial in predicting success needed in occupation like drafting. Further researches have shown that the test is also very helpful in predicting successful performance in mechanical work and engineering shop-work Bennett Mechanical Comprehension test assesses the mechanical principles involved in the various diagrams of machinery presented *The Purdue Mechanical Adaptability test* is another mechanical aptitude test that assesses the principles of mechanical nature. Besides these, *Differential Aptitude test* (DAT) and *General Aptitude test Battery* (GATB) are also frequently used in predicting success in various jobs.

c) Interest tests :

Although interest tests are of great importance in vocational guidance and career counselling, they have also been the preferred choice of some industrial organizations. Such interest tests include items about many daily activities and objects from among which the examinees (or test-takers) select their preferences. The rationale behind this is that if the individual exhibits the same pattern of interests and preferences as those who are successful in any given occupation, the probability is that the individual would tend to be satisfied in that occupation. Two widely and frequently used interest inventories are the *Strong-Campbell Interest inventory* and the *Kuder Occupational Interest Survey*.

d) Tests of Psychomotor abilities :

Psychomotor abilities are considered very important in work situation. Psychomotor abilities include both the perceptual-motor abilities and physical efficiency abilities. Psychomotor abilities include control precision, finger dexterity, manual dexterity, reaction time, response orientation, wrist-finger speed, rate control (timing), etc. Physical proficiency abilities include external flexibility, gross-body coordination, equilibrium, stamina, etc.

The *Minnesota Rate of Manipulation test* is one popular measure used for predicting success on task requiring psychomotor abilities. It consists of two parts. In the first part, the examinee's task is to place 60 cylindrical blocks in 60 wells in the board and in second task, he is to turn all the blocks over. The score is determined on the basis of time taken to complete each task *McQuarrie Test for Mechanical ability* and *O'connor Finger and Tweezer Dexterity tests* are also very popular measure of psychomotor abilities. Unlike most measures of psychomotor abilities that utilize special equipment, the MacQuarrie test utilizes only paper-and-pencil task and covers seven subtests reach as *Tracing, Tapping, Dotting, Copying, Location, Blocks* and *Pursuit*. These tests have proved to be valid in predicting success required in aviation mechanic and stenographer. The O'connor tests require the testee to fill the 100 small holes with pins using fingers or in some cases, tweezers. The score is the amount of time required to complete the task. The tests have been found most important for predicting success for the job of power sewing machine operators as well as jobs that require manipulative tasks. For the test of coordination, the most common measure is the pursuit motor test which establishes aiming skill or motor coordination.

Another popular measure of psychomotor abilities frequently used in *Purdue Pegboard Test*, which measures dexterity for two types of activity: gross movement of hands, fingers and arms and fingertip detexterity needed in assembly tasks. This test has been considered most fruitful in predicting jobs requiring various kinds of manual labor.

e) Personality tests :

Researchers have shown that some personality characteristics are important for some kinds of occupations and are very important for job success. In business and industrial organizations several approaches to personality assessment have been reported. They are self-report inventories and projective techniques.

The self-report inventories include several items that deal with specific symptoms, feelings and situations. The test-takers are requested to indicate how well each item describes themselves or how much they agree with the content of the item. In industrial organization *Guilford-Zimmerman Temperament Survey* and *Minnesota Multiphasic Personality Inventory* (MMPI) are popularly used for predicting likely success in different jobs. The Guilford-Zimmerman Temperament Survey yields *ten* independent personality traits: *General activity, Restraint, Ascendance, Sociability, Emotional stability, Friendliness, Objectivity, Thoughtfulness, Personal relations* and *Masculinity*. MMPI is another very popular measure of personality that assesses psychopathological traits. Besides these inventories, Eysanck personality, Questionnaire (EPQ) and Edward Personal Preference Schedule are also used in personnel selection and predicting job success.

Projective Techniques, though mainly used in clinical psychology, are also used for assessing candidates for high level executive positions. *Rorschach test* and *Thematic Apperception Test* are the two most popular projective tests of personality. Responses to 10 standardized inkblots in Rorschach test provides some clue to the personal selection, though researchers have indicated low predictive value of Rorschach test for personal selection. In TAT, the applicant's responses are recorded in terms of a story because he is instructed to construct a story based upon ambiguous pictures. In TAT there are nineteen ambiguous pictures and the applicant is to write story for each picture. Stories are analyzed. TAT has been found to yield good predictions of success in entrepreneurial occupations as well as in line management.

Besides these psychological tests, some other tools/devices are used for predicting success in various jobs. Important ones are as under.

Application Blank:

Standard application blank consisting of questions dealing with the applicant's personal life and work history as well as some questions about his present status, that is, age marital status address have been used in industrial organization since 1894 (Owens, 1976). For high level jobs, the applicants are requested to describe their interests and hobbies, their career goals and reading habits. An application blank is usually limited to questions that provide very useful guides for selection as well as for predicting success in job. However, some industrial organisations use excessively long application blanks in an attempt to gather information on every possible facets of the applicant's life.

One major problem with application blank has been that sometimes applicants tend to provide misleading or fraudulent information. The common distortions occur on questions relating to previous job title, pay and level of responsibility.

• Biographical Inventories :

Biographical inventory or biographical information blank is typically much longer than this application blanks and tends to cover information on the applicant's life in greater detail. The rationale behind this is that on-the-job behaviour is related very much to past behaviour in a variety of situations as well as to various attitudes, values and preferences. Owens and Schoenfeldt (1979) have reported that the biographical inventories have proved a valid predictor of success in a variety of jobs such as scientist, middle-level manager, production worker, salesperson and office worker. Biographical inventories have a great predictive value than any other selection procedures except for tests of cognitive abilities (Anderson & Shakleton, 1986)

- **Employment Interview :**

    Almost all employers want to meet the applicant in person. Interview provides one such means. Interview is the face-to-face interaction between employers and the applicants aspiring for a job. Interviews range in length from cursory meetings of five to ten minutes to elaborate affairs lasting for some hours only. There are two traditional kinds of interview-*unstructured interview* and *structured interview* or *patterned interview*. The unstructured interview is characterized by a lack of structure or any advance planning. The format to questioning as well as questions asked are left entirely to the discretion of the interview. Due to this. it is just possible that three interviewers conducting separate unstructured interviews with the same applicant may receive three different impressions of the same applicant.

    The structured interview is opposite to the unstructured interview, uses a predetermined list of questions to be asked of each applicant. The questions usually deal with the prior work experience, etc. The responses are recorded. Structured interviews, as compared to the unstructured interviews, have a high potential for predictive validity. Indeed, a carefully prepared structured interview have been found to be as valid predictors of job success as a typical battery of employment tests. (Campion et al., 1988) Most of the industrial organizations prefer to use interview in collaboration with test scores.

- **Work sample tests :**

    Such tests include measures developed by sampling from the work actually performed on the job as we find with various stenographic and typing test. Work sample tests are correct predictors only when the role prescriptions for a job form a rather homogeneous unit. If the nature of job is complex requiring many different types of activities all of which are equally important to success, the use of work sample test would not be normally expedient. Several industrial organizations have developed work sample tests for skilled and semi-skilled positions. In some instances, special equipment simulating those actually used in job, has been constructed. Work sample tests have also been popular in predicting success in clerical jobs. Managerial work is not much suitable for work sample tests. However, some aspects of work have been simulated with considerable success. Truly speaking, the *managerial games*, *in-basket exercises*, and *discussion groups* are often used to represent work samples (Thornton & Byham, 1982, Fredericksen, 1962)

    Work sample tests range from those that involve relatively simple tasks such as 6-minute typing sample to some complex samples of performance. In fact, there are two common features to all work sample tests that must be kept in mind. *First*, every work sample test puts the applicants in a situation that is more or less similar to the actual work situation and tends to assess performances on tasks similar to those that make up the task. *Second*, every work sample *differs* in some ways from the job in which it will be used. Ordinarily, the applicant tries to impress their prospective employers by showing high levels of motivation in work sample than they will show on the job. In light of this, it is reasonable to regard work sample test as a measure of *maximal performance* rather than a measure of *typical performance*. (Murphy & Davidshofer, 1988)

- **Assessment Centers :**

Initially called situational testing, the assessment center is a method that places the applicants in a simulated job situation so that their behaviour under stress can be evaluated. In fact, assessment center, as used today, is the lineal descendent of the multiple assessment procedures commonly used in World War II by German and British psychologists and adapted by the American Office of Strategic Services (OSS) as one of the primary instruments in selecting various agents and operatives. It has since been adopted by about 2000 organizations and are popular in countries like Britain, Brazil, Australia, Japan, Western Europe, etc. Although assessment centres used in different organizations differ widely in terms of content and organization, there are some commonalities among them that constitute their basic features.

Those points of communality are as under :

- Use of multiple methods: The activities of the assessment centers include the use of cognitive ability tests situation tests, personality tests, interviews, peer evaluations and performance tests. The rationale behind this is that each test has its strengths and weaknesses and therefore, a combination of diverse tests is considered necessary for capitalizing on the strength of each test.
- Assessment in groups: In each assessment canter, small groups of participants (usually 6 to 10 participants) are assessed simultaneously.
- Assessment by groups: At the assessment canter, the assessment is done usually by a group of managers, consultants, psychologists or some mix of these three groups. The behaviour of each participant is assessed by a number of different assessors and the final ratings represents the assessment teams consensus regarding the evaluation of the individual.
- Use of situation tests: Nearly every assessment canter uses same type of work sample or situational test. Here some of the popular techniques used are *in-basket test*, *leaderless group discussion*, *oral presentations* and *role playing*.
- Assessment along multiple dimensions: At the assessment centres the assessment is done for different dimensions and the end result of an assessment centre is, in fact, a consensus rating along each of several dimensions.

Assessment centers are considered to be fair and relatively unbiased methods of making selection and promotion. Evidences for the validity of assessment centers have been consistently positive (Borman, 1982; Hinvichs, 1978). For example, Borman (1982) reported a correlation of 0.50 between composite assessment ratings obtained at the beginning of an army personnel training and later performance in training. Henrichs (1978) reported similar levels of validity in predicting advancement within organization. In fact, what is more important is that the validity of assessment centers rating for predicting promotion tends to increase as one goes into higher level of management Huck and Bray (1976) have reported that assessment centers demonstrated identical levels of validity for both white and Black employees in predicting both job performance and potential for advancement.

2. Psychological Testing : Evaluation issues :

In organization and industry the measurement of present and past job performance is considered as critical as the successful prediction of future performance. In general, there are three classes of data that are frequently used in measuring the performance of the workers. They are : *data arising from production count, personnel data* and *judgemental methods*.

In the organization, the production count may be the basis for assessing the performance of the workers. For example, the number of calls attended by an operator, number of parts produced by the mechanist by the deliveryman may be used as basis for assessing the job performance Personnel data, that is, relevant data related to the workers or personnel, such as the absenteeism by the workers, frequent leave taken by workers, ill-health of the workers also force them to produce less as compared to the workers who work regularly for eight hours per day. In judgement methods *ratings rankings* or *peer nominations* are done to evaluate the performance of workers. Of these three, judgemental methods are the most popular ones.

There are two most general types of judgement, which are commonly used in measuring the performance of the workers. They are *ranking methods* and *rating methods*. A discussion follows:

a) Ranking methods :

There are several ranking methods available for evaluating the effectiveness of the employees. The following are the three important ranking methods commonly used: *forced-distribution scale, full ranking method* and *pair-comparison method*.

In force distribution scale the supervisor or managers are forced to place a fixed number of workers as belonging to the different categories such as a fixed number of workers into superior category, a fixed number of workers into average category and a fixed number of workers into poor category. For example, the management may ask the superiors to assign 20% in the superior category, 20% in the poor category and 60% in the average category. In such scale, the judgements about the performance of the workers are made on relative rather than absolute basis, that is, the workers are assessed as poor, for example, relative to their performance in particular group in which they work.

In full ranking method the supervisor or manager assigns rank to each worker rather than group them into assigned categories. As a consequence, in this method the number of categories of ranks is equal to the number of workers. For example, if 10 workers are to be ranked on the basis of their effectiveness of performance, one worker will get rank of 1, another 2 and so on 10th rank to the last worker. The workers are ranked from highest to lowest or best to worst in terms of overall job effectiveness.

In paired-comparison method each worker is compared with every other worker in the section or group. A comparison is made between two people at a time and subsequently a judgement is made as to which person within the pair is better. When all possible comparisons have been completed, an objective rank ordering is obtained on the basis of the worker's score in each comparison. For example, suppose that a manager or supervisor is asked to evaluate 10 workers by the method of paired comparison technique. This means that a total of 45 comparison must be made because there will be 45 possible pairs obtainable with 10 workers. The formula is as under :

$$\frac{N(N-1)}{2} = \frac{10(10-1)}{2} = \frac{10 \times 9}{2} = \frac{90}{2} = 45$$

In this formula N means the number of workers to be evaluated.

In a nutshell, it can be said that in a forced-distribution scale a relative crude measure is obtained, that is, workers in top category are judged to be more effective than those in the bottom category. A full ranking method, no doubt, provides a more exact comparison of workers but it takes considerable effort. A paired-comparison method provides an interval-level measure of relative effectiveness, but it involves a large number of comparisons, which is cumbersome.

Which is the best ranking method? In fact, the choice of any method depends upon the way in which rankings will be used. For some promotional decisions or for allocating cutbacks, a forced distribution ranking is considered to be appropriate. But for evaluating the extent to which the workers differ in their levels of performance, a paired-comparison method would be more appropriate.

b) Rating Scales

Rating scales require the managers or supervisors to evaluate each worker with regard to a particular standard such as 'good', 'average', 'poor' etc, or in terms of some other standards or they may be described in some concrete behavioural terms, etc. Thus rating scales are different from ranking methods. In fact, ranking methods require supervisors or managers to decide on a relative basis that is, which of the two workers is a better performer. In contrast to this, rating scales require supervisors to evaluate the level of each worker's performance.

There are different types of rating scales. Of these various scales, two rating scales are frequently used in evaluating the performance of the workers—*graphic rating scale* and *behaviour-based scales*.

The Graphic rating scale is the simplest form of rating instrument. A graphic scale is one in which a response of the supervisor is indicated by marking a designated position on a line such as one anchored by *extremely rapidly* at one end to *extremely slowly* at the other or *strongly agree* on one end to *strongly disagree* on the other end or some other analogue. In rating the quality of work based on observations of the workers' performance, the supervisor or manager might express a judgement in terms of the position marked by him An illustration is provided here :

Mohan's performance as a machine operator in the department of mechanical engineering is

| Consistently Unsatisfactory | Below Average | Above Average | Consistently Superior ✔ |

In the above example, Mohan as a machine operator has been rated by the supervisor as having consistently superior performance. Such scale is often used for assessing broad dimension of performance such as Quality of work, Planning or Oral communication.

The obvious advantage of the graphic rating scale is its simplicity because once it is decided what is to be assessed, the construction of graphic scale is very simple. But at the same time, the simplicity of graphic rating scale is also a source of ambiguity. There is often a good deal of ambiguity in the interpretation of anchors. Supervisors have often different interpretation of what is meant by 'average' performance, 'unsatisfactory performance', etc. To remove such source of ambiguity, behaviour-based scales, which include *mixed standard scales, behaviourally anchored rating scales* and *behavioural observation scales* were developed.

Behaviour-based scales are those rating scales on which the supervisors have a clear-cut behaviourally defined anchors and they tend to indicate whether the workers being evaluated typically exhibits the performance reflected by the anchor. There are three types of behaviour-based Scales- *Mixed standard scales* (MSS), *Behaviourally anchored rating scales* (BARS) and *Behavioural observation scale* (BOS) A discussion follows.

The Mixed standard scale (MSS) is a scale that presents the supervisor with concrete behavioural examples of average, good and poor performance. Rather than producing a vague anchored dimension, here the supervisor using this scale is asked to indicate whether the worker's being evaluated typically exhibit performance, which is better than, about equal to or worse than the level of performance shown in the behavioural example. The MSS has two basic features. *First* it presents a very concrete, behavioural standards for evaluating good, average and poor performance. *Second*, this scale requires a very simple judgements on the part of the supervisor, that is, supervisor is simply required to judge the worker's performance as good, better or worse than the level prescribed in the behavioural example. However, there are some problems with MSS and the most serious problem is that the supervisors who act as raters, are often inconsistent. Due to this, it is just possible that a worker who is rated as better may be rated as worst on the same dimension.

Behaviourally anchored rating scales (BARS) have been developed to deal with the ambiguity problem of the graphic rating scales. BARS present a concrete behavioural statements, which are used to identify the levels of performance and which taken together provide a series of clear behavioural exemplars of the performance dimension. BARS attempts to evaluate the performance of the workers in terms of some specific behaviours (rather than in terms of general traits or attitudes). These specific behaviours are called *critical incidents* for the job considered critical to success on the job. Supervisors who are familiar with job and workers, observe the performance of the workers and record those behaviours, which are considered critical incident behaviours. Some of these critical incident behaviours are associated with good performance and some are associated with poor performance. Subsequently these behaviours are used as standards for judging the performance of the individual workers. In BARS much depends upon the observation skill of the supervisors in determining the kinds of behaviour that are truly critical to successful or unsuccessful performance on the job. Moreover, if the list of critical incident behaviours is inadequate, any evaluation based on these behaviours would be misleading Moreover, some researchers have shown the BARS approach to the evaluation is no more objective as it has been claimed and also not free of bias than scales that don't have behavioural anchors (Murphy & Coustans, 1987)

Latham and Waxley (1977) have pointed out that both the graphic rating scale as well as BARS don't serve the real purpose of evaluation. The graphic rating scales require the supervisors to make vague and an unanchored judgements. Likewise, BARS require supervisors to indicate what sort of behaviours they would expect, although in some cases none of the behavioural anchors included in the scale have been observed. Due to these problems, a simpler alternative called *Behaviouralobservation scale* (BOS) has been developed.

A BOS consists of behaviours believed to be critical for effective performance on job. Here also the workers are evaluated on the basis of critical incidents of behaviour on the job. In this sense, it is similar to BARS. However, it differs from BARS in the sense that here supervisors rate workers on the *frequency* of these critical incidents as they are observed to occur in a given of time. One example of a BOS in evaluating the performance of a routine clerk in government office is presented below:

Table 19.1: A Behavioural Observation Scale for a Routine Clerk

Rate the frequency with which each of the following behaviours have occurred during the last one month. Use the following scale to rate frequency

1	2	3	4	5	6	7
Never	Almost never	A few times	About half of time	Often	Most of the Time	Always

1. Reports office in time
2. Prepares a good draft
3. Speaks clearly
4. Tries to complete task in time
5. Quarrels with his colleagues

The rater who uses BOS is requested to indicate how frequently each of the critical incidents in the list has occurred over a specific period of time. The evaluation yields a total score for the employee determined by adding the scores or rating for each critical incident. Finally, a worker who frequently performs effective critical behaviours and who very rarely performs ineffective behaviours receive a high score whereas a worker who frequently performs ineffective behaviours receives a low rating.

The process involved in completing a BOS seems considerably simpler and more straightforward as compared to other types of scales. Rather than asking supervisors or managers to make complex and unreliable judgement about workers, in BOS, he is merely asked to report how frequently different behaviours have occurred in a given period. However, BOS is not without problem. Murphy and Balzer (1981) have reported that BOS is, in fact, trait-rating scale in disguise. Supervisors (or raters) infer that the employee he perceives as 'good' must have carried out many effective work behaviours rather than making an inference that a worker is good because he has carried out the specific behaviours included in the BOS.

In a nutshell, psychological testing in business and industrial organization have several fields. In whatever fields, they are applied, they prove to be of immense importance.

Application of Psychological Testing in Sports

As we know sport psychology is a new branch of psychology in which the principles of psychology are applied in sports and exercise setting. This branch is fully dedicated to the enhancement of athletic performance as well as social-psychological aspect of human environment (Cox, 1998). In recent years two sport psychologies have emerged (Martens, 1987). One is *academic sport psychology* and another is *applied sport psychology*. The Academic sport psychology throws light on the disciplinary or research-oriented aspects of field. On the other hand, *applied sport psychology* focuses on the professional or applied aspect of the field of sport psychology.

The psychological testing plays a significant role especially in applied sport psychology. Since 1992 there have been significant advances in doing various types of psychological testing in sport and exercise. According to significant review done by Ostrow (1996) there are about 314 self-report measures used in sport, exercise and physical activity studies published in about 45 journals during previous 30 years. Here a brief discussion of some of those self-assessment measures is being provided.

- **Test of Performance Strategies :**

Thomas et al., (1999), developed Test of Performance Strategies (TOPS). It assesses athletes' use of psychological skills in competition and during practice. The test assesses dimension like emotional control, automaticity, goal-setting, imagery, activation, self-talk, relaxation and attentional control in both during competitive games and exercises. Taylor et al., (2008) reported that the TOPS subscales for both competitive games as well as practice significantly tend to make discrimination between medallists and non-medallists among American athletes in 2000 Sydney Olypics. According to Weinberg and Forelanza (2012) TOPS is the most widely used test by sport psychology consultants. It is a most popular choice for assessing the effectiveness of psychological skills training interventions. (Woodcock et al., 2012)

- **Athletic Motivation Inventory :**

    The Athletic Motivation Inventory (AMI) has been developed by Tutko, Ogilvie and Lyon at the Institute for the study of athletic motivation at San State College (Tutko & Richards, 1971, 1972). The AMS assesses traits associated with high athletic achievement such as aggression, drive, determination, leadership, responsibility, emotional control, self-confidence, mental toughness, conscience development, coachability, and trust. The test has proved to be a good predictor for athletic success.

- **Profile of Mood States :**

    The Profile of Mood States (POMS) originally consists of 65-item unipolar inventory, measures six affective states: *depression*, *anger*, *tension*, *vigor*, *fatigue* and *confusion*. (McNair, Lorr & Droppleman, 1971). Subsequently, a 72-item bipolar version of the original POMS has been developed by Lorr and McNair (1988). The bipolar version has also six sub-scales. The examples of bipolar form are agreeable-hostile, composed-anxious, etc. Sport psychologists, for ease of administration, have developed some shortened versions of original unipolar POMS. For example, a 37-item version was developed by Schacham (1983) and a 40-item version was developed by Grove and Prapavessis (1992). A 27 item version was developed by Terry et al., (1996) for use with child athletes.

- **Test of Attentional and Interpersonal Style :**

    This is a self-report method for assessing attentional focus developed by Nideffer (1976a, 1976b). The Test of Attention and Interpersonal Style (TAIS) is based upon two-dimensional structure of attentional focus. The two dimensions are: *width* and *direction*. The width dimension of the athlete's attentional focus ranges from broad to narrow whereas the direction dimension tends to vary from internal to external. Some athletes seem to be internally directed while others seem to be externally directed. The TAIS tends to assess both these dimensions of athlete's attention. Several sports related versions of TAIS have also been developed. (Albrecht & Feltz, 1987)

- **Measures of Mental toughness :**

    Mental toughness is considered to be one of the important characteristics of athletes. Guncciardi et al., (2009, 2011) have developed sport-specific measures of mental toughness, which is, in fact, a multidimensional construct. The mental toughness consists of specific values, attitudes, cognitions and emotions of the athletes and are common among various sports. These are personal values, self-confidence, attention control, positive and tough attitudes, resilience, sport intelligence, self-motivation, etc. Tests are available for assessing mental toughness in football and cricket. Also there are evidences that reveal that various psychological skills training programmes are effective in enhancing the mental toughness of the athletes.

    Besides these tests, several other self-report measures have been developed. For example, Marsh and Cheng (2012) have developed the physical self-description Questionnaire for assessing participant's physical self-concept. They also develop the Elite Athlete self-Description Questionnaire (EASDQ) for assessing elite athlete's skill level, aerobic and anaerobic fitness, body suitability, mental competence, etc. Jackson and Eklund (2012) have developed tests to assess dispositional and state 'flow' or some balanced challenges or skills. All these various traits/skills can be used to enhance self-awareness and are urgently needed to gain the edge of peak performance. (Ravizza, 2010) Still another important step is the development of *Sport Personality Questionnaire*, which assesses twenty dimensions in four key areas of athletes. Popularly called SPQ 20, it consists of 168-item and uses standard Ten (Sten) scoring system. The test provides information about athlete's personality and mental skills. The four key areas are: *achievement and competitiveness, confidence and resilience, interaction and sportsmanship* and *power and aggressiveness*. The first covers traits like achievement, adaptability, competitiveness, conscientiousness, visualization, intuition and goal setting. The second area covers managing pressure, self-efficacy, fear of failure control, flow, stress management, emotions, self-talk, and self-awareness. The third area covers traits like ethics, empathy and relationship. The fourth area covers aggressiveness and power.

Recently, there has been a renewed interest in knowing the relationship between sport performance and various personality characteristics such as perfectionism, narcissism and resilience. However, American Psychological Association has not favoured this approach and has expressed the view that use of personality inventories alone in selecting athletes to a team or to drop them from a team would be an abuse of testing that should not be tolerated.

Application of Psychological Testing in Military Services

Military psychology is a subdiscipline of psychology which studies the behaviour of people within the context of armed forces. Military psychology is broadly defined in terms of application of psychological principles and theories to military context of the different subbranches of military psychology. Military selection is one very important branch. So far as the psychological assessments and testing are concerned, they play a very vital role in military selection.

In fact, during the First World War and the Second World War, there emerged an urgent need to test a large number of candidates within a short span of time so that they may be assigned their jobs in which they would function in a most effective way during military operations. Until World War I, psychological tests were individual tests, which used to take more than an hour. Such testing was not practical from the point view of war fields. As a consequence, there was an urgent requirement for the tests which can be easily administered in group. Army Beta test which were group tests of intelligence were developed. These tests were developed by a group of US psychologists working under the leadership of R.M. Yerkes including L.M. Terman. The Alpha Army test heavily bear on verbal skills whereas the Army Beta test was a performance test, which was for illiterates as well as for those emigrants to America whose English was poor. For armed force services, subsequently, the Armed Forced Qualification Test (AFQT) was developed. This test acted as a preliminary screening instrument and provided a single score based on an equal number of arithmetic, vocabulary spatial relations, and mechanical ability items. Subsequently the

Armed Services Vocational Aptitude Battery (ASVAB) was developed for use in all wings of armed forces as a composite selection and classification batteries. During Second World War Army General classification Test (AGCT) was also developed and was considered suitable for literate adults. Examples of modern group tests are *Armed Services Vocational Attitude Battery* (ASVAB) and the Scholastic Assessment Test [SAT (Formerly known as Scholastic Aptitude test)]. During Second World War, group observational exercises for selection and placement of soldiers formed an innovation in military services.

There are *three* ways in which military selection using various types of tests is organized. One common way is to make all candidates go through the same psychological tests, irrespective of the nature of jobs. This is *uniform procedure technique*. Another way to assign the candidates to different types of tests keeping in view the nature of job they have applied. This is *diversified selection procedure*. A third way is one which *mixes* these two procedures.

Under uniform selection policy since all candidates whether they apply for soldier or for officer, are administered the same test. The cognitive test battery is very popular test to be administered here. It measures different cognitive qualities such as mathematical reasoning, verbal reasoning as well as the use memory. In American army, all candidates take AFQT, which produces an index for trainability. Here, it is easily assumed that since all recruits or candidates start with the same basic training, they are all thought to possess the same level of qualities.

Under diverse selection policy, the candidates for soldiers take test battery meant for soldier and the candidates applying for officer, take the tests meant for officer. Here the basic assumption is that within different categories, the candidates have the same qualities but between categories, they have to depend upon different qualities.

In mixed selection policy, the military academics responsible for selection of armed forces through various types of test adopt both uniform and diversified policy. Belgian Defence Forces use a mixed system of selection.

So far as the Indian context is concerned, it was around the time of Second World War that Military psychology found its beginning and that too on a low key basis. It was, then, mainly concerned with solution of personnel's problem in armed forces. In 1943 officer selection Board was established at Dehradun for applying psychological method of selection. It was renamed at *Psychological Research Wing* (PRW) in 1949 and was established at Delhi. In 1962, PRW was renamed as *Directorate of Psychological Research* (DPR). It became a full-fledged institute called as *Defence Institute of Psychological Research* (DIPR) in 1982 working under *Defence Research and Development Organization* (DRDO) at Delhi. DIPR has developed several psychological tests, which have been successfully used for selection of armed force personnels. It has also developed aptitude tests for assessing abilities to work in inhospitable conditions. In fact, DIPR has 66 psychological tests to its credit.

A review of literature reveals that various types of psychological tests are used for selection of armed forces. This selection is performed by trained selection officers, psychologists as well as by military specialists. Five categories of psychological tests are commonly applied for selection and promotion of defence personnels.

- **Aptitude tests :**

    Aptitude tests are very popular tests employed for the selection of armed forces personnels. Aptitude test batteries contain various sub-tests providing indication about different cognitive abilities such as numerical reasoning, verbal reasoning, mechanical comprehension, spatial ability, general reasoning, etc. Sometimes aptitude tests are also called as cognitive mental ability test or intelligence test. In India a *Comprehensive Battery of Cognitive Abilities* (CBCA) has been constructed for selection of officers in the armed forces by DIPR (2007). Several existing tests of intelligence have been carefully reviewed and a new model has been developed. This new model attempts to asses different aspects of cognition, namely, decision-making, attention, memory, reasoning ability, problem solving ability, and the like. Based on PASS model developed by Das, Naglieri and Kirby (1994), the CBCA intends to assess the individual competence at three levels of cognitive functioning— *registration, processing* and *higher order processing*. A in-depth review of literature shows that the Royal Air Force of the United Kingdom, the Royal Norwegian Air Force, the Swiss Air Force, the Turkish Air Force as well as Portuguese Air Force use the following sub-tests (or some of them) for assessing aptitudes of the pilots.

    - The Instrument comprehension test is used for assessing general and spatial reasoning.
    - The Control and Velocity Test is used for tracking moving targets using a joystick
    - Psycho-motor test is used for assessing coordination between hand, eye and foot.
    - Test of short term memory such as Digital recall test is used in which the candidates are required to reproduce the memorized set of numbers using a keyboard after it has disappeared from the candidate's sight.
    - The vigilance test is used as a test of attention power and capacity.

- **Academic tests :**

    Academic tests or also called as *knowledge tests* intend to provide information about knowledge in specific domain. These tests are, infact, a sort of traditional examinations like essay test or open-ended test. Such tests have also been developed as a large scale multiple choice test batteries. Since such tests provide a detailed information about educational qualification, they are popularly applied for selecting military personals not only in USA but also in Netherlands and Belgium.

- **Assessment Scales:**

    Assessment scales have also been frequently used as a part of psychological testing in selection of defence personnels. There are three types of assessment–*self assessment, supervisor assessment* and *peer assessment*. In self-assessment scales or tests, the candidates describe themselves directly or indirectly by answering the various questions. An indirect self-assessment in done when the candidates give a self-description through their friends or through the eyes of enemy. In supervisor assessment, the manager or supervisor evaluates the people usually through using various types of psychological tests or scales. Often we find that many armed forces organize the selection boards for evaluating candidates who want promotion. This illustrates supervisory assessment. People can also be evaluated by their colleagues. This is called peer assessment. During an interview following the group exercises, the candidates may be asked to describe the most important and influential person during the exercise.

- **Personality tests :**

There is a growing realisation that the personality structure of the candidate is a very important and useful indication about the candidate's adaptation skills to military training and military way of life. Personality tests are of two types–*projective tests* and *psychometric test orself-report personality inventory*. The projective tests for selection of officers in Indian armed forces have been successfully used. Most of these projective tests have been successfully used. Most of these projective tests have been developed by defense scientists. However, due to some inherent limitations, it is not feasible to use projective tests in all situations. Hence, attempt was made to use self-report personality inventories for selection in defense services. Some researchers have made an extensive review in which for selection, training and promotion, personality inventories like the NEO Personality Inventory (Revised), the Minnsota Multiphasic Personality Inventory (MMPI), the California Psychological Inventory (CPI), the Inward Personality Inventory (IPI), the Commander Traits Inventory (CTI) etc., have been extensively used. The MMPI and IPI were developed to diagnose the psychopathological cases. The NEO Personality inventory and the CPI have been developed to assess normal personalities and have been successfully used in basic selection procedures. The CPI has been widely used in military research and selection of defense personnel. (Gough, 1987) The CPI is one of the important tests included in Belgian Defense Forces as well as it has also been incorporated by USA Navy. Likewise, NEO is a popular test in USA and Belgium's Air Force Selection team. It is based upon the Jung's personality theory and measures cognitive style. The CTI is very popular among Swedish military services.

In India at DIPR (2008) a selection battery consisting of a cognitive test and a personality test has been developed for selection of *Other Ranks* in the Indian Army. The cognitive test consists of non-verbal format using matrices types of items in view of educational level and diversity in population of applicants. The personality tests aims at locating the basic data to assess the candidate's suitability to perform a given task, the willingness and ability to learn and being trained, and tendency to learn and upgrade his knowledge and skills. The personality test has been developed in bilingual form (Hindi and English) having situation judgemental format.

- **Biographical data Inventory :**

Such inventory aims at providing information regarding the candidate's past experience and personal details like age, sex, gender, birth place, home address, father's education, mother's education, parents' annual income, etc. Military services board tries to make prediction on the basis of these personal informations. According to one popular review, biographical data have been more successfully used in military settings than in civil settings. Such biographical data inventory has also been known variously as *Rational bio-data scales* or *Biographic inventory*.

In a nutshell, it can be said that in military services, the psychological testing plays a significant role not only in selection of defense personnels but also in their training and promotion.

Summary and Review

- As we know, psychological tools are used in various kinds of assessments. Psychological testing has to do with the instrumentation part of the assessment where psychometric instruments are used for gathering information.
- Psychological testing is more common in clinical setting. Clinical psychologists have used psychological tests for diagnostic, prognostic as well as for thereauptic decision in various mental health settings. Tests like MMPI, Rorschach test, TAT, Halstead-Reitan Neuropsychological Battery, Leuria-Nebraska Neuropsychological Battery, etc. are very popular in clinical settings.
- Psychological testings have also been applied in educational settings. Here different types of educational tests are applied in institutional settings as well as in individual decision-making. At individual decision-making level, these tests have been applied under two major heads–career planning and educational planning.
- Psychological testing has also been applied in industry, organisation and business. In these situations psychological testings have been applied to deal with prediction issues as well as evaluation issues.
- Psychological tests have also been successfully applied in the field of sports, which is covered by sport psychology particularly by applied sport psychology. Various specifically designed psychological tests such as Test of performance strategies, Athletic Motivation Inventory, Profile of Mood states, Measures of Mental toughness, Test of attentional and interpersonal style, etc. have been used for promoting mental health and awareness among athletes.
- Psychological tests have also been applied in defense services for selection, training and promotion of military personnels. The most popular tests applied here are the Cognitive ability tests, Personality tests, Biographical-data inventory, Aptitude test, Academic tests, etc.

Review Questions

1. Discuss how psychologists apply psychological tests in clinical setting.
2. Explain the importance of psychological testing in educational setting.
3. Do you think that psychological tests are of any help in counselling and guidance?

4. Discuss the role of psychological tests in the field of applied sports psychology.
5. Explain how psychological tests fulfill their responsibilities in the field of industry, organization and business.
6. Explain the role of psychological testing in defense services.

Part-III : Basic Statistical Concepts

Chapter – 20

Basic Ideas in Statistics

Learning objectives :

- Meaning of Statistics
- Brief History of Statistics
- Population and Sample
- Parameter and Statistics
- Variable and Constant
- Levels of measurement scale
- Parametric statistical test Vs. Non-parametric statistical test
- Types of statistical methods
- Degree of freedom
- Null hypothesis, Alternative hypothesis and Research hypothesis
- Level of Significance
- Type I error and Types II error
- One-tailed test Vs. Two-tailed test
- Some dirty words about statistics
- Summary Questions
- Review Questions

Key Terms :

Alternative hypothesis, Statistics, Descriptive statistics, Inferential statistics, Nominal scale, Ordinal scale, Interval scale, Ratio scale, Degree of freedom, Type I error, Type II error, Independent variable, Dependent variable, Null hypothesis.

In the present chapter some basic facts about statistical principles, which a student is expected to understand and know have been discussed. These basic principles will enable the readers to know about both elementary statistics and advanced statistics. A brief historical review of the statistics would enable the students and teachers know about the gradual growth of statistics from the beginning to the present.

Meaning of Statistics

By a common observation, statistics is facts and figures. In reality, the word statistics has some common meanings. In the oldest sense, it referred to any sort of facts (numerical or otherwise) that reflect the conditions and prospects of society or state. However, this meaning of statistics has been narrowed so that when the word is used to mean any facts relating to society and the physical environment, numerical facts exclusively are implied. Another meaning of statistics has been developed during the letter part of the nineteenth century. In this sense, the term statistics refer to theories and techniques involved in collecting, summarizing and interpreting numerical facts. Clarifying it further, Ferguson and Takane (1989) have pointed out that the statistics is the branch of scientific methodology and it basically deals with the collection, classification, description and interpretation of data obtained by conduct of experiments and surveys. Likewise Elifson et al. (1990) have defined statistics as a method for dealing with data resulting from organisation and analysis of numerical facts or observations that are collected in accordance with systematic plan. In a nutshell, it can be concluded that statistics is mathematical process of gathering, organising, analyzing and interpreting numerical facts or data.

If we review the literature relating to the various meanings of statistics, we shall come to the conclusion that today, statistics has been encountered in four different contexts.

- Statistics can mean *applied statistics*, which is the science of describing, organizing and analyzing a set of quantitative data.
- Statistics can also mean a *set of indices* such as average (Mean), which is the outcome of application of statistical procedures.
- Statistics can also mean *statistical theory* and in this, it is regarded as the branch of mathematics, especially owing to the theory of probability.
- A fourth meaning much similar to the above second meaning, is that it is an *index descriptive of a sample*. Thus average of a sample is *statistics* whereas the same average of the population is called as parameter.

Brief History of Statistics

Statistics has a long and venerable historical background. Perhaps the earliest use of statistics started when the Ancient Chief counted the number of effective soldiers that he had or the required number be needed to defeat his enemies. Later on, statistics was used to report death in some diseases and in the study of natural resources.

In the seventeenth and eighteenth century some important mathematicians were requested by gamblers to develop such principles based on numerical facts that may ensure the chance of winning at cards and dice. The two most outstanding mathematicians of that time named Bernoulli and DeMoivre were involved in developing such principles of probability. In 1730s Demoivre was able to develop the equation for normal curve LaPlace and Gauss, two other important mathematicians of that time, also joined hand to work on probability. They worked to apply the principles of probability to the field of astronomy.

In the early nineteenth century a Belgium statistician Quetelet applied statistics to investigation of social and educational problems. Francis Galton in England also applied statistics to social sciences. Both Quetelet and Galton perceived statistical methods as a quantitative and powerful tool for dealing with the mass data obtained from society and human beings. In fact, beyond any doubt, Galton made an unparallel impact upon the introduction and use of statistics in social science. Throughout his life, he made a venerable contributions to the fields of heredity, psychology, anthropometry and statistics. Our basic understanding of correlation is largely credited to him. Among Galton's basic contributions, development of centiles or percentiles was worthmentioning. Later on, another mathematician Karl Pearson collaborated with Galton and was instrumental in developing many of correlation and regression formulas, which are even today used. During the first-third of the twentieth century owing to the work of Pearson and R.A. Fisher, an English statistician, both theoretical and applied statistics expanded rapidly.

The famous American psychologist James Mckeen Cattell contacted Galton and other European statisticians. Cattell and his student E.L. Thorndike began to apply statistical methods to various psychological and educational problems. As a consequence of their efforts and work of Cattell and Thorndike, both theoretical and applied statistics courses started being taught in American Universities in next few years.

In the twentieth century new statistical techniques and methods relating to the study of small samples were developed. In this field the most important contribution was made by R.A. Fisher. As it is known to all, his most of the statistical methods and techniques were developed in an agricultural or biological setting. Social scientists also adopted the Fisher's method and made use of his significant ideas.

Population and Sample

Population in, everyday language, is defined as the entire group of individuals that a researcher wishes to study. By entire group is meant every single individual. For example, we speak of population of India, population of Bihar, Population of New Delhi, etc. This is one use of the term population based on geographical region at a particular time. However, statisticians use the term population in a more general sense to refer not only to a defined groups of individuals but also to defined groups of animals, objects, materials, things or happenings of any kind. Thus there may be population of doctors, students, teachers, laboratory animals, birds, insets, fishes in the river, etc.

Keeping in view what has been elaborated for classifying meaning of population, it can be said that the term population as used in psychology and education, can best be described as the complete set of observation (or measure) about which the researcher likes to draw some conclusions. Analyzing this way of defining population consists of two main features.

- In this way of defining the term population, it does not refer to people rather it refers to some observed characteristics. Thus, if the researcher's interest lies in the intelligence scores of all those students who have been enrolled in class II of Patna St. Xavier's High School, it is this set of intelligence scores that will constitute the population and not the set of enrolled students in class I. The single observation of the population is technically called as *element*. Likewise, if the interest of the researcher is in observation of the number of schools in the concerned district, it will be population and any one school will be called as element of population. Statistically speaking, it is also possible to exist population even though there is just one person in observation. For example, if the researcher is interested in studying reaction time of a person under various conditions, here population would consist of a set of large number of possible reaction times of the same person that was measured in various conditions.

- The second feature of the above definition of population also clearly implies that the very set of observations, which constitutes the population is determined by the specific interest of the researcher. Thus defining population is not a statistical question rather a precondition to the application of the statistical procedure. Generally, the researcher would like to select his observation so that adequate statistical treatment would easily be done within the legitimate condition.

Sometimes a distinction is made between *finite* population and *infinite* population. The number of students enrolled in a particular university, number of books in a library, number of schools in a particular district are the examples of finite population because they can be easily counted to arrive at a finite number. The number of fishes in a river, the possible rolls of a die and possible observations in some scientific experiment are examples of infinite population. Most populations studied by social scientists are finite in nature.

Sample is the subset of the population. In other words, the sample is the portion of the population that is selected for observation. Numerically, a sample could be as small as a one element or it could consists of all but one element of the population. A sample could be drawn from all female students of the university or from all teachers of the university. Let us take an example. Suppose the researcher is going to conduct an opinion poll in the city of Delhi. The researcher is going to ask people if they believe the present government is doing a good job. Since the poll is likely to be conducted by telephone, the researcher defines his population as people who live in the city and are listed in telephone directory or listed with any of the local telecom company. Realizing that it would take more time than he can spend to call everyone listed in phone book (the entire population), he decides to call only 500 people. The example clearly illustrates that it is necessary to study a sample because the population is so large that it is impractical to call every person listed in telephone book. If the researcher is more ambitious and optimistic, he can take a sample of even 1000 people. It would be here worthmentioning to state that larger sample has some obvious benefits over the smaller sample.

Parameter and Statistics

It is desirable to distinguish between parameter and statistics. A parameter is a measurement that describes characteristics of the population such as population mean. In other words, a parameter is a property descriptive of the whole population. A statistic is a measurement that describes the characteristics of the sample such as sample mean. The sample value is presumed to be an estimate of the corresponding population parameter. Usually, population values or parameters are not known and the researcher estimates them from the sample values. Let us take example. Suppose the researcher randomly selects a sample of 1000 girls of 20 years old from Mumbai City. He measures the height of these 1000 girls, adds them and divide the total by 1000 and gets the average or mean height of five feet three inches (5'3"). This mean is the sample value and will be called as *statistic*. This value is an estimate of population parameter (population mean), which would have been obtained had it been possible to measure all the members (girls) in the population. Statisticians frequently use different symbols for a parameter and a statistic. By using different symbols, it can be readily said that if a characteristic such as mean is describing a population or a sample.

Variable and Constant

Variable is definded as any characteristics of a person, group or environment that can vary or denote some kind of difference. In other words, variable is a characteristic, which may take on different values. Some of the examples of variables are weight, sex, group cohesion, political ideology, intelligent test scores, number of errors in spelling test, rank of student in the class, etc. In fact, the concept of variable does not emphasize that each observation must differ from all the others. What is necessary is that the possibility of difference exists. Let us take an example to illustrate this fact. Suppose the researcher is interested in knowing the mean weight of class 10th students in a large school. Accordingly, he selects five students and measure their weight and finds that all the five students are of the same weight. It is still here proper to refer weight as a variable since possibility of getting students of different weight exists. On the other hand, if he decides to study weight only among the class Xth students, clearly it means that level of class is not a variable in this research.

On the other hand, if it is not possible for a characteristic to have other than single value, such characteristic is called as *constant*. In other words, a value is considered as *constant* if it never changes. For example, pi whose value is equal to 3.14 is one illustration of constant. Likewise, if the researcher is studying only female delinquents, sex would be the example of a constant. In fact, in a particular study, there are several variables and constants to which attention should be paid. Suppose the researcher is interested in political attitude among students in a college. Here political attitude is the prime variable but other variables such as age, sex political affiliation of parents may have to be taken in interpreting the findings. At the same time, membership in specific college and the time when the data were collected are example of constants. Infact, different results may be obtained if the study is conducted in different college or in the same college twelve years later.

The particular value of the variable is called as *variates* or *variate values*. For example, in studying the weight of the adult males, weight is a variable but weight of any particular person is a *variate* or *variate value*. There are many ways of classifying the variable. One popular way of classifying the variable that bears a functional relationship to one another, distinguishes between *dependent variable* and *independent variable*. In experiment the variable subject to manipulation is called the independent variable and the variable under observation is called dependent variable.

The dependent variable (symbolized by letter Y), is thus a variable being predicted or explained. In fact, it is dependent upon the independent variable (symbolized by letter X). The functional relationship between the dependent variable and the independent variable is usually expressed as $Y = f(X)$. This expression states that a given variable Y is some unspecified function of another variable X. The symbol f expresses that generally a functional relationship between the two variables exist. In experimental design, the dependent variable is referred to as the *criterion variable* and independent variable as the *experimental* or *treatment variable* or *predictor variable*. Suppose the researcher wants to study the impact of instruction technique (Lecture Vs. demonstration method) upon the classroom achievement of the students. In this example, the instruction technique would be the example of independent variable (X) and the achievement would be the example of dependent variable (Y). According to the above expression, that is, $Y = f(X)$, the prediction of Y (achievement) depends upon the value of X and the known functional relationship.

Another popular classification of variables is *discrete variable* vs. *continuous variable*. A discrete variable (also called as discontinuous variable) is defined as variable that has specific values and can have no value between two specific values. Thus a discrete variable is one which can take only certain values and none in between. For example, the number of members in the family, number of books in the library, the number of rooms in the schools, kinds of occupations are discrete variable. A person has 4 or 5 members in his family but not 4.7 member. A person might have visited the doctor's clinic 6 times or 7 times but not 6.5 times. All these illustrate discrete variable. Continuous variable is defined as variable that can take on an infinite number of values. The basic characteristic of continuous variable is that within whatever limits its value may range, any value at all is possible. For example, length is a continuous variable. It is just possible for an object to be 4'1", 4'2", 4'3" or any conceivable amount in between. Other examples of continuous variable are age, weight, temperature, intelligence, etc.

A variable that is really a continuous one may give the appearance of being discrete because of the measurement scale used. In fact, continuous variables are often expressed as whole number and therefore, appear to be discontinuous. Scores on intelligence tests, for example, are reported as whole numbers and there is maximum and minimum scores, which can be obtained. Such provision, at a glance, gives the appearance of a discrete variable, it is really the result of restrictions imposed by the test. Theoretically, intelligence could be expressed in any amount and therefore, it is continuous variable.

Another important classification of variables has been made on the basis of the nature of information made available by the measuring operation. Here four class of variables have been identified : *nominal variable*, *ordinal variable*, *interval variable* and *ratio variable*. Nominal variable is a primitive type of variable and is basically the property of a group that permits the making of statements only of *equality* or *difference*. Such variable places individual or objects into categories, which are homogeneous, mutually exclusive and make no assumption about any ordered relationship between categories. For example, if the researcher wishes to classify individuals on the basis of marital status, the convenient categories would be Single, Married, Separated and Divorced. Other examples of nominal variables are blood type, sex, religious affiliation, etc. The fundamental principle of nominal variable is that of *equivalence*, that is, all persons or objects placed in the same category are considered to be equivalent. In no sense, member of one category can be said to be 'better' or 'more' than the member of the other category. Ordinal variable is another important variable. It is the property of members of group defined by their rank ordering. In other words, with respect to the members of the group, not only are statements of equality and difference but also statements of the kind such as *greater than* or *less than* are legitimately made. However, statements regarding the number of times one member is 'greater than' or 'less than' another or about equality of difference between members are not possible. For example, we can rank order the university degree from higher to lower that is, 1. Doctorate 2. Master degree 3. Bachelor degree but it can't be said that the Ph.D. is twice as high as the Bachelor degree or that a particular difference exists between these categories.

An interval variable is a property defined by the operation that reflects differences in magnitude, that is, here measuring differences in size or amount of the statements of equality of intervals, in addition to sameness or difference or greater than or less than. One of the basic features of interval variable is that it does not have a *true zero point* and a zero point is only arbitarily defined for convenience. Temperature measurement (both Fahrenheit and Celsius) and calendar time are its examples. Let us take example. Suppose the temperature of three objects A, B and C are 20°, 40° and 60° respectively. In this example, it is appropriate to tell the difference between temperature of A and B is exactly equal to the difference in temperature B and C. It is also appropriate to tell that the difference in temperature A and C is twice the difference between temperature A and B or B and C. However, it is *not* appropriate to tell that B has twice the temperature of A or C has thrice the temperature of A. This is because the zero point is not true rather arbitrary. If today is 40° temperature and yesterday it was 20°, we cannot say that today is twice as hot as yesterday.

Ratio variable is defined as the property of members of the group by an operation that permits statement of equality of ratios in addition to all the statements made with respect to above three variables. Here ratios do reflect ratio of magnitudes. Such variable has a meaningful or true zero point. Here numbers used reflect distances from natural origin. Length, weight, temperature measured on Kelvin scale, measure of elapsed time such as age, years of experience and reaction time are examples of ratio variable. Here not only the difference between 30" and 40" is considered to be same as difference between 60" and 70" but it is also true that 60" inch is twice as long as 30". All these happens because of a true zero point on the measurement scale.

Of these four variables, in psychological and educational researches, ordinal variables and interval variables are most common. Nominal variables are less frequently used. Ratio variables are used in natural sciences.

Levels of Measurement Scale

In general, measurement involves either categorizing events or objects (qualitative measurement) or using numbers to categorize the size of the event according to some rules (quantitative measurement). There are four types of scales that are associated with measurement. These are as follows :

- **Nominal scale of measurement :**

In nominal scale of measurement observations are labelled and categorized. The term 'nominal' means 'having to do with names'. Therefore, measurements that are done on this scale involves naming things. Since nominal scale does not involve highly complex measurement and simply involves rules for placing individuals or objects into categories, the measurement exists here at the *weakest* level. A nominal scale is also known as *categorical scale* or as a *classificatory scale*. For example, when the researcher wants to know the sex of a person responding for a questionnaire, it would be measured on a nominal scale consisting of two categories—male and female. Likewise, when a psychiatrist classifies the patients as 'paranoid', 'schizophrenic', 'obsessive-compulsive neurotics', he is using nominal scale.

Since in a nominal scale the measurement operations partition a class into a set of mutually exclusive subcases, only relationship involved is that of *equivalence*. By the relation of equivalence is meant that members of any one subclass must be equivalent in property being measured. For example, all patients belonging to the subclass of 'schizophrenic' are equal with respective to the symptoms of schizophrenia. Likewise, all persons placed under the subclass of 'female' are equal to each other. This equivalence relationship is *reflexive* (a = a for all values of a), *symmetrical* (if a = b, then b = a) and *transitive* (if a = b, and b = c, then, a = c).

The admissible statistical operations on nominal measurement are frequency counts made, chi-square test and test based upon binomial distribution.

- **Ordinal scale of measurement :**

This is the next higher level of measurement where categories are not only homogeneous and mutually exclusive rather they stand in some kind of *relation* to one another. The types of relationship encountered in an ordinal scale are 'greater', 'higher', 'more preferred', 'more difficult', 'more', 'disturbed', 'more prejudiced', 'more healthier', 'more prestigious', etc. The example of ordinal scaling includes rank ordering: socio-economic status, final merit list of the students' academic achievement, academic institution in a state or district according to prestige, names of leaders in terms of their leadership qualities, etc. Many social scientists believe that all the attitude scales employed in the social sciences are examples of ordinal scales. Therefore, responses to Likert type scale in terms of 'strongly agree', 'agree', 'neutral', 'disagree' and 'strongly disagree' constitute ordinal scale.

The formal properties of ordinal scale is that it not only incorporates relationship equivalence but also of relationship of rank ordering, that is, greater than or less than. This relationship of rank ordering is *irreflexive* (it is not time for any a that a > a), *asymmetrical* (if a > b, then b > a) and *transitive* (if a > b and b > c, then a > c).

The admissible statistical operations in ordinal measurement are median, Spearmen Rank order correlation, Kendall rank correlation coefficient, Mann-Whitney U test, Kruskal-Wallis one way analysis of variance by rank and other *ranking statistics* or *ordering* statistics. On ordinal measurement median is considered most appropriate statistic for describing central tendency of scores because median is not affected by the changes of any scores, which are above or below it as long as the number of scores above and below remains the same.

- **Interval scale of measurement :**

When the measurement scale has all the characteristics of ordinal scale plus equality of intervals (or differences) between any two numbers on the scale, it is said to constitute what is called interval scale. In other words, in an interval scale the difference (or the interval) between number on the scale reflects a difference in magnitude. However, ratios of magnitudes are not meaningful. One of the basic features of interval scale is that hereboth zero point and the unit of measurement are arbitrary. Examples of interval scale are calendar years and degrees of temperature on Centigrade scale or Fahrenheit. Through an example it can be easily shown that the ratios of temperature (or intervals) are independent of unit of measurement and of the zero point. As we know, the freezing point on Celsius scale is 0° and boiling point is 100°. On the other hand, on Fahrenheit scale freezing occurs at 32° and boiling occurs at 212°. Some other readings of the same temperature[1] on two scales are as under :

| Celsius | : | 0 | 20 | 40 | 100 |
| Fahrenheit | : | 32 | 68 | 104 | 212 |

In the above readings the ratio of difference between temperature on one scale is equal to the ratio between equivalent differences on the other scale. For example, on Celsius scale the ratio of differences between 40 and 20 and 20 and 0 is (40–20)/ (20–0) = 1. For same readings on Fahrenheit scale the ratio is (104 – 68)/(68 – 32) = 1. Thus ratios in both cases are the same, namely, 1. Thus on interval scale, the ratio of any two intervals is independent of the unit used and the zero point. Both are arbitrary in such scale. It is due to arbitrary zero that we cannot say that a day which has temperature of 40° Celsius is twice as hot as a day having temperature of 20° Celsius.

The formal properties of interval scale includes *equivalence* (like nominal scale), *greater-than* relation (like ordinal scale) as well as specification of ratios between two intervals or differences.

---

1. Temperature on Celsius has been converted to Fahrenheit by a formula:

$$F = \frac{9}{5}C + 32$$

Since interval scale is the first truly quantitative scale, all common parametric statistics like mean, standard denotion t test, F test, product-moment correlation etc. are easily applied to the measurements done by interval scale

- **Ratio scale of measurement:**

The ratio scale of measurement possess all the features of interval scale plus has the property of absolute zero or true zero point. Measurement obtained with respect to height, weight, temperature or Kelvin scale, etc are the examples of such measurement In ratio scale, the ratio of any two scale points is independent of unit of measurement. Let us take an example. A weighs 40 Kilogram and B weighs 20 Kilogram. We can say that A is twice heaver than B just because there is true zero point in such measurement that reflects 'no' weight or zero weight. The weight is measured on the scale of pound and on the scale of gram also. The ratio between any two weights is independent of unit of measurement. Thus if two objects are weighed not only an pounds but also in grams, it cannot be said that ratio of two weights in pounds is equal to the ratio of two weights in grams.

Ratio scales are frequently encountered in natural sciences. The formal properties of a ratio scale includes *equivalence, greater than, known differences* between two intervals (or differences) and *known ratio* of any two scale values. Here number associated with the ratio scale values are true numbers with a true zero point. However, the unit of measurement is arbitrary. Any parametric or nonparametric statistical test can conveniently be used with measurements obtained from ratio scale.

Of these four scales, nominal scale and ordinal scale are most popular in behavioural sciences, followed by interval scale. Ratio scales are most preferred in physical or natural science such as physics, chemistry, etc.

Parametric Statistical tests vs. Non-parametric Statistical Tests

Parametric statistics are those statistics, which make some assumptions about the nature of the population from which samples were drawn. Since the population values are called parameter, these statistics are called parametric statistics. t test and F test are common examples of parametric tests. The meaningfulness of the results of any parametric tests depends upon the validity of those assumptions. Since these assumptions are not ordinarily tested, they are assumed to hold. The important assumptions of parametric tests are as under :

- The population from which samples were drawn must be normally distributed. In other words, the observations must be drawn from normally distributed population.
- The observation must be independent one. In other words the selection of any one case from population should not bias the selection of other case. Similarly, the score assigned to any one case must not bias the score, which is assigned to any other case. Thus independence of observation includes two things : selection of samples from population must be random and assignment of scores to the observation must be unbiased.
- The variables involved must have been measured on an interval scale or ratio scale so

that arithmetical operations may be easily carried out.
- The populations from which samples are drawn must exhibit the homogeneity of variance or same variance or in some cases must have known ratio of the variance.
- The population from which samples are drawn must have *homoscedasticity* of variances which means that there must be equality of variances between columns and rows.

Non-parametric tests are those tests, which make very few assumptions or no assumptions either about the parameters or about the shape or nature of the population from which sample is assumed to be drawn. For this reason, nonparametric statistics are also called as *distribution-free statistics* or loosely *short-cut statistics*. Examples of non-parametric statistics are chi square test, Mann-Whitney U test, Kendall's Rank difference correlation, Spearman's Rank difference correlation etc. The major assumptions of non-parametric tests are as under:

- The variable involved must have been measured on nominal scale or ordinal scale. In other words, the obtained data must be in terms of categories (frequency) or rank.
- The observations should be independent. In other words, the selection of samples from population must be random as well as the assignment of scores to the observation must be unbiased.
- The variable under study (dependent variable) must have underlying continuity. It means that variable under study must be present in any amount in each and every element of population from which sample is drawn.

Advantages of Nonparametric statistical tests :

There are some obvious advantages of the nonparametric tests. Some of these are as under:
- For analyzing data which are categorical, that is, measured on nominal scale, nonparametric tests are the only choice. No parametric statistical tests apply to such data.
- If the size of the sample is very small, there is no alternative to using a non parametric test unless the nature of the population distribution is exactly known.
- Since nonparametric tests typically make very fewer assumptions about the data, they are considered more relevant to a particular situation and are less susceptible to violation. These violations are frequently seen with parametric tests.
- Nonparametric statistical tests are easy to apply and that they are most appropriate for many analyses of dichotomous data.
- Nonparametric statistical tests are usable when observations come from several different populations. Parametric tests cannot handle such data without making some unrealistic assumptions or requiring cumbersome computations.
- Nonparametric statistical tests are available to treat data, which are inherently in ranks as well as data whose apparently numerical scores have the strength of ranks. If the data are inherently expressed in ranks or even if they can be categorized only as plus or minus (more or less, better or worse), they can be easily treated by non-parametric statistical

tests whereas they cannot be treated by parametric statistical tests unless some unrealistic assumptions are spelled out about the underlying distributions.
- Nonparametric statistical tests are comparatively much easier to learn and apply than parametric tests. Naturally then, their interpretation is also more direct than the interpretation of parametric tests.

Disadvantages of Nonparametric Statistical tests

Despite several advantages of nonparametric statistical tests, there are some obvious disadvantages as under :

- Nonparametric statistical tests generally don't have as much statistical power as parametric tests. By statistical power is meant a probability that a test will reject a false null hypothesis. Besides, nonparametric tests are more likely to fail in detecting the real difference between two treatments. Therefore, wherever the researcher has choice between a parametric test and a nonparametric test and when most of the assumptions of parametric tests are met in data, he should always choose the parametric test because in such situation, nonparametric tests would be considered as 'wasteful'.
- Another disadvantage of nonparametric test is that their table of significant values are scattered widely and appear in different formats. Therefore, they are inaccessible to the behaviour scientists.
- Nonparametric statistical tests generally so operate as to throw away part of the information contained in data and that their labour-saving claims become illusory especially in large samples.

Difference between parametric statistical tests and Nonparametric statistical tests :

A comparative study of parametric statistical tests and nonparametric statistical tests reveals the following points of distinction.

- Parametric statistical tests are applied when data have been obtained on interval scale or ratio scale whereas nonparametric statistical tests are applied when data have been obtained on nominal scale or ordinal scale.
- A parametric statistical test is dependent upon population characteristics or parameter for its use whereas a nonparametric statistical test does not depend so much on the population characteristics or parameters for its use.
- Nonparametric tests have fewer assumptions about population characteristics as compared to the parametric statistical tests.
- Parametric tests are more powerful in rejecting a false null hypothesis as compared to the nonparametric tests.
- For parametric statistical tests the shape of the distribution should be normal one whereas for nonstatistical tests, the shape of the distribution is not relevant.

- Nonparametric statistical tests are typically much easier to learn and to apply in comparison to the parameteric tests.
- When both parametric and nonparametric statistical tests are not applicable, it is almost automatic that the parametric tests are inappropriate and nonparametric tests should be used (Kurtz & Mayo, 1980)

Types of Statistical Methods

There are two basic types of statistical methods or procedures : descriptive statistics and inferential statistics. A brief description of these two statistics is presented below :

Descriptive statistics :

Descriptive statistics are statistics, which provide a meaningful and convenient techniques for *describing* the major features of data that are of interest. In other words descriptive statistics are used in describing the properties of samples or of population on the basis of data available. For example, if the researcher measures the intelligence of all students of a university, and compute the mean of the scores, it will be an example of descriptive statistics because it describes the features of the population. On the other hand, if he measures the intelligence of a sample of students, say only 500, taken from the student population of the university, this sample mean would also be example of descriptive statistics because they describe the characteristics of the sample. In a nutshell, descriptive statistics summarize, organize and simplify the data. Examples of descriptive statistics are mean, variance, standard deviation, range standard error of mean, skewness and kurtosis of distribution of scores. Most statisticians also regard correlation between variables as descriptive statistics. Thus descriptive statistics may belong to three categories according to the properties of sample describe by them : *statistics of location, statistics of dispersion and statistics of correlation.*

Interferential statistics :

Inferential statistics are another popular statistical methods used in researches of behaviour sciences. Also known as *inductivestatistics* or *sampling statistics*, such statistics allow the researcher to draw an inference about the population on the basis of sample data. In other words, the objective of inferential statistics is to draw inference about conditions, which exist in a larger set of observations (population) from the study of the part of that set (sample). Suppose the researcher wants to study the prevalence of drug use in Indian College campus. Here population is defined as full-time enrolled students in Indian colleges and universities. Students are to fill out a questionnaire, which asks about the type and frequency of drug used. Since population is too large to make a contact to every student, a sample of students, say, 1000, is randomly selected and they are requested to respond to the questionnaire. After analysis of the data for the sample, the researcher is likely to make general statement about the population. Here sample data are being used to infer something about the population. Statistics used here are called as inferential statistics. Examples of inferential statistics are t-test, F test etc. Inferential statistics are also used in estimately the sampling error and the confidence intervals of population parameters. Interferential statistics are considered a very crucial part of conducting experiments in these statistical techniques, which are often used to assess whether or not a treatment effect has occurred.

Degree of freedom

Degree of freedom means freedom to vary. The origin of the term degree of freedom is linked to the advanced statistical theory underlying the probability distribution. The number of degrees of freedom of an estimate of a parameter refers to the number of independent values, which contribute to the estimate. Suppose we have seven scores and the mean of these seven scores is to be 8. Here the seventh score makes adjustment in the variation brought about by the first six scores and assures that the mean of the scores will be 8. For example, suppose six scores are ; 15, 6, 4, 12, 4, 13. In order for mean to be equal to 8, the seventh score must be 2. Here number of independent values is 6 ($N - 1 = 7 - 1 = 6$), which is the degree of freedom. The number of degrees of freedom of an estimate of a parameter is not laways one less than the number of original observations. In fact, the decrease depends upto the number of independent restrictions required in arriving at the estimate or (what is the same thing) the number of constants that must be determined from the observations. The principles underlying them is always the same. The principle is that a statistic, as an estimate of parameter, has degree of freedom equal in number to the number of independent observations contributing to its value. This principle maybe stated in any of the following ways :

(i) The number of observations contributing to the value of the statistic that are free to vary. In determining mean of the distribution, $N-1$ observations are free to vary and only one is fixed.

(ii) The number of observations minus the number of constants determined from them used in calculating the statistic. In determining mean, we use N observations and one constant from the observation.

(iii) The number of observations less the number of independent restrictions placed upon them in calculating the statistic. For example, in calculating standard deviation, the researcher has N observations (N deviation scores), which have the single restriction that they must sum to zero.

Null hypothesis, Alternative Hypothesis And Research Hypothesis

The first step in decision-making procedure based on statistical tests is to formulate a null hypothesis (H_0). The reality is that this statistical hypothesis is usually stated in null form, which is nothing but assertion of the fact that there exists no difference. For example, in case of a single sample mean and a known population mean, the null hypothesis states that there exists no difference between population parameter and that particular sample mean. In case of two sample means which are the estimates of two population means, the null hypothesis would state that no difference exists between two population means. An alternate way of stating the null hypothesis is simply to state that two samples are drawn from populations having the same mean.

The null hypothesis is not limited to zero differences nor to differences between statistics. Null hypothesis can also be asserted in other forms. For example, other form of the null hypothesis may assert that the results found in an experiment don't differ significantly from the results to be obtained on some probability basis or according to some stipulated theory. Researchers are of view that the hypothesis other than null can surely be stated. For example, we may assert that the experimental group receiving training would exhibit performance 10 points above the performance of the control group receiving no training. But it is very difficult to formulate such precise expectation in many situations. Hence, the researcher always prefer to proceed by formulating a null hypothesis which is exact (Garrett, 1981).

A null hypothesis (H_0) is usually formulated for the express purpose of being rejected. If a null hypothesis is rejected, its opposite called *alternative hypothesis* (H_1) is supported. In fact, alternative hypothesis is the operational statement of the experimenter's *research hypothesis*, which is simply the prediction derived from theory under consideration (Siegal & Castellan, 1988). According to Elifson et al., (1990) alternative hypothesis is somtimes referred to as the *research hypothesis*. Let us take an example to illustrate the fact. Suppose a theory in social psychology permits us to predict that two specified groups of people will differ in the expression of aggression. This is an example of a research hypothesis. To test this research hypothesis, we can state it in operational form, which will be our alterative hypothesis. For example, we may state that mean of Group A will be higher on the measure of aggressiveness than the mean of Group B or that the mean of group A and the mean of group B will be the same on the measure of aggressiveness. If the null hypothesis is rejected, the alternative hypothesis would be accepted.

The alternative hypothesis is stated in *directional way* or *nondirectional way*. It is the nature of research hypothesis that determines how the alternative hypothesis would be stated. If the research hypothesis simply permits to state that two groups will differ from each other, then alternative hypothesis would be stated in non-directional way stating that mean of Group A and the mean of Group B would be the same. However, if the research hypothesis would permit to state the direction of difference by emphasizing the fact that one specified group would have either the larger or the smaller mean than the other, then the alternative hypothesis will be stated in directional way.

Level of Significance

After the null hypothesis and the alternative hypothesis have been properly stated, the next step is to specify a *level of significance* or also called as *alpha level*. In behavioural research the common practice is to reject null hypothesis in favour of alternative hypothesis if the statistical test yields a value where associated probability of occurrence under the null hypothesis is equal to or less than some small probability. This probability is technically known as the *level of significance*. The most common values of level of significance are .05 and .01 level. Thus if a null hypothesis is rejected at .05 level of significance, it means that 5 times in 100 replications of the experiment, the null hypothesis will come to be true and 95 times, it will come to be false. This, in turn, indicates that 95% probability exists that the obtained experimental results are due to the experimental treatment rather than due to some chance factors. Likewise, .01 level of significance indicates that one time in 100 replications of the experiment, the null hypothesis is true and 99 times, it is false. Therefore, rejection of null hypothesis at .01 level of significance provides greater confidence in the experimental results. There is no strict rule to choose a level of significance. Researcher chooses according to precision and confidence he wants to impose.

Type I Error and Type II Error

In taking decision about the null hypothesis, two types of error may be made. These decision errors reflect situations in which *right procedures* lead to *wrong decisions*. Rejection of null hypothesis (H_0) when, in fact, it is true constitutes Type I error denoted by Greek letter a. In such situation, the alternative hypothesis (H_1) is accepted when the null hypothesis is true. On the other hand, acceptance of null hypothesis when, infact, it is false constitute Type II error denoted by Greek letter b. Here, despite the fact that an alternative hypothesis is true, the null hypothesis is accepted. Type I error is called as *alpha error* and Type II error is also called as *beta error*.

The probability of committing a Type I error is related to the level of significance. The larger the probability of level of significance, the more likely it is that the null hypothesis will be rejected falsely, that is, higher is the probability of committing type I error. That is reason why, the rejection of null hypothesis at .05 level (larger than .01 level) produces more type I error than when null hypothesis is rejected at .01 level. There is an inverse relationship between livelihood of making Type I error and Type II error. A decrease in Type I error will lead to increase in Type II error for any given N. In order to reduce Type I error whenever the researcher makes the level of significance more stringent (by making the level of significance smaller such as going from .05 to .01), he succeeds no doubt but he increases the number of Type II error. On the other hand, whenever the researcher uses a more relaxed level of significance in order to reduce Type II errors, he succeeds but he increases the number of Type I errors. If the researcher wants to reduce both type of errors, he must increase the sample size N.

The whole facts as outlined above may be summarized as follows :

	H_0 is true	H_1 is true
Acceptance of H_1	Type I error	Correct Decision
Acceptance of H_0	Correct decision	Type II error

The above summary clearly shows that when we accept H_1 where H_0 is true, it constitute Type I error and when we accept H_0 when H_1 is true, it constitutes Type II error. However, when we accept H_1 where H_1 is true, and accept H_0 where H_0 is true, constitutes correct decisions.

In any statistical inference there always exists danger of committing any of these two types of error. Therefore, a good researcher always wants to maintain a balance between the probabilities of making these two types of errors. Different types of statistical tests offer the possibility of different balances. In achieving such balance, the notion of power of statistical test is very important. By power of statistical test is meant its probability of rejecting H_0 when in reality it is false. In terms of equation, it can be said

$$\text{Power of statistical test} = 1-\beta$$

Where, β is the Type II error

It has been shown that for a particular test, the probability of committing a Type II error decreases as the sample size N increases. As a consequence, the power of a statistical test increases as the size of N increases. The power of the statistical test is also related to the nature of alternative hypothesis. The directional alternative hypothesis (one-tailed test) is more *powerful* than a non-directional test (two-tailed test). The power is also influenced by variance, level of signification and other variables.

One-tailed test vs. Two-tailed test

When a directional alternative hypothesis is stated, the resulting test is called as *one-tailed test*. On the after hand, when a non-directional alternative hypothesis is stated, the resulting test is referred to as *two-tailed test*. When a researcher makes a directional alternative hypothesis, the null hypothesis is also directional. A one-tailed test can be one-tailed in either direction or tail of the sampling distribution. One-tailed test and two-tailed tests differ in terms of location (but not in terms of the size) of the region of rejection. In one-tailed test the region of rejection lies entirely at one tail of the sampling distribution whereas in two-tailed test the region of rejection is located at both tails of one distribution.

Let us take an example to illustrate the basic idea contained in one-tailed test and two-tailed test. Suppose the researcher wants to study the impact of a training programme upon the classroom achievement of the students. For this, he prepares two groups—control group and experimental group. The control group receives no training whereas the experimental group receives the designated training. Suppose the researcher hypothesizes that training would make a difference in classroom achievement of the two groups of students. The null hypothesis here would be that training does not make a difference in the achievement of the students or there will be no difference in the classroom achievement of two groups of students. Such framing of hypothesis is non-directional and hence, it is an example of two-tailed test. Here the researcher is merely interested in knowing if there were a difference and therefore, either large increase or large decrement in achievement would lead to the rejection of null hypothesis. Since difference in either direction would lead to rejection of null hypothesis, it is two-tailed test. If the level of significance is set at .05 level, the size of region of rejection comprises 5 percent of the entire area included under the curve of sampling distribution and is equally divided into two tails (see Figure 20.1).

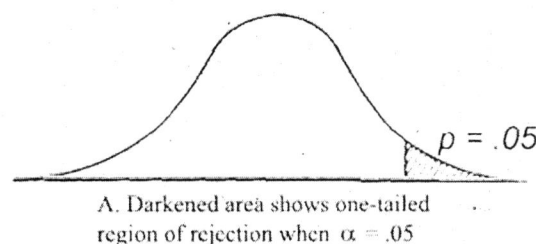

A. Darkened area shows one-tailed region of rejection when α = .05

Figure 20.1 Region of Rejection for One-tailed Test.

On the other hand, if the researcher in the above experiment formulates an alternative hypothesis stating that the training would enrich the classroom achievement of the students and accordingly, the classroom achievement of experimental group would be better than the classroom achievement of control group, this illustrates one-tailed hypothesis because here the researcher is interested in determining whether the training would lead to increased achievement. Here only the large increases in performance would lead to the rejection of null hypothesis. The null hypothesis here, would be that the performance of experimental group would not be higher than the performance of the control group or the training will not improve the performance of the experimental group. If the level of significance is set at .05, then the size of the region of rejection comprises 5 percent of the entire area included under the curve of the sampling distribution and this entire area will lie at the upper end or tail. (see Figure 20.2). If somehow the researcher has reason to believe that the experimental group would score lower than the control group, he may set up the directional hypothesis that the mean of the experimental group is equal to or greater than that of the control group. Rejecting this would naturally support the alternative hypothesis that the mean of the experimental group was lower than that of the control group. In this case if the level of significance is set at .05 level, this 5% area will lie at the lower and of the curve.

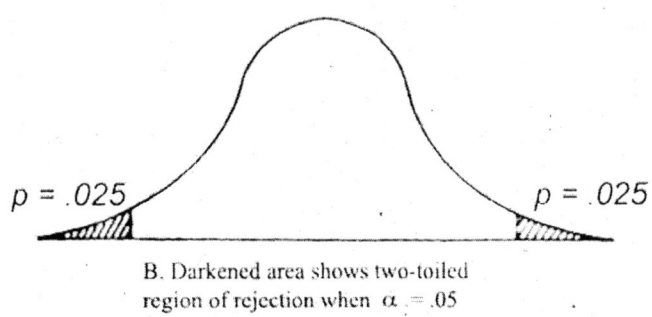

B. Darkened area shows two-toiled region of rejection when α = .05

Figure 20.2 Region of Rejection for Two-tailed Test.

The important point to be noted is that at .05 level, the two regions of one-tailed test and two-tailed test differ in location but not in size.

A two-tailed test is considered most appropriate if the researcher's concern is with the absolute magnitude of difference, that is, with the difference only. He is not concerned with the sign of the difference. When a directional or one-tailed test is to be used? In general, if the researcher has pre-knowledge about the likely effect of the treatment, he may use a directional test. Of course, there are many circumstances in research where the direction of difference is of substantial interest. There are only few situations where direction of the difference is not of much importance. According to Ferguson and Takane (1989) one-tailed test should be used more frequently than two-tailed test.

However, there are some problems with one-tailed test. Most researchers are of view that one-tailed test makes it too easy to reject null hypothesis and therefore, make it too easy to commit type I error. In other words one-tailed test allows the researcher to reject the null hypothesis even when the difference between sample and population is relatively small. Another problem with one-tailed test is that such test looks for a treatment effect in one direction only. For example, in the above illustration suppose the training given to the group proves to be boring and monotonous to the experimental group and it caused decrement in academic achievement. Since the one-tailed test has no critical region on the left-hand side (or lower tail) of the sampling distribution, it would not detect a significant decrease in classroom achievement. The test is looking only for an increase. The non-directional test (or two-tailed test) is sensitive to a change in either direction. For these reasons, many researchers disagree about using the one-tailed test, even when there is clearly directional hypothesis (Aron, Aron & Coups, 2006).

Some Dirty words about Statistics

There are some complaints against statistics. Some of the common complaints are are under:
- Statistics is said to be dry. Any statistical result is nothing but statement about the condition of data. It is bound to be dry unless the individual takes interest in the data and understands the significance of the statistical results.
- It is depersonalising and misleading. Any statistical statement is usually made about a group rather than about an individual. When it is said that the mean height of students of class XII is 5 feet, it does not provide any idea about the height of any one particular student whose height may be greater or less than 5 feet. Thus it is depersonalizing and mislead. Therefore, it has been rightly commented, *"There are three kinds of lies ; lies, damned lies and statistics."*
- It is also said that the statistical methodology dictates the natures of investigations. It is true that some investigators lean more toward conducting such researches, which are objective and where things are easily measurable than towards researches which are plain and meaningful. But here really, the statistics is not the culprit rather the problem lies in the attitude of the investigator.

- It is said that statistics is too mathematical to be understood by the common people. Although this is true that a number of propositions in statistics have been derived from mathematical derivations, a good working understanding of statistics can be easily acquired by a simple knowledge arithmetic and elementary principles of algebra commonly taught in high schools.

Summary and Review

- Statistics is a branch of scientific methodology and it deals basically with the collection, classification, description and interpretation of data obtained by conduct of experiments and surveys.
- Population is defined as complex set of observation (or measures) about which the researcher likes to draw some conclusions. Sample is a subset of population. More specially, it is that portion of the population that is selected for observation.
- Parameter is a measurement that describes a characteristic of the population whereas the statistics is the measurement that describes the characteristic of the sample.
- Variable is defined as any characteristic of a person, group or environment that can vary or denote some kind of difference. A value is considered as constant if it never changes. The popular classifications of variables are dependent variable and independent variable, discrete variable and continuous variable, nominal variable, ordinal variable, interval variable and ratio variable.
- There are four levels of measurement scale: nominal scale of measurement, ordinal scale of measurement, interval scale of measurement and ratio scale of measurement.
- Psychologists and educators have distinguished between parametric statistics and nonparametric statistics. Parametric statistics refer to statistics, which is based upon some assumptions about population from which sample data have been derived. Non parametric statistics does not specify such assumptions or specifies a fewer assumptions about the population. Hence, it is also called as distribution-free statistics.
- There are two basic types of statistical methods : descriptive statistics and inferential statistics. Descriptive statistics are one which conveniently describe the major features of data that are of interest. Mean and Standard deviation are examples of descriptive statistics. Inferential statistics are one which allow the researcher to draw inference about the population on the basis of samples data. Inferential statistics is also known as *inductive statistics* or *sampling statistics*. Examples of inferential statistics are t ratio and F test.
- Degree of freedom refers to the number of independent values, which constitutes to the estimate. The origin of degree of freedom is linked to the advance statistical theory underlying probability distribution.
- Null hypothesis (H_0) refers to the hypothesis of no difference. Research hypothesis is the hypothesis derived from a known theory regarding variable under study. Alternative

hypothesis (H_1) is the operational statement of the research hypothesis. On rejection of H_0, H_1 is accepted by the researcher.
- Level of significance (also called as alpha level) refers to some selected levels (such as .05 and .01) where null hypothesis is rejected in favour of alternative hypothesis if the statistical test yields value where associated probability of occurrence is equal to or less than some small probability.
- Type I error is the rejection of null hypothesis when, infact, it is true and Type II error is the acceptance of null hypothesis when, in fact, it is false.
- When a directional alternative hypothesis is stated, it is called one-tailed test and when a non-directional alternative hypothesis is stated, it is called as two-tailed test. In one-tailed test the critical region of rejection lies at one tail or end of the sampling distribution whereas in two-tailed test the critical region of rejection is located at both ends or tails of the sampling distribution.
- There are some dirty words about statistics. Commonly, it is said that the statistics is dry, depersonalizing, and misleading. It is also said that statistics dictates the nature of investigations and is too mathematical to be understood by common people.

Review Questions

1. Define statistics. Make distinction between population and sample.
2. Citing example discuss the different types of variables.
3. Discuss the different levels of measurement scales.
4. Citing examples make distinction between parametric statistical test and non-parametric statistical tests.
5. Make distinction between description statistics and inferential statistics.
6. Citing examples make distinction between null hypothesis, research hypothesis and alternative hypothesis.
7. Make distinction between one-tailed test and two-tailed test.
8. Write short notes on the following :
    (a) Level of significance
    (b) Degree of freedom
    (c) Interval scale of measurement scale
    (d) Alpha error and Beta error.

Chapter – 21

Measures of Central tendency

Learning objectives :
- Meaning of Central Tendency
- Arithmetic Mean
- Median
- Mode
- Comparison of Mean, Median and Mode
- Which of these three measures of central tendency is most appropriate?
- Summary Questions
- Review Questions
- Key Terms

Key Terms:
Arithmetic Mean, Central tendency, Median, Mode, Open-ended distribution, Skewed distribution, Grouped data, Ungrouped data

The researcher has constructed an Anxiety scale and has before him the scores of a group of men and a group of women. Now he wants to know whether one group score is better than the other and if so, to what extent. He may set the two distributions of scores side by side but this will present before him an approximate number. In stead, he would prefer to find *average* score of each group and would like to compare them. This average is called as *central tendency*.

Meaning of Central Tendency

The central tendency of a distribution refers to a single summary figure, which describes the high or low level of a set of observations. In other words, the central tendency is the central reference value, which is usually close to the point of the greatest concentration of scores of the distribution and may, in some sense, be thought to typify the whole set of distribution. In common use, there are three measures of central tendency: *Arithmetic Mean*, *Median* and *Mode*. The average is the popular term for arithmetic mean. There are two basic situations in which use of the measure of central tendency is of importance :

- When the researcher wants to compare the level of performance under two or more conditions or of two or more groups, he calculates a value of central tendency. For example, the researcher may be interested in knowing whether level of concentration of the students is higher in the morning than in the evening or whether Group A excels than Group B on the measure of intelligence. These can be easily done by resorting to some measures of central tendency.
- The researcher may like to compare the level of performance of a group with that of standard reference group. For example, if a teacher finds that the average IQ of students is 110 and he may further like to be interested in knowing whether this obtained average is above or below the standard set up by the group of educational institutions.

We shall discuss all the three measures of central tendency one by one.

Arithmetic Mean

The arithmetic mean or briefly 'mean' is defined by the sum of the separate scores (SX) divided by their number or N. For example, if there are five set of scores or measurements like 10, 20, 12, 13, 30, their sum is 85 and if it is divided by N, that is, 5, its arithmetic mean would be 85 divided by 5 is equal to 17. Its symbol is M or $\bar{X}$ (spoken as X bar) which is here equal to $\frac{\Sigma X}{N}$. This illustrates calculation of mean from ungrouped data.

Calculation of Mean from group data (or from frequency distributions)

There are two ways to calculate mean from frequency distribution: *short method and long method*. Let us illustrate each of these two methods.

Calculation of Mean by short method or code method :

When mean is to be calculated by short or code method from the scores arranged in frequency distribution, its formula becomes :

$$\bar{X} = M = AM + Ci \qquad (21.1)$$

where, M = arithmetic mean
AM = assumed Mean (or Midpoint of the class interval to which the coded score of zero is assigned.
C = correction
i = length of class interval

The calculation of Mean by Formula 21.1 is illustrated in Table 21.1.

Table 21.1: Frequency distribution for computing the mean of 100 scores by short method

1	2	3	4	5
Class Interval	Midpoint	Frequency (f)	X'	fx'(3 × 4)
85-89	87	2	5	10
80-84	82	3	4	12
75-79	77	4	3	12
70-74	74	6	2	12
65-69	67	5	1	05 (+51)
60-64	62	7	0	0
55-59	57	3	−1	−3
50-54	52	2	−2	−4
45-49	47	6	−3	−18
40-44	42	2	−4	−8 (−33)
		N = 40		$\Sigma fx' = 18$
				$\Sigma(\text{Positive})fx' = 51$
				$\Sigma(\text{Negative})fx' = -33$
				$\Sigma fx' = \text{Difference} = 18$

$$M = AM + Ci$$

$$AM = 62; \; C = \frac{\Sigma fx'}{N} = \frac{18}{40} = 0.45; \; i = 5$$

$$\therefore M = 62 + (0.45)(5) = 62 + 2.25 = 64.25$$

As per requirement of formula 21.1, AM (Assumed Mean) that is the midpoint of the class interval in which the mean is assumed to lie, has been calculated. There is no rule to assume in which class interval mean can be expected to lie but it should be preferably be done in the middle of the distribution and in the class interval which has the highest frequency. And we assume that the midpoint of the interval is the mean of scores in that interval. Subsequently, the correction is calculated by dividing the sum of the product of frequency and deviations by N. In Table 21.1, $C = \frac{\Sigma fx'}{N} = \frac{18}{40} = 0.45$. The length of class interval is 5. When all these have been calculated, Mean is finally calculated by formula 21.1. The entries in column 4 indicates deviation of the scores (X = Midpoint) from above and below the assumed mean. All deviations above the assumed mean are positive and all deviations below the assumed mean are negative. The entries in Column

2 has been obtained by adding the upper and lower value of the class interval and diving it by 2 such as $\dfrac{60+64}{2} = \dfrac{124}{2} = 62$.

Calculating Mean by Long method :

The arithmetic mean can also be calculated by somewhat a *longer method*. Here formula becomes

$$\overline{X} = M = \dfrac{\Sigma fx'}{N} \qquad (21.2)$$

where,
$\overline{X}$ or M = Mean
Σ = sum of
f = frequency
X = scores (Midpoint)
N = total Number of subjects or scores

The calculation of mean by long method has been illustrated in Table 21.2 where the data have been taken from Table 21.1.

Table 21.2: Calculation of Mean by long method

1	2	3	5
Class Interval	X(Midpoint)	f	fx(2 × 3)
80-89	87	2	174
80-84	82	3	246
75-79	77	4	308
70-74	74	6	444
65-69	67	5	335
60-64	62	7	434
55-59	57	3	171
50-54	52	2	104
45-49	47	6	282
40-44	42	2	84
		N = 40	Σfx = 2582

$$\bar{X} = M = \frac{\Sigma fx}{N} = \frac{2582}{40} = 64.55$$

(The answer by short method and long methods are same. The difference in fraction is due to rounding errors.)

The entries in column 1, 2 and 3 have been calculated as in Table 1. The entries in column 4 is the product of column 2 and column 3 and its total sum is $\Sigma fx = 2582$. We get mean after dividing $\Sigma fx' = 2582$ by N that comes to 64.55.

Averaging Means

Often in research the researcher gets mean of two or three or more than three groups and he wants to have average of all these means. In such situations, ordinarily there are two conditions: when N of the groups are different and when N of the groups is the same. Let us see how to average mean in these two conditions.

When N of the groups are unequal :

Suppose that the researcher wants to average the means of three groups. Group A, Group B and Group C. Their details are as under :

	Mean	N
Group A	60	20
Group B	70	10
Group C	50	30

As it is know, mean = $\frac{\Sigma X}{N}$. If the researcher has mean and number of cases, he can have the sum of scores by solving this equation :

$$\Sigma X = Mean\ (N)$$

Thus, the researcher gets the sum of scores for each group and they are added and divided by the total number of scores.

Mean	N	ΣX
60	20	1200
70	10	700
50	30	1500
	$\Sigma N = 60$	$\Sigma X = 3400$

$$\therefore M_T = \bar{X}_T = \frac{3400}{60} = 56.66$$

It should be clearly noted that the mean cannot be obtained by averaging three sample means.

When N of the groups are equal :

When N of all the groups are identical, the researcher can directly average the means of all the concerned groups for getting an overall average.

Basic Properties of Mean :

There are some important properties of mean as under :
- Mean, unlike Median and Mode, is responsive to the exact position of each score in the distribution. If a score is increased or decreased by any amount, the value of mean will reflect that change.
- The mean is more sensitive to the presence (or absence) of extreme scores in the distribution than median and mode.
- The mean is amenable to arithmetic and algebraic manipulation in a way the other measures are not. In fact, when further statistical calculation is to be done, the mean is the most preferred statistic.
- If the individual scores in the distribution are all multiplied or divided by a constant number, say K, the mean is also respectively multiplied or divided by the same number.
- If a constant number, say K, is added to or subtracted from each individual score of a sample, the mean also gets respectively increased or decreased by the same number.

There are some negative features or *drawbacks* of arithmetic mean.
- The mean cannot be calculated from open-ended class interval such as 'below 20' or 'above 80' because in such situation midpoint (X) cannot be determined.
- The mean cannot be determined by inspection nor it can be located graphically.
- The mean cannot be determined if a single observation in the series is somehow missing.
- In an extremely skewed distribution, the mean is no longer considered as the true representative of the distribution.
- The mean cannot be used when the researcher is dealing with qualitative characteristics such as beauty, honesty and so on.

Median

Median (Symbol *Mdn.*) is another important measure of central tendency. Median is defined as the *point* on the scale of scores below which 50% of the scores lie and above which 50% of the score lie. By virtue of its middle position, median characterises the central tendency of the scores. In this way, median is that point above which half of observations fall and below which half of observations fall. Median can be calculated from both ungrouped data and group data.

Calculation of Median from ungrouped data :

When data are ungrouped and median is to be calculated, ordinarily *three* types of situation arises
- When N is odd :

When N is odd, the scores are arranged in order of size and the middle score becomes the

median. For example, suppose seven students got the score on spelling test as under :
$$8, 7, 15, 6, 3, 11, 9 \ (N = 7)$$
According to equation, median is :
$$\text{Median } \frac{n+1}{2} = \frac{7+1}{2} = \frac{8}{2} = 4\text{th score}$$
After arranging the scores in order of size :
$$3, 6, 7, 8, 9, 11, 15$$
The fourth score is 8. Hence, 8 is the median.

- When N is even :

If N is even (that is, divisible by 2), the $\frac{n+1}{2}$ the score falls midway between two consecutive middle score. For example, if we drop the last score in the above example, N becomes even (N = 6) and then, median is calculated as under :
$$3, 6, 7, 8, 9, 11$$
By applying same formula, that is, $\frac{n+1}{2} = \frac{6+1}{2} = \frac{7}{2} = 3.5$th score in order of size. Hence, here median falls in between 7 and 8. Here the score at 3.5th point is 7.5, which is the midpoint of scores 7 and 8 $\left(\frac{7+8}{2} = \frac{15}{2} = 7.5\right)$. Thus median is 7.5.

The third situation which is slightly complex one arises when the median or $\frac{n+1}{2}$ th score falls within a set of identical scores in the distribution. In such situation median is calculated by the *method of interpolation*. Suppose the scores are :
$$7, 8, 10, 15, 15, 15, 16, 17, 19 \ (n = 9)$$
In this example Median lies at $\frac{n+1}{2} = \frac{9+1}{2} = 5$th score. But in counting off four score from the lowest one in the distribution, first of three identical scores, that is, 15 gests included in those of four scores. The three identical scores of the set of 15 is then assumed to occupy one unit interval ranging from 14.5 (lower limit of 15) to 15.5 (upper limit of 15) and each score occupies 1/3 = 0.33 of this interval. On counting off as only one set of these three identical scores for calculating median, the upper limit of that score is arrived at 15.5 + 0.33 = 15.88. Therefore, the median becomes 15.88.

Calculation of Median from Grouped data :

When data are grouped into frequency distribution, median can be calculated by an equation. The Median is calculated with the help of the following equation.

$$\text{Median} = l + \left(\frac{N/2 - F}{fm}\right) \times i \qquad (21.3)$$

where,

l	=	The exact lower limit of the class interval upon which median lies
N	=	Total number of frequencies
$N/2$	=	One-half of the total number of frequencies
F	=	Sum of frequencies on all intervals below l
fm	=	Number of frequencies within the interval upon which the median falls
i	=	length of class interval

Let us illustrate the calculation of median by equation method.

Table 21.3: Computing Median from grouped data by equation method

Class Interval	Frequency(f)	Cumulative frequency (cf)
20-21	2	60
18-19	4	58
16-17	6	54
14-15	3	48
12-13	4	45
10-11	12	41
8-9	6	29
6-7	8	23
4-5	6	15
2-3	9	9
	$N = 60$	

In Table 21.3, $N/2=30$, which lies in the class interval 10-11. Therefore, $l=9.5$, $F=29$, $fm=12$ and $i=2$. Accordingly, median by equation 21.3 becomes

$$\text{Mdn} = 9.5 + \left(\frac{30-29}{12}\right) \times 2$$

$$= 9.5 + (1/12) \times 2 = 9.5 + (.08) \times 2 = 9.66$$

Basic Properties of Median :

There are some basic properties or characteristics of the median as under :
- Since median is sensitive to the number of scores below it but not to how far they are below, the median is less sensitive as compared to mean to the presence of a few extreme scores.
- If the distribution is open-ended, mean cannot be calculated but calculation of median remains possible.
- If the distribution is strongly asymmetrical (that is, skewed) median is the better choice to represent the central tendency of the distribution.
- Median stands next to mean in resisting the influence of sampling distribution in ordinary circumstances.
- Median can be calculated while the researcher is dealing with qualitative characteristics that cannot be measured quantitatively but can be arranged in ascending or descending order. Examples of qualitative characteristics are High honesty, Average Honesty and Low Honesty.

Mode

The mode (Symbol : Mo) is another measure of central tendency. For ungrouped data, the mode is defined as that score (or datum), which occurs most frequently. But when the data have been grouped into frequency distribution, the mode is defined as the midpoint of the class interval containing the highest number of frequencies or cases.

In situations where the values of score (X) occur with equal frequency and where that frequency is equal to or greater than 1, mode cannot be calculated. For example, for the set of scores such as 8, 10, 12, 16, 18, 20, 19, no mode can be calculated. Likewise, for the set of scores such as 4, 4, 4, 8, 8, 8, 12, 12, 12, scores occur with a frequency of 3. In case where two adjust values of X occurs with the same frequency and which is larger than the frequency of occurrence of other values of X, the mode is defined as mean of two adjacent values of X. For example, in set of observations or scores 7, 7, 9, 9, 10, 10, 10, 10, 11, 11, 11, 11, 15, 15, 17, 18, 18, 18, the values of 10 and 11 both occur with a frequency of 4, which is greater than the frequency of occurrence of the remaining values of scores. Therefore, mode may be here calculated as midpoint of 10 and 11, that is $\frac{10+11}{2}$ = 10.5. Likewise, when two adjacent values of score occur in such a way that the frequencies of both are greater than the frequencies found in adjacent intervals, then each value of X may be taken as a mode and such set of distributions would be said to have two modes or bimodal in nature. For example, in set of observations like 8, 8, 7, 7, 9, 9, 9, 9, 9, 10, 10, 12, 12, 12, 12, 12, 12, 14, 14, the score 9 occurs five times and 12 occurs six times. Both these frequencies of occurrence are greater than the frequencies of occurrence of the adjacent values. Thus this set of observation has two modes, that is, 9 and 12.

In case of grouped data, the mode is the midpoint of the class interval with highest frequency. Thus is Table 1 the mode is the midpoint of class interval 60-64 having highest frequency. Thus mode is $\frac{60+64}{2} = 62$. In Table 3, the mode is the midpoint of class interval $\frac{10+11}{2} = \frac{21}{2} = 10.5$ because this class interval contains highest frequency.

The mode is also calculated by the following formula.

$$\text{Mode} = 3 \text{ Median} - 2 \text{ Mean} \qquad (21.4)$$

The calculation of Mode by formula 21.4 obviously requires that both mean and median must be calculated before the mode is calculated. As a consequences, this takes both time and labour.

The mode may directly be computed from a frequency distribution by using formula 21.5 which is as under:

$$\text{Mode} = X_L \left(\frac{d_1}{d_1 + d_2} \right) i \qquad (21.5)$$

where,

X_L = lower limit of the class interval that contains highest frequency (called modal class)

d_1 = Difference between the frequencies of the modal class and class interval directly below it

d_2 = Difference between the frequencies of the modal class and class interval directly above it

i = Length of the class interval

Applying formula 21.5 to Table 1, we get the following values:

$X_L = 59.5$
$d_1 = 7 - 3 = 4$
$d_2 = 7 - 5 = 42$
$i = 5$

The value 59.5 is the lower limit of the class interval 60–64, which contains the highest frequency, that is, 7. The class interval above 60–64 is 65–69 and it contains the frequency of 5 and the class interval below 60–64 is 55–59, which contains the frequency of 3. Accordingly $d_1 = 7 - 3 = 4$ and $d_2 = 7 - 5 = 2$. The length of the class interval is 5. Therefore, mode is calculated as under by formula 21.5.

$$\text{Mode} = 59.5 \left(\frac{4}{4+2} \right) 5 = 59.5 \, (.66) 5 = 59.5 \times 3.33 = 62.83$$

Basic Properties of Mode

Like mean and median, there are also some basic features of mode as under:
- Although mode is the easiest measure of central tendency, it is not very stable from sample to sample and often, more than one mode is found in the same distribution.
- Mode is of limited practical value. It does not lend itself to algebraic manipulation.
- The mode is considered as a measure of central tendency only for distributions that taper off systematically toward the extremities.
- The mode can be used as a central location for both qualitative as well as quantitative data.
- When the distribution contains two, three or more modes, it becomes difficult to interpret and compare the data.
- The mode is not suitable for further statistical calculation.

Comparison of Mean, Median and Mode

If the researcher is dealing with interval and ratio variables, the arithmetic mean is the most appropriate measure of central tendency. The calculation of mean incorporates all the particular values of the variable. On the other hand, median is the ordinal statistics. Its calculation directly rests upon the ordinal properties of the data. If the scores are arranged in order of size, median is the middle value. Its calculation does not incorporate all the particular values of the variable rather it incorporates fact of their occurrence above or below the middle value. Mode is the nominal statistics. Its calculation does not incorporate all particular values of the variable or their order. Rather it merely incorporates their frequency of occurrence.

Moreover, a good comparison of mean, median and mode may be done when all these three have been calculated from the same frequency distribution and represented graphically. When represented graphically, mean would be point on X axis that corresponds to the center of gravity of the distribution. The median would be the point on the X-axis where ordinate divides the total area into two equal-halves. Half of the areas under curve will fall to the left and half of the areas will fall on the right of the ordinate at median. The mode would be point on X-axis which corresponds to the highest point of curve. Moreover, if the plotted frequency is symmetrical, mean, median and mode will be the exact same. However, if the frequency distribution is skewed, these three measures would not coincide. If the distribution is positively skewed, mean may be greater than median, which, in turn, would be greater than mode. In case of negative skewness, the reverse may be true.

Which of these three measures of Central tendency is more appropriate?

Since the arithmetic mean is rigorously defined, easily calculated and readily amenable to algebraic treatment, it is preferred to median and mode. It also provides a good estimate of population parameter as compared to median and mode. However, when the distribution contains some atypical or extreme score, median is to be preferred. For example, consider the set of distribution of such as 5, 6, 3, 8, 4, 60. The mean after incorporating all the particular values is 86/6 = 14.33 whereas the median of the distribution 3, 4, 5, 6, 8, 60 is 5.5. Under such Circumstances, it is recommended to use a statistical procedure that is based on ordinal properties of the data in preference to the procedures which incorporates all the particular values of the variable and be grossly affected by the extreme or atypical values. Thus if the distribution shows some assymmetry (such as skewness) median is the preferred statistic because in such situation, it can always easily be interpreted as the middle value. When the distribution has emerged strictly from nominal variable, where the most frequently occurring class or value is preferred, the mode becomes the most appropriate statistic. Mode is rarely used in interval, ordinal and ratio variables where mean and medians are preferred ones.

Summary and Review

- The central tendency of the distribution is defined as the central reference value or single summary value, which is usually close to the point of the greatest concentration of scores of distribution.
- There are three measures of central tendency : Arithmetic Mean, Median and Mode.
- Arithmetic mean or simply mean is defined as the sum of separate scores divided by their the number of the scores or measurement. It is most stable and preferred measure of the central tendency. It is calculated by both short method and long method. It is most suitable to interval and ratio variable.
- Median is defined as the point on the scale of score above which lies 50% of cases and below which lies 50% of the cases. It is a preferred statistic when the researcher is dealing with ordinal variable.
- Mode is defined as the most frequent score in the ungrouped data. However, when the data have been grouped into frequency distribution, mode is defined as the midpoint of class interval containing maximum frequency. When the distribution of scores has two modes, it is called as bimodal distribution. Mode is most suited when the researcher is dealing with a nominal variable.
- Of these three measures of central tendency Mean is the most preferred statistic and mode is the least preferred one, Median comes in between these two. The reason of least preference to mode is that it is crude and unstable measure of central tendency.

Review Questions

1. What do you mean by the central tendency of distribution. Discuss the features of mean.
2. Define mean. Make difference between mean and median.
3. Make a comparative study of mean, median and mode.
4. Define median. How does it differ from mode.

5. Citing examples discuss in what situations the three measure of central tendency are most appropriate.
6. Calculate Mean, Median and Mode from the following set of data :

Class Interval	Frequency
110-119	4
100-109	8
90-109	6
80-89	2
70-79	5
60-69	18
50-59	3
40-49	4
30-39	2
20-29	6
	N = 60

7. Calculate Mean, Median and Mode from the following set of ungrouped data
60, 20, 30, 10, 40, 32, 40, 38, 40, 60, 72 (N = 11)

■ ■ ■

Chapter – 22
Measures of Variability

Learning objectives :
- Meaning of Variability
- Measure of Variability
- Range
- Semi-interquartile range or Quartile deviation
- Average deviation or Mean deviation
- Standard deviation
- Variance
- Comparing measuring of variability
- Summary and Review
- Review Questions
- Key Terms

Key Terms :

Average deviation, Range, Quartile deviation, Standard deviation, Variance, Open-ended distribution, Deviation.

The variability is one of the important concepts in statistics. Variability is understood as a quantitative measure of the degree to which scores in the distribution are spread out or clustered together. Very simply, if the scores in the distribution are all the same, then there is no variability. On the other hand, if there are small differences between scores, then the variability is small and if there are big differences between scores, then the variability is large. In the present chapter a discussion of all those methods will be done through which variability in the distribution is measured. Besides, the concept of variability and their relevance for psychological and educational researches will also be discussed.

Meaning of Variability

Measures of variability express quantitatively the extent to which scores in the distribution scatter about or cluster together. Whereas the measures of central tendency provides a summary description of the level of performance of a group, a measure of variability provides a value that reflect scatter or spread of score of separate score around their central tendency (Garrett, 1981). There are various measures of variability and usefulness of this statistic can be illustrated through an example. Suppose an intelligence test has been administered to a group of 60 girls and 60 boys. The mean score of boys is 70.5 and the mean score of girls is 70.3. In terms of means, there is no difference between boys and girls. But further suppose that the range of scores of boys is from 45 (lowest) and 80 (highest) whereas the range of scores of girls is from 65 to 75. The difference in the range shows that the boys are more variable than girls and this greater variability may be of more interest than lack of difference in mean. If there is little variability in the group, the group is said to be more or less *homogeneous* and if the group contains greater variability, the group is said *heterogeneous* one.

Measures of Variability

Ordinarily, there are five important measure of variability: *Range*, *Semi-interquartile range*, *Average deviation* or *Mean deviation*, *Standard deviation* and *Variance*. Each of these measures will be discussed in separate section.

Range

It is said that range is the rough-and-ready measure of variability. The range is defined as high score in distribution minus low score plus one. In terms of formula

$$\text{Range} = (\text{Highest score} - \text{Lowest score}) + 1 \quad (22.1)$$

For example, in a distribution if the highest score is 70 and the lowest score is 30, then range is equal to 70-30 = 40 + 1 = 41. In formula of range, 1 has been added in determining the range. The range extends from the extreme lower real limit to the extreme upper real limit. When the data have been grouped into frequency distribution, range is defined as distance between upper exact limit of the top class interval and the lower exact limit of the bottom class interval. Thus, range like all other measures of variability, is the *distance* and not, like measures of central tendency, a *location*.

Of all the measures of variability, the range is rough and the most unstable measure. The unstability of range is due to the fact and this measure varies from sample to sample. An example will illustrate why it is so. Suppose in a distribution of score, the lowest score is 40 and the highest is 100. The next score below 100 happens to be 80. By using formula 20.1, the range is formed to be (100-40)+1=61 but the 19 of the points making up this range are due to the high high score of 100. There is high probability that the next sample may not certain this high deviate score and hence, the range may be much smaller. Garrett (1981) has rightly pointed out that the range takes into account of the extremes scores only and is unreliable when N is small or when there are larger gaps in the frequency distribution.

Researchers are of view that the range as a measure of variability, can be used when the investigation is in haste and has no time to compute other measures of variability.

Semi-Interquartile Range or Quartile deviation

The Quartile deviation (Symbol Q) also known as semi-interquartile range, is defined as the half of distance between Q_1 (25th percentile) and Q_3 (75th percentile). In terms of equation, it is written as :

$$Q = \frac{Q_3 - Q_1}{2} \quad (22.2)$$

where,
Q = Quartile deviation
Q_1 = first quartile on score scale or 25th percentile
Q_3 = third quartile on the score scale or 75th percentile

The reality is that the quartile points are the three score points, which divide the distribution into four parts, each containing an equal number of cases. These points are symbolized as Q_1, Q_2 and Q_3, which are equivalent to P_{25}, P_{50} and P_{75} respectively. The difference between Q_1 and Q_3 is called as *interquartile range* and when this difference is divided by 2, it becomes *semi-interquartile range*.

Formula no. 22.2 shows that for calculating Q, it is first necessary to calculate Q_1 and Q_2. Q_1 can be calculated by the formula 22.3 and Q_2 by formula 22.4.

$$Q_1 = 1 + i \left(\frac{N_4 - Cumf_1}{fq} \right) \quad (22.3)$$

$$Q_3 = 1 + i \left(\frac{3N/4 - Cumf_1}{fq} \right) \quad (22.4)$$

where,
l = exact lower limit of the interval in which the quartile falls
i = length of class interval
$cumf_i$ = Cumulative frequency up to the interval which contains the quartile
f_q = frequency of the class interval containing the quartile.

In Table 22.1, calculation of Q has been shown. In Table 22.1, the scores on Numerical reasoning test of 100 students have been grouped into frequency distribution and Q has been calculated.

Table 22.1: The calculation of Q from scores of 100 students on Numerical reasoning test

Class Interval	Frequency (f)
90-99	2
80-89	6
70-79	10
60-69	20
50-59	30
40-49	15
30-39	8
20-29	6
10-19	3
	N = 100

$N/4 = \dfrac{100}{4} = 25$

$\dfrac{3N}{4} = \dfrac{300}{4} = 75$

For Q_1,

$l = 39.5$, the exact lower limit of the class interval that contains Q_1 or $N/4 = \dfrac{100}{4} = 25$. This is reached by counting off 25 scores from the lowest interval of the distribution which is, here, 10-19. Thus 3+6+8+15 = 32 in which 25 lies and 15 belongs to class interval 40-49, the lower limit of which is 39.5.

$cumf_i$ = 17, cumulated frequency upto the interval containing Q_1
fq = 15, the frequency of the interval on which Q_1 falls
i = 10, the length of class interval

Accordingly, Q_1 will be:

$$Q_1 = 39.5 + 10 \ \dfrac{25-17}{15}$$

$$= 39.5 + 10\left(\dfrac{8}{15}\right) = 39.5 + 10(0.53)$$

$$= 39.5 + 5.33 = 44.83$$

For Q_3,
l = 59.5, the exact lower limit of the class interval that contains Q_3 or $3N/4 = \frac{100 \times 3}{4} = 75$

$cumf_i$ = 62, the cumulated frequency upto the interval containing Q_3
f_q = 20, the frequency of the interval on which Q_3 falls
i = 10, the length of class interval

Accordingly, Q_3 will be :

$Q_3 = 59.5 + 10 \frac{75 - 62}{20}$

$= 59.5 + 10 \frac{13}{20}$

$= 59.5 + 10 (1.43) = 59.5 + 0.6 = 60.1$

$\therefore Q = \frac{Q_3 + Q_1}{2} = \frac{60.1 - 44.83}{2} = 7.6$

Since quartile deviation is associated with the median, it follows that whenever median is used as a measure of central tendency, the Q is the most appropriate measure of variability. It must be noted that median is the statistic to be used as a measure of central tendency when the distribution is *skewed*.

Since Q focuses only on the middle 50% of the distribution, it is less likely to be influenced by extreme scores and therefore, is considered as a better and more stable measure of variability than the range. However, Q does not take into account the actual distances between the individual scores and therefore, it does not give a complete picture of how scattered or clustered the scores are. Like range, Q is also considered as a crude measure of variability. Moreover, since Q ignores 25 per cent of the data at the upper end and 25 per cent of data at the lower end, it is not regarded as a reliable measure of variability. Q is also not affected by scores other than Q_3 and Q_1. It gives no idea about the distribution and variability of other scores both within the interquartile rage and beyond it.

Average Deviation or Mean Deviation

Another measure of variability is the average deviation (AD) or Mean eviation (MD) as it is sometimes called. The average deviation is defined as the mean of deviations of all of the separate scores in the series taken from mean (and occasionally from median and mode). In other words, AD is simply the arithmetic mean of the absolute values of the deviations from a measure of central tendency. By absolute value is meant the size of a value or quantity taken together ignoring it signs. Although AD can be calculated from any measure of central tendency, in practice, deviations are mearly always measured from mean or from the median. Since the sum of deviations from mean is equal to zero, it follows that we can obtain no average deviation unless the procedure is changed. Therefore, in actual practice, all the deviations from mean is summed *disregarding* the signs. In other words, in averaging deviations to find the AD, no account of sings is taken into consideration and all the signs whether plus or minus are taken to be positive ones.

AD is calculated from ungrouped data, by the following formula :

$$AD = \frac{\Sigma |x|}{N} \quad (22.5)$$

when,

AD = Average deviation
x = Deviation of a score from mean
| | = Bars enclosing x indicates that the signs have been ignored.

On the other hand, when data have been grouped into frequency distribution, AD is calculated by the following formula

$$AD = \frac{\Sigma |f_x|}{N} \quad (22.6)$$

where,

fx = Deviation multiplied by the frequency of the interval concerned

The rest of the symbols are defined as it has been done in formula 22.5.

An example will make clear how to compute AD. Suppose there are a set of five scores as under :

Scores(X)	Deviation(X–M) = x
10	–10
15	–5
20	0
30	+10
25	+5
$\Sigma X = 100$	$\Sigma x = 30$ (Disregarding signs)

$$\text{Mean} = \frac{\Sigma X}{M} = \frac{100}{5} = 20$$

$$AD = \frac{\Sigma |X|}{N} = \frac{30}{5} = 6$$

The above example illustrates the calculation of AD from ungrouped data. The mean of a set of five scores has been calculated and it is 20. Subsequently, deviation of each score from mean that is X–M has been calculated and these deviations have been summed by ignoring the signs called $\Sigma |X|$.

If the sum of plus signs and the sum of minus signs differ by as much as N/2, recompute it until the errors are located. This comes to be 30, which has been divided by N, that is, 5 and then, AD = 6.

The next example illustrates calculation of AD from a frequency distribution. Table 22.2 shows the calculation of AD from grouped data.

Table 22.2: Calculation of AD from a frequency distribution

1	2	3	4	5
Class Interval	Midpoint(X)	f	x^i	fx'
80-84	82	2	+4	8
75-79	77	4	+3	12
70-74	72	6	+2	12
65-69	67	5	+1	5
				+37
60-64	62	16	0	0
55-59	57	5	−1	−5
50-54	52	3	−2	−6
45-49	47	5	−3	−15
40-44	42	4	−4	−16
		N = 50		$\Sigma fx' = -42$
				$\Sigma fx' = 37-42 = -5$

Mean = AM + Ci

$AM = 62, C = \dfrac{\Sigma fx'}{N} = \dfrac{-5}{50} = -.1; i = 5$

∴ Mean = 62 + (−.1)(5) = 62+(−.5) = 62 − .5 = 61.5

After calculating mean, now AD will be calculated as under

Class Interval	Midpoint(X)	f	Deviation (X − M) or x	fx
80-84	82	2	20.5	41
75-79	73	4	15.5	62
70-74	72	6	10.5	63
65-69	67	5	5.5	27.5

60-64	62	16	.5	8
55-59	57	5	−4.5	−22.5
50-54	52	3	−9.5	−28.5
45-49	47	5	−14.5	−72.5
40-44	42	4	−19.5	−78
		N = 50		Σfx=403

(*Ignoring the sign*)

$$\text{By formula 22.6, AD} = \frac{|\Sigma fx|}{N} = \frac{403}{50} = 8.06$$

In computation of deviation from mean, it is always important to remember that the mean is always subtracted from the midpoint (X). Thus $X-M$ equals x (deviation). Here the computation is algebraic, which means that plus and minus signs are recorded. When the midpoint or X is numerically greater than mean, the value of X will be plus and when numerically less than the mean, the value of X will be minus.

As we know, when the distribution is normal, the AD includes middle 57.50 of cases, that is, 28.75 of cases fall above and 28.75 of cases fall below the mean. Therefore, AD is always somewhat larger than the Quartile deviation, which includes the middle 50% of cases.

Average deviation (*AD*) can also be calculated from median and then, its formula is modified as under :

$$AD \text{ from Median} = \frac{\Sigma |X - Mdn|}{N} \qquad (22.7)$$

When, the symbol | | means the 'absolute value of' which means that signs are ignored completely and all deviations are added together. Thus in a distribution of 5, 2, 9, 6, 3, 8, 7, the median is 6 and deviations from median are (5−6) = −1, (2−6) = −4, (9−6) = 3, (6−6) = −0, (3−6) = −3, (8−6) = 2 and 7−6 = 1. The sum of these deviations ignoring sign is 14 and if divided by N = 7, we get 2 as AD. Since AD is a minimum (and hence, a unique value) when taken from median, AD from median is slightly preferable to AD from mean.

At a glance, AD appears to be an attractive measure of variability. But it has proven to have a mathematical intractability that severely limits its use. For example, it is of no use in statistical inference. It has seldom been used in research in the past 50 years. AD cannot be a satisfactory measure when taken about mode or while dealing with a skewed distribution. Moreover, due to disregard of signs in finding the AD, all negative deviations have the same influence upon amount of variation as positive. Consequently, it results in a nonalgebraic quantity, which makes it unpopular in mathematical operations and is left with very limited use in statistical theory.

Standard Deviation

The standard deviation symbolized by Greek letter sigma (s) is one of the most popular and depedendable measures of variability. Another name for standard deviation is *root mean square deviation*, which is seldom used today but may be encountered in the older literature. The standard deviation uses the mean of the distribution as a reference point and measures variability by considering the deviation or distance between each score and the mean. Obviously, it determines whether the scores are generally near or far from the mean. In other words, the standard deviation approximates the average distance from the mean. Therefore, roughly speaking, the standard deviation is the average about that scores differ from the means no doubt, but it is not simply arithmetic mean.

There is clear distinction between Standard deviation and Average deviation. In computing AD, the signs of plus and minus are ignored, whereas in computing SD difficulty associated with signs is avoided by squaring the each deviations. In computing AD the deviation may also be taken from median or mode apart from mean but in case of SD, the deviation is always taken from mean.

The standard deviation (SD) can be computed from both *ungrouped data* and *grouped data*. Let us see how it is done.

Calculation of SD from ungrouped data

When the data are ungrouped, the SD can be calculated by the following formula.

$$\sigma = \sqrt{\frac{\Sigma X^2}{N}} \quad (22.8)$$

$$\sigma = \sqrt{\frac{\Sigma X^2}{N-1}} \quad (22.9)$$

In both formulas 22.8 and 22.9, X means deviation from mean (that is, $X - \bar{X}$) and N = size of sample. When N is equal to 30 or greater than 30, formula 22.8 is applied but when N is small, that is, less than 30, formula 22.9 is applied. The N–1 in the formula 22.9 is used to make a correction for bias. The reality is that the sample standard deviation is a biased estimate of population standard deviation and this bias is downward one. In other words, the sample standard deviation tends to be smaller than the population standard deviation. Hence, the sample standard deviation always be corrected for bias. Such correction becomes ideal when N < 30. Let us take an example: when N is 8, and the researcher is to divide SX^2 by N, it makes a difference in the value of SD if he divides it by 8 or 7. But if N is 200 whether the researcher divides SX^2 by 199 or 200, it makes a trivial difference. Table 22.3 presents the calculation of SD from ungrouped data.

Table 22.3: Calculation of SD from ungrouped data

X	x(X–M)	x²
40	+9	81
30	–1	01
20	–11	121
18	–13	169
12	–19	361
40	+9	81
10	–21	441
35	+4	16
55	+24	576
50	+19	361
ΣX = 310	Σx = +65–65 = 0	Σx² = 2208

N = 10

$$\text{Mean} = \frac{310}{10} = 31$$

$$\sigma = \sqrt{\frac{\Sigma x^2}{N-1}} = \sqrt{\frac{2208}{10-1}} = \sqrt{\frac{2208}{9}} = \sqrt{245.33} = 15.66$$

The computation of SD from the above formula (formula 22.9) incorporates steps as under:

Step 1 : Add the scores and divide it by N in order to find Mean
Step 2 : Find deviation of each score from Mean which equals X– or X–M.
Step 3 : Square each deviations to find on x^2
Step 4 : Sum the square deviations to find out Sx^2
Step 5 : Divide this sum by N–1 to find out $Sx^2/N–1$
Step 6 : Find out the positive square root of the result obtained in step 5 and the outcome is the desired value of SD.

There is also a *raw score formula* for computation of SD. When a calculating machine is available, it is much easier to use the raw score formula for computing the standard deviation. Here the formula for calculating SD is :

$$SD = \frac{1}{N}\sqrt{N\Sigma X^2 - (\Sigma X)^2} \qquad (22.10)$$

Here, N = Number of scores and X is the original score.

If we apply formula 22.10 to the data shown in Table 22.3, SD would be calculated as shown in Table 22.4.

Table 22.4: Computation of SD from original scores by raw-score formula

1 Scores (X)	2 X²
40	1600
30	900
20	400
18	324
12	144
40	1600
10	100
35	1225
55	3025
50	2500
ΣX = 310	ΣX² = 11818

$$\sigma = \frac{1}{N}\sqrt{N\Sigma X^2 - (\Sigma X)^2}$$

$$= \frac{1}{10}\sqrt{10(11818) - (310)^2}$$

$$= \frac{1}{10}\sqrt{118180 - 96100}$$

$$= \frac{1}{10}\sqrt{22080} = 1/10 \times 148.593 = 14.859 = 15.00$$

(The difference in real value of SD of the same data by two formulas is due to rounding error. (cf table 22.3 and Table 22.4)

There is still another way of calculating Standard deviation from original scores (ungrouped data) in which the researcher takes assumed mean at zero. This is illustrated in Table 22.5.

Table 22.5: Calculation of Standard deviation from original scores (ungrouped data) when assumed mean is taken at zero

X	x'	x'²
40	40	1600
30	30	900
20	20	400
18	18	324
12	12	144
40	40	1600
10	10	100
35	35	1225
55	55	3025
50	50	2500
$\Sigma X = 310$	$\Sigma x' = 310$	$\Sigma x'^2 = 11818$

Assumed Mean $(AM) = 0$

Actual Mean $= \dfrac{\Sigma X}{N} = \dfrac{310}{10} = 31$

Correction or $C = 31 - 0 = 31$ $\quad C^2 = 961$

$$\sigma = \sqrt{\dfrac{\Sigma X'^2}{N} - C^2} = \sqrt{\dfrac{11818}{10} - 961} = \sqrt{1181.8 - 961} = \sqrt{220.8} = 14.859 = 15.00$$

Again, we get more less same value of Standard deviation that was found in Table 4. Minor difference in value of standard deviation is due to rounding errors. In Table 5 entries in column 2 has been obtained by taking deviation from assumed mean which is zero. Thus $X-AM = 40-0 = 40$ and so on other entries have been obtained.

Computation of SD from grouped data :

When data have been grouped into frequency distribution, the SD is computed by two methods–*long method* and *short method* or *coded method*. The computation of SD by long method is done by the following formula

$$\sigma = \sqrt{\dfrac{\Sigma f X^2}{N}} \qquad (22.11)$$

where,

Σfx^2 = sum of the square deviation of each class interval multiplied by their respective frequency

N = Number of cases in sample

As an example, data of Table 22.2 is being presented in Table 22.6 for computation of SD by formula 22.11

Table 22.6: Computation of SD by long method

1	2	3	4	5	6
Class Interval	Midpoint X	f	Deviation X	fx	fx^2
80-84	82	2	20.5	41	840.5
75-79	77	4	15.5	62	961
70-74	72	6	10.5	63	661.5
65-69	67	5	5.5	27.5	151.25
60-64	62	16	.5	8	4
55-59	57	5	−4.5	−22.5	101.25
50-54	52	3	−9.5	−28.5	270.75
45-49	47	5	−14.5	−72.5	1051.25
40-44	42	4	−19.5	−78	1521
		$N = 50$			$\Sigma fx^2 = 5562.5$

$$\text{Mean} = 61.5; \quad SD = \sqrt{\frac{\Sigma fx^2}{N}} = \sqrt{\frac{5562.5}{50}} = \sqrt{111.25} = 10.55$$

The entries in column 6 has been calculated by multiplying entries in column 4 and 5, that is, $(X)(fx)$. The entries in the remaining columns are as is Table 22.2.

The SD can also be calculated by short method or coded method from the frequency distribution. In such case, the formula for computation of SD is as under

$$\sigma = i \sqrt{\frac{\Sigma fx'^2}{N} - C^2} \quad (22.12)$$

where,
i = length of class interval
c = correction
fx' = deviation from assumed mean multiplied by frequency of the respective class interval
N = number of cases or scores

To illustrate the computation of SD by formula 22.12, we are taking the data of Table 22.2 and this is being presented in Table 22.7.

Table 22.7: Computation of SD by short method

Class Interval	Midpoint X	f	x'	fx'	fx'²
80-84	82	2	4	8	32
75-79	77	4	3	12	36
70-74	72	6	2	12	24
65-69	67	5	1	5	5
60-64	62	16	0	0(+37)	0
55-59	57	5	−1	−5	5
50-54	52	3	−2	−6	12
45-49	47	5	−3	−5	45
40-44	42	4	−4	−16(−32)	64
		N = 50		Σfx' = −5	Σfx'² = 223

$$c = \frac{\Sigma fx'}{N} = \frac{-5}{50} = -.1$$

$$i = 5$$

$$\therefore \sigma = \sqrt{\frac{\Sigma fx'^2}{N} - c^2} = 5\sqrt{\frac{223}{50} - (-.1)^2}$$

$$= 5\sqrt{4.46 - .01} = 5\sqrt{4.45} = 5 \times 2.11 = 10.55$$

In statistics there are two such terms which are closely related to SD. They are: *sum of squares* and *variance*. The sum of squares of a series is the sum of deviations (taken from mean) squared. In fact, sum of squared deviations is called as sum of squares in brief. The variance is the sum of squares divided by N. More simply, variance is SD². Accordingly, SD is the positive square root of variance. (Any number has both positive and negative square root. For example, the square root of 9 is both +3 and −3).

Major features or characteristics of SD :

There are some important characteristics of SD, which a student must know. They are as under
- Like mean, the standard deviation is responsive to the exact position of every score in the distribution. If the researcher shifts any score, which is more deviant from mean, the standard deviation will be larger than before. Likewise, if the shift is to a position much close to mean, the standard deviation is likely to be reduced.

- The standard deviation is very much sensitive to the presence or absence of extreme scores in the distribution. In fact, the standard deviation, in such situation, is affected more than semi-interquartile range or Q. Due to this reason, the standard deviation may not be the best choice among the measures of variability when the distribution contains a few very extreme scores or when the distribution is badly skewed.
- The standard deviation shows adequate resistance to its sampling fluctuation. It means that if repeated random samples are drawn from the same population, the exact numerical value of the standard deviation would tend to vary less than the numerical values of other measures of variability computed on the same samples.
- In general, the properties of standard deviation are related to those of mean.
- If all scores have identical value in the distribution, the standard deviation will be zero.
- Addition or subtraction of a constant number to or from each individual score in the distribution, does not make any difference in the value of standard deviation but the multiplication or division of each score by a constant number produces an identical change in the value of standard deviation.
- In a normal distribution the quartile deviation and standard deviation have a constant relationship. $Q = 0.6745\sigma$
- The more homogeneous the scores are, the smaller will be standard deviation and conversely, the more heterogeneous the scores are, the larger would be the value of standard deviation.

Variance

The variance is another measure of variability though not directly used as a measure of dispersion. The variance is defined as the sum of squared deviations from the mean divided by N. The symbol for variance is σ^2 or SD2. Accordingly, variance is also defined as standard deviation squared. The sum of squared deviations, in brief, called as sum of squares has its own symbol Ss. The formula for variance is :

$$\sigma^2 = \frac{\Sigma(X - \bar{X})^2}{N} \qquad (22.13)$$

where,

σ^2 = variance
X = score
$\bar{X}$ = mean of scores
X – M = deviation of score from mean
N = number of scores

In terms of symbol of the sum of squared deviations (that is, SS), the variance formula can be written as under:

$$\sigma^2 = \frac{SS}{N} \quad (22.14)$$

In formula 22.14 SS replaces the $\Sigma(X-M)^2$ of formula 22.13.

Let us take an example to illustrate the computation of σ^2 from a set of 5 scores

X	$X - \bar{X}$	$(X - \bar{X})^2$
10	−8	64
20	+2	4
30	+12	144
14	−4	16
16	−2	144
$\Sigma X = 90$		$\Sigma(X-M)^2 = 232$

$$\bar{X} \text{ or Mean} = \frac{\Sigma X}{N} = \frac{18}{5} = 18$$

$$\sigma^2 = \frac{\Sigma(X-\bar{X})^2}{N} = \frac{232}{5} = 46.4$$

Since Standard deviation (σ) is the positive square root of σ^2, here $\sigma = \sqrt{46.4} = 6.81$

Variance is not ordinarily used as a direct measure of variability but it is used in some other ways such as it is basic to analysis of variance (Kurtz & Mayo, 1980). Thus variance is a statistic, which finds its greatest use in inferential statistics. However, at descriptive level, it has a flaw. Since its calculated value is expressed in terms of squared units of measurements, it is little used in descriptive statistics. However, this defect is remedied by taking the square root of the variance and thus, returning to a measure in the original units of measurement.

Comparing Measures of Variability

We have discussed several measures of variability such as range, Q, SD, etc. In the present section a comparative study of these various measures would be undertaken. Honestly speaking, there are two primary considerations that tend to determine the value of any statistical measurement.

1) The measure must provide a stable and reliable description of the scores. Specifically, the measure should not be greatly affected by minor details in the set of distribution.
2) The measure should have a consistent and predictable relationship with other statistical measurements.

The first consideration throws light upon the factors that affect variability and the second consideration throws light upon the relationship with other statistical measures. These two are being discussed below

1. Factors affecting variability :

A review of literature shows that there are four factors that do influence the variability. They are: extreme score, size of the sample, stability under sampling and open-ended distribution.

- **Extreme scores :**

Range is highly affected by the extreme score. In fact, range is determined by only two extreme scores—highest score and lowest score in the distribution. Therefore, any extreme score profoundly affect the range. Standard deviation and variance are also influenced by the presence or absence of extreme score. Since these two measures are based upon the squared deviations, a single extreme score can have a profound impact. For example, a score that is 15 points away from mean will contribute to $15^2 = 225$ points to the sum of squared deviation. For this reason, variance and standard deviation should be treated very carefully in such situation where there exists some extreme score. However, quartile deviation or Q focuses upon the middle of the distribution and therefore, it is least affected by the extreme score. Therefore, Q is the best choice as a measure of variability whenever the distribution is skewed.

- **Size of the sample :**

Range is readily influence by the size of the sample because if the researcher increases the number of individual in the sample, it also tend to increase the range because each additional score has the potentiality to replace the higher or lowest score. However, SD, Q, and Variance are relatively unaffected by size of the sample and therefore, they provide better measure of variability.

- **Stability under sampling :**

When several different samples are taken from the same population, it is expected that they should be similar due to some family resemblance. When standard deviation and variance are used to assess variability, these samples will tend to have more or less similar variability and thus, they are said to be stable under sampling. Quartile deviation also provides a reasonable stable measure of variability. But range is likely to change from sample to sample and thus, is unstable under sampling.

- **Open-ended distributions :**

When the nature of distribution of scores is such that there is no specific boundary for the highest and lowest score, it is called *open-ended distribution.* Such distribution usually results when there is possibility of infinite or undetermined scores. For example, when a subject who cannot solve the problem, has taken an undermined or infinite amount of time to arrive at a solution. In open-ended distribution, the standard deviation and the variance cannot be calculated. However, in such case Q is the only appropriate and best measure of variability.

2. Relationship with other statistical measures :

The standard deviation and variance have a direct relation to mean. In fact, they are based on deviations from mean. As a consequence, mean and standard deviation tend to be reported together. Since mean is the most common and stable measure of central tendency, the standard deviation is the most common and stable measure of variability. Likewise, quartile deviation and median both are based upon percentiles and therefore, they share a common foundation and tend to be associated.

In case the researcher is using median as a measure of central tendency, the appropriate choice as a measure of variability would be Quartile deviation. The range has no direct relationship to any statistical measure. For this reason, range is rately used in relation to other statistics.

Moment About the Mean

The mean and standard deviation are very closely related to a family of descriptive statistic known as moments. In fact, the term moment originates in mechanics. It refers to a measure of force with respect to its tendency to provide rotation. The strength of the tendency depends on two factors—amount of the force and distance from the origin of the point at which force is applied. If a number of forces such as F_1, F_2, F_3 at distances X_1, X_2 and X_3 are applied, the moment of the first force about the origin is $F_1 X_1$, the momen of the second force is $F_2 X_2$ and so on. These moments are additive in nature so that SFX is the total moment about the origin. If the total moment is divided by the total force, the quotient is termed as 'moment'. The whole thing can still better be explained with the example of a lever supported by fulcrum. If a force f_1 is applied to the lever at a distance x_1 from the origin, then $f_1 x_1$ is called the moment of the force. Likewise, if a second force f_2 is applied at a distance x_2, the total moment is $f_1 x_1 + f_2 x_2$. If distance x is squared, it is called second moment, and if the distance is cubed, we get a third moment and so on. In case of frequency distributions, the origin is the analog of fulcrum and the frequencies in various class intervals are analogous to forces operating at various distances from the origin. The first four moments about the arithmetic mean are as under :

$$m_1 = \frac{\Sigma(X - \bar{X})}{N} = 0 \quad (22.14)$$

$$m_2 = \frac{\Sigma(X - \bar{X})^2}{N} = \sigma^2 \quad (22.15)$$

$$m_3 = \frac{\Sigma(X - \bar{X})^3}{N} = \frac{\Sigma x^3}{N} \quad (22.16)$$

$$m_4 = \frac{\Sigma(X - \bar{X})^4}{N} = \frac{\Sigma x^4}{N} \quad (22.17)$$

In general, rth moment about the mean is given by the following formula

$$m_r = \frac{\Sigma(X - \bar{X})^r}{N} \quad ; r = 0, 1, 2, 3, 4 \ldots \quad (22.18)$$

With the help of these moments, we can measure the central tendency of a set of observations, their variability, their symmetry and peakedness. The points to be remembered are

- The first moment about origin measures *mean*.
- The second moment about the mean measures *variance*.
- The third moment about the mean measures *skewness*.
- The fourth moment about the mean measures *kurtosis*.

The first moment (m_1) about the mean is computed from sum of deviations raised to power 1 and amounts to zero for all symmetric or asymmetric distributions. A we know, as the algebraic sum of deviations of a given set of observations from their mean is zero, the first moment about mean is always zero. The second moment about the mean is the arithmetic average of the sum of squared deviations from mean and is idential with variance. The third moment about mean is calculated to obtain skewness in the distribution and the fourth moment is calculated to obtain a measure of peakedness or kurtosis in the distribution.

Summary and Review

- The measures of variability express quantitatively the extent to which scores in the distribution scatter about or cluster together.
- There are five measures of variability: range, quartile deviation, average deviation, standard deviation and variance.
- The range is the most roughest measure of variability. It reflects difference between high score and low score in the distribution.
- The quartile deviation is based upon the middle 50% of the distribution. It is defined as one half distance between Q_1 (25th percentile) and Q_3 (75th percentile).
- The average deviation or also called as mean deviation is the arithmetic mean of the deviations of all of the separate scores in the series from mean (sometimes taken from median and mode). All the deviations from mean are summed disregarding the signs. It is not in much use.
- The standard deviation is the most stable measure of variability and is directly related to mean. The standard deviation is defined as the square root of the arithmetic mean of the squared deviations of the scores from their mean. That is why, it is called as *root-mean-square deviation*.
- Variance is another measure of variability but it is commonly not used directly. It is very common statistic in analysis of variance. It is defined as standard deviation squared (s^2) and accordingly, the standard deviation is defined as the positive square root of variance.
- When mean is the preferred statistic as a measure of central tendency, the standard deviation should be the preferred measure of variability. When median is used as the measure of central tendency, quartile deviation should be the preferred measure of variability. In case of open-ended distribution, median rather than mean is the appropriate statistic as a measure of central tendency and quartile deviation is the appropriate measure of variability.

- The concept of moment is related mean and standard deviation. There are four common moments about mean. The first moment about origin measures mean, the second moment about mean measures variance, third moment about mean measures skewness and fourth moment about mean measures kurtosis.

Review Questions

1. What is meant by the measures of variability? Discuss some important measures of variability.
2. Citing examples, make a comparative study of average deviation and standard deviation.
3. Define standard deviation. Discuss its major features.
4. Point out the major features of quartile deviation.
5. Citing example, illustrate the relationship between standard deviation and variance.
6. Find out the average deviation, quartile deviation and standard deviation from the following set of data.

Class Interval	Frequency
90-94	2
85-89	3
80-84	5
75-79	2
70-74	16
65-69	2
60-64	2
55-59	4
50-54	1
45-49	3
	N = 40

7. Define the concept of Moment. Discuss the significance of first four moments used in description statistic.
8. Calculate first moment and second moment from the following set of scores :
   8, 10, 16, 12, 14, 10, 2, 8 (N = 8)

Chapter – 23

Normal Curve

Learning objectives:
- Meaning of Normal Curve
- Historical background of Normal Curve
- Equation of Normal Curve
- Properties of Normal Curve
- Normal Curve as a statistical model
- Applications of Normal Curve
- Measuring divergence from normal distribution
- Why do frequency distributions deviate from Normal Curve?
- Summary and Review
- Review Questions
- Key Terms

Key Terms:

Normal curve, Normal distribution, Skewness, Positive Skewness, Negative Skewness, Kurtosis, Leptokurtic distribution, Platykurtic distribution, Mesocurtic distribution, Floor effect, Ceiling effect.

The normal curve is a very important concept in statistics and serves many functions in statistical procedures. It is now high time to know it more closely. We need to know what a normal curve is, what are its features, how it is useful as a statistical model and how to put it to work in answering some questions. A historical consideration of normal curve is due to be discussed in the present chapter.

Meaning of Normal Curve

A frequency curve, which is symmetrical and bell-shaped is called as normal probability curve or normal curve. The normal curve represents a normal distribution of many physical, psychological and biological measurements are observed to approximate normal distribution. What happens in a normal distribution? In normal distribution scores or measures are concentrated closely around the center and taper off from this central high point or crest to the left and the right.

There are relatively few scores at low end of the scale, the increasing number up to maximum at the middle position and a progressive falling-off toward the high end of the scale. If the area under the curve is marked by a line drawn perpendicularly through central high point to the base line, resulting two parts of curve will be similar in shape and very nearly equal in area. Thus a perfectly symmetrical curve or the *bell-shaped curve* emerges and is called as *normal probability curve* or *normal curve*. A normal curve has been presented in figure 23.1

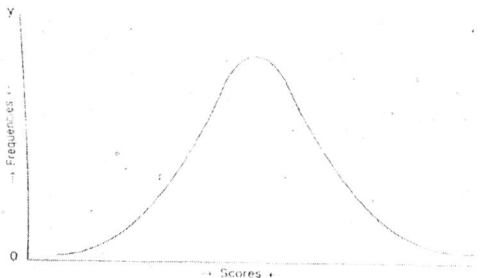

Figure 23.1 Normal Distribution Curve.

Historical Background of Normal Curve

In the eighteenth century gamblers were interested in studying the chances of beating various gambling games and they requested some mathematicians to help them out. In 1733 Abraham De Moivre was first to develop a mathematical equation of the normal curve. Later some significant development occurred in the beginning of the 19th century. Indeed, it seems that Pierre-Simon (the Marquis de Laplace) and Carl Friedrich Gauss tried to rediscover the normal curve independently of De Moivre's work. Gauss's primary interest was associated with problems in astronomy such as the determination of the orbit of a planet from a number of measurements or observations, which are subject to error. These problems led to the consideration of a *theory of errors of observation* and it became so popular that at that time normal curve came to be known as the *normal law of error* (This name is now obsolete). Due to the significant contribution done by Gauss, this curve is also known as *Gaussian Curve*. In the middle part of the 19th century Adolphe Quetelet, a Belgian researcher, promoted the applicability of normal curve. He was basically a teacher of mathematics and astronomy and originator of the basic method of physical anthropometrics. He was of view that the normal curve could be better extended to apply to problems in meteorology, anthropology and human affairs. It was his viewpoints that mental and moral traits when measured would conform to the normal distribution.

In the latter part of the 19th century Sir Francis Galton did champion work in the field of individual differences, a domain considered very important in the field of psychology and education. He also reported that many mental and physical traits conformed to the normal curve and was greatly impressed by the applicability of normal curve to the various natural phenomena. Today normal curve is variously referred to as the *bell-shaped Curve*, the *Gaussian Curve*, the *De Moivre's Curve*, the *Curve of error*.

Equation of Normal Curve

The general equation that describes mathematically the normal curve is often expressed as under:

$$Y = \frac{N}{\sigma\sqrt{2\pi}} e^{\frac{-x^2}{2\sigma^2}} \quad (23.1)$$

Y = height of the curve above X-asix, that is, frequency of a given x-value
N = number of observation or cases
σ = Standard deviation of the distribution
x = scores (expressed as deviation from the Mean) laid off along the base line or X axis.
π = 3.1416 (the ratio of the circumstance of a circle to its diameter). It is constant
e = 2.7183 (base of Napierian system of logarithms). This is also constant.

The close look at formula 23.1 suggests some features of the normal curve. Since deviation that is, x is squared, either positive or negative values of it have the same effect and yield the same value of Y. That is the reasons why the curve is bilaterally symmetrical. As x increases and because x^2 has a negative sign, the exponent of e decreases and therefore, Y also decreases. Y is at maximum when x is equal to zero, for that is when the exponent of e is a maximum. All the measures of the central tendency, that is, mean, median and mode are zero.

By using formula 23.1 and substituting appropriate values of different points along the base line, the researcher can solve for the ordinates of these points and then, draw the curve from the highest of these ordinates. In doing so, the researcher should take enough points along the base line to cover the range adequately. Suppose the researcher selects the first point at mean. For the Mean, the deviation value, that is, x is equal to zero. Since x is equal to 0, the exponent of e is also 0 and quantity raised to 0 power is equal to 1. Then Formula 23.1 becomes:

$$Y = \frac{1}{\sqrt{2\pi}} = \frac{1}{\sqrt{2(3.1416)}} = \frac{1}{\sqrt{6.2832}} = \frac{1}{\sqrt{2.5066}} = .3989$$

Other values can then be taken for x, the ordinates can be computed and subsequently, normal curve could be draw using Y value. But these calculations are rarely done because Tables are available from which information may be readily obtained.

Properties of Normal Curve

There are some obvious characteristics or properties of normal curve as under:
- The curve is symmetrical, that is, 50% cases lie left to mean and the remaining 50% lie right to mean (see Figure 23.2)

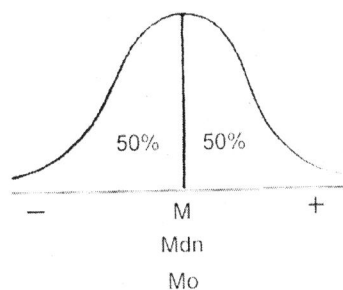

Figure 23.2 Normal curve showing location of Mean, Median and Mode as well as Symmetrical two halves.

- In a normal curve mean, median and mode coincide.
- The normal curve is asymptotic. It comes near the X axis but does not meet X-axis and tends from minus infinity to plus infinity. However, for practical purposes, the curve may be taken to end at points $+3\sigma$ to -3σ sigma scores from mean (Garrett, 1981)
- The maximum height of ordinate of curve occurs at mean, that is, where $z = 0$ and in the unit of normal curve, it is equal to .3989.
- About 68.26 percent area of curve falls at ± 1 standard deviation unit from the mean, about 95.44 percent area falls within $\pm 2\sigma$ and about 99.73 falls within $\pm 3\sigma$ and beyond $\pm 3\sigma$, only 0.28% of the cases lie (see Figure 23.3).
- In the unit normal curve, the limits $\sigma = \pm 1.96$ include 95% and the limits ± 2.58 include 99% of the total area of curve and 1% the area respectively falls beyond these limits (see Figure 23.3).
- The points of inflection of the curve occur at points plus or minus one standard deviation unit above or below the mean. Thus the curve changes from convex to concave in relation to the horizontal axis at these points under study.
- Since normal distribution is bilaterally symmetrical, its coefficient of skewness is always zero.

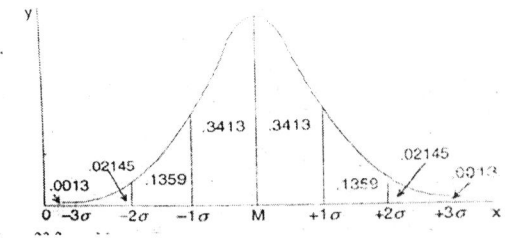

Figure 23.3 Normal curve showing areas at difference sigma score distances from the Mean.

(495)

Normal curve as a Statistical Model

As a statistical model, normal curve provides an ideal distribution of populations of observations. Researchers are of view that the normal curve appears to be a rather good fit to many (but not all) distributions of measurements done in psychology, education, biology, anthropology, etc. For example, several measures of mental traits (such as intelligence) tend to be normally distributed. Mathematical astronomers have reported that the normal curve fits the errors of observations in many situations. In homogeneous populations, the distribution of stature follows the normal curve only.

In taking normal curve as a model for distributions of populations of observations, there are however, some precautions that must be taken. *First*, the normal curve as a model describes an infinity of observations, which are on a continuous scale of measurement. As we know, the real populations may be large but they are not of infinite size. Recorded observations are discrete rather than continuous, and so far as the concrete data are concerned, there is no infinity of observations. Although normal curve approximates distributions of many kinds of data, no real variable is exactly normally distributed. Despite all these, it does not matter much because it is the utility of the model that counts. *Second*, most of the variables are not normally distributed. Variables like size of family income, human body weight, frequency of accidents and reaction time possess various degrees of skewness. Not only this, even variables which ordinarily exhibit a normal distribution in homogeneous populations may fail to do so under changed circumstances. For example, the distribution of stature of a mixed group of men and women is bimodal and test of intelligence which might yield a normal distribution of scores if properly constructed, may produce positive or negative skewness if it consists of too easy or too difficult items for the group being studied.

Another way in which normal curve functions as a model is for the distributions of sample statistics rather than for raw observations. Let us take an example. Suppose the researcher draws a large number of random samples from the same population and compute mean of each sample. It would be found that the distribution of this large number of means tends to approximate normal curve. This is a very important property in statistical inference because the shape of such distributions must be known in order to provide appropriate and necessary inference.

There are also some other roles of normal curve in relation to data. *First*, normal curve serves as a *relative frequency distribution*, which describes events that have occurred. In this sense, normal curve may provide information that half of a population have obtained intelligence test scores above the mean. *Second*, normal curve may serve as a *probability distribution* providing information about expected values. In this second sense, the normal curve may provide an accurate basis for stating that if any person is selected at random from the population, the chances are one out of two or three that he will score above mean.

Applications of Normal Curve

If the researcher is able to assure that the obtained distribution is normal or approximately normal, he would be able to solve a number of problems. In other words, the normal curve is applied to many areas for solving many problems. In the present section, a discussion of those applications would be done.

1. In determining the percentage of cases in a normal distribution within given limits :

With the help of normal curve or normal distribution, the researcher can easily compute the percentages of cases falling between the given limits.

Example: If a distribution of scores is normal with the Mean of 20 and Standard deviation of 5, what percentages of cases will fall in between 15 and 25. What percentage of cases lie above 23 and below 11?

A score of 25 is 5 points above mean (20) and a score of 15 is 5 points below mean. If these distances of 5 scores units is divided by σ (5), it is clear that 25 is $+1\sigma$ above mean and 15 is -1σ below mean (see Figure 23.4). In a normal distribution 34.13% cases lie within mean and $+1\sigma$ and same percentages of cases, that is, 34.13% lie within mean and -1σ. Thus in between 15 and 25 in this distribution, 34.13%+34.13% = 68.26% cases lie. This result may also be stated in terms of chances. Since 68.20% of cases fall between 15 and 20, the chances are about 68 in 100 that any score in the distribution, will be found between these two points.

So far as the score of 23 is concerned, its upper limit is 23.5 and it is 3.5 points above mean. If this difference is divided by standard deviation, that is, 5, we get 3.5/5 = .7σ. It means that the score of 23 lies .7σ above mean (see Figure 23.4). In normal distribution in between mean and .7σ lies 25.80% cases. Accordingly, 50.00–25.80 or 24.20% cases must lie above the upper limit 20 or 20.5 in order to fill out 50% cases in the upper half of the normal curve. In terms of chances there are about 24 chances in 100 that any score in the distribution will be larger than 20.

Likewise, the lower limit of score 11 is 10.5, which falls 9.5 score point below the Mean. If it is divided by standard deviation of 5, we get $-9.5/5 = -1.9\sigma$ from the Mean. In a normal distribution in between mean and -1.9σ lies 36.21%. Accordingly, 13.79% (50–36.21) cases must lie below the lower limit of 11 (that is, 10.5) in order to fill out the 50% of cases in lower half of the normal curve (see Figure 23.4). In terms of chances, there are about 13.79 or 14 chances in 100 that any score in the distribution will be less than 11.

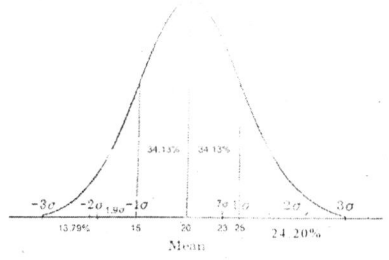

Figure 23.4

2. In finding out the limits in any distribution that include a given percentage of cases :

Another important application of normal curve is in knowing the limits that contain a given percentages of cases.

Example : Suppose Mean of a distribution is 12 and the Standard deviation is 4. Assuming that this distribution is normal, what limits will include the middle 70% of the cases?

In normal distribution the middle 70% cases include 35% of just above and 35% just below the mean. From Table of showing the percentages of cases in normal distribution, what limits will include the middle 70% of the cases?

In normal distribution the middle 70% cases include 35% of just above and 35% just below the Mean. From Table of showing the percentages of cases in normal distribution, it is clear that 34.85% or 35% lies between mean and $+1.03\sigma$ and similarly 35% of the distribution fall between the mean and -1.03σ. Therefore, the middle 70% of cases fall between mean and $\pm 1.03\sigma$. Since $\sigma = 4$, the middle 70% of the cases fall between Mean and $4 \times 1.03 = \pm 4.12$ score units. If we add ± 4.12 to Mean = 12, it is found that the middle 70% of cases in a given distribution lie between 16.12 to 7.88, the two score limits (see Figure 23.5).

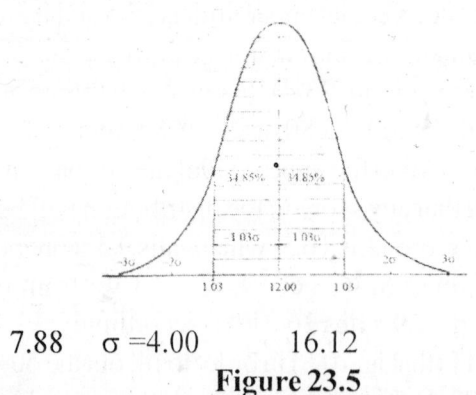

7.88 $\sigma = 4.00$ 16.12
Figure 23.5

3. In comparing two distributions of scores in terms of overlapping :

The normal curve is also used in comparing the two large distribution of scores in terms of overlapping.

Example : A test of intelligence was administered on a group of boys (M = 400) and on a group of girls (M = 350). The Mean score of boys group is 30.55 and the Standard deviation is 6.52. The Mean score of girl's group was 30.65 and the Standard deviation is 7.21. The Mean of boys is 30.01 and Median of girls is 32.62. What percentage of boys exceed the median of girls' distribution?

In this example, the girls' median is 32.62 − 30.55 = 2.07 score points above the boys' mean. When we divided 2.07 by 6.52, the standard deviation of boys group, we find that girls' median is .32σ above the Mean of boys' distribution. Now, between mean and .32σ there lies 12.55% cases of the normal distribution and therefore, 37.45% (50% – 12.55%) of the boys exceed the median of the girls.

4. In determining the relative difficulties of the question, problems and test items :

The researcher often uses normal curve in determining the relative difficulties of the various questions, problems or items of the test.

Example : Suppose that 15% of a large unselected group passed first test item, the second test item was passed by 20% of the same group and the third item was passed by 30% of the same group. If the test capacity measured by test problems is assumed to be normally distributed, find out the relative difficulty of test items 1, 2 and 3.

In the above example test item No. 1 is passed by 15% of the group. Therefore, the task is to find out a cut in the distribution such that 15% of the entire group (percent passing) lies above and 85% (percent failing) lies below the given point. The highest 15% in a normally distributed group has 35% of cases in between its lower limit and mean (see Figure 23.6). It is clear that in normal distribution 34.85% or 35% cases lie between mean and 1.03σ. It means that test item 1 belongs at a point on the baseline of the curve which has distance of 1.03σ from mean. So, 1.03σ is set down as the difficulty value of test item 1. Test item 2 is passed by 20% of the same group. In a normal distribution, the highest 20% has 30% cases below its lower limit and Mean (see Figure 23.6) from table.

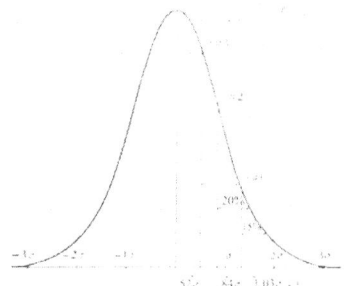

Figure 23.6

Showing fractional parts of the normal curve at different sigma score, it is clear that 29.95% or 30% cases lies between Mean and .84σ. So the difficulty value of test item 2 is taken to be 0.85σ.

The third test item is passed by 30% of the same group. The highest 30% in normally distributed group has 20% of the cases between its lower limit and the Mean (see Figure 23.6). It is clear that 19.85% or 20% cases lies between mean and 0.852σ. Hence, the difficulty value of test item 3 is 0.82σ. Let us summarize the results as follows:

Test items	Passed by %	σ values	σ difference
1	10	1.03σ	—
2	20	0.84σ	0.19
3	30	0.52σ	0.32

The σ difference in difficulty between test items 2 and 3 is 0.32 and between test items 1 and 2, is 0.19. These differences provide an index of relative difficulties of the test items.

5. In separating a given group into subgroups according to the capacity when trait is assumed to be normally distributed :

The normal curve is also applied in separating a given group into subgroups according to some capacity when the measured trait is assumed to be normally distributed.

Example: Suppose that the researcher has administered the verbal reasoning test to 200 college students. He wishes to classify this group into five subgroups—A, B, C, D and E according to ability, the range of ability to be equal in each subgroup. On the assumption that the trait measured by the verbal reasoning test is normally distributed, how many students should be placed in groups A, B, C, D and E.

To solve above problem, let us first represent the position of six subgroups diagrammatically on a normal curve as shown in Figure 23.7. If the baseline of the curve extends from +3σ to –3σ, that is, over the range of 6σ, we have to reduce it to 5 categories because there are only five groups into which the total students are to be divided. If we divided 6σ/5, we would get 1.2σ as the baseline extent to be dotted to each subgroup. There intervals are laid down on the base line as shown in Figure 23.7.

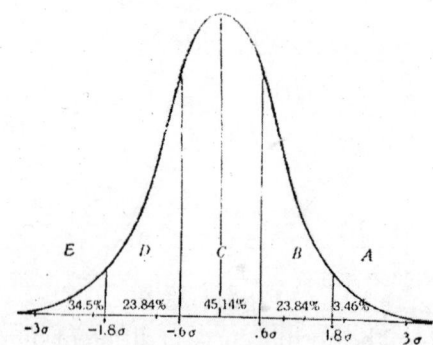

Figure 23.7

Each subgroup has been demarcated by the perpendicular erected on the base line Group A covers the upper 1.2σ, Group B the next 1.2σ, Group C covers .6σ right and 6σ left of the mean, thus again constituting (+.6σ) + (–6σ) = 1.2σ, Group D next 1.2σ and Group E next 1.2σ. Thus Group A and B occupy position in upper half of the curve where as D and E occupy positions in the lower half of the curve.

For finding what percentage of whole group belongs to A, it is necessary to find out what percentage of a normal distribution falls between 3σ (upper limit of Group A) and 1.8σ (lower limit of Group A). From table showing areas under normal curve, it is found that 49.86% of the normal distribution is said to lie between Mean and 3σ and 46.41% lie between Mean and 1.8σ. Therefore, 3.45 or 3.5 of the total area (49.86% – 46.41%) under the normal curve falls between 3σ and 1.8σ. Thus Group A consists of 3.5% of the whole group. Similar procedure is applied in calculating the percentage of other four groups. In this way in case of Group B, we have to find out percentages of cases in between 1.8σ and .6σ. In between Mean and 1.8σ, 46.41% of the normal distribution falls and 22.57% of the normal distribution falls between Mean and .6σ. Subtracting, we find that 46.41% – 22.57% = 23.84% of our distribution belongs to group B. Group C lies from .6σ above to –.6σ below the mean. In between Mean and .6σ lies 22.75% of the normal distribution and the same percentage lies between Mean and –.6σ. Thus Group C consists of 22.75% + 22.75% = 45.14% of the normal distribution Group D lies between –.6σ and 1.8σ and contains exactly the same percentage that group B contains, that is, 23.84%. Finally, Group E which lies between –1.8σ and –3σ, contains the same percentage of the whole distribution as group A contains, that is, 3.5. Now, the percentage and number of men in each subgroup can be presented as under:

	Group				
	A	B	C	D	E
1. Percentage of total in each group	3.5	23.8	45	23.8	3.5
2. Number in each group	7	48	90	48	7

Thus assuming the normal distribution that the measured capacity follows, 7 students out of 200 students should be placed in group A, the superior group, 48 in group B, high average ability group; 90 in group C, the average ability group, 48 in group D, the low average ability group and 7 in group E, the inferior group.

Measuring Divergence from Normal distribution

As we know a normal distribution is symmetrical and bell-shaped. However, such as perfect symmetrical curve rarely exists in reality as we ordinarily don't measure the entire population rather we take a sample from it. Therefore, in actual practice, a slightly distorted well-shaped curve is accepted as normal curve.

Now the question is how do we measure distortion or divergence from normality? There are two important measures of divergence from normality: *Skewness* and *Kurtosis*. In fact, Skewness and Kurtosis are the two types of errors of normal distribution. A discussion follows :

Skewness :

Skewness reflects the degree of asymmetry in the distribution. It is defined as the extent to which a frequency distribution departs from a symmetrical shape, in which mean, median and mode all coincide. Any distribution will be said to be skewed when mean and median fall at different points in the distribution and the center of gravity or balance shifts to one side or the other, that is, to the left or right of the distribution. In bell-shaped distribution mean becomes exactly equal to median and skewness is, of course, zero. In this way, skewness refers to symmetry or asymmetry of the frequency distribution (Ferguson & Takane, 1989). Thus one desirable characteristic of skewness is that it should be zero when the distribution is symmetrical. Another characteristic is that in skewed distributed mean is always pulled toward the skewed (or pointed) end of the curve.

There are two types of skewness—*positive skewness* and *negative skewness*. A distribution is said to be negatively skewed or to the left when scores are massed at the high end of the scale (right end) and are gradually spread out toward the lower end (see Figure 23.8). Thus in the negative skewness, the scores are skewed towards the lower end of the scale. A distribution is said to be positively skewed or to the right when scores are massed at the low (or left) end of the scale and gradually spread out toward the right or high end of the scale (see Figure 23.9).

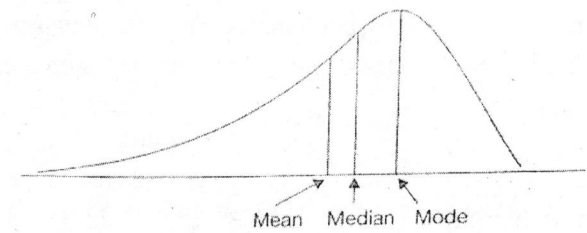

Figure 23.8 Negative Skewness (the curve inclines more to the left)

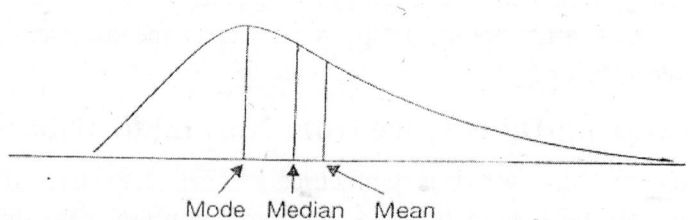

Figure 23.9 Positive Skewness (the curve inclines more to the right)

If, in any distribution, there is greater gap between mean and median, the higher will be the degree of skewness. In all skewed distribution mean is pulled towards the skewed end of the distribution. As a consequence, in a negatively skewed distributed, mean lies to the *left* of the median and in positively skewed distribution, the mean lies to the *right* of the median. Thus negative and positive skewness are opposite of each other.

The skewness can be measured by various formulas, which express skewness in pure numbers, free from unit of variable:
(i) Pearson's first coefficient of skewness

$$Sk = \frac{\text{Mean} - \text{Mode}}{\text{Standard direction}} \quad (23.2)$$

(ii) Pearson's second coefficient of skewness

$$Sk = \frac{3(\text{Mean} - \text{Median})}{\sigma} \quad (23.3)$$

where,
 Sk = a skewness
 σ = Standard deviation

(iii) Percentile coefficient of skewness :

$$Sk = \frac{(P_{90} + P_{10})}{2} - P_{50} \quad (23.4)$$

Formula 23.4 measure skewness in terms of percentiles.

(iv) T.L. Kelley's coefficient of skewness :

$$Sk = \frac{\frac{P_{90} + P_{10}}{2} - Mdn}{D} \quad (23.5)$$

where, D is the 10–90 percentile range. The simplicity of the Kelley's formula is obvious. If the distribution is symmetrical, the average of the 10 the and 90th percentile will equal to the median, and the value of skewness will be zero. If P_{90} is above the median then P_{10} is below the median, the skewness will be positive. Negative skewness is, of course, the exact opposite of positive skewness.

(v) Moment's coefficient of skewness :

The skewness can also be calculated by making use of third moment about mean and is defined as :

$$Sk = \frac{m_3}{m_2\sqrt{m_2}} \quad (23.6)$$

Where, m_3 = third moment to be calculated by formula 22.16 and m_2 = second moment to be calculated by formula 22.15.

(vi) Bowley's Quartile coefficient of skewness :

Here skewness is measured in terms of Q_1, Q_2 and Q_3 and Q. The formula is :

$$Sk = \frac{(Q_3 - Q_2) - (Q_2 - Q1)}{2Q} \quad (23.7)$$

where, Q is the quartile deviation

If the distribution is symmetrical, $Q_3 - Q_2 = Q_2 - Q_1$. In such situation, skewness will be zero. Let us take example to illustrate the calculation of skewness by these different formulas.

Suppose Mean of the distribution is 15 and Mode is 14 and standard deviation is 3, then skewness by formula 23.2 would be

$$Sk = \frac{15-14}{3} = \frac{1}{3} = .33$$

It is an example of positive skewness. Further suppose that mean of distribution is 50, median is 51 and standard deviation is 5, then skewness by formular 23.3 would be calculated as under:

$$Sk = \frac{3(50-51)}{5} = \frac{-3}{5} = -.6$$

This distribution is negatively skewed.

In computing skewness by formula 23.4, we need the value of P_{90}, P_{10} and P_{50}. Suppose in a distribution, $P_{10} = 20$, $P_{50} = P_{40}$ and $P_{90} = 70$, then skewness by formula 23.4 would be

$$\text{Skewness} = \frac{70+20}{2} - 40$$

$$= \frac{90}{2} - 40 = 45 - 40 = 5$$

This distribution is positively skewed.

Let us illustrate, now, computation of skewness by formula 23.5. Suppose in a distribution, N = 100, $P_{10} = 20$, Median = 50 and P_{90} is 70. Substituting these values in formula 23.5, we get

$$\text{Skewness} = \frac{\frac{70+20}{2} - 50}{70 - 20} = \frac{45-50}{50} = \frac{-5}{50} = -.10$$

The distribution is slightly negatively skewed.

We can also compute skewness with the help of third moment by using formula 23.6. As an illustration consider the zero of scores A and B distribution.

A:	X	x or $X - \bar{X}$	x^2	x^3
	9	−1	1	−1
	8	−2	4	−8
	10	0	0	0
	12	+2	4	+8
	11	+1	1	+1
	$\Sigma X = 50$		$\Sigma x^2 = 10$	$\Sigma x^3 = 0$

$$\bar{X} = \frac{\Sigma x}{N} = \frac{50}{5} = 10$$

$$m^2 = \frac{\Sigma x^2}{N} = \frac{10}{5} = 2$$

$$m^3 = \frac{\Sigma x^3}{N} = \frac{10}{5} = 2, \text{ Skewness} = \frac{m_3}{m_2\sqrt{m_2}} = \frac{0}{2\sqrt{2}} = 0$$

Hence, the set of score in distribution A is symmetrical because skewness is zero.

B:	X	x or $X-\bar{X}$	x^2	x^3
	10	−15	225	−3375
	15	−10	100	−1000
	20	−52	25	−125
	30	+5	25	125
	50	+25	625	15625
	$\Sigma X = 125$		$\Sigma x^2 = 1000$	$\Sigma x^3 = 11125$

$$\bar{X} = \frac{\Sigma X}{N} = \frac{125}{5} = 25$$

$$m^2 = \frac{\Sigma x^2}{N} = \frac{100}{5} = 200$$

$$m^3 = \frac{\Sigma x^3}{N} = \frac{11125}{5} = 22.25, \text{ Sk} = \frac{m_3}{m_2\sqrt{m_2}} = \frac{22.25}{200\sqrt{200}} = .786$$

Distribution of score in set B is positively skewed.

The skewness can also be computed by using formula 23.7 where it is done through quartile deviation or Q. Suppose in a distribution of scores $Q_1 = 15$, $Q_2 = 30$ and $Q_3 = 48$, then, by formula 23.7, the skewness would be :

$$Sk = \frac{(Q_3 - Q_2) - (Q_2 - Q_1)}{2Q}$$

In the above example, Q by formula 20.2 would be

$$Q = \frac{Q_3 - Q_1}{2} = \frac{48 - 15}{2} = \frac{33}{2} = 16.5$$

After applying formula 23.7,

$$Sk = \frac{(48-30) - (30-15)}{2 \times 16.5} = \frac{18-15}{33} = \frac{3}{33} = .09$$

The distribution is slightly positively skewed.

The distribution is slightly positively skewed.

The question of how much skewness a distribution must exhibit before it can be considered as *significantly* skewed cannot be answer satisfactorily, unless we have a standard error for the index of skewness. Standard errors for above different formulas except formula 23.5 will not be discussed here because they are not satisfactory.

In the above example, applying formula 22.2, we get skewness of –.10. The standard error for this formula can be easily calculated by applying formula 23.8 as under :

$$S_{sk} = \frac{.5188}{\sqrt{N}} \qquad (23.8)$$

where, S_{sk} = standard error of skewness

$$= \frac{.5185}{\sqrt{100}} = \frac{.5185}{10} = .051$$

Dividing the skewness –.10 by its standard error (that is .051) we get –.10/.051 = –1.96 as the value of z. Since this is exactly 1.96 corresponding to .05 level of significance, it is considered significant. Had it been less than 1.96, it would not be then considered significant. So, we conclude that there is evidence to indicate that the scores in the distribution are not distributed symmetrically.

One important question relating to skewness is often asked by psychologists or educators is : *What are the situations that lead to skewness?* There are *five* situations, which produce skewness in the distribution of scores. Those five situations are as under :

- There may be natural restriction at the low end of the scale and this would tend to produce positive skewness. This is called as *floor effect*. For example, population of city, minimum speed attained by car at stop signals, number of offspring, time required by a six year old child to run 50 meters, etc. may lead to such floor effect, causing positive skewness in the distribution.
- There may be some natural restriction at the high end of the scale, resulting in negative skewness. This is called *Ceiling effect*. The number of persons that can live in two-bed room flat is one such example.
- There may be some artificial restriction at the low end of the scale resulting in positive skewness. For example, the distribution of scores made by students on a very difficult task may produce positive skewness.
- There may be some artificial restriction at the high end of the scale resulting in negative skewness. This can be illustrated by the distribution of scores of student on a test that is too easy.
- Lastly, skewness in the distribution of scores may be the function of the measuring system itself. One example is the record of time required to complete a task when the researcher is really interested in speed.

Kurtosis

The term Kurtosis comes from the Greek word *Kyrtos*. Kurtosis reflect how much the shape of a distribution differs from a normal curve in terms of whether its tails are heavier (thicker) or lighter (thinner) than the normal distribution. In other words, Kurtosis refers to 'peakedness' or flatness of a frequency distribution as compared with the normal distribution. A frequency distribution more peaked than normal distribution is called as *leptokurtic distribution* and more flatter than the normal distribution is known as *platykurtic distribution*. A normal distribution is known as *mesokurtic* (meso means middle or medium), which means that it fall between leptokurtic and platykurtic distributions. In Figure 23.10 the three types of curves, namely, leptokurtic, mesokurtic and platykurtic curves have been drawn.

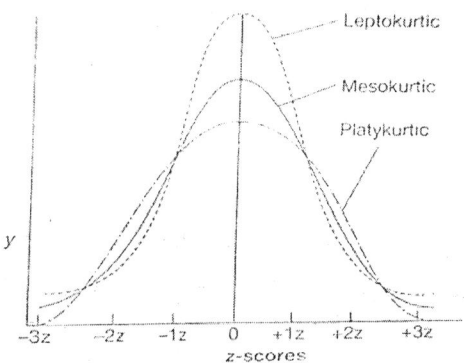

Figure 23.10 Different Forms of Kurtosis – Leptokurtic, Mesokurtic, Platykurtic.

When the value of kurtosis is zero, the shape of distribution is mesokurtic, which is the shape of normal curve. When the value of kurtosis is negative, the curve is platykurtic and when its value is positive, the curve is leptokurtic.

The value of kurtosis can be computed by different formulas. Important ones are as under:

(i) *Kurtosis in terms of percentiles*:

The formula for computation of kurtosis is:

$$Ku = \frac{Q}{(P_{90} - P_{10})} \qquad (23.9)$$

When,

Ku = Kurtosis

Q = Quartile deviation

The formula 23.9 may also be written as:

$$Ku = \frac{P_{75} - P_{25}}{2(P_{90} - P_{10})} \qquad (23.10)$$

(ii) *Kurtosis in terms of moment coefficient*:

$$Ku = \frac{m_4}{m_2^2} - 3 \quad (23.11)$$

Where, m_4 and m_2 are the fourth and second moments about mean.

Let us illustrate the computation of kurtosis by the above formulas.

Suppose in a distribution of 50 scores $P_{10} = 15$, $P_{25} = 23$, $P_{75} = 40$, $P_{90} = 60$, then what will be its kurtosis? In this example, Q would be calculated as under:

$$Q = \frac{P_{75} - P_{25}}{2} = \frac{40 - 23}{2} = \frac{17}{2} = 8.5$$

Therefore, K_u by formula 23.9 would be

$$Ku = \frac{Q}{(P_{90} - P_{10})} = \frac{8.5}{60 - 15} = \frac{8.5}{45} = .188$$

K_u by formula 23.10 would be

$$K_u = \frac{P_{75} - P_{25}}{2(P_{90} - P_{10})} = \frac{40 - 23}{2(60 - 15)} = \frac{17}{90} = .188$$

Both formulas 23.9 and 23.10 yields the same value of kurtosis. Since the value of kurtosis is positive, curve for the distribution is said to be leptokurtic. It is well to remember that if the numerical value of K_u is obtained by formula 23.10 or 23.11 it must always lie in between .00 and .50 though most values will fall within the limits of .21 and .31 (Kurz & Mayo, 1980).

The value of kurtosis can also be computed through moment coefficient. Let us take an example:

Scores (X)	$X - \bar{X}$ or x	x^2	x^4
8	0	0	0
7	−1	1	1
6	−2	4	16
9	+1	1	1
10	+2	4	16
$\Sigma x = 40$	$\Sigma x = 0$	$\Sigma x^2 = 10$	$\Sigma x^4 = 34$

$$\bar{X} = \frac{\Sigma X}{N} = \frac{40}{5} = 8$$

$$m^2 = \frac{\Sigma x^2}{N} = \frac{10}{5} = 2$$

$$m^4 = \frac{\Sigma x^4}{5} = \frac{34}{5} = 6.8$$

Now K_u by formula 23.11 would be

$$Ku = \frac{m_4}{m_2^2} - 3 = \frac{6.8}{2 \times 2} - 3 = \frac{6.8}{4} - 3 = -1.3$$

Since the value of the k_u is negative, the curve for this distribution would be called as platykurtic curve.

The computation of the by formula 23.9 or by 23.10 is very simple and easy. When k_u is computed by formula 23.9 or 23.10, the standard error of K_u (Sk_u) can also be easily computed by formula 23.12 for knowing the stability of kurtosis.

$$Sk_u = \frac{.2778}{\sqrt{N}} \qquad (23.12)$$

$$\therefore Sk_u = \frac{.2778}{\sqrt{50}} = \frac{.2778}{7.07} = .039$$

To determine whether the obtained value of kurtosis, that is, .188 is significantly different from .2632, signifying mesokurtosis or normal distribution, we simply take its deviation from the fixed value, .2632 and divide the difference by the standard error.

Thus,

$$\frac{.1880 - .2632}{.039} = \frac{-.0752}{.039} = -1.928$$

Since this obtained quotient is smaller than 1.96, corresponding to the .05 level of significance, it is concluded that there is no evidence to disapprove the assumption that the curve was leptocurtic.

The reader must *remember* that for the normal distribution curve, the formula 23.9 gives K_u = .2632. If the obtained K_u is greater than .2632, the distribution is considered playkurtic and if less than .2632, the distribution is considered leptocurtic.

Why do frequency distributions deviate from normal distribution?

There are various reasons behind deviation of a frequency distribution from normal distribution. When deviation occurs from normal distribution, skewness and kurtsis result. The important causes of such asymmetry are as under :

- **Selection :**

Selection is one of the important causes of deviation of distribution from normal form. If the researcher selects a group of 50 superior students or a group of 50 inferior students and administer intelligence test on them, the resulting distribution would hardly be normal. Not only this, selection is likely to produce skewness and kurtosis in the distribution even when a well-standardized test has been developed. For example, if the selected group of individuals are bilinguals, from high or low socio-economic status, from over age groups for a grade are likely to produce skewed distribution etc. Likewise, it has been reported that the scores made by small and homogeneous groups are likely to yield leptokurtic distributions and scores from large and heterogeneous groups are likely to be broad and yield platykurtic distribution. Likewise, distributions of physical traits such as height, weight, etc. sometimes yield skewed distributions. For example, the distribution of height of 16-year old girls of high socio-economic status or distribution of weight from children from slum areas or very poor families are likely to yield skewed distribution.

- **Unsuitable or poorly developed tests :**

  The tests should be appropriate for the abilities of the sample being studied. If the test is too difficult, most of the scores will pile up at the lower end of the scale and the resulting distribution will be positively skewed. Likewise, if the test is made up of too easy items, most of the scores will pile up at the higher end of the scale and the resulting distribution will be negatively skewed. Likewise, a poorly developed test having poor reliability and validity is also likely to yield a skewed distribution. Similarly test having ambiguous or poorly made items tend to yield nonnormal distribution.

- **Non-Normal distributions :**

  There are various types of nonnormal distributions that result when there is real lack of normality in the trait being measured. Examples of nonnormal distributions are *skewed distributions, platykurtic distribution, leptokurtic distribution, J-shaped distribution, rectangular distribution, U-shaped distribution,* etc. Such nonnormality of distribution usually arises when some of the hypothetical factors determining the strength of a trait are very dominant over others and hence, are present more often than chance will allow. Figure 23.11 presents common types of non-normal distributions : J-shaped distribution, rectangular distribution, bimodal distribution and U-shaped distribution.

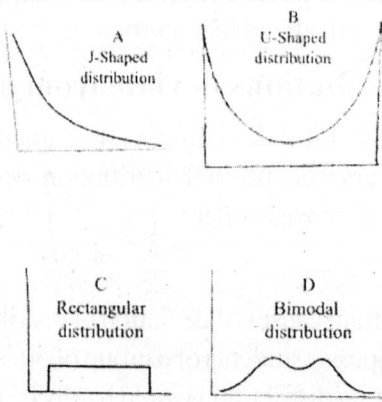

Figure 23.11 Shapes of some major Non-Normal distribution.

The J-shaped distribution often occurs to describe some forms of social behaviour. For example, if the researcher notes down the number of students appearing at the lecture hall in time, the number who arrived 5 minutes, 10 minutes and 20 minutes late, he would get J-shaped of distribution. If such frequency is plotted on a graph with X-axis showing the time and frequency of arrival with Y-axis, the distribution will be highest at zero (on time) on X-axis and will fall off rapidly on we go to the right, producing J-shape curve. The U-shape distribution is sometimes obtained in the measurement of social and personality traits especially when the group is extremely heterogeneous with respect to some attitude or trait being measured is likely to be present or absent in all-or-none manner. The bimodal distribution may result from measuring the strength of the grip in a group having both men and women. A rectangular shape of the distribution may result from a set of ranks resulting from ordinal measurement. Non-normal curves often occur in medical statistics. For example, the likelihood of death due to some kind of degenerative diseases is the highest during maturity and old age and minimal during the early years of life.

- **Errors in construction and administration tests :**

If items of the test are very easy at the beginning of the test and much difficult later on, an increment of one point of score at the upper side of the test will be much greater than an increment of one point at the low end of the scale. Such unequality is likely to reduce the spread of the distribution. The scores tend to pile up at some intermediate point and to be stretched out at the low end of the scale. Not only this, various kinds of errors such as errors in timing or in giving instructions, large difference in motivation may promote some individuals to score higher and some other individuals to score lower than normally would. As consequence, there would occur skewness in the distribution.

Summary and Review

- Normal curve also known as Gaussian curve, DeMoivre curve, the curve of error is a curve, which results from the normal distribution of scores. A normal distribution is a distribution in which the majority of scores is centered at the middle of the distribution and very few cases lie at the upper end and lower end of the scale.
- There are several important properties of normal curve. For example, normal curve is bilaterally symmetrical, asymptotic, its skewness coefficient is zero etc. It is extended to $\pm 3\sigma$ including both sides of the distribution. Very few cases lie beyond $\pm 3\sigma$.
- Normal curve as a statistical model serves various functions.
- There are five major applications of normal curves commonly done in behavioural sciences.
- Skeweness and Kurtosis are the popular measures of divergence from normality. Skewness refers to symmetry or asymmetry of the distribution. There are two types of skewness: positive skewness and negative skewness. There are various formulas for computing skewness. Kurtosis refers to the flatness or peakedness of distribution in relation to that of normal one. Kurtosis is of three types : leptocentic, mesokurtosis and platykurtic. Mesokurtic is the other name of normal distribution.

- There are several reasons of asymmetry in the distributions. Important ones are selection, unsuitable or poorly made tests, nonnormal distributions and errors in construction and administration of tests.

Review Questions

1. Define normal curve. Discuss its important characteristics.
2. Discuss normal curve as a statistical model.
3. What is skewness? Discuss, with example, the type of skewness.
4. What is kurtosis? Discuss, with examples, the different types of kurtosis.
5. What are the reasons of asymmetry in the distribution?
6. If mean is 50.00, median is 52.32 and the standard deviation is 10.40, find out its skewness.
7. In a distribution if $P_{10} = 10.00$, $P_{25} = 15.00$, P_{75} is 40.00 and P_{90} is 50.00, then find out its kurtosis.

■ ■ ■

Chapter – 24

Percentile and Percentile Rank

Learning Objectives
- Meaning of Percentile and Percentile rank
- Computation of Percentile
- Computation of Percentile rank
- Uses and limitation of Percentile
- Uses and limitations of Percentile rank
- Concept of Decile and Quartiles.
- Summary and Review
- Review Questions
- Key Terms

Key Terms:
Percentile, Percentile rank, Decile, Quartile.

Transformation of scores are done by changing observed or raw scores in some systematic way. There are several purposes of such transformation. One such important purpose is to make scores from different tests instruments comparable. Such transformations became necessary when the researcher wishes to compare or combine an individual scores on different tests or instruments since their raw scores are not directly comparable. There are various kinds of transformations of raw scores to comparable scores. However, we shall stick in this chapter to some most popular kinds of transformations like percentile, percentile rank, decile, etc.

In educational and psychological literature especially that which is concerned with mental tests, these two terms percentile and percentile rank are very popular and frequently used. Percentile and percentile rank are two different terms signifying two different things. But the reality is that in many situations they simply indicate two alternative ways of expressing essentially the same idea that is the relation between a test score and the percent of individual who secures scores below it. In the present chapter, a discussion regarding these two concepts and their uses, computation and comparison would be discussed in detail.

Meaning of Percentile Rank

The percentile or centile system is widely used in the educational and psychological measurement for reporting the standing of the individual relative to the performance of a known group. A percentile is defined as a score point below which a specified percent of scores in the distribution falls. In other words, a percentile is a score point below which a given percentage of the cases occur. The specified percent is called as *percentile rank (PR)*. These two should not be confused due to the similarity in expression. The percentile rank may take any value between zero and 100 whereas a percentile point may have any value that scores may have. For example, suppose Mohan has obtained a score of 165 in college entrance test and his this score is such that 80 percent of the applicants score below it. In this example, Mohan's percentile point is 165 and his percentile rank is P_{80}. In a nutshell, the percentile rank of a particular score is defined as the percentage of individuals in the distribution with scores at or below the particular value. When a score is identified by its percentile rank, the score is called percentile. Notice that these are but two different ways of expressing the relation between a test score and the percentage of individuals who have obtained scores below it. These differences in orientation lead to two different computational methods.

Computation of Percentile

In computation of percentile, the more or less the same method is used as it is used in case of computation of median. As we know, median is P_{50}, Q_1 is P_{25} and Q_3 is P_{75}. Median is the point below which lies 50% of the cases, Q_1 is the point below which lies 25% of cases and Q_3 is the point in the distribution below which lies 75% of the cases. The computational procedures of these statistics is very similar.

The method of computing percentile is much similar to that of median. It is done through formula 24.1

$$P_P = l + \left(\frac{PN - F}{f_P}\right)i \quad (24.1)$$

where,
- P_P = percentage of distribution wanted such as 30%, 40%, etc.
- PN = part of N to be counted off in arriving at P_P.
- l = exact lower limit of the class interval upon which P_P lies
- F = sum of all scores upon intervals below l
- f = number of scores (or frequency) within the interval that contains P_P
- i = length of class interval.

The formula 24.1 is used in computation of percentile when the data have been grouped into frequency distribution table. Let us illustrate the calculation of percentile from data presented in Table 24.1

Table 24.1

Class interval x	Frequency f	Cum.f
100-109	2	50
90-99	3	48
80-89	5	45
70-79	1	40
60-69	6	39
50-59	27	33
40-49	2	6
30-39	1	4
20-29	3	3
	N = 50	

Sohan has obtained a percentile rank of 70 in the above distribution. What would be his percentile point or score? This question can be answered with the help of applying formula 24.1.

$$P_P = l + \left(\frac{PN - F}{f_P}\right) i$$

Here, $PN = 35$ (70% of 50), F is 6, fq is 27, $l = 49.5$ and $i = 10$. Thus substituting these values in formula 24.1, we get

$$P_P = 49.5 + \left(\frac{35 - 6}{27}\right) 10$$

$$= 49.5 + \left(\frac{29}{27}\right) 10 = 49.5 + 10.74 = 60.24$$

Thus, Sohan's percentile rank of 70 entitles him to a percentile score of 60.24

Computation of Percentile Rank (PR)

There is also a way of finding an individual's percentile rank (PR) or the position of the scale of 100 to which his score entitles him. What is the percentile rank of the individual who scores 85 in the distribution presented in Table 24.1? This problem can be tackled with the help of formula 24.2.

$$P_R(X) = \frac{100}{N}\left[F + \frac{(X-1)f}{i}\right] \quad (24.2)$$

where,

$P_R(X)$ = desired percentile rank of the given score (X)
N = number of scores in the distribution

F = cumulative frequency upto the class interval containing X
l = exact lower limit of the class interval containing X
f = frequency in the class interval containing X
i = length of the class interval

Now, by substituting the values in formula 24.2, we get

$$PR(85) = \frac{100}{50}\left[45 + \frac{(85-79.5)5}{10}\right] = 2\left[45 + \frac{(.5 \times 5)}{10}\right]$$

$$= 2[45 + .25] = 2 \times 45.25 = 90.5$$

It means that the individual having a score of 85 entitles him to a PR of 90.5 which would be reported as 91.

The student should carefully note the converse relationship between percentile and percentile rank. The percentile is a score below which a specified percentage falls and the percentile rank is the percentage below a specified score.

Percentile rank can also be calculated from data expressed in terms of rank. There are many occasions in which persons or objects are ordered in rank rather than measured on some trait. For example, N students may be ranked by teacher from rank 1 to N lowest with respect to punctuality. In such situation Percentile rank may be calculated with the help of formula 24.3.

$$PR = 100 - \left(\frac{100R - 50}{N}\right) \quad (24.3)$$

Suppose a student has obtained 4th rank on the measure of punctuality in a group of 20. His PR by formula 24.3 would be:

$$PR = 100 - \left(\frac{100 \times 4 - 50}{20}\right) = 100 - \left(\frac{350}{20}\right) = 100 - 17.5 = 82.5$$

Thus a student whose score entitles him 4th rank has a percentile rank of 82.5 or 83.

Uses and Limitations of Percentile

There are three major common uses of percentiles.
- The most frequent application of percentile is in psychological and educational testing. Standardized tests usually report norms in terms of percentile values of raw scores. If a person obtains a score on a standardized test and such score is at 60th percentile, it is clear that he or she has obtained score higher than 60 percent of the group of individuals on whom the standardized test was administered. In a nutshell, percentiles are often used in interpreting test scores or other measures.
- Some small number of percentiles may be used in determining the accurate picture of the shape of a frequency distribution. Thus if a researcher knows P_{10}, P_{25}, P_{50}, P_{75} and P_{90}, he can formulate a basic idea regarding a complete understanding of the shape of the distribution.

- Percentiles are also used in computing some other statistics. For example, skewness and kurtosis may be measured by statistics that are combinations of two or more percentiles. It is also worth mentioning to note that the most popular statistic for median is nothing but 50th percentile.

Despite these uses, percentiles have some serious limitations. Percentiles are unequal units of measurements and cannot be treated arithmetically. In other words, there is no justification in adding them, averaging them, combining them or treating them in any arithmetical fashion. Nothing can be done with them. As statistics go, they are dead ends. Since percentile are unequal units of measurements, many researchers believe that it is better to proceed without them or with some substitutes. In fact, at the center of distribution, the use of percentile scores tends to exaggerate differences that are actually not in existence. This can be better understood if we pay attention to Figure 24.1. This figure clearly shows that a raw score of 33 is equivalent to fiftieth percentile, a raw score of 36 to the percentile point of 60 and a raw score of 30 to a percentile point of 40. Thus a change in six raw scores units become equivalent to a change in 20 percentile units. In this way, there is a piling up of percentile points at the center of the distribution and the differences between them at these points of the curve have little meaning.

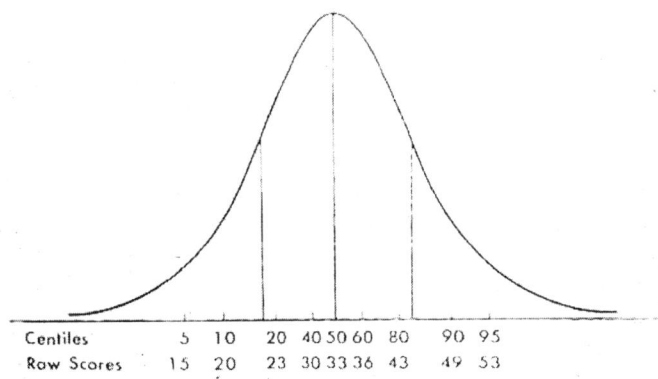

Figure 24.1 Centile and Raw Score equivalents for a set of Data.

Uses and Limitations of Percentile Rank

The most frequent and common application of percentile rank is in testing. Standardized tests often report norms in terms of percentile rank. If a person obtains a score on a standardized test, which has a percentile rank of 75, it becomes obvious that the person exceeds 75 percent of the group on which the test was administered. Likewise, when obtained scores or ranks are changed to percentile ranks, a person's performance on tests or his position in ordered data can be brought into comparison, irrespectives of the dissimilar nature of original or raw scores. As we know, the various individual profiles as needed in counselling and placement, frequently use percentile ranks as the common unit.

Despite these uses, percentile rank has two serious *limitations*. *First*, percentile ranks are not subject to algebraic treatment and therefore, cannot logically be used when two or more scores are to be combined into a composite score. *Second*, percentile ranks are not in proportional relationship to the raw scores. A look at Figure 24.1 reveals that the difference between raw score of 23 and 20 (which is 3) is represented by corresponding difference in percentile rank 20 – 10 = 10 whereas difference in the raw score of 95 and 90, that is, 5 is represented by the corresponding difference in percentile rank 49 and 53, that is, only of 4 percentile ranks. No general statement can be made regarding the nonproportionality in the percentile scale and the consequences of obscuring through its use reveal relatively greater differences near the top and bottom of the distribution.

Concept of Deciles and Quartiles

Since the concepts of deciles and quartiles are related to percentile, it is essential to throw a light upon these two related concepts. A decile is a percentage rank that divides the distribution into 10 equal parts. The 1st decile (specified as D_1) is the 10 percentile and the 9th decile (specified as D_9) is the 10 percentile and the 9th decile (specified as D_9) is the 90th percentile. D_{10} is the 100th percentile. A quartile, already discussed in chapter 22, is the percentile rank that divides the distribution into four equal parts. There are four quartiles in any scale of percentile rank. There are four quartiles in any scale of percentile ranks. They are first quartile (Q_1) which is 25th percentile, Q_2 which is equivalent to the 50th percentile (or 50th decile) and third quartile (Q_3) which is equivalent to 75th percentile and Q_4 is the 100th percentile. Q_2 is also known as median of the distribution. Q_{-1} can be calculated by formula 22.3, (see chapter 22), Q_2 can be calculated by formula 21.3, Q_3 can be calculated by formula 22.4 and Q_4 can be calculated by formula 22.4.

$$Q_4 = l + \left(\frac{N-F}{f_q}\right)i \quad (24.4)$$

Obviously, this formula is similar to the formulas for Q_1, Q_2 and Q_3 and therfore, its notations are defined as used.

Deciles can be calculated with the help of following formulas :

$$1\text{st Decile } (D_1) = l + \left(\frac{N/10 - F}{f_d}\right)i \quad (24.5)$$

$$2\text{nd Decile } (D_2) = l + \left(\frac{2N/10 - F}{f_d}\right)i \quad (24.6)$$

$$3\text{rd Decile } (D_3) = l + \left(\frac{3N/10 - F}{f_d}\right)i \quad (24.7)$$

$$\text{4th Decile } (D_4) = l + \left(\frac{4N/10 - F}{f_d}\right)i \quad (24.8)$$

$$\text{5th Decile } (D_5) = l + \left(\frac{5N/10 - F}{f_d}\right)i \quad (24.9)$$

$$\text{6th Decile } (D_6) = l + \left(\frac{6N/10 - F}{f_d}\right)i \quad (24.10)$$

$$\text{7th Decile } (D_7) = l + \left(\frac{7N/10 - F}{f_d}\right)i \quad (24.11)$$

$$\text{8th Decile } (D_8) = l + \left(\frac{8N/10 - F}{f_d}\right)i \quad (24.12)$$

$$\text{9th Decile } (D_9) = l + \left(\frac{9N/10 - F}{f_d}\right)i \quad (24.13)$$

$$\text{10th Decile } (D_{10}) = l + \left(\frac{10N/10 - F}{f_d}\right)i \quad (24.14)$$

where,

N = number of scores in the distribution
l = exact lower limit of the class interval upon which the required decile lies
F = sum of all scored upon interval below l
f_d = number of scores within the interval upon which the needed decile falls
i = length of class interval

To illustrate, let us compute D_1 from Table 24.1 by formula 24.5. In this situation, $N/10 = 50/10 = 5$, $l = 39.5$ (because class interval 40–49 contains $N/10$, that is 5, $F = 4$, $fd = 2$ and $i = 10$. Thus substituting these values in formula 24.5, we get

$$D_1 = 39.5 + \left(\frac{50/10 - 4}{2}\right)10$$

$$= 39.5 + \left(\frac{1}{2}\right)10 = 39.5 + 5 = 44.5$$

Summary and Review

- Percentile point or score and Percentile rank are the two different ways of expecting the relation between a test score and the percentage of individuals who have obtained scores below it. The percentile rank of a particular score is defined as the percentage of individuals in the distribution with scores at or below the particular value. When a score is identified by its percentile rank, the score is called percentile.
- Specific formulas are available to compute percentile and percentile rank.
- Percentile rank can also be computed from ranked data.
- There are different uses of percentile and percentile rank. However, there are also some limitations of these two statistics.
- The concept of deciles and quartiles are related to percentile. A decile is the point on the score scale below which a given decile of the cases lie. In fact, deciles for any given distribution are ten in number and are described as D_1 to D_{10}. D_1 is the 10th percentile and D_{10} is 100th percentile. A quartile is the percentile rank, which divides the distribution into four-equal parts. There are four quartiles : Q_1, Q_2, Q_3 and Q_4. Q_1 is the 25th percentile, Q_2 is the 50th percentile (also called as median), Q_3 is the 75th percentile and Q_4 is the 100th percentile.

Review Questions

1. Citing examples make distinction between percentile and percentile rank.
2. Define percentile, percentile rank, decile and quartile. Discuss their mutual relationship.
3. Find out the value of 75th percentile rank, D_4 and D_3 from the following data

Scores	f
90-94	2
85-89	3
80-84	4
75-79	5
70-74	17
65-69	3
60-64	6
55-59	6
50-54	4
	N = 50

Chapter – 25

Correlation and Regression

Learning objectives :
- Meaning and Types of Correlation
- Pearson Product-Moment Correlation
- Major Properties of Pearson r
- Assumptions of Pearson r
- Interpretation of Pearson r
- Correlation and Causation
- Rank difference Correlation
    - Spearman's Rank difference Correlation
    - Kendall's Rank difference Correlation
- Two Important Correlation Methods involving More than Two variables
- Meaning of Regression and Regression equation
- Obtaining a and b coefficient
- Limitations of Regression Analysis
- Comparison between Regression and Correlation
- Summary and Review
- Review Questions
- Key Terms

Key Terms:

a coefficient, b coefficient, Correlation, Positive Correlation, Negative Correlation, Pearson r, Spearman's rank difference correlation, Kendall's rank difference correlation, Linear relationship, Regression equation, Partial correlation, Part correlation, Multiple correlation, Regression Analysis.

Scientists believe that the world is understandable. In search of understanding of man and his affairs, various types of questions arise regarding the possible relationship between phenomena. Such questions like whether intelligence is related to academic achievement, whether parental income is related to the child continuation in school, whether broken homes are related to delinquency, whether supply of economic foods is related to price etc, are very common. The major task of any branch of science is that of discovering and measuring relationships through comparisons of sets of data. As new relationships are discovered, understanding of the world is increased. Moreover, when existing relationship permits prediction of events, control over the environment is strengthened. Variables that are not related to one or more other variables are of little importance. In a nutshell, knowledge of relationships has always been the fundamental one for understanding and controlling the environment. In the present chapter some important measures of relationship such as Pearson r, Spearman Rank difference correlation, Kendall's Rank difference correlation and their computation and uses would be discussed. There are many other measures of correlation but only the above three that are most common would be discussed. Simultaneously, the related concept of regression and regression equations would also be discussed.

Meaning and Types of Correlation

The term correlation is often used to refer to any sort of relationship between objects or events. Such assertions like correlation between fact and theory, correlation between height and weight of children, correlation between crime and poverty refer to this kind of relationships. However, in statistics, the term correlation refers exclusively to the relationship between variables that can be quantified. The situation in which statistical correlation is applicable is *always* one in which there is pair of measures for each individual in the group. For example, in order to apply the method of correlation to determine whether intelligence (X) and academic achievement (Y) are related, we must have the measures of two for a number of individuals. Thus correlation method can be applied to data only where a basis exists for pairing one measurement with other. The statistic that describes the degree of relation between two variables is known as *correlation coefficient*. Thus the coefficient of correlation is a *numerical* index which expresses quantitatively the *magnitude* and *direction* of the relationship. In other words it is a kind of ratio that reflects the extent to which changes in one variable are accompanied by changes in the other variable.

If the relationship between two variables, that is, X and Y is such that large values of one variable tend to be accompanied by the large values of the other, the correlation is said to be *positive one*. Thus in position correlation, there is a similar trend of increasing or decreasing. When large values of one variable tend to be associated with small values of other variable (that is, opposite trend), the correlation is said to be *negative* one. When data consists of pairs of measures, they are called as *bivariate data*. When the two variables consisting of bivariate data are correlated, either may be spoken of as the *correlative* of other.

Obviously, there are many situations in which more than two variables are of interest. For example, scores on three or four psychological tests may be obtained for a group of students and these scores may be related to academic performance. In fact, the study of relation between variables can be extended to any number of variables. Methods which are used in the study of more than two variables are called *multivariate* statistical methods.

Pearson Product-Moment Correlation

The most widely used method of correlation is the Pearson product-moment correlation coefficient. This is a statistic which is applied on interval and ratio measurement. This statistic is descriptive of the magnitude of the relation between two variables. The symbol r which is short for regression (a concept closely related to correlation), is used to indicate the Pearson Product-moment correlation. Hence, it is also referred to as *Pearson r*. It is named after Karl Pearson, a student of Francis Galton. In fact, Pearson along with Galton, played a major role in developing the correlation coefficient.

The size of Pearson r varies from +1 through 0 to –1. Most correlation coefficients tell us two things. *First*, it gives an indication of magnitude of the relationship between two variables. It is worth noting at this point that a correlation of –0.67 is of the same size as that of +0.67. The sign has nothing to do with size of the relationship but it does give information about the direction of the relationship. The former value (–0.67) indicates negative correlation—one increases (or decreases) while other decreases (increases) and the latter value (+0.67) indicates positive correlation—one increases (or decreases) the other also increases (or decreases). The absence of relationship is denoted by a correlation coefficient .00 or thereabouts.

Pearson r can be computed by different formulas. For example, Pearson r may be calculated from deviations occurring from means, by a method of difference, directly from raw scores or from the standard scores. In each situation, the formula differs. The basic formula for computation of Pearson r is as under :

$$r_{xy} = \frac{\Sigma_{xy}}{N\sigma_x\sigma_y} \qquad (25.1)$$

where,
Σ_{xy} = sum of the product of x and y deviations taken from actual mean
σ_x = standard deviation of X scores
σ_y = standard deviation of Y scores
N = Number of pairs

The above basic formula for the Pearson Product-moment coefficient is commonly designated by r_{xy}. Since $\sigma_y = \sqrt{\Sigma x^2 / N}$ and $\sigma_y = \sqrt{\Sigma y^2 / N}$, the basic formula may also be written as

$$r_{xy} = \frac{\Sigma xy}{\sqrt{(\Sigma x^2)(\Sigma y^2)}} \qquad (25.2)$$

Where Σx^2 and Σy^2 are the sum of deviations.

When the researcher is interested in calculating Pearson r directly from raw scores, the formula becomes.

$$r = \frac{N\Sigma XY - (\Sigma X)(\Sigma Y)}{\sqrt{[N\Sigma X^2 - (\Sigma X)^2][N\Sigma Y^2 - (\Sigma Y)^2]}} \qquad (25.3)$$

where,

X = Scores on X test or variable
Y = Scores on Y test or variable

Let us illustrate the calculation of Pearson r by both formula 25.2 and 25.3.

Suppose an X test and Y test have been administered to a group of 10 students. Their scores have been presented in Table 25.1.

Table 25.1: Scores of ten students on X and Y test

1	2	3	4	5	6	7
X	Y	x	y	x^2	y^2	xy
10	5	+1	−7	1	49	−7
15	4	+6	−8	36	64	−48
10	6	+1	−6	1	36	−6
12	12	+3	0	9	0	0
8	14	−1	+2	1	4	−2
5	14	−4	+2	16	4	−8
9	18	0	+6	1	36	0
6	12	−3	0	9	0	0
5	25	−4	+13	16	169	−52
10	10	+1	−2	1	4	−2
$\Sigma X = 90$	$\Sigma Y = 120$	$\Sigma x = 0$	$\Sigma y = 0$	$\Sigma x^2 = 91$	$\Sigma y^2 = 366$	$\Sigma xy = -125$

$$\text{Mean}_X = \frac{\Sigma X}{N} = \frac{90}{10} = 9 \qquad \text{Mean}_Y = \frac{\Sigma X}{N} = \frac{120}{10} = 12$$

Now r by formula 25.2 becomes :

$$r_{xy} = \frac{\Sigma xy}{\sqrt{\Sigma x^2 \times \Sigma y^2}}$$

$$r_{xy} = \frac{-125}{\sqrt{91 \times 366}} = \frac{-125}{182.49} = -0.684$$

If we apply formula 25.3, Pearson r can be calculated in this manner. The data in Table 25.1 is being reproduced in Table 25.2

Table 25.2

X	Y	X²	Y²	XY
10	5	100	25	50
15	4	225	16	60
10	6	100	36	60
12	12	144	144	144
8	14	64	196	112
5	14	25	196	70
9	18	81	324	162
6	12	36	144	72
5	25	25	625	125
10	10	100	100	100

$\Sigma X = 90$ $\Sigma Y = 120$ $\Sigma X^2 = 900$ $\Sigma Y^2 = 1806$ $\Sigma XY = 955$

Now r by formula 23.3 can be calculated as under

$$r_{xy} = \frac{N \Sigma XY - (\Sigma X)(\Sigma Y)}{\sqrt{N \Sigma X^2 - (\Sigma X)^2 \left[N \Sigma Y^2 - (\Sigma Y)^2 \right]}}$$

$$= \frac{(10)(955) - (90)(120)}{\sqrt{[(10)(900) - (90)^2][(10)(1806) - (120)^2]}}$$

$$= \frac{9550 - 10800}{\sqrt{[9000 - 8100][18060 - 14400]}}$$

$$= \frac{-1250}{\sqrt{[9000 - 8100][18060 - 14400]}} = \frac{-1250}{1814 - 94} = -0.688$$

Thus the value of Pearson r coefficient of correlation is –0.668 of X text and Y test.

Major Characteristics of Pearson *r*

The major characteristics of Pearson r are as under :

- *Pearson r is a product moment r:*

As we know in physics or mechanics, a moment is defined as a tendency to produce motion about a point. Thus in statistics the first moment about zero as an origin is mean, which is equal to $\Sigma x/N$. Also the first moment about mean is always zero since $\Sigma x/N = 0$. If the researcher raises the scores to the second power, he gets second moment and therefore, $\Sigma x^2/N$ is the second moment about the mean. This is similar to the formula of variance, that is σ^2. If a moment involves the product of two variables, it is called as product moment thus, the fact that the product moment leads r, has led statisticians to conclude *r* as the Pearson product moment coefficient of correlation.

- *Pearson can be either positive or negative*:
  - Pearson r is confined within limits of –1.00 to +1.00.
  - In Pearson r_{xy} equals r_{yx}. It does not matter whether it is (8)(4) or (4)(8) because their product remains unchanged. If each product remains unchanged by such a reversal, it follows that if one reverses all of the products, their total would remain the same and hence $r_{xy} = r_{yx}$.
  - Pearson r may be regarded as arithmetic mean. In computing *r* standard deviation is computed with N (sometimes $N-1$) as denominator, and therefore, *r* is exactly the mean of such standard score products.

Assumptions of Pearson *r*

There are certain assumptions of Pearson *r*. For yielding effective result, it is essential that these assumptions must be fulfilled before applying the test. Some of these assumptions are as under:

- One popular assumption of Pearson r is that the units of measurement are equidistant throughout the range of scores involved in correlation and this is true for both variables. This assumption presents no difficulties. It is quite obvious that a difference in height from 5 feet 8 inches to 6 feet is the same as difference from 4 feet 8 inches to 5 feet. One is also correct in assuming a difference of 20 IQ points on a standardized intelligence test means the same thing, whether the change is from Mohan's IQ of 50 to Sohan's 70 or from Ram's IQ of 80 to Shyam's IQ of 100.
- In computing Pearson *r*, it is assumed that the regression is *linear*. In fact, linearity refers to the tendency on the part of data, when plotted, to follow a straight line as closely or more closely than some other curve. In a scatter diagram the regression is linear if the means of successive columns lie on straight line and means of successive rows lie on another straight line. If the line connecting the means of the columns and the line connecting the means of rows does not deviate systematically from straight line, we have sufficient evidence for linear regression. To make it more clear, if each time the X score increases by one unit, the Y-score increases (or decreases) by a constant amount, the regression is linear. Thus if each time the researcher selects a group of males who are one inch taller and he finds that their average weight is 3 pounds higher than the preceding group, it constitutes the example of linear regression. If regression is not linear, Pearson is inapplicable (Kurtz & Mayo, 1980).
- Pearson *r* does not assume normality although many people believe that both variables must be normally distributed for justification of computation of *r*. In some interpretation of *r*, the normal distribution within each column and row is assumed but in most interpretations, no such assumptions are made. Thus no matter what the shape of distribution of either variable is we are always justified in computing *r* for determining the strength of linear relationship between two variables under study.

- Pearson *r* does *not* assume that an increase in score on one variable is or is *not* related to increase or decrease in the other variable. Pearson *r* only measures the strength of amount of the relationship between two variables (if it exists). In a nutshell, the researcher does not assume anything about either the direction or the amount of the relationship. Instead, he uses Pearson r to determine the size and sign of the existing correlation which may be positive, negative or zero.

Interpretation of Pearson *r*

The interpretation of Pearson *r* is very important. When should *r* be called high, when low and when medium? A Pearson *r* of .60 between two variables indicate low relationship or high relationship? What should be the level of *r* for making prediction from one variable to another? These questions can be satisfactory answered only when we are able to correctly interpret Pearson *r*.

There are various ways of interpreting Pearson *r*. Some of the important and popular methods of interpretation are as under :

- **Pearson *r* may be interpreted in terms of simple verbal description :**

The obtained Pearson *r* may be interpreted verbally by saying that *r* from .00 to ± 0.20 indicates negligible relationship, from ± 0.20 to ± 0.40 indicates low correlation, from ± 0.40 to ± 0.70 indicates a substantial or marked relationship and from ± 0.70 to ± 1.00 indicates high or very high (or perfect) relationship. There is a fair agreement among the researchers regarding these interpretation. This is only a broad and somewhat tentative interpretation. This is because of the fact that Pearson *r* is subjected to influence by many factors such as nature of variables, variability of the group, the reliability coefficients of the tests used for assessing the variables as well as by the purpose for which *r* was computed. In a nutshell, Pearson *r* is always judged with references to the conditions under which it was obtained as well as by the objectives or purpose of the research. For example, if the purpose of the research is to make prediction regarding the standing of an individual on one of the two variables being correlated, a high value of Pearson *r* would prove to be better than when the purpose of the research is simply to know about the likely achievement of the persons.

- **Pearson *r* may be interpreted in terms of coefficient of alienation :**

The Pearson *r* can also be interpreted in terms of coefficient of alienation symbolized as *k*. The coefficient of alienation may be thought as measuring the absence of relationship between *X* and *Y* in the same sense as *r* measures the presence of relationship between *X* and *Y*. As a consequence, when $k = 1.00$, $r = .00$ and when $k = .00$, *r* is 1.00. The larger the value of *k*, the smaller the extent of relationship and therefore, there will be less precise the prediction from *X* to *Y*. The formula for computation of *k* is as under :

$$k = \sqrt{1 - r^2}$$

Take an example suppose r between X and Y is .60, then, its $k = \sqrt{1-(60)^2} = \sqrt{.64} = .80$. Hence, the dispersion of score averages about regression line is about 80 percent as much as their dispersion about their own mean. The quantity 1–12 us sometimes also called as an *index of efficiency of prediction*. In a nutshell, the quantity $\sqrt{1-r^2}$ is a good index of the predictive value of the correlation coefficient and gives us a measure of the extent to which the errors in our estimate have been reduced. When $r = \pm 1.00$, k is equal to .00 and it obviously means that we have made no error in predicting one score from other. When $r = .00$, $k = 1.00$ and therefore, the standard error of estimate is the same as the standard deviation of the marginal frequencies. On the other hand, where $r = 0.50$, $k = 0.87$ and it obviously meant that our error of estimates are .87 as large as, or 13% smaller than, they would be if the two variables were not correlated.

Interpretation of r in terms of k (or reduction in the standard error of estimate) makes several assumptions. Besides equality of units and linear regression, it assumes that scores in each column and each row in the scatter diagram are normally distributed about the regression line. It also assumes that the standard deviation in various rows and columns are equal.

- **Interpretation of Pearson r in terms of coefficient of determination :**

As we know, variations of scores around sample mean is referred to as the *total variation* or *total sum of square* (TSS). This total variation is a measure of the variation in the Y scores (usually the dependent variable). It is considered basic to the determination of the variance and the standard deviation. This total variation consists of two components that may be added together. These two components are *explained variation* and *unexplained variation*. Thus

Total variation = unexplained variation + Explained variation

or, $\Sigma(Y-\bar{Y})^2 = \Sigma(Y-Y')^2 + \Sigma(Y'-\bar{Y})^2$

where, Y' = predicted score and $\bar{Y}$ is the mean of Y scores. When $r = 0.00$, $\Sigma(Y'-\bar{Y})^2$ is also 0.00. Consequencely, the total variation is equal to unexplained variation. In simple words, when $r = 0.00$, all variation is unexplained. On the other hand, when $r = 1.00$, $\Sigma(Y-Y')^2 = 0.00$ since all of the scores are on the regression line. In such situation, the total variation equals with explained variation. In simple words, when $r = 1.00$, all variation is explained :

There ratio of the explained variation to the total variation is called coefficient of variation and is symbolized by r^2. Thus equation for r^2 is

$$r^2 = \frac{\text{explained variation}}{\text{Total variation}} = \frac{\Sigma(Y'-\bar{Y})^2}{\Sigma(Y-\bar{Y})^2} \qquad (25.4)$$

Thus, when explained variation is 150.00 and the total variation is 200, then $r^2 = 150/2 = 0.75$. Then $r = \sqrt{r^2} = 0.866$. The statistic coefficient of determination is so called presumably because it is the proportion of variation of the one variable that is determined by the variation in the other. The value of r^2 is always *positive* and varies from 0 to 1.

In general, in interpreting the magnitude of relation between two variables (X and Y) the coefficient of determination is more informative. If $r = 0.70$, $r^2 = (0.70)^2 = 0.49$, we can state that 49 percent of variation of one variable *(Y)*, is predicted from the variation of the other *(X)*. The remaining 51 per cent of variation of one variable is unexplained. When a $r = .90$, the unexplained variance is 19 percent and explained variance is 81 percent.

Since r^2 represents the proportion of variation accounted for, $1 - r^2$ represents proportion of variation that is not explained in terms of the correlation between X and Y. This concept is known as the *coefficient of indetermination* and symbolized as K^2. Thus K^2 represents proportion of variation in Y that must be explained by variable other than X. Since $r^2 + k^2 = 1.00$, the proportion of the variance in Y not accounted by X is given by k^2 when $r^2 = 0.50$, $k^2 = 0.50$.

- **Interpretation of Pearson r in terms of coefficient of forecasting efficiency :**

Pearson r can also be interpreted in terms of the *coefficient of forecasting efficiency* symbolized as E or also known as *coefficient of dependability* or *the predictive index*. E is derived from k as under :

$$E = 1 - \sqrt{1 - r^2} \qquad (25.5)$$

or

$$E = 1 - k$$

when, r between X and Y scores is 0.60, from formula 25.5, $E = 1 - \sqrt{1 - (.6)^2} = 1 - \sqrt{1 - .36} = 1 - \sqrt{.64} = 1 - .80 = .20$, which means test's forecasting efficiency is 20%. When $r = .96$, $E = -.56$ that means the forecasting efficiency of the test is 56%. Likewise, when $r = 0.98$ and E becomes .80, the test is said to be 80% efficient.

Correlation and Causation

The very fact that correlation depends upon the extent to which one variable is affected by another squarely brings up the question of causation. When we talk about *cause*, we usually refer to a sufficient reason for the occurrence of an event and thus, we think about an orderly and invariable sequence *effect preceded by cause*. We ordinarily interpret causal relationship like this : A and B are related as cause and effect if B occurs after A and if B does not occur, when A is absent. Thus in any causal relationship an invariable sequence of events is involved. Such relationships are said to be *sine qua non* of understanding and controlling the environment because they enable us to explain and also predict events.

So far as the correlation is concerned, it does not demonstrate invariable *sequence* and hence, does not indicate which of the two related variables is cause and which is the effect. Thus causation implies an invariable sequence whereas correlation is simply the mutual association between X and Y. Thus in psychology and education, the presence of correlation between two variables can rarely be interpreted as implying a direct causal relationship. When X and Y are correlated, they may be correlated with an underlying variable or a set of variables and not because of one is cause and other is the effect. For example, in a group of children of varying ages, a correlation may be found between height and weight. Such a correlation may come about because both height and weight are correlated with age. If the effect of age is somehow removed, the correlation may vanish. Besides, we are constantly bombarded with reports of relationship : Cigarette smoking is related to lung cancer; alcohol consumption is related to birth defects; carrot consumption is related to a better eyesight. Do these relationships mean that cigarette causes lung cancer or alcohol consumption causes birth defects or carrot consumption causes a better eyesight? The answer is definitely no. Although there may be a causal relationship but a simple correlation does not prove it.

What we are emphasizing is that correlational studies simply do not allow inference of causation. Correlation is a *necessary* but not a sufficient condition for establishing cause and effect relationship between two variables. According to Huff and Geis (1954) the faulty causal inferences from correlation data are called *post hoc fallacy*.

Despite this limitation, correlation is extremely important in premilinary investigation of causal relationship. It is often found that the variables which are causally related show correlation and that variables that don't demonstrate correlation are not relate to casualty (Tate, 1948).

Rank Difference Correlation

In psychological and educational researches, it is often the case that variables whose relationships the researcher wishes to investigate are available in order of merit, importance or on some other quality. In such situations differences among individuals are expressed by ranking them in 1-2-3-4 order. For example, the person may be ranked in order of merit of punctuality, honesty, salesmanship, academic achievement, social adjustment, etc. Thus in these situation data are obtained on ordinal measurement. There may be situations where data have been obtained in scores such as in interval measurement and ratio measurement, but due to some conveniences, the researcher prefers to convert these data in terms of ranks. For calculating correlations between X and Y variable, data obtained in either of the above situation, statisticians have provided what is called *rank difference correlation*.

There are two popular methods of rank-difference correlation : *Spearman Rank-order correlation* and *Kendall Rank-order correlation*. A brief description of these two methods is as under :

• Spearman Rank-order correlation :

Spearman proposed a method of computing correlation between two sets of rank data. This method is known as Spear Rank-order Correlation and is designed as P (read *rho*). The formula for calculating P is as under :

$$P = 1 - \frac{6\Sigma d^2}{N(N^2 - 1)} \quad (25.6)$$

where,

- P = rank difference correlation coefficient
- Σd^2 = sum of the squares of differences in rank
- N = number of pairs

The calculation of P is simple. We rank the data (if they are given in score), find the difference between the paired ranks, square them, sum to obtain Σd^2 and then apply formula P. Table 25.3 illustrate the calculation of P.

Table 25.3: Calculation of P from a set of scores obtained on two academic achievements by 6 students

Persons	1 X	2 Y	3 Rank$_X$	4 Rank$_Y$	5 D	6 D²
A	10	27	6	5	+1	1
B	15	28	4	4	0	0
C	18	20	2	6	−4	16
D	16	30	3	3	0	0
E	14	32	5	2	3	9
F	20	36	1	1	0	0
						$\Sigma D^2 = 25$

P by formula 25.6:

$$P = 1 - \frac{6 \times 25}{6(6^2 - 1)} = 1 - \frac{150}{6 \times 35} = 1 - \frac{150}{210} = 0.29$$

The correlation between Ranks of X and Y variable is positive but low.

In the example of Table 25.3 the scores obtained on X and Y test have been ranked from 1, largest, to 6, smallest. They might equally be ranked from 1, smallest to 6, largest. Had we ranked the scores from 1, smallest, to 6, largest, P would have been −0.29 but in view of the changed meaning of the ranks, the interpretation would have been the *same*.

The coefficient P varies from −1.00 to +1.00. The more P departs from 0, the stronger is the relationship. In fact, P is a product-moment coefficient of correlation between ranked variables. Had we treated the ranked data of Table 25.3 as raw scores and applied formula 25.3, we would have obtained the same result. At this point one thing worthmentioning is that when N is below 25, it is often advisable to rank these scores and to compute the correlation coefficient P by rank difference method rather than by computing r by the longer and more laborious product-moment method. In such cases where N is small, P yields as adequate result as that obtained by Pearson r. Moreover, P is easier to compute than r. Researchers have revealed that when ranks are treated as scores, and there are no ties among the ranks, $P = r$. But when there are tied ranks, a correction is needed in P to be equivalent to r (Edwards, 1954). In fact, the presence of ties in either or both sets of ranks has no influence upon the computation of D^2 but it does decrease P. Kendall (1955) have also introduced a correction for P between ranking having ties but in practice this correction is rarely used.

The significance of P can be tested in two ways. One way is to test the significance of P against the null hypothesis by means of a Table with a given N (Siegel & Castellan, 1988, p. 360). In Table 25.3, $N = 6$ and $P = 0.29$. This obtained P is not significant at .05 level because at this level the required value of P is 0.886. Therefore, we accept the null hypothesis, which states that the two variables under study are not associated (that is, independent) in the population and the observed value of P differs from zero only by chance.

Some statisticians recommend the testing of the significance of P by Student's t and the formula is:

$$t = P\sqrt{\frac{N-2}{1-P^2}} \qquad (25.7)$$

Thus, substituting the values, we get

$$t = 0.29\sqrt{\frac{6-2}{1-(.29)^2}} = 0.29\sqrt{\frac{4}{1-.08}}$$

$$= 0.29\sqrt{\frac{4}{0.92}} = 0.29\sqrt{4.35} = 0.29 \times 2.08 = 0.60$$

Here $df = N-2 = 6-2 = 4$. At df 4, the required value of t for being significance at .05 level is 2.776 (two-tailed test). Since obtained t is less than 2.776, we conclude that P is not significant. Hence null hypothesis is accepted with conclusion that two set of ranked data are not correlated or are independent whatever correlation has been found, that is by chance only.

When N is larger than about 20 to 25, the significance of P under the null hypothesis may be tested by computing z by the following formula.

$$z = P\sqrt{N-1} \qquad (25.8)$$

If $P = 30$ and $N = 25$ we get

$$z = .30\sqrt{24} = 1.47$$

Since the value of z is less than 1.96, it is concluded that the obtained P is not significant even at .05 level.

Kendall's Rank difference correlation

Kendall Rank difference correlation coefficient symbolized as T or τ (read as tan) measures correlation with the same sort of data for which Spearman's Rank difference correlation coefficient is suitable. In other words, Kendall's rank difference correlation is applicable to two sets of rank data or to two sets of scores when converted into rank. One obvious advantage of Kendall's rank difference correlation coefficient (T) over Spearman's rank difference correlation coefficient (P) is that T can be generalized to a partial correlation coefficient. T can be calculated by the following formula

$$T = \frac{2S}{N(N-1)} \quad (25.9)$$

where,
N = number of objects or individuals ranked on both X and Y variable
S = total number of agreements in ordering minus number of disagreement in ordering

Let us take an example to illustrate its calculation. Below is the calculation of T from set of ranked data of seven subjects.

Table 23.4: Calculation of T from two sets of ranked data

Individuals	:	A	B	C	D	E	F	G
X test Rank	:	3	2	4	6	1	5	7
Y Test Rank	:	5	4	7	3	2	1	6

For computing T, we shall rearrange the order of the individuals so that ranking on X test becomes in natural order.

Individuals	:	E	B	A	C	F	D	G
X test Rank	:	1	2	3	4	5	6	7
Y Test Rank	:	2	4	5	7	1	3	6

Ranks on X test appears in natural order and corresponding rank of the individual appear on Y test. Having arranged the ranks on variable X in their natural order, now we determine the value of S for the corresponding order of ranks on the variable Y. On Y rank, first rank from left is 2. On the right of rank 2, five ranks are larger than 2 and one rank is smaller than 2. Therefore, its contribution to S would be 5 – 1 = 4. The next rank on Y test is 4. There are three ranks larger than 4 and two ranks are smaller than 4. So its contribution to S would be 3 – 2 = 1. The next rank on Y is 5. Right to 5, there are two ranks larger than 5 and two ranks are smaller than 5. So its contribution to S would be 2 – 2 = 0. In this way, the contribution of each rank on Y variable has been determined in this way :

S = (5 – 1) + (3 – 2) + (2 – 2) + (0 – 3) + (2 – 0) + (1 – 0)
= (4) + (1) + (0) + (–3) + (2) + (1) = 8 – 3 = 5

Now by substituting the values in formula 25.9, we get

$$T = \frac{2 \times 5}{7(7-1)} = \frac{10}{42} = 0.238$$

The significance of T can be tested by using a Table (Siegal & Castellan, 1988, p. 362) when $N \leq 10$; this Table can be used. If the obtained $p < \alpha$, null hypothesis may be rejected. In the above example $T = 0.238$, $N = 7$ and when we enter table with these entries, we get $p = 0.281$. Since obtained T is less than p value, that is, 0.281, null hypothesis is rejected and therefore, the two sets of ranked data are correlated.

When N is larger than 10, T is approximately normally distributed and therefore, its significance can be tested with the help of z to be calculated by formula 25.10

$$z = \frac{3T\sqrt{N(N-1)}}{\sqrt{2(2N+5)}} \quad (25.10)$$

Suppose $N = 20$, $T = 0.67$, then its z value would be

$$z = \frac{3 \times .67\sqrt{20(20-1)}}{\sqrt{2(2 \times 20 + 5)}}$$

$$= \frac{2.01 \times 14.50}{\sqrt{2 \times 45}} = \frac{29.145}{9.48} = 3.07$$

Associated with $T = 0.67$, $z = 3.07$ which permits us to reject null hypothesis at the significance level of .001 (Siegel & Castellan, 1988, p. 319).

A comparative study P and T:

As we know both P and T can be applied for the same set of ranked data. The numerical value of P and T will not be equal for the same set of rank data. This happens because P and T have different underlying scales and therefore, numerically, they cannot be directly compared.

P and T also differ from the pointview of their interpretation. P, as it has already been explained, is the same as Pearson r computed from ranked data. T, on the other hand, is nothing but the *difference* between the probability that, in the observed data, X and Y are in the same order and the probability that X and Y data are in different order.

Despite all these, both P and T tend to utilize the same amount of information in the data and both tend to detect the existence of association in the population.

Two Important correlation coefficients Involving more than Two Variables

Here in this section an attempt will be made to discuss two important correlation methods, which involve *more than* two variables. They are Partial correlation and Multiple correlation coefficient (R).

Partial correlation :

Partial correlation may be regarded as an extension of theory and technique of simple two

variable linear correlation to such problems, which involve three or more variables. The correlation between two variables is sometimes misleading if there is dependence of these two variables upon a third variable. For example, correlation between height and weight of children of varying ages will be positive because both height and weight of children are related to increasing age. As age of children grows, both height and weight also increase. Thus the correlation between height and weight of children may be misleading if age is not controlled. The factor of age can be controlled in two ways :
- experimentally selecting all children of the same age
- Statistically by eliminating the impact of age by partial correlation.

In former method the researcher has to restrict drastically the sizes of the sample. But in case of partial correlation since the researcher uses all of the data, it is preferred to the experimental method, that is, the former method.

Many attributes increase regularly with advancement in age from about 5 to 19 years of age. Examples of such attributes are height, weight, physical strength, intelligence, vocabulary, general knowledge, reading skills, etc. Let us take an example. Suppose the researcher wants to correlate height and physical strength in varying age group of children. In this situation, the correlation between these two attributes is likely to be high and positive because of the age factor which is highly correlated with age variable. Therefore, the effect of age has to be partialled out or removed or eliminated. Partial correlation is one such statistic which allows to calculate the correlation between X_1 (one variable) and X_2 (second variable) while eliminating the effect of X_3 (third variable).

Let us now see what is meant by eliminating or removing or partialling out the effect of third variable. To explain its meaning let X_1, X_2 and X_3 stand for three variables and X_1 and X_2 are to be correlated eliminating the impact of X_3. In fact, all or part of correlation between X_1 and X_2 may be due to the fact that both are correlated with X=3. As we know, a score on X_1 may be divided into two parts : One part is the score predicted from X_3 and the other part is the residual or error of estimate in predicting X_1 from X_3. Likewise, a score on X_2 may be divided into two parts: one part is predictable from X_3 and a residual or error of estimate in predicting X_2 from X_3. In fact, the correlation between these two sets of residuals (or error of estimate) in predicting X_1 from X_3 and in predicting X_2 from X_3 is called as partial correlation coefficient. It is the part of correlation, which remains when the effect of third variable is partialled out.

The formula for calculating the partial correlation eliminating the third variable is as under :

$$r_{12.3} = \frac{r_{12} - r_{13}r_{23}}{\sqrt{(1 - r_{13}^2)(1 - r_{23}^2)}} \quad (25.11)$$

where,
r_{12} = correlation between X_1 and X_2
r_{13} = correlation between X_1 and X_3
r_{23} = correlation between X_2 and X_3

$r_{12.3}$ = correlation between X_1 and X_2 (or residuals) when X_3 has been removed.

The correlation between two variables is called as *zero-order correlation* and the correlation between two variables when the impact of third variable is eliminated is called *first order partial correlation*. Take an example to illustrate the calculation of first-order correlation. Suppose a researcher obtains the following zero-order correlation between two variables:

X_1 and X_2 = 0.83
X_1 and X_3 = 0.78
X_2 and X_3 = 0.80

In this situation, partially out the likely impact of third variable (X_3) upon X_1 and X_2, it would be by formula 25.11

$$r_{12.3} = \frac{.83-(.78)(.80)}{\sqrt{(1-.78^2)(1-.78^2)}} = \frac{.83-.62}{\sqrt{(.39)(.36)}} = \frac{.21}{.37} = -0.57$$

Partial correlation may be used to remove the effect of more than one variable. When the partial correlation between X_1 and X_2 is calculated with the effect of both X_3 and X_4 eliminated, it is called as *second-order partial correlation*. The formula for calculating second-order partial correlation is as under

$$r_{12.34} = \frac{r_{12.3} - (r_{14.3})(r_{24.3})}{\sqrt{(1-r_{14.3}^2)(1-r_{24.3}^2)}} \qquad (25.12)$$

where,

$r_{12.34}$ = partial correlation between X_1 and X_2 eliminating the impact of X_3 and X_4
$r_{12.3}$ = First order partial correlation between X_1 and X_2
$r_{14.3}$ = First order partial correlation between X_2 and X_4

Since partial correlation deals with three or more variables simultaneously, it is classified as a multivariate method and then, obviously it is related to the method of multiple correlation and factor analysis. In a very real sense, use of partial correlation is a substitute for experimental controls. There are two basic assumptions underlying partial correlation:

- There exists linearity of regression of the two variables upon the third variable
- There should be equal scattering of the values of the two variables for the different values of the third variable.

In reality, neither of these assumptions is ordinarily satisfied. Therefore, the partial correlation should be regarded as a sort of average value.

The obtained value of partial correlation may be tested for it significance by using t test. The formula for the required ± is as under:

$$t = \frac{r_{12.3}}{\sqrt{(1-r_{12.3}^2)(N-3)}} \qquad (25.13)$$

where, N = Number of subjects and the rest of the symbols are defined as it has been done in formula 25.11.

Here $df = N-3$ and the obtained value of t is further referred to the Table of t for interpretation. The significance of partial r may also be determined by way of the z transformation (Garrett, 1981).

Sometimes a situation is encountered where the impact of X_3 (third variable) is eliminated from one variable only say for example X_1 and X_2 is then correlated with the residual of X_1. This type of situation calls for what is called *semi-partial correlation* or also known as *part correlation*. The formula for part correlation between X_1 and residuals of X_2 with X_3 eliminated is as under :

$$r_{1(2.3)} = \frac{r_{12} - (r_{13})(r_{23})}{\sqrt{1 - r_{23}^2}} \quad (25.14)$$

If in a situation $r_{12} = 0.80$, $r_{13} = 0.75$ and $r_{23} = 0.60$, then by formula 25.14, part correlation by formula would be

$$r_{1(2.3)} = \frac{.80 - (.75)(.60)}{\sqrt{1 - 0.60^2}} = \frac{.80 - .45}{\sqrt{0.64}} = \frac{.35}{.80} = .437 = .44$$

Generally, the situation for part correlation arises when the measurements have been obtained prior to an experimental treatment and subsequent measurements on the same participants under the condition of experimental treatment.

Multiple correlation

In some situation the researcher wants to estimate the relationship between one variable (X_1) and a combination of two other variables. This situation warrants for what is called multiple correlation designated as letter R. In other words, correlation between a set of obtained scores and the same scores predicted from multiple regression equation is called as a coefficient of multiple correlation. Suppose the researcher has the following three zero-order correlation coefficients based upon the three variables (N = 200).

X_1 = Academic achievement
X_2 = Intelligence test score
X_1 = General knowledge test scores

The zero-order correlations are as under :

$r_{12} = 0.40$
$r_{13} = 0.50$
$r_{23} = 0.30$

The researcher wants to compute multiple correlation coefficient between academic achievement and the combined effect of the two tests. The formula is as under :

$$R_{1.23} = \sqrt{\frac{r_{12}^2 + r_{13}^2 - (2r_{12}r_{13}r_{23})}{1 - r_{23}^2}} \quad (25.15)$$

$$R_{1.23} = \sqrt{\frac{.40^2 + .50^2 - (2)(.40)(.50)(.30)}{1 - .30^2}}$$

$$= \sqrt{\frac{.16 + .25 - .12}{.91}} = \sqrt{\frac{.29}{.91}} = \sqrt{.31} = .56$$

Multiple R has the following important characteristics:
- It is always positive
- It is always less than 1.00
- It is always greater than the zero-order correlation coefficients such as r_{12}, r_{13} and r_{23}.

A multiple R coefficient of 0.56 indicates that the scores in variable 1 (academic achievement) *predicted* from a multiple regression equation containing variables 2 (intelligence test scores) and variable 3 (general knowledge test scores) correlate .56 with score *obtained* on variable 1. In other words, R gives correlation between a criterion (1), which is dependent variable here and a team of test scores (2 and 3), which are independent variables or called as *predictors*. How many scores on the two predictors be combined to predict academic achievement?

The significance of R can be estimated by calculating its standard error. The formula is as under:

$$SE_R = \frac{1 - R^2}{\sqrt{N - m}} \qquad (25.16)$$

where,
N = size of sample
m = number of variables being correlated
$(N-m)$ = degrees of freedom

In the above example, $R = .56$ 3 and $N = 200$, then SE_R by formula 25.16 would be

$$SE_R = \frac{1 - .56^2}{\sqrt{200 - 3}} = \frac{1 - .31}{\sqrt{197}} = \frac{.69}{14.03} = .049 = .05$$

The .95 confidence interval for the population R is $.56 \pm 1.96 \times .05$ or from 0.462 to 0.658. Therefore, the obtained R is significant.

Meaning of Regression and Regression Equation

The concept of regression is very closely related to the concept of correlation. As it is well-known, the idea of regression came first and subsequently, the correlation method was evolved. It began with the scientific work of Sir Francis Galton in 1885 who was making very important studies on heredity where he did a study of the relationship between heights of children and the heights of their parents. His final observation was that the mean height of the offspring deviated less from their general mean height than the heights of the parents from which they came. This falling back of heights of offspring toward general mean is called as the law of *filial regression* or more popularly known as *regression towards the mean*. His conclusion was that the offspring of tall parents tended on the average to be shorter than their parents and the offspring of short parents tended on the average to be taller than their parents. He used the term 'regression' for such phenomenon. In statistics the term regression is used to refer to the problem of predicting one variable from the knowledge of another or possibly many other variables. In fact, Galton wanted a single value which could express the amount of this regression phenomenon in any particular relationship problem. Karl Pearson solved this problem by developing a method of correlation to which his name is attached (Pearson r).

In statistical prediction attempt is made to predict the theoretical score of an individual in one variable after knowing his score in a related variable. In prediction situation, the dependent variable Y is customarily referred to as the *criterion variable* and the independent variable X is known as the *predictor variable*. The task of regression is accomplished through what is called *regression equation*. In a simply word, a regression equation is an equation for predicting the most probable value of one variable from the known value of anther variable. The obvious purpose of regression equation is simply to tell the researcher what score he may expect to attain on one variable if he already knows that person's score on another variable. Regression can be computed only when the X (predictor or independent variable) and Y (the criterion variable or dependent variable) possess correlation other than zero. In fact, regression translates the association between two variables (or between one variable and a combination of two or more variable) into a mathematical pattern into which one variable is expressed as a function of other (s).

Regression is of two types : *simple regression* and *multiple regression*. In simple regression only two variables—criterion variable and predictor variable exists and here scores on the criterion variable are predicted from the given scores of the single predictor variable. For example, when on the basis of intelligence score (predictor variable) of a student, academic achievement (criterion variable) is predicted, it is called as simple regression. On the other hand, in multiple regression more than two variables are involved. A prediction about the criterion variable is made on the basis of a combination of more than two predictor variables. For example, when the academic achievement of a student is predicted on the basis of intelligence score and the amount of time devoted for the study, it forms an example of multiple regression.

In this chapter our concern is only with the data that are linearly related, if related at all. Therefore, the regression equation of interest is the equation of a straight line. This is the simplest of the pattern which characterizes many of the bivariate data of psychological and educational research. The general equation for the straight line is :

$$Y' = bX + a \qquad (25.17)$$

where, Y' is the predicted value of Y and Y' is not usually the same as Y because the score predicted from this equation will not be exactly the same as one actually obtained. The value of predicted Y's will, in general, be closer to the mean of Y scores than are the values of the observe Y's.

Y' = the predicted score of Y
b = the slope of line
a = Y intercept (the value of Y when X is zero)

When b and a have been calculated, the equation describes one and only one straight line. For example, if $b = 4$ and $a = 7$, we have

$$Y' = 4X + 7$$

If Mohan has 10 on X variable, his likely score on Y would be

$$Y' = (4)(10) + 7 \text{ (by formula 25.17)}$$
$$= 47$$

Thus we have $X = 10$ and $Y' = 47$. These points form a straight line and all values which satisfy the equation 24.17 will lie on the straight line.

Ordinarily, two conditions are involved in using the regression equation in prediction
- The scores on the criterion variable (Y variable) are a linear function of the scores on the predictor variable (X variable).
- Both the criterion variable and the predictor variable should have unimodal or symmetrical distribution in population.

The logic underlying statistical prediction as well as under the estimation of accuracy of statistical prediction involves the following major points :
- There exists correlation between criterion variable and predictor variable in the sample from the specified population.
- The relationship holds true in further samples taken from the same population.
- The criterion scores to be observed, in fact, pair with the predictor scores is essentially the same way as they were in the past.
- The accuracy of predicted criterion scores can be easily estimated from the extent to which past criterion scores have scattered about the regression line.

Obtaining a and b coefficient

In linear regression equation such as equation 25.17, there are two important coefficients: b coefficient and a coefficient. It is expected that the students must know how to obtain there two coefficients :

b coefficient :

As explained earlier, b coefficient indicates the relationship between the changes in Y in reference to change in X. This ratio of change in one variable to change in another variable is referred to s the *slope* of the line. The slope of any line is nothing but simple the ratio of distance in vertical direction to the distance in horizontal direction (see Figure 25.1). The slope of a line, then, clearly measures the rate of change and in equation form,

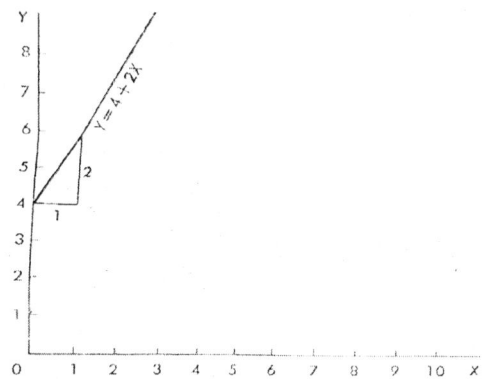

Figure 25.1 A Line with Positive Slope.

it clearly tells us how much of Y score changes (that is, either increases or decreases) for each unit of change in X score. In figure 25.1 for one unit of X score, there occurs two units of changes in Y score. Therefore, *b* coefficient is 2.

There are several ways of calculating *b* coefficient from the given data. Two popular ways areas under :

• **Based on raw scores :**

b coefficient can be directly calculated from X and Y raw scores by using the following formula :

$$b_{YX} = \frac{\Sigma XY - \frac{(\Sigma X)(\Sigma Y)}{N}}{\Sigma X^2 - \frac{\Sigma X^2}{N}}$$

(25.18)

where,
- *b* = coefficient in predicting Y on the basis of X
- X = X scores
- Y = Y scores
- N = number of subjects

Take an example to illustrate the calculation of *b* coefficient. Suppose the scores of 5 students on X test and Y test have been obtained as shown in Table 25.4. We need to computer *b* coefficient using formula 25.18.

Table 25.4: Scores of 5 students on X and Y test

X	Y	XY	X²
4	10	40	16
6	16	64	36
9	12	108	81
2	15	30	4
5	11	55	25
$\Sigma X = 26$	$\Sigma X = 64$	$\Sigma XY = 207$	$\Sigma X^2 = 162$

$$b_{yx} = \frac{\Sigma XY - \frac{(\Sigma X)(\Sigma Y)}{N}}{\Sigma X^2 - \frac{\Sigma X^2}{N}}$$

$$= \frac{207 - \frac{26 \times 64}{5}}{162 - \frac{(26)^2}{5}} = \frac{207 - 333.8}{162 - 135.2} = \frac{-126.8}{26.8} = -4.73$$

b coefficient can also be calculated by using Pearson r :

If the Pearson r and the standard deviations of the variables are already calculated, b coefficient can be calculated by using formula 25.19 for predicting Y on the basis of X scores :

$$b_{yx} = r_{xy} \frac{\sigma_y}{\sigma_x} \qquad (23.19)$$

Where r_{xy} is the Pearson r between X and Y scores and σ_y is the standard deviation of Y scores and σ_x is the standard deviation of X scores. If $r_{xy} = .70$, $\sigma_x = 5.00$ and $\sigma_y = 6.00$, then, b coefficient by formula 25.19 would be

$$b_{yx} = .70 \frac{6}{5} = 0.84$$

a coefficient :

The quantity of *a* is a constant in straight line regression equation. It is *Y intercept* which is the value of Y where the line intersects the Y axis. In Figure 25.1 *a* is equal to 4 because the straight line touches at this point on Y axis.

The a coefficient is calculated using formula 25.20 in predicting Y on the basis of X

$$a_{yx} = \bar{Y} - \bar{X}(b_{yx}) \qquad (25.20)$$

Where, $\bar{X}$ and $\bar{Y}$ are the means of X and Y scores and b_{yx} is the b coefficient.

To illustrate its calculation, suppose $\bar{Y} = 25$, $\bar{X} = 20$ and $b_{yx} = 0.4$, then a coefficient would be
$$a_{yx} = 25 - (20)(.4) = 25 - 8 = 7$$

When a_{yx} and b_{yx} have been computed, their values are put in the following equation (25.21) to give the regression equation of Y on X.
$$Y' = a_{yx} + b_{yx} X \qquad (25.21)$$

For drawing the linear regression line of Y on X, the Y' score for each of the X score is computed using the regression equation. Each Y' score is subsequently plotted graphically against the corresponding X score.

Limitations of Regression Analysis

There are some limitations of regression analysis as under :
- The effectiveness of regression analysis is dependent upon the assumption that the relationship has not changed since the regression equation has been computed.
- It is also important to keep in mind while doing regression analysis that the relationship shown by the scatter diagram may not be the same if the equation is somehow extended beyond the values used in computing the equation.

Comparison between Regression and Correlation

There are some similarities and differences between correlation and regression. The important point of similarities between these two are as under :

Similarities :
- The coefficient of correlation (r) takes the same sign as the regression coefficients such as b_{yx} and b_{xy}.
- Both correlation and regression can be simple as well as multiple. The simple correlation and simple regression are called bivariate statistics because only two variables are involved. Multiple correlation and multiple regression are called as multivariate statistics because more than two variables are involved.
- If the value of b coefficient becomes significant at a given level of significance, r also becomes significant at that level of significance.

Differences :
- The correlation coefficient need not imply cause-and-effect relationship between dependent and independent variable. On the other hand, the regression coefficient clearly establishes the cause-and-effect relationship. The variable representing the cause is the independent or prediction variable whereas the variable representing the effect is taken as dependent or criterion variable. Thus regression very clearly indicates the cause-and-effect relationship.

- Correlation coefficients are *descriptive statistics* for assessing the relationship between variables whereas the regression is the *predictive statistics* for predicting the most likely value of a variable based upon one or more other correlated variables.
- Correlation indicates relationship between two or more variables, which vary so that the changes in one variable are accompanied by changes in other variable. On the other hand, by regression is meant stepping back or returning to the average value and, in fact, it represents a mathematical measure expressing the average relationship between the two or more than two variables.
- A correlation coefficient such as r_{xy} or r_{yx} is a relative measure of the linear relationship between X and Y variable and is independent of the unit of measurement involved. On the other hand, regression coefficients such as b_{xy} or b_{yx} are absolute measures representing changes in the amount of variable for a unit change in the value of other variable.
- A correlation coefficient that measure, the direction and degree of the linear relationship between two variables, is actually mutual and symmetric, that is, $r_{xy} = r_{yx}$. Moreover, it is immaterial which of X and Y is dependent variable and which is independent variable. In regression the picture is different. Here the major purpose is to establish functional relationship between X variable and Y variable and on the basis of such relationship attempt is made to predict the value of dependent variable for any given value of the independent variable. It means that the regression emphasizes upon the fact that which variable is dependent one and which is independent one. That is the reason why here there are two regression lines Y on the basis of X and X on the basis of Y.
- Correlation is limited only to the study of the linear relationship between the variables and therefore, its application is limited. Regression, on the other hand, studies both linear and nonlinear relationship between the variables under study and therefore, has comparatively under applications.

Summary and Review

- The coefficient of correlation is a numerical index, which expresses quantitatively the magnitude and direction of the relationship between two or more than two variables. The correlation may be positive, negative or zero.
- The most widely used method of correlation is Pearson r or product-moment correlation. There are different methods of computing Pearson r.
- There are several assumptions of Pearson r. One very important assumption is that of linearity, which refers to the tendency on the part of data, where plotted, to follow a straightline as close as or more closely than some other curve.
- There are various ways of interpreting correlation coefficient. For example, it can be interpreted in terms of simple verbal description, it can be interpreted in terms of coefficient of alienation (K) it can be interpreted in terms of coefficient of determination (r^2) and it can also be interpreted in terms of coefficient of forecasting efficiency (E).

- Rank difference correlation is another method of correlation based on ranked data or data obtained on ordinal measurement. There are two popular methods of rank difference correlation: Spearman's rank difference correlation and Kendall's rank difference correlation.
- There are two such common methods of correlation in which more than two variables are involved. One such correlation is partial correlation and the other is multiple correlation. In partial correlation the two variables are correlated by partialling out the effect of third variable. In multiple correlation the correlation is computed between one variable and a combination of two or more than two variables.
- In psychological and educational statistics regression is used to indicate the problem of predicting one variable from the knowledge of another or possibly many other variables. The task of regression is accomplished through what is called *regression equation*. In predicting one variable from the known other variable, the regression equation for straight line is $Y' = bX + a$, where b is the slope of line and a is the Y intercept. The a coefficient and b coefficient can be obtained with the help of different formulas.
- Although there are some similarities between regression and correlation these two terms differ in many ways. Obviously, there are also some limitations of regression.

Review Questions

1. Define correlation. Discuss its type.
2. Discuss the major properties of Pearson r. Also discuss its major assumptions.
3. Make a comparative study of K and r^2.
4. Discuss the ways Pearson r is interpreted.
5. Write short notes on the following :
   (a) Correlation and causation
   (b) a coefficient
   (c) b coefficient
   (d) Regression equation for straight line.
6. What is rank difference correlation? Discuss the various ways through rank difference correlation is computed.
7. Make a comparative study of regression and correlation.

Chapter – 26

z test, t-test and F test

Learning objectives :
- Significance of difference between two sample means using z test
- Significance of difference between two sample means using to t test
- z test vs. t test
- Robustness of the t test
- F test : Analysis of variance
- Summary and Review
- Review Questions

Key Terms:
F test, t test, z test, Robustness of t test, Homogeneity of variance, Posterior test, Tulley's HSD test, Scheffe's test.

z test, t-test and F test : Significance of Mean differences

In psychological and educational researches there often arises situations in which the researcher is interested in knowing the fact that whether the obtained difference between two or more than two sample means reflect a true difference between the population means or they are simply due to sampling fluctuations or some chance factors. Usually in such situation the researcher is in need of some statistical tests, which can confidently tell him about the reality. Statisticians have provided some tools like z test, t-test and F test, which easily provide information about the significance of the difference between these sample means. All these three tests are parametric statistical tests. In the present chapter attempt would be made to discuss these various statistical tests in detail.

Significance of the Difference between two sample means using z-test

With regard to the significance of difference between two means, usually two situations arose :
- When means are independent or uncorrelated and samples are large ($N \geq 30$)
- When means are correlated and sample size is small ($N < 30$)

Means are independent or uncorrelated when computed from different samples or from uncorrelated tests administered to the sample sample.

The z test is used to test the significance of difference between two means when sample is large ($N \geq 30$). A z test is based upon a standard score called as z score, which was introduced in 1924 by R.A. Fisher. Any score expressed in units of standard deviations of the distribution of scores in the population with mean set at zero and the standard deviation is equal to 1, is known as z score. A z score tells us how many standard deviations we are away from mean (either above or below). The z test is calculated by dividing the difference between two sample means by the standard error of the sample mean. If the obtained value of z test exceeds the ± 1.96, it is said to be significant at .05 level of significance and if it exceeds ± 2.58, it is said to be significant at .01 level.

Let us illustrate the calculation of z test from two uncorrelated or independent samples. Suppose that we have computed the following statistics for two samples a group of boys ($N = 100$) and a group of girls ($N = 120$) on an intelligence test.

Boys	Girls
$N = 100$	$N = 120$
$\overline{X}_1$ or Mean$_1$ = 50.55	$\overline{X}_2$ or Mean$_2$ = 40.60
Population standard deviation or $\sigma_1 = 10.66$	Population standard deviation or $\sigma_2 = 9.58$

For applying z test, we start with a null hypothesis (H_0) stating that difference between Mean$_1$ and Mean$_2$ is 0. Therefore, we are saying that the obtained difference between 50.55 − 40.60 = 9.95 is merely due to chance fluctuations. The formula for z test is under :

$$z = \frac{\overline{X}_1 - \overline{X}_2}{\sigma_{dx}} \quad (26.1)$$

where,

$z = z$ test; $\overline{X}_1 =$ Mean 1, $\overline{X}_2$ Mean 2 and $\sigma_{dx} =$ standard error of difference between two means.

The standard error of difference between two means (σ_{dx}) is calculated by the following formula

$$S_{dx} = \sqrt{S_{\overline{x}_1}^2 + S_{\overline{x}_2}^2} \quad (26.2)$$

where, $S_{\overline{x}_1}$ is the Standard error of Mean 1 and $S_{\overline{x}_2}$ is the standard error of Mean 2.

The Standard error of each sample mean is calculated as under :

$$S_{\overline{x}_1} = \frac{\sigma_1}{\sqrt{N_1}} \quad (26.3)$$

$$S_{\bar{x}_2} = \frac{\sigma_2}{\sqrt{N_2}} \qquad (26.4)$$

where, σ_1 and σ_2 are the population standard deviation

Now,

$$S_{\bar{x}_1} = \frac{\sigma_1}{\sqrt{N_1}} = \frac{10.66}{\sqrt{100}} = \frac{10.66}{10} = 1.066$$

$$S_{\bar{x}_2} = \frac{\sigma_2}{\sqrt{N_2}} = \frac{9.58}{\sqrt{120}} = \frac{9.58}{10.95} = 0.87$$

The standard error of difference between the two Means can be calculated by formula 26.2

$$S_{dx} = \sqrt{(1.066)^2 + (.87)^2} = \sqrt{1.14 + .76} = \sqrt{1.9} = 1.38$$

Now, we convert each sample mean to z score by using the following formula.

$$z = \frac{\bar{X}_1 - \bar{X}_2}{S_{dx}} \qquad (26.5)$$

Thus,

$$z = \frac{50.55 - 40.60}{1.38} = \frac{9.95}{1.38} = 7.21$$

Since the value of z score (or z test) exceeds 2.58, it is taken to be significant .01 level of significance. Therefore, the null hypothesis (H_0) is rejected and alternative hypotheses (H_1) which states that boys and girls differ on the measure of intelligence, is accepted.

The major shortcoming in using z test as an inferential statistics is that z scores require that *the researcher knows the value of population standard deviation*, which is needed to compute the standard error. Most often the standard deviation of the population is not known, the standard error for sample means can't be calculated. Without standard error, there is no means to quantify the expected amount of distance (or error) between sample mean and population mean. Therefore, we are left with no means of making precise quantitative inferences about the population based on z scores.

Major *assumptions* in using z test are as under :
- Each score in the sample occurs at random and is independent of all other scores
- The size of sample should be larger one, that is, N >30 so that the sample means and therefore, the differences between them should have normal sampling distribution in the population.
- The dependent variable is a continuous measurement variable.
- The population standard deviation should be known to the researcher.

Significance of Difference between Two Means Using *t* Test

The t-test is similar in structure to z test except that it uses the estimated standard error because the value of population standard deviation is not known.

The *t*-test is used when the size of the sample is small ($N < 30$) or large ($N \geq 30$) and the significance of different between two sample means is to be estimated. Thus for large sample both t test and z test can be used but for small sample ($N < 30$), only *t* test should be used because its distribution conform to leptokurtic t distributions only instead of the mesokurtic normal distribution.

So far as the history is concerned, the distribution of t was developed originally in 1908 by W.S. Gossett who wrote under the pen name "Student". That is why, it is also called as *Student's t-test*. Gossett spent his entire working life with a firm of Guinness, first in Dublin and later in London. Since his service rule prohibited him to publish an article in his own name, he published his research article, anonymously using the name 'Student'.

The t test is very *similar* to the z test. The t test can be calculated both for independent a correlated sample means as well as for correlated sample means. In both cases of computation, the numerator and dominator of t test formula contains the same thing, that is, $t = \frac{\text{Mean difference}}{\text{Standard error of mean difference}}$ or $\frac{M_1 - M_2}{\sigma_{d\bar{x}}}$. What makes primary difference when faced with an independent or correlated sample means is the way standard error of mean difference is calculated by the following formula

$$\sigma_{d\bar{x}} = \sqrt{\sigma_{\bar{x}_1}^2 + \sigma_{\bar{x}_2}^2} \qquad (26.6)$$

It is clear that formula 26.6 is same as formula 26.2. Accordingly, various symbols are defined in the same. For calculating the standard error of mean ($\sigma_{\bar{x}}$), we need to calculate standard deviation of the distribution, which is calculated by formula 26.7 when N_s of the two independent samples are less than 30.

$$\sigma = \sqrt{\frac{\Sigma x^2}{(N-1)}} \qquad (26.7)$$

where,
- σ = Standard deviation
- x = deviation of scores from actual mean
- N = number of persons

Now let us take an example of calculating t from independent samples. Suppose the researcher has obtained the following data after using anxiety scale on boys and girls.

	N	Mean ($\bar{X}$)	SD (σ)
Boys =	25	10.25	2.11
Girls =	20	7.26	2.33

Do boys differ from girls with respect to their mean based upon an anxiety scale?

First, we shall calculate the standard error of mean of the group of boys ($N_1 = 25$) and the standard error or mean of the group of girls ($N_2 = 20$) by formula 26.8.

$$\sigma_{d\bar{x}} = \frac{\sigma}{\sqrt{N-1}} \quad (26.8)$$

Where σ_x = standard error of mean; σ = standard deviation and N = Number of individuals in the sample. When N < 30, it is advisable to use N – 1 in the denominator (Garrett, 1966). Thus

$$\sigma_{d\bar{x}} = \frac{2.11}{\sqrt{25-1}} = \frac{2.11}{459} = 0.43 \text{ (for Boys)}$$

$$\sigma_{\bar{x}_2} = \frac{2.33}{\sqrt{20-1}} = \frac{2.33}{4.36} = 0.53 \text{ (for girls)}$$

Now, the standard error of mean difference can be calculated by formula 26.6

$$\sigma_{d\bar{x}} = \sqrt{\sigma_{\bar{x}_1}^2 + \sigma_{\bar{x}_2}^2}$$

$$\sigma_{d\bar{x}} = \sqrt{(0.43)^2 + (0.53)^2}$$

$$= \sqrt{.18 + .28} = \sqrt{.46} = .678 = .68$$

Now *t* test by formula 26.9 would be computed

$$t = \frac{\bar{X}_1 - \bar{X}_2}{\sigma_{d\bar{x}}} \quad (26.9)$$

Where, $\bar{X}_1$ = Mean of one sample, $\bar{X}_2$ = Mean of second sample and $\sigma_{d\bar{x}}$ = significance of difference between means and t = t test.

Thus the value of t by formula 24.9 would be

$$t = \frac{10.25 - 7.26}{.68} = \frac{2.99}{.68} = 4.39$$

Entering Table of *t* distribution with $df = (N_1 - 1) + (N_2 - 1) = (25 - 1) + (20 - 1) = 43$, we find that the obtained *t* is significant at .01 level and therefore, the null hypothesis (H_0) is rejected in favour of alternative hypothesis (H_1). The conclusion is that boys and girls differ much respect to the anxiety level.

When the *t* test is to be calculated from correlated sample means, the formula calculating standard error of mean difference ($\sigma_{d\bar{x}}$) is changed and it is written as

$$\sigma_{d\bar{x}} = \sqrt{\sigma_{\bar{x}_1}^2 + \sigma_{\bar{x}_2}^2 - 2r_{12}\sigma_{\bar{x}_1}\sigma_{\bar{x}_2}} \quad (26.10)$$

Where r_{12} = correlation between two test scores and other symbols are defined as it was done in formulas 26.6 and 26.8.

Let us illustrate the calculation t test from the correlated mean samples.

Suppose a group of 25 students have been tested twice on an academic achievement test. The initial test yielded a mean score of 50.00 with the standard deviation of 6.75 on the test scores. Subsequently, the group underwent a training for improving their overall academic achievement for about 2 months. Then, the group was again tested on the same academic achievement test. The second administration yielded a mean score of 62.00 with the standard deviation of 4.56. The correlation between initial scores and final scores was 0.72. Now the question is : has the group made significant improvement in their academic achievement over 2 months? This question can be answered by using t test. The above data may be tabulated as under :

	Initial test	Final test
Number of students	25	25
Mean	50.00	62.00
Standard deviation	6.75	4.56

Correlation between initial and final test scores : $r = 0.72$

$$\sigma_{\bar{x}_1} = \frac{\sigma_1}{\sqrt{N_1 - 1}} = \frac{6.75}{\sqrt{25-1}} = \frac{6.75}{4.9} = 1.37$$

$$\sigma_{\bar{x}_2} = \frac{\sigma_2}{\sqrt{N_2 - 1}} = \frac{4.56}{\sqrt{25-1}} = \frac{4.56}{4.9} = 0.93$$

$$\sigma_{d\bar{x}} = \sqrt{(1.37)^2 + (.93)^2 - 2 \times .72 \times 1.35 \times .91}$$
$$= \sqrt{1.87 + .86 - 1.77} = \sqrt{0.96} = 0.98$$

$$t = \frac{62 - 50}{0.98} = \frac{12}{0.98} = 12.24$$

Since obtained t is significant at .01 level with $df = (N-1) = 25 - 1 = 24$, the null hypothesis (H_0) is rejected in favour of alternative hypothesis (H_1) and it is concluded that students' academic achievement has significantly improved with training.

z test vs t test

Many textbooks recommend that z test should be used for larger sample ($N \geq 30$) and t test should be used for small sample ($N < 30$). This distinction of small and large sample is arbitrary. Moreover, this distinction results from the fact that in testing statistical hypotheses when samples are large, the normally distributed z ratio can be meaningfully used. For larger N, the difference between z and t may be viewed as trivial. However, for small N the difference between z and t should not be viewed as trivial (Ferguson & Takane, 1989). The arbitrary distinction of above small and large sample may be easily viewed by noting that the values of t change gradually around $N = 30$ in Table of t test rather than suddenly. Those researchers who use z for the large sample must resort to t at some point. So that they may be able to prevent their results from being biased due to the effect of small samples. Therefore, t is always preferred and thus, the researcher is protected from being biased. As df gets larger, the t distribution gets closer in shape to a normal z-score distribution. One obvious advantage of using t test as compared to z test is that in case of t test the researcher does not need to know the value of population standard deviation. This obviously means that one can proceed with hypothesis testing even though we have little or no information about the population. One consequence of this extra versatility of t test is that it is possible to use t test even in those circumstances where hypothesis testing with z-score would not even allow for the statement of null hypothesis.

Robustness of the t test

A statistical test may or may not be a robust test. Let us see what is meant by robustness of statistical test. By *robustness* of a statistical test is meant that the probabilities statements resulting from its use are insensitive or not affected by the violations of assumptions of the concerned test. The t test for the mean of independent samples have three fundamental assumptions.

- The variables are normally distribution in the population from which samples were drawn
- The variances of the two samples are homogeneous that is $\sigma_1^2 = \sigma_2^2$
- Sample should be randomly selected and this condition applies to ensure that the sample is representative of population so that valid generalizations can be done from sample to population.

Many statisticians have suggested that t test is robust and it is least affected by any departure from the assumption of normality. In fact, t test is more likely to be robust to violations of normality when N > 30. When the assumption of homogeneity of variance is violated, the degree of robustness of t test depends upon both the size and the relative size of the two samples as well as upon the differences between the two variances.

F test : Analysis of Variance

Analysis of variance (ANOVA) is a statistical test that is used to compare two or more populations (or treatments) to determine whether there are any mean differences among them. The common practice is that when there are only two treatments or two sample means, t test is applied and ANOVA is applied to the situation that warrants more than two sample means but ANOVA can also be applied to test the mean differences of only to sample means.

For understanding the logic of analysis of variance, we must consider variances. In analysis of variance, hypothesis testing is about whether the means of samples differ more than the researcher expects if the null hypothesis were true. Surprisingly, this question about means is answered by analyzing *variances* and hence, the name analysis of variance. Thus for understanding the logic of analysis of variances, we consider variances and we shall begin with *two* different ways of estimating variances—*between groupvariance* and *within group variance*. Between group variance refers to the variability in scores of the different groups of subjects. It reflects variation in means of the various groups under study. Grammatically, it should be *among group variance* but *between groupvariance* is traditional. Within group variance reflects variability inside each of the groups because scores within a group are not all the same rather they differ from each other. Thus the *average variability* of the scores within each group is called as *within group variance*. Both between group variance and within group variance together constitutes '*total variance*'.

Let us take an example to illustrate between group variance and within group variance. Suppose that there are four different samples (or four groups of subjects) with N = 5 in each group. They have been tested under four difference temperature conditions on arithmetical problem solving task. Their scores have been shown in Table 26.1. Now the simple question is do these groups differ with the respect of problem solving ability? This is a single-factor experiment. In ANOVA, a factor is called as an *independent variable*. In the above experiment, temperature is an independent variable and the *dependent variable* is the number of problems solved correctly. This experiment represents results of an independent measures experiment comparing the performance under four different temperature conditions.

Table 26.1: Hypothetical data from an Experiment comparing Arithmetical problem solving skills under four different temperature conditions (Celsius)

Sample 1 30° Temperature	Sample 2 40° Temperature	Sample 3 45° Temperature	Sample 4 50° Temperature
4	8	7	3
5	10	14	0
3	9	20	1
2	12	18	2
0	10	20	5
$\Sigma X = 14$	$\Sigma X = 49$	$\Sigma X = 79$	$\Sigma X = 11$

Mean of Sample 1 : $\dfrac{\Sigma X^1}{N_1} = \dfrac{14}{5} = 2.8$

Mean of Sample 2 : $\dfrac{\Sigma X^2}{N_2} = \dfrac{49}{5} = 9.8$

Mean of Sample 3 : $\dfrac{\Sigma X^3}{N_3} = \dfrac{79}{5} = 15.8$

$$\text{Mean of Sample 4:} \quad \frac{\Sigma X^4}{N_4} = \frac{11}{5} = 2.2$$

In the above example we find that scores of four groups differ from each other and not only this, the average variability of within each group also differs.

Now let us concentrate upon the process of analyzing the variability of these two types of variability—*between group variability* and *within group variability* (cf. Figure 26.1)

- **Between group variability :**

If we compare the mean of the sample 1 (Mean = 2.8) with mean of sample 2, (Mean = 9.8) say for example, they are different. Why are they different? There are three possible answer to this question :

a) Individual differences :

The subjects enter the experiment with different backgrounds, abilities, aptitudes and attitudes. As a consequence, their scores are likely to differ from each other and consequently, the means of the sample may also differ.

b) Treatment effect :

The mean of sample 1 may differ from the means of other remaining three samples due to the treatment imposed. The sample 1 has been tested in room temperature of 30°, sample 2 in room temperature of 40°, sample 3 in room temperature of 45° and sample 4 in room temperature of 50°. If temperature affects performance, the means of the four samples are likely to differ from each other.

c) Experimental error :

Whenever measurement is done, there is a chance of error. The error may be caused by lack of attention in subjects, poor motivation, poor tests/equipment or by some unpredictable changes in event being measured. Even though the researcher is measuring the same individual under same condition, the scores may be different. Such kind of unexplained and uncontrolled difference is technically known as experimental error. All researchers try to keep experimental error as small as possible but it cannot be eliminated altogether.

- **Within group variability :**

This reflects variability within each group. A close examination of these four groups indicates that the scores within a group are different. The score of one subject differs from the score of the other subject of the same sample. Why are they different? There are two possible answers to this question :

a) Experimental error : There is always a pretty chance that the difference in scores of the subjects of the same group is caused by the experimental error occurring due to some fault in instrument/test, lack of motivation and attention, etc.

b) Individual differences : Since scores came from different subjects all independently and all subjects are different from each other, there is high probability that scores become different for different individuals.

Treatment effect does not apply to explain the within group variability because all subjects within a sample receive similar treatments. Therefore, it does not account for the variability. The details of analysis of between and within group variances have been presented in Figure 26.1.

Once the total variance or variability is analyzed into two basic components (between variance and within variance), the researcher simply compare them. The comparison is made by simply computing a statistic called *F-ratio* (F is for Sir Ronald Fisher, a prominent statistician who developed analysis of variance.) For independent measures, single factor analysis of variance, F ratio (or F test) is computed by dividing between group variance by within group variance

$$F = \frac{\text{Between group variance}}{\text{Within group variance}} \quad (26.11)$$

Explaining between variance and within variance in terms of sources (see Figure 26.1), the formula 26.11 may be written as under :

$$F = \frac{\text{treatment effect} + \text{individual difference} + \text{experimental error}}{\text{individual difference} + \text{experimental error}} \quad (26.12)$$

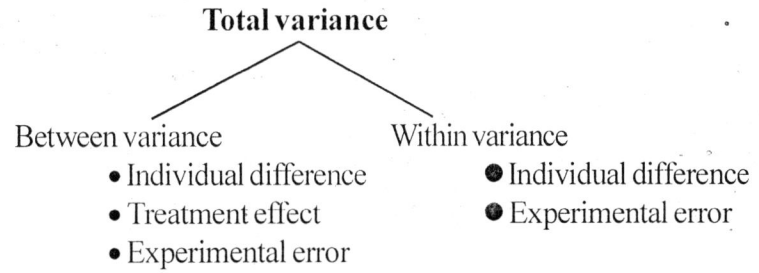

Figure 26.1 Details of analysis of Total variance, Between variance and Within variance

A simple inspection of formula 26.2 reveals that between variance and within variance differ in only one respect—variability (or mean differences) caused by only the treatment effect. This single difference between the numerator and demonstrator of F-ratio is very important in determining if a treatment effect has really occurred. The basic purpose of the doing the experiment was to find out whether or not the treatment has any effect.

Assumptions of F test

There are four important assumptions in using analysis of variance or F test for the solution of problem of the type discussed in this chapter. The unfortunate part is that sometimes we cannot tell whether these assumptions hold but the fortunate part is that it sometimes does not matter much. Those four assumptions are as under :

- It is assumed that the participants or subjects are assigned at random to the difference groups.
- It is also assumed that the distribution of variable in the population from which the different samples have been drawn is normal. The reality is that the researcher does not know whether or not this assumption holds but various empirical studies have shown that even when the distribution departs appreciably from normality, it has very little effect upon the results of F test. However, if the sample is markedly skewed or leptokurtic, it is advised that such data must be transformed and then, analysis of variance should be started.
- It is also assumed that all samples have the same variances. This is called *homogeneity of variance*. In other words, σ^2 of sample A = σ^2 of sample B = σ^2 of sample C. In practice, however, if one sample has a standard deviation even two or three times as great as another group, this is considered sufficiently close to equality to have little effect upon final result.
- In ANOVA attempt is made to divide the total variance into between group variance and within group variance. As a consequence, it is assumed that the factors, which account for the variations in individual's score are *additive*.

How to compute F ratio from independent or uncorrelated measures or scores in one criterion of classification?

Suppose the researcher wishes to study the effects of three experimental conditions designated as A, B and C upon the performance of a problem solving task. There are 30 subjects and 10 are assigned at random to each of three group and the same test is administered to all. The question is : Do the mean scores achieved under three experimental conditions differ significantly? The data as well as calculation of F test have been presented in Table 26.2.

Table 26.2: Hypothetical experiment in which 30 subjects are assigned at random into three experimental conditions, each condition having 10 subjects. Three groups have been tested under three different conditions.

Conditions

A	B	C
8	2	8
4	3	7
7	5	9
5	4	9
3	3	8
6	6	9
5	9	9
8	3	7
5	2	8

	9	3	6
(Σ)	60	40	80
Means	6	4	8

(ΣX) Grand sum = 60 + 40 + 80 = 180

Step 1: Correction (C) = $\dfrac{180}{30} = 6$

Step 2: Total sum of squares : $(8^2 + 4^2 + 7^2 ... 7^2 + 8^2 + 6^2) - C = 1266 - 6 = 1260$

Step 3: Between sum of squares of Means of A, B and C =

$$\dfrac{(60)^2}{10} + \dfrac{(40)^2}{10} + \dfrac{(80)^2}{10} - 6$$

= (360 + 160 + 640) – 6 = 1160 – 6 = 1154

Step 4: Sum of squares within conditions A, B and C :
Total sum of squares – Between sum of squares
= 1260 – 1154 = 106

Summary : Analysis of Variance

Sources of variation	df	Sums	Mean square (variance)
Between Means of condition	k – 1 = 3 – 1 = 2	1154	577
Within means of conditions	N – k = 30 – 3 = 27	106	3.92
Total	29	1260	

$$F = \dfrac{\text{Between variance}}{\text{within variance}} = \dfrac{577}{3.92} = 147.19$$

From Table of F distribution :
df = 2 and df_2 = 27
F at .05 level = 3.35
F at .01 level = 5.49

Since obtained value of F (147.19) far exceeds the value of F at .01 level, it is taken to be significant one. Therefore, null hypothesis (H_0) is rejected in favour of alternative hypothesis (H_1). In this example, the null hypothesis was that the three sets of scores are in reality random samples drawn from the same normally distributed population and therefore, the means of the three conditions did not differ through the fluctuation of sampling. This null hypothesis is rejected and therefore, we conclude that the means of three conditions do differ in reality.

Tests after ANOVA or F test :

Since in the above example the researcher has found significant difference between means, the next task is to locate where the difference or differences are. This is because F test only indicates about overall differences among the means. Winer (1962) has summarized about a half dozen methods for doing this task. Some of these methods are very rigorous than others and they reduce the probability of making a Type I error. One such test is known as *Tukey's Honestly Significant Difference (HSD) test*. Another such test has been developed by Scheffe (1957) and this *Scheffe technique* is both simple as well as regorous. Tests used after F test are called as *posteriori test*. In the present example the *Scheffe technique* would be applied.

In the above example of Table 26.2 since there are three means, three comparisons may be made :

A vs. B, B vs. C and A vs. C

For each comparison, an F ratio like the following is computed :

$$F = \frac{(\bar{X}_1 - \bar{X}_2)^2}{S_w^2(N_1 + N_2)/N_1 N_2} \qquad (26.13)$$

Where, $\bar{X}_1$ = Mean of scores of one condition; $\bar{X}_2$ = Mean of scores of another condition, S_w^2 = within group variance or mean square; N_1 = Number of cases in one condition and N_2 = Number of cases in another condition.

Now, for distribution A and B :

$$F = \frac{(6-4)^2}{3.92(10+10)/(10)(10)} = \frac{4}{78.4/100} = \frac{4}{.784} = 6.10$$

For distribution B and C :

$$F = \frac{(4-8)^2}{3.92(10+10)(10)(10)} = \frac{16}{78.4/100} = \frac{16}{.784} = 20.4$$

For distribution A and C :

$$F = \frac{(6-8)^2}{3.92(10+10)/(10)(10)} = \frac{4}{78.4/100} = \frac{4}{.784} = 6.10$$

The .05 level of significance of F for $df(2, 27)$ is 3.35. This value is multiplied by $K - 1$ where k is the number of groups or treatment. In this example, we have $(3 - 1)(3.35) = 6.70$. Each of these three computed F is compared with 6.70. Only one of them is larger than 6.70 and this is 20.4, the F computed between B and C. Therefore, it follows that there is significant difference between means of B and C at .05 level of significance and there is no significant difference between each of other two comparisons.

F test for testing the significance of difference between two sample means :

F test can also be applied for testing the significance of difference between only two sample means although in such situation, the most appropriate statistic is *t* test provided its assumptions are fulfilled. In such situation, the relationship between *t* test and *F* test is straightforward :

$$F = t^2 \qquad (26.14)$$
$$\text{or } t = \sqrt{F}$$

and here *df* for between group variance is 1.

Summary and Review

- The significance of difference between two sample means can be calculated by either z test or t test. z test is used when there is a larger sample ($N \geq 30$) and t test is used when there is smaller samples ($N < 30$). However, for larger samples t can be used because for larger sample, difference between *t* and *z* test is trivial. One obvious advantage of using t test over z test is that in case of t test the researcher does not need to know the value of population standard deviation.
- *t* test is a robust test because *t* test is least affected by violation of its assumptions.
- *t* test has some assumptions : variables are normally distributed, the variances of the two samples are homogeneous and samples should be randomly selected.
- *F* test named after its discoverer R.A. Fisher, is a statistical test used for testing the significance of the mean differences between two or more than two sample means. In F test, the total variance is broken into two important categories—*between variance* and *within variance*. In *F* test between variance and within variance are compared.
- Between variance is accounted for in terms of individual differences, treatment effect and experimental error. Within variance is accounted for in terms of experimental error and individual differences only.
- When there are two sample means, *t* and *F* bears a certain relations. In this situation $t = \sqrt{F}$ and $F = t^2$.
- A significant F yields about the overall difference among the sample means under study. Such situation warrants the application of some *posteriori tests* like Scheffe test and Tukey's HSD test for locating the exact difference between two sample means. There are many other such posteriori tests which are available.

Review Questions

1. Citing examples make distinction between *z* test and *t* test.
2. Discuss the major assumptions of *t* test. Do you consider *t* test as a robust statistic?
3. What is *F* test? Discuss the major assumptions of *F* test.
4. What do you mean by posteriori tests? Discuss some of such popular posteriori tests.

Chapter – 27

Some Important Non-Parametric Statistics

Learning objectives :

- Meaning and Nature of Chi-square test
- Major characteristics of Chi-square test
- Assumptions of Chi-square test
- Major uses of Chi-square test
- Chi-square as a test of Equal Probability hypothesis
- Chi-square as a test of Independence
- Chi-square test for Goodness of fit.
- Chi-square test for a 2 × 2 table
- Chi-square test for a 2 × 2 table with small expected frequencies
- Chi-square from a Table larger than 2 × 2 table with small expected frequency
- Major sources of Errors in chi-square test
- Advantages and Disadvantages of Chi-square test
- Nominal Measures of Association based upon X^2
  - Phi coefficient
  - Coefficient of contingency
- Mann-Whitney U test
- Median Test
- Kendall's Coefficient of Concordance : W
- Summary Questions
- Review Questions

Key Terms :

Chi-square Test, Phi-Coefficient, Coefficient of contingency, Mann-Whitney U test, Median test, Coefficient of Concordance.

While a few parametric statistics like Spearman rank-difference correlation and Kendall rank-difference correlation have been discussed in chapter 23, still some more important non-parametric statistics will be discussed in this chapter. As we know in nonparametric tests, no assumptions about the precise form of the sample distribution is made. That is why, they are also called as *distribution-free statistics*. However, sometimes certain assumptions are made such as that a distribution is continuous or that the sample populations having identical shapes or distribution are symmetrical about the same point. As discussion about the nature of non-parametric tests have already been done in chapter 18, we shall not repeat those things here again. In the present chapter a discussion about only five important nonparametric statistical test such as chi-square test (X^2) phi-coefficient (ϕ) coefficient of contingency (C), Mann-Whitney U test and coefficient of concordance (W) would be undertaken.

Meaning and Nature of Chi-square Test (X^2)

Chi square test (X^2) is a statistical test of significance when we have data that are expressed in frequencies or data that are in terms of percentages or proportions or that can be reduced to frequencies. The chi-square test was originally *developed* by Karl Pearson in 1900 who is sometimes hailed as the *father* of science of statistics. That is why, chi-square test is sometimes also called as *Pearson chi-square test*. Chi-square test (X^2) is, in fact, a descriptive measure of the magnitude of discrepancies between the observed frequencies (f_o) and expected frequencies (f_e). The chi-square is often used to determine whether a set of observed frequencies is such that it might have arisen as a consequence of chance fluctuations from the frequencies which would be expected on the basis of a theory of any kind. Mathematically, chi-square test may be defined as the sum of ratios of squared deviations of observed frequencies (f_o) divided by the expected frequency (f_e). The formula for chi-square test is as under :

$$X^2 = \Sigma \frac{(f_0 - f_e)^2}{f_e} \qquad (27.1)$$

where,

f_o = observed frequencies
f_e = expected frequencies
Σ = Sum of values after dividing the square the differences between f_o and f_e.

The major steps in computation of chi-square is very straight forward and simple :
- Find the value of expected frequency (f_e)
- Find the difference between f_e and f_o
- Square the differences of f_o and f_e
- Divide the squared value by f_e
- Finally, add the resulting values from all categories.

Another formula to compute chi-square which is much easier than formula 27.1, is as under :

$$X^2 = \Sigma \left(\frac{f_0^2}{f_e} \right) - N \qquad (27.2)$$

Formula 27.2 is easier than formula 27.1 because here we need not calculate the difference between $f_o - f_e$.

Major characteristics of Chi-Square test (X^2)

There are certain characteristics of chi-square (X^2). The major ones are as under :

- X^2 is always positive because the difference between f_o and f_e is squared.
- Since X^2 is a descriptive measure of the magnitude of the differences between f_o and f_e, the greater the difference between f_o and f_e relative to f_e, the larger will be the value of X^2. If there is no difference between f_o and f_e, X^2 will be zero.
- The magnitude of X^2 will vary from series to series owing to the variations in the observed frequencies as well as due to the fact that the expected frequencies being constant over the categories. For example, a researcher tosses a coin 100 times and gets 65 heads and 35 tails. Another researcher tosses the coin again 100 times and he gets 45 heads and 55 tails. In both cases the expected frequencies are same that is, 50 in heads and 50 in tails. But the amount of X^2 will be different in both cases because of different values of observed frequencies in heads and tails.
- Different categories of responses are not independent. It means that if frequency of one category is known, the frequency of other category is fixed and determined. For example, if a coin is tossed 100 times and one gets 60 heads, then frequency in tails is fixed, that is, 40. Thus these frequencies are not independent.
- The distribution of chi-square is positively skewed. The exact shape and form of the distribution of chi-square depends upon the *df* of the chi-square. As the *df* approaches 30, the shape of X^2 approaches that of normal curve.
- The chi-square test is a non-directional test. Since X^2 is arrived at by squaring the difference between observed and expected frequencies, it has no sign.
- The chi-square test is insensitive to the effects of order (Siegel & Castellan, 1988). Thus when either response categories or groups or both are ordered, the chi-square is not a best statistic.
- The chi-square is not in itself a measure of association when applied to a contingency table but that it is possible to derive measure of association from it. The phi co-efficient and the coefficient of contingency are the examples.

Assumptions of Chi-square test (X^2)

There are some assumptions of chi-square. Important ones are as under :

- It is assumed that the sample drawn is a random sample from the population about which some inferences are likely to be made. However, in practice this requirement is seldom fully met because of the difficulty in recognizing each element of the target population and therefore, in giving each element an equal opportunity for inclusion in the sample.
- It is also assumed that observations are independent. For correct use of X^2, it is essential

that all N observations are made independently and don't influence each other. By independence of observations is meant that each observed frequency is *generated* by a different subject or participant. Thus one observed frequency per subject or respondent should be generated. A chi-square test would be inappropriate if a person produces responses that can be classified in more than one category or contribute more than one frequency count to a single category. In general, the set of observations will not be completely independent when their number *exceeds* the number of subjects. Thus although X^2 makes no assumptions about normal distributions of their variables but it requires that no individual should be counted in more than one cell.

- The variable under study should be a continuous variable, that is, nowhere in the sample or population the trait should be absent.
- The data should be obtained on nominal (or categorized) scale of measurement. In other words, subjects are required to provide their responses in different categories such as 'Agree', 'Disagree', 'Neutral', 'Favourable', 'Unfavourable' etc.
- The data should be expressed in terms of frequencies, percentages or proportions.
- The X^2 test is dependent on the assumption that the observed frequencies (fo_y) will be normally distributed about their expected frequencies (fe_y) (Minium, 1970). When fe is quite small, the distribution of fo's remains no longer normal rather it becomes skewed and theoretical chi-square model will not be adequate. This matter becomes of greater consequence when df is small rather than large.

Major Uses of Chi-square Test

There are many uses of chi-square test. Of those uses, the following four are most common ones :

- The chi-square is used to determine whether a certain distribution differs from some predetermined theoretical distribution. For example, suppose a researcher tosses a coin 100 times and each time he records the results of the toss. He is studying these observed results against the frequencies he world expect by chance. If he get 70 heads and 30 tails, the expected frequency based on chance would be 50 heads and 50 fails. This is called as *equal probability hypothesis*.
- A second use of chi-square is in testing the *independence of two variables*, which are usually the nominal variables. The basic question here is to decide whether the two variables are independent of each other. The data are arranged in form of a table called *contingency table* which may compose of any number of rows and any number of columns. Two variables are said to be independent when the distribution for one of the variables is not related to nor dependent on the categories of the second variable. The null hypothesis here states that the two variables are not related (and, therefore, independent of each other) and make predictions for the expected frequencies for all categories.

- The chi-square test is used for the *goodness-of-fit*. When goodness-of-fit is done, the researcher is trying to find out if the distribution of the observed data is similar to that of the normal or some other distribution. Thus in this use of chi-square, the researcher is testing a hypothesis about the shape of a population frequency distribution. Here the null hypothesis predicts a specific type of distribution (normal or some other) for the population. The chi-square test compares the frequency distribution obtained for the sample to the frequency distribution specified by the null hypothesis. The chi-square test determines how well the obtained sample data *fit* the hypothesis, hence the name of the test-*goodness of fit*.
- The chi-square is also used in testing the significance of various statistics. For example, it is used in testing the significance of coefficient of contingency, coefficient of concordance, Phi-coefficient, etc.

Chi-square As a test of Equal Probability Hypothesis

One of the popular uses or application of the chi-square is in testing the differences in observed results or frequency from those expected on the hypothesis of equal probability, which becomes the null hypothesis. Let us take an example to illustrate calculation of chi-square in this situation.

Suppose 120 students were asked to expressed their attitude towards a proposition: Should India attack Pakistan on Kashmir issue? by marking 'Yes', 'No', 'Cannot say'. Of these 120 students, 40 marked 'Yes', 70 marked 'No' and 10 marked 'Cannot say'. Do these results indicate a significant trend of attitude opinion?

The observed data (fo) have been presented in the first row of Table 27.1 and the expected frequencies (fe) on the null hypothesis of equal probability have been presented in the second row of the same Table. Below are the details of calculation of chi-square.

Table 27.1: Chi-square test from data to be expected on Equal probability hypothesis

	Answers			
	Yes	No	Cannot Say	N
Observed (fo)	40	70	10	120
Expected (fe)	40	40	40	120
($fo-fe$)	0	30	−30	
($fo-fe$)²	0	900	900	
$\frac{(fo-fe)^2}{fe}$	0	22.5	22.5	

$$X^2 = \Sigma \frac{(f_0 - f_e)^2}{f_e} \quad \text{(by formula 27.1)}$$

$$= 0 + 22.5 + 22.5 = 45$$

X^2 by formula 27.2, : $\quad X^2 = \Sigma\left(\dfrac{f_o^2}{f_e}\right) - N$

$$= \left(\dfrac{40^2}{40} + \dfrac{70^2}{40} + \dfrac{10^2}{40}\right) - 120$$

$$= \left(\dfrac{1600}{40} + \dfrac{4900}{40} + \dfrac{100}{40}\right) - 120$$

$$= (40 + 122.5 + 2.5) - 120 = 45$$

$$df = (r-1)(c-1) = (2-1)(3-1) = 1 \times 2 = 2$$

(r stands for row in Table and c stands for column in Table. There are two rows and three columns in the Table.)

Entering Table[1] of chi-square at $df\ 2$, we find that the obtained value of X^2, that is, 45 far exceeds the value of chi-square given at .01 level ($X^2 = 6.64$). The obtained value of chi-square even exceeds the chi-square value given at still smaller level, that is, at .001 ($X^2 = 10.83$). As a consequence, X^2 value is taken as significant one and null hypothesis which is, here, expressed in terms of equal probability hypothesis is rejected. Accordingly, we conclude that the group really expresses the view that India should not attack Pakistan on Kashmir issue.

Chi-square as a Test of independence

Another important use of chi-square is in testing the extent to which two variables are related or independent (not related). In such situation, the researcher wishes to investigate the relationship between traits or attributes, which can be classified into two or more categories. For example, a researcher may wish to determine if attitudes about career differ as a function of academic year in the college. Thus in test of independence two variables are involved and observed and expected frequencies are compared. Here the expected frequencies are those the researcher would expect to get if two variables are independent of each other. Let us take an example to illustrate the calculation of X^2 in such situation.

Suppose that the researcher is going to determine if attitude of a group of students towards career differ as function of academic year in a college. A random sample of college students is selected. The students are asked to fill out a questionnaire, specifying year in college and what they intend to do when they complete their college education. Their responses are classified and recorded in Table 27.2 and calculation of X^2 have been done accordingly. The basic question here is : Is there a relationship between academic year in college and attitude about career?

---

1. Table can seen at page 323 (Table 6) in the book entitled *Non-parametric statistics* (second edition, 1988) by Siegel & Casteltan.

Table 27.2: Chi-square test from data to be expected on Independence hypothesis Plans after college

Year	Public Sector Work	Private Sector Work	Undecided	Total
Intermediate	25	30	5	60
Bachelor	40	50	15	105
Master	70	30	10	110
Ph.D.	60	10	5	75
Total	195	120	35	350

I. Calculation of Independence value (fe):

$$\frac{195 \times 60}{350} = 33.4 \quad \frac{120 \times 60}{350} = 20.6 \quad \frac{35 \times 60}{350} = 6.0$$

$$\frac{195 \times 105}{350} = 58.5 \quad \frac{120 \times 105}{350} = 36.0 \quad \frac{35 \times 105}{350} = 10.5$$

$$\frac{195 \times 110}{350} = 61.3 \quad \frac{120 \times 110}{350} = 37.7 \quad \frac{35 \times 110}{350} = 11.0$$

$$\frac{195 \times 75}{350} = 41.8 \quad \frac{120 \times 75}{350} = 25.7 \quad \frac{35 \times 75}{350} = 7.5$$

II. Difference between fo and fe ($fo - fe$):

25 – 33.4 = 8.4	30 – 20.6 = 9.4	5 – 6 = –1
40 – 58.5 = 18.5	50 – 36.0 = 14	15 – 10.5 = 4.5
70 – 61.3 = 8.7	30 – 37.7 = 7.7	10 – 11 = –1
60 – 41.8 = 18.2	10 – 25.7 = 15.7	5 – 7.5 = –2.5
$\Sigma = 0$	$\Sigma = 0$	$\Sigma = 0$

check: $fo - fe = 0$ (The difference between fo and fe should be zero. But sometimes fractional values may be obtained due to rounding errors.)

III. Calculation of X^2:

$$\frac{(-8.4)^2}{33.4} = 2.11 \quad \frac{(-9.4)^2}{20.6} = 4.28 \quad \frac{(-1)^2}{6} = 0.16$$

$$\frac{(-18.5)^2}{58.5} = 5.85 \quad \frac{(14)^2}{36} = 5.44 \quad \frac{(4.5)^2}{10.5} = 1.92$$

$$\frac{(8.7)^2}{61.3} = 1.23 \quad \frac{(7.7)^2}{37.7} = 1.57 \quad \frac{(-1)^2}{11} = 0.09$$

$$\frac{(18.2)^2}{41.8} = 7.9 \quad \frac{(15.7)^2}{25.7} = 9.59 \quad \frac{(-2.5)^2}{7.5} = 0.83$$

$\Sigma = 17.09 \quad \Sigma = 20.88 \quad \Sigma = 3.00$

$X^2 = 17.09 + 20.88 + 3.00 = 40.97$

If one likes to calculate X^2 with the help of formula 27.2, one will proceed like this :

$$\frac{(25)^2}{33.4} = 18.7 \quad \frac{(30)^2}{20.6} = 43.7 \quad \frac{(5)^2}{6} = 4.2$$

$$\frac{(40)^2}{58.5} = 27.3 \quad \frac{(50)^2}{36} = 69.4 \quad \frac{(15)^2}{10.5} = 21.4$$

$$\frac{(70)^2}{61.3} = 79.9 \quad \frac{(30)^2}{37.7} = 23.9 \quad \frac{(10)^2}{11} = 9.09$$

$$\frac{(60)^2}{41.8} = 86.1 \quad \frac{(10)^2}{25.7} = 3.9 \quad \frac{(5)^2}{7.5} = 3.3$$

$\Sigma = 212 \qquad \Sigma = 140.9 \qquad \Sigma = 37.99$

$X^2 = (212.00, 140-90 + 37.99) = 350$
$= 390.89 - 350 = 40.89$

We get the same value of X^2 by both formulas. The fractions difference is due to rounding errors.

Entering Table of X^2 with $df = (r-1)(c-1) = (4-1)(3-1) = 3 \times 2 = 6$, we find that at .01 level of significance, a chi-square value of 16.81 is needed. Since our obtained value (40.97) exceeds this critical value (16.81), the X^2 is taken to be significant one. As a consequence, the independence hypothesis (null hypothesis Ho) is rejected and we conclude that there is relationship between year in college and attitude about career in students.

Chi-square Test for Goodness-of-fit

The chi-square test for goodness-of-fit is one the most popular applications. Here the researcher usually tests the divergence of observed results from those expected on a hypothesis of normal distribution or some other distribution. In other words, the major aim of the researcher is to test whether the obtained results fits with the data expected on normal distribution (or some other distribution).

There can be two types of situations here. One situation may be that when the obtained results available in frequency (category wise) and other situation may be one where obtained results are available in frequency distribution table. Let us see how calculation of X^2 in both situation is done.

Calculation of X^2 when obtained results are available in frequency only (category wise) :

Suppose 80 assistant mangers have been classified into three groups: Excellent, Good and Satisfactory by a consensus of excentive directors. Now the question is: Does this distribution of ratings differ significantly from that to be expected if managerial ability is normally distributed in our population of assistant managers? The obtained data and calculation of X^2 is being done in Table 27.3.

Table 27.3: Chi-square test for divergence of observed data from those expected on the hypothesis of a normal distribution

	Excellent	Good	Satisfactory	
Observed (fo)	50	20	10	80
Expected (fe)	12.8	54.4	12.8	80
(fo-fe)	37.2	–34.4	–2.8	$\Sigma fo-fe=0$
(fo-fe)²	1383.84	1183.36	7.84	
(fo-fe)²/fe	108.11	21.75	0.61	

$$X^2 = 108.11 + 21.75 + 0.61 = 130.47$$
$$df = (r-1)(c-1) = (2-1)(3-1) = 2$$
or
$$df = c - 1 = 3 - 1 = 2$$

In Table 27.3, the value of fe has been calculated as under:

The entries in second row of the Table 27.3 have been found by first dividing the base line of a normal curve (taken to extend over 6σ) into 3 equal segments. Thus each segment consists of $6\sigma/3 = 2\sigma$. Entering Table[1] showing fractional parts of the total area under the normal probability curve, we find the proportion of the normal distribution to be found in each of these segments as under:

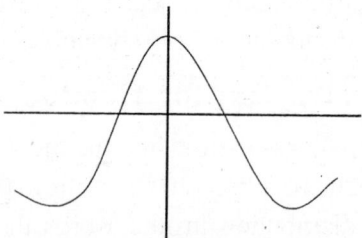

$-3\sigma\ -2\sigma\ -1\sigma\ \ 0+1\sigma\ +2\sigma\ +\ 3\sigma$

Figure 27.1: Normal distribution curve with $\pm 3\sigma$

Between $+3\sigma$ and $1.\sigma = .4986 - .3413 = .1573 = .16$
Between $+1.\sigma$ and $1.\sigma = .3413 + .3413 = .6826 = .68$
Between -1σ and $-3\sigma = .4986 - .3413 = .1573 = .16$
                                                              1.00

Now .16 or 16% of 80 = $\dfrac{16}{100} \times 80 = 12.8 (fe)$

---

1. Table can be seen in the book entitled *Statistics in Psychology and Education* by H.F. Garrett, Tenth Indian Print (1981) at page 458.

.68 or 68% of 80 = $\frac{68}{100} \times 80 = 54.4(fe)$

.16 or 16% of 80 = $\frac{16}{100} \times 80 = 12.8(fe)$

Entering Table of distribution of chi-square at $df=2$, we find that a critical value of X^2 is 9.21 at .01 level. The obtained X^2 value is 130.47, which far exceeds the above critical value. It means that X^2 is significant and we reject the hypothesis of normal distribution. X^2 is significant and we reject the hypothesis normal distribution of managerial ability (null hypothesis Ho) and accordingly, conclude that the obtained results does not fit the normal distribution hypothesis.

Calculation of X^2 when obtained results are available in frequency distribution table :

Suppose 30 students were randomly selected and were administered anxiety scale. Their scores have been arranged in a frequency distribution table as it has been presented in Table 27.4. Now the question is : Do these observed data conform to widely known distribution such as normal distribution?

Table 27.4: Frequency distribution table based on scores of 30 students on Anxiety Scale

Class Interval	Frequency (f)
85 – 89	1
80 – 84	1
75 – 79	2
70 – 74	3
65 – 69	9
60 – 64	8
55 – 59	2
50 – 54	1
45 – 49	1
40 – 44	2
	N = 30

For calculating X^2 from 27.4, the first step would be to compute fe (frequency expected on the hypothesis of normal distribution). For calculating fe, there are several steps to be followed as under:
- Find out the mean of the distribution, which is 65.5.
- Find out the standard deviation of the distribution, which is 9.98.
- Find out z-score $\left(\frac{X-M}{\sigma}\right)$ of each class interval separately.

- Then, find not the cumulative proportion (*CP*) of each *z*-score with help of a Table[1] showing areas and ordinates of the normal curve in terms of *z*-score.
- Subsequently, cumulative frequency is found out multiplying each *Cp* by *N*.
- At last, by the process of successing subtraction starting from the bottom of the Table 27.5, *fe* of each class interval is calculated. For example, the lowest class interval 40-44 has cumulative frequency of .69 so, its *fe* is also the same, that is, .69. But for finding the *fe* of next class interval, that is, 45-49, cumulative frequency of class interval 40-44, that is, .69 is subtracted from the cumulative frequency of class interval 45-49. Thus 2.01 – .69 = 1.32 becomes the '*fe*' for the class interval 45-49. This procedure is repeated till the last and uppermost class interval, that is, 85-89.

The above entire procedure for calculating *fe* has been demonstrated in Table 27.5.

Table 27.5: Calculation of *fe* from data shown in Table 27.4

Class Interval (1)	*f* (2)	Upper limit (3)	Deviation (X–M) (4)	z Score $\frac{(X - M)}{\sigma}$ (5)	Cumulative Proportions (CP) (6)	Cumulative Frequency (CP × N) (7)	*fe* (8)
85 – 89	1	89.5	+25	+2.5	.9938	29.81	.49
80 – 84	1	84.5	+20	+2.0	.9772	29.32	1.33
75 – 79	2	79.5	+15	+1.5	.9332	27.99	2.75
70 – 74	3	74.5	+10	+1.0	.8413	25.24	4.5
65 – 69	9	69.5	+5	+.5	.6915	20.74	5.74
60 – 64	8	64.5	0	0	.5000	15.00	5.75
55 – 59	2	59.5	–5	–.5	.3085	9.25	4.49
50 – 54	1	54.5	–10	–1.0	.1587	4.76	2.76
45 – 49	1	49.5	–15	–1.5	.0668	2.00	1.32
40 – 44	2	44.5	–20	–2.0	.0228	0.68	.68
	N = 30					Σ*fe* = 29.81	

Mean = 65.5; SD(σ) = 9.98

There is difference of 0.19 between total *fo* and total *fe* and this is due to *rounding error*. After calculating *fe*, the X^2 may be computed through the procedure shown in Table 27.6.

---

1. Table can be seen at page 503 of the book entitled *Fundamental statistics in Psychology and Education* by J.P. Guilford & B. Fruchter (1981), Mc-Graw Hill Publication.

Table 27.6: Calculation of X² from data taken from Table 27.4 and Table 27.5

Class Interval (1)	fo (2)	fe (3)	fo Rearranged (4)	fe (5)	fo–fe (6)	(fo–fe)² (7)	(fo–fe)²/fe (8)
85 – 89	1	0.49 ⎤					
80 – 84	1	1.33 ⎟					
75 – 79	2	2.75 ⎟					
70 – 74	3	4.50 ⎦	7	9.07	–2.07	4.68	0.51
65 – 69	9	5.74	9	5.74	+3.26	10.63	1.85
60 – 64	8	5.75	8	5.75	+2.25	5.06	0.88
55 – 59	2	4.49 ⎤	6	9.25	–3.25	10.56	1.14
50 – 54	1	2.76 ⎟					
45 – 49	1	1.32 ⎟					
40 – 44	2	0.68 ⎦					
	N = 30	Σ=29.81			Σfo–fe=.19		Σ= 4.38

df = Number of categories – 3
= 4 – 3 = 1

The critical value of X^2 at df = 1 for .05 level of significance is 3.841. Since the obtained value of X^2, that is, 4.38 exceeds the critical value, the X^2 is taken to be significant one. Accordingly, null hypothesis (Ho) that is, the hypothesis of normal distribution, is rejected. Therefore, it is concluded that the obtained results don't fit to the expected the normal distribution.

Chi-square test for 2 × 2 Table

In psychological and educational researches a frequently occurring type of contingency table is the 2 × 2 or four fold contingency table. Let us represent the four cells of 2 × 2 table and marginal frequencies by the following notation:

A	B	A + B
C	D	C + D
A + C	B + D	N = (A+B+C+D)

Chi-square may then be calculated by the following formula.

$$X^2 = \frac{N(AD-BC)^2}{(A+B)(C+D)(A+C)(B+D)} \quad (27.3)$$

Here, the terms in the numerator $AD – BC$ is simply the difference between the two cross products and the terms in the denomination is the product of the four marginal totals.

Let us take an example: Suppose a researcher would like to see if there is relationship between handedness and eye preference. A random sample of 300 subjects is selected. For each subject, the researcher determines two things : (1) whether the person is left-handed or right-handed and (2) which eye the person prefers to use when looking through a camera viewfinder. The question is: Is there relationship between hand preference and eye preference? The observed frequencies are as follows :

Hand preference

		Left	Right	
Eye Preference	Left	40^A	80^B	120
	Right	20^C	160^D	180
		60	240	300

X^2 by formula 27.3 would be as under

$$X^2 = \frac{300[(40)(160) - (80)(20)]^2}{120 \times 80 \times 60 \times 240}$$

$$= \frac{300 \times 23040000}{311040000} = 22.22$$

$df = (r - 1)(c - 1) = (2 - 1)(2 - 1) = 1$

For $df = 1$, the critical value of X^2 at .01 level of significance is 6.64. The obtained value far exceeds this critical value of X^2. Therefore, X^2 is significant at .01 level. The null hypothesis which states that hand preference and eye preference are not related is rejected. Accordingly, it is concluded that there is relationship between these two variables.

Chi-Square test for 2 × 2 Table with small expected frequencies

As we know, the theoretical distribution of X^2 may be plotted as a continuous curve for any given df. But actual samples of real data may show marked discontinuity. This is particularly true for 2 × 2 tables, which, of course, have only one degree of freedom. Therefore, some correction for continuity becomes necessary. In such situation, a correction known as *Yates' correction for continuity* is applied.

When a chi-square is to be calculated from a problem having 1 df (such as 2 × 2 or 1 × 2 table) and when any expected frequency is less than 5 (some textbooks prefer less than 10), Yates' correction for continuity is applied in calculating X^2. This correction consists of reducing .5 each obtained frequency that is greater than expected and increasing by the same amount each observed frequency that is less than expected frequency. Such correction has the impact of reducing the amount of each difference between obtained and expected frequency to the extent of .5. As a consequence, the size of the chi-square is reduced. The formula for calculating X^2 with Yates' correction for continuity is as under :

$$X_c^2 = \frac{N(|AD - BC| - N/2)^2}{(A + B)(C + D)(A + C)(B + D)} \quad (27.4)$$

The term $|AD - BC|$ is the absolute difference, the difference taken regardless of sign. For applying Yates' correction for continuity the following points should be kept in mind:

- A correction of .5 is applied to all cells in the table even through only one or two frequencies are small.
- The decision to apply Yates' correction for continuity is done on the basis of low expected frequency (fo) and not on the basis of low observed frequency.
- Yates' correction for continuity is applied to instances of 1 df only such as in 2 × 2 and 1 × 2 table. It should not be used with 2 × 3, 3 × 3 table, etc.

Let us take an example to illustrate the calculation of Yates' correction for continuity. Suppose an Academic achievement test was administered to a group of 25 boys 25 girls all randomly selected from class IV standard. Of the 25 boys, 23 boys achieved at or above the norm established by the school authority and 2 were below that norm. Among girls, 18 girls achieved at or above the norm and 7 were below that norm. The question is: Are the boys really better than girls in academic achievement? The data have been summarized in Table 27.7.

Table 27.7: Yates' Correction in X^2

	At or above Norm	Below Norm	
Boys	23^A	2^B	25
Girls	18^C	7^D	25
	41	9	50

In the Table 27.7, it is proper to get an idea about actual value of fe so that a decision about Yates' correction may be taken. Calculation of fe:

Cell A: $\frac{41 \times 25}{50} = 20.5$ cell $B = \frac{9 \times 25}{50} = 4.5$

Cell C: $\frac{41 \times 25}{50} = 20.5$ cell $B = \frac{9 \times 25}{50} = 4.5$

Since two cells contain fe less than 5, the situation warrants for application of Yates' correction for continuity. Now applying formula 27.4, we get

$$X^2 = \frac{50(161 - 36 - 25)^2}{25 \times 25 \times 41 \times 9} = \frac{500000}{230625} = 2.168 = 2.17$$

Entering Table of chi-square distribution with 1 df, is we find that at .05 level the critical value of X^2 is 3.84. Since obtained value of chi-square is lower than this critical value, X^2 is not significant one and null hypothesis (Ho) is accepted. As a consequence, it is concluded that the boys are really not better than girls in academic achievement or that is no difference between boys and girls in terms of academic achievement. It is important point to note here that if the X^2 is computed without Yates' correction from Table 27.7 by formula 27.3, its value becomes 3.39. Thus when Yates' correction is applied, the actual value of X^2 is reduced from 3.39 to 2.17. The X^2 of 3.39 slightly misses the critical value of X^2 given at .05 level (that is, 3.84).

Chi-square from a Table larger than 2 × 2 with small expected frequency

When X^2 is to be competed from 2 × 2 table or 1 × 2 table with small expected frequency (less than 5), we apply Yates' correction of continuity and proceed ahead. But suppose the table is large than 2 × 2 such as 3 × 2, 3 × 3, 4 × 3, 4 × 4 etc. (that is, $df > 1$) the expected frequency is small (less than 5) in cells (this will be especially true when the number of participants is small), what should be done? In such tables Yates' correction for continuity cannot be honestly applied. One solution is to combine the related categories to increase the expected frequency and then, the total number of cells would be reduced. But this is the solution of the last resort because such adjustment is based upon the results of the experiment. In such situation, the best solution is to add more participants to the study. If somehow, it is not feasible, an alternative test called *Fisher's exact test* way be employed. The safe rule is that when $df \geq 1$, the expected frequency should be equal to or greater than 5 in 80% of the cells. When these requirement are not met, other statistical tests are recommended (Siegel & Castellan, 1988; Elitson et al., 1990)

Major Sources of Error in Chi-square test

The chi-square appears to be very simple, clearly understood and easy-to-use statistic. It has been widely used no doubt, but also has been widely misused. Lewis and Burke (1949) have pointed out in their research article that there are nine major sources of errors, which make the use of X^2 incorrect. Such sources of error are as under :

(i) **Lack of independence among observations :** The use of X^2 is most appropriate if all N observations are made independently and don't influence each other. Thus if we have only 10 subjects and decide to build our N up to 100 by measuring subjects ten times, this would not yield an independent observation and the use of X^2 here will be incorrect. According to Lewis and Burke, this lack of independence is one of the most common cause of error in chi-square.

(ii) **Small theoretical frequencies :** There is a disagreement among experts regarding the minimum expected or theoretical frequency. Some say it should be 5 or some say it should be 10. This creates confusion when X^2 is to be computed from 2 × 2 table. In such situation Yates' correction for continuity is applied. But even regarding Yates' correction for continuity, there is some differences of opinion among the experts (Kurtz & Mayo, 1988).

(iii) **Use of infrequency data :** The chi-square is generally legitimately applied only to data which are in form of frequencies. It is not correct to insert the scores of test or other measurements in the cells and compute chi-square.

(iv) **Failure to equalize Σfo and Σfe**: In computing X^2 by formula 27.1 or 27.2, it is essential that Σfo and Σfe should be equal and therefore, its difference should be zero. But it has been found that some researchers ignore it or don't pay much attention to it and therefore, error creeps into the calculation of X^2.

(v) **Neglect of frequencies of non-occurrence**: Sometimes the researchers neglect or don't pay attention to a fact that why some events are coming up more often than it should come up. Due to this some other events don't occur or are blocked. Due to the neglect of frequencies of non-occurrences, some errors creep into the calculation of X^2. Suppose the researcher tosses a coin 120 times and gets 46 heads, against the expected or theoretical occurrences of 60. If he computes X^2 by comparing the observed frequency 46 with expected frequency, 60, he gets

$$X^2 = \Sigma \frac{(fo-fe)^2}{fe} = \frac{(46-60)^2}{60} = \frac{196}{60} = 3.26$$

with 1 df, a X^2 of 3.26 is not significant even at .05 level. Here one error has been made. The error is that the researcher has failed to take any account of 54 times that the coin did not turn up heads.

(vi) **Incorrect determination of df**: In a table having two rows and two columns (2×2) or in some other tables of varying rows and columns, the usual formula for determining df is $(r-1)(c-1)$. Thus in table having four rows and three columns (4×3), we have df $(4-1)(3-1) = 6$. In a formal language, df is always equal to the number of cells in table minus the number of restrictions placed during the theoretical calculations. In 4×3 table, there are 12 cells and we have restrictions on all but one of the column totals, on all but one of row totals and on N. Thus

$$df = 12 - (4-1)(3-1) - 1 = 12 - 3 - 2 - 1 = 6$$

But the common formula for determining df, that is, $(r-1(c-1)$ is not applicable when X^2 is used as goodness of fit. In such situation df = Number of categories -3. That is why in Table 27.6 df is equal to the number of categories minus 3. Reality is that in such situation, we forced our data and theoretical curve to agreed exactly on N, Mean and standard deviation Thus there are three restrictions.

(vii) **Incorrect computations**: In calculation X^2 some common mistakes result from the use of formulas based on percentage or proportion or from some incorrect understanding of Yates' correction for continuity.

(viii) **Incorrect categorizing**: If the researcher is required to subjectively classify or place data in one of the several categories, there exists some possibility of errors. Such errors are likely to occur most frequently in situations, when drawings are to be classified according to the subject-matter categories or when the researcher is trying out some new methods for grouping the responses made to the projective techniques like Rorschach test or

TAT. The best way to minimize or avoid such errors especially when working with such data is to set up categories on the basis of external criteria in advance.

(ix) **Indeterminate theoretical frequencies :** This source of error is very infrequent. This source of error arises when the *fo* or observed frequencies are in any way related or some mutually contradictory assumptions regarding the expected frequency can be made with about equal justification. Thus this is a technical source of error and therefore, does not often arise.

Advantages and Disadvantages of Chi-square test

The greatest advantage of X^2 is that it may be used to test the consistency between any set of frequency data and any hypothesis or theoretical values, irrespective of what the latter may be. Another related advantage of X^2 is that it is a very useful statistic in curve fitting, providing the researcher with a definite figure, which tells the researcher whether or not he may regard any given fit of theoretical curve to a set of distribution as a good one.

Despite these advantages, there are some limitations or disadvantages of chi-square test as under:
- The use of chi-square test, for all practical purposes, is limited to the use of data in form of frequencies, percentage or proportions.
- The chi-square is insensitive to the order in which categories are arranged because such ordering has no effect upon the size of the chi-square. This may be a disadvantage for one or both variables.
- The chi-square tells us *only* about the difference between our sample results and those expected on a theory. For example, in test of independence X^2 tells us whether or not the classification on one variable is related to the classification of on the other variable in non-chance manner. *The X^2 test does not tell anything about the strength of that relationship* (Kurtz & Mayo, 1980).
- Another limitation of X^2 stems from the fact that the value of X^2 is proportional to the sample size. For example, if the sample in Table 27.8 were doubled to 240 and the observed frequencies were increased proportionately (that is, 56, 64, 74 and 46), the value of X^2 would be enhanced from 2.72 to 5.44. As a consequence, the researcher would reject the null hypothesis (Ho) rather than accept it despite the fact that the relationship has *not* changed. Due to this reason, many researcher want to avoid calculation of X^2 with large sample.

Table 27.8: Yates' Correction in X^2

Response	Age		Total
	Adult	Young	
Agree	28(32.5)	32(27.5)	60
Disagree	37(32.5)	23 (27.5)	60
	65	55	120

$X^2 = 2.72$. In parenthesis there appears the value of fe:
- Another limitation of X^2 arises when it is to be calculated with small N or when the expected frequency is small say less than 5. When the contingency table is 2×2 and any expected frequency is below 5, we proceed with applying Yates' correction for continuity. However, some propose that Yates' correction for continuity should be applied when fe is below 10. Moreover, the chi-square statistic becomes highly distorted when fe is very small. Let us take an example. Suppose a cell has the value of $fe = 1$ and $fo = 5$. The difference between 'fo' and 'fe' is 4. The contribution of this cell to the total X^2 value is:

$$X^2 \text{ for this cell} = \frac{(fo - fe)^2}{fe} = \frac{(5-1)^2}{1} = \frac{16}{1} = 16$$

Now, consider another instance where $fe = 10$ and $fo = 14$. The difference between $fo - fe$ is still equal to 4 but contribution of this cell to the total value of X^2 would be:

$$X^2 \text{ for cell} = \frac{(fo - fe)^2}{10} = \frac{(14-10)}{10} = \frac{16}{10} = 1.6$$

Clearly, a small fe especially when its value is less than 5, has a great influence on the chi-square value. Thus X^2 is too sensitive when fe values are extremely small. A chi-square test preferably should not be performed when the expected frequency of any cell is less than 5. One ideal way to avoid small expected frequencies is to use large samples. In fact, X^2 test can best be used only if fewer than 20% of cells have an expected frequency of less than 5 and no cell has an expected frequency of 1 or less than 1.

Nominal Measures of Association based upon X^2

As we know that the chi-square in itself is not a measure of association when applied to a contingency table but that it is possible to derive measures of association from it. The phi-coefficient (ϕ) and the coefficient of contingency (C) are two such important derivations.

Phi-efficient (ϕ)

The phi-coefficient is a measure of the extent of association or relation between two sets of attributes measured on *nominal or categorical scale*, each of which may take on only two values. In other words, phi coefficient is a measure of association between two attribution measured on nominal scale and arranged into 2×2 table. The value of ϕ ranges from to ± 1.00. ϕ may be calculated directly from the 2×2 table with the help of formula 27.5.

$$\phi = \frac{AD - BC}{\sqrt{(A+B)(C+D)(A+C)(B+D)}} \quad (27.5)$$

Where,

ϕ = Phi-coefficient
A, B, C&D = Four cells of 2 × 2 table

Let us take an example to illustrate its calculation on a sample of 200 students to determine if there is a relationship between socio-economic status (SES) and attitudes towards demonetisation. The results are listed in 2 × 2 table as under:

Table 27.9: Responses of 200 students towards demonetisation

Attitude		Lower	Middle	Total
	Approve	70^A	30^B	100 (A+D)
	Disapprove	30^C	70^D	100 (C+D)
		100	100	200
		(A+C)	(B+D)	(A+B+C+D)

$$\phi = \frac{(70)(70) - (30)(30)}{\sqrt{(100)(100)(100)(100)}}$$

$$= \frac{4900 - 900}{10000} = \frac{4000}{10000} = .40$$

The phi-coefficient in related to chi-square from a 2 × 2 table by very simple and easy equation.

$$X^2 = N\phi^2 \qquad (27.6)$$

Likewise, phi-coefficient is derived from chi-square by the following equation:

$$f = \sqrt{\frac{X^2}{N}} \qquad (27.7)$$

The significance of ϕ can be tested through its relationship with X^2. When $\phi = .40$, its chi-square value with the help of formula 27.6 is equal to $200 \times (.40)^2 = 32.00$. If chi-square is significant, phi is also taken to be significant one. At $df = (r-1)(c-1) = (2-1)(2-1) = 1$, the critical value of X^2 is 3.841 on .05 level of significance. Since the obtained value of X^2, that is, 32 far exceed this critical, it is taken to be significant one. Accordingly phi is also taken to be significant one. Thus the null hypothesis which states that the socio-economic status and attitude towards demonetisation are not related (or independent) is rejected and therefore, it is concluded that there is relationship between socio-economic status and attitude towards demonetisation.

Assumptions of phi-coefficient:

The following are some of the major assumptions of X^2
- The measurements must be obtained on two nominal or categorical variables.
- Both the variables must have genuine dichotomies, that is, two classes or categories of the variables must be separated by a real gap between them. However, ϕ can also be used to data that are measurable on continuous variables if certain allowances for the

continuity with appropriate corrections and modifications are made (Guilford & Fruchter, 1981)
- The data in 2 × 2 table should be in terms of frequency or proportions.
- There is some associations or relationship between the variables.

Features of Phi-coefficient

The following are some of the important features of phi-coefficient:
- The phi-coefficient is a product moment correlation coefficient (Edwards, 1967)
- The values of phi-coefficient varies from –1.00 to +1.00 but seldom reaches this extreme range.
- The phi-coefficient is a very useful and desirable statistical tool in test construction and item analysis. Phi-coefficient is especially useful when item-item correlations are needed.
- The phi-coefficient is a close relative of the chi-square. The relation between phi and chi is clearly observed through formula 27.6 and 27.7.
- The phi-coefficient is the appropriate statistic when traits are truly dichotomous

Limitations of the phi-coefficient

One of the major limitations of the phi-coefficient is that the size of the coefficient is related to the way in which the two variables are split. When both variables are evenly divided a we find in Table 27.10 (Part I and Part III), the maximum limit of correlation coefficient ±1.00 may be achieved. If the marginal totals are unequal, the maximum value of the phi-coefficient will vary but in any case will be less than ±1.00 (see Table 27.10 Part II & IV).

Table 27.10: 2 × 2 Table showing how the size of Phi-coefficient varies with Marginal values

Part I

0	100	100		
100	0	100	$\phi = -1.00$	(Two variables are evenly distributed)
100	100	200		

Part II

40	30	70		
30	0	30	$\phi = -.43$	(Two variables are distributed on 70-30 basis)
70	30	100		

Part III

100	0	100		
0	100	100	$\phi = -1.00$	(Two variables are evenly distributed)
100	100	200		

Part IV

30	40	70		(Two variables are
0	30	30	$\phi = +.43$	distributed on 70-30
30	70	100		basis)

Co-efficient of Contingency (C):

Another measure of relationship based on nominal variables is coefficient of contingency symbolized by C. The coefficient of contingency was developed by Karl Pearson primarily for use with square tables having two or more than two equal number of rows and columns such as 3×3, 4×4, 5×5, etc. The efficient of contingency is computed through chi-square with the help of formula 27.8.

$$C = \sqrt{\frac{X^2}{X^2 + N}} \qquad (27.8)$$

Where, X^2 = chi-square; N = Number of cases.

A simple algebraic manipulation enables us to write the formula for X^2 based upon formula 27.8 as under

$$X^2 = \frac{NC^2}{1 - C^2} \qquad (27.9)$$

We can also calculation C directly from contingency table by the following formula

$$C = \sqrt{1 - \frac{N}{S}} \qquad (27.10)$$

where,

$$S = \Sigma \left(\frac{fo^2}{fe} \right) \qquad (27.11)$$

The coefficient of contingency has some features as under:
- The value of C becomes equal to zero when the variables are independent.
- The maximum value of C is always less than 1 and is determined by the number of rows and columns in the table. When the number of rows is equal to the the number of Columns, the maximum value of C is given by $\sqrt{(K-1)/K}$. Thus for 2×2 table the maximum value of C would be $\sqrt{(2-1)/2} = \sqrt{1/2} = .707$ and for a 3×3 table, the maximum value would be $\sqrt{(3-1)/3} = .816$ and for 5×5 table, the maximum value would be .89 and for 10×10 table, the maximum value would be .95.

Calculation of the coefficient of contingency

Let us illustrate calculation of C from a 3×3 table (Table 27.11) which represents the bivariate data regarding marital status and depression. The question is whether the two variables, that is, marital status and depression are related.

Table 27.11: Coefficient of Contingency between Marital status and Depression

		Married	Never Married	Formerly Married	
	High	15	20	18	53
Depression	Moderate	16	20	30	66
	Low	30	16	15	61
		61	56	63	180

For computing C, we shall first calculate X^2 by formula 27.1

Calculation of fe:

$$\frac{61 \times 53}{180} = 17.96 \quad \frac{56 \times 53}{180} = 16.49 \quad \frac{63 \times 53}{180} = 18.55$$

$$\frac{61 \times 66}{180} = 22.37 \quad \frac{56 \times 66}{180} = 20.53 \quad \frac{63 \times 66}{180} = 23.10$$

$$\frac{61 \times 61}{180} = 20.67 \quad \frac{56 \times 61}{180} = 18.98 \quad \frac{63 \times 61}{180} = 21.35$$

fo–fe

```
15 – 17.96 = –2.96    20 – 16.49 = 3.51     18 – 18.55 = –.55
16 – 22.37 = –6.37    20 – 20.53 = –.51     30 – 23.10 = 6.9
30 – 20.67 =  9.33    16 – 18.98 = –2.98    15 – 21.35 = –6.35
Σfo–fe =       0                  0                     0
```

$$X^2 = \frac{(-2.96)^2}{17.96} + \frac{(-6.37)^2}{23.37} + \frac{(9.33)^2}{20.67} + \frac{(3.51)^2}{16.49} + \frac{(-.53)^2}{20.53} + \frac{(-2.98)^2}{18.98}$$

$$+ \frac{(-.55)^2}{18.55} + \frac{(6.9)^2}{23.10} + \frac{(-6.35)^2}{21.35}$$

$= .487 + 1.736 + 4.21 + .747 + .01 + .467 + .016 + 2.06 + 1.888 = 11.621$

Now, C by formula 27.8, would be

$$C = \sqrt{\frac{11.621}{11.621 + 180}} = .246 = .25$$

The significance of C is tested through X^2. For 3×3 table, the $df = (3-1)(3-1) = 4$. At df

4, the critical value of X^2 at .05 level of significance is 9.488. Since the obtained value of X^2 (11.621) exceeds this critical value, X^2 is taken to be significant. Consequently, C is also significant. The null hypothesis that the marital status and depression are not related (or independent) is rejected. Thus it is concluded that marital status and depression are related.

Limitations of coefficient of contingency :

One of the limitations of C is that it does not remain constant for the same data when the number of categories varies. For example, C computed from 4×4 table cannot be compared to the C computed from the same data arranged into 5×5 table. Also, the maximum value which C can take, depends upon the fineness of the grouping. Where $k = r$, the upper limit of C depends upon $\sqrt{\frac{k-1}{k}}$. Another limitation is that the value of C is always positive and it ranges between 0 and +1.0 but it cannot attend the upper limit or unity. Still another limitation is that for computation of C, data must be amenable to the computation of X^2. As we know that X^2 test can best be used only if fewer than 20% of cells have an expected frequency of less than 5 and no cell has expected frequency of 1 or less than 1, it is also said that C is not directly comparable to some other measures of correlation like Pearson r, Kendall's τ and Spearman's P.

Mann - Whitney U Test

The Mann-Whitney U test is a non-parametric substitute of parametric t test. In brief, it is called as *Wilcoxon test*. This test is most suited to data obtained on ordinal measurement or are converted into ordinal measurement. The test was developed by Austrian born US mathematician H.B. Mann and US statistician D.R. Whitney who had jointly published this test in 1947. The Mann-Whitey U test is sometimes referred to as the rank test or *Sum-of-Ranks test* and if N of the two groups are equal, it is also known as the *Wilcoxson T-test*.

The U test can be applied to both continuous and discrete measurement variables. The test is particularly applied to test the significance of difference between unpaired observations of two independent samples or groups either of equal or unequal sizes. It is also applied when the researcher wants to avoid the assumptions of t test or when the researcher finds that the measurement is weaker on interval scaling. The U test can be applied to both small sample and large sample. In a nutshell, the Mann-Whitney U test is used when there are scores for two groups such that all scores may be placed in one overall ranking and the researcher wants to test the differences in central tendency between these two groups.

Computation of Mann-Whitney U values :

In Mann-Whitney U test there are always two independent groups and therefore, two N_s. Let n_1 be the number of cases in the smaller of the two groups and n_2 be the number of cases in the larger one. If both samples are equal in size, that is, $n_1 = n_2$ either N may be used as n_1 or n_2. The

U is calculated by the following formulas.

$$U = n_1 n_2 + \frac{n_1(n_1+1)}{2} - R_1 \quad (27.12)$$

$$U = n_1 n_2 + \frac{n_2(n_2+1)}{2} - R_2 \quad (27.13)$$

we will get two separate values by these two formula. The smaller of the two is called U and the larger values is called U'. The sum of $U + U'$ must be equal to $n_1 n_2$. For calculating U value, we first combine the observations or scores from both groups, rank then in order of increasing size. Rank 1 is assigned to the smallest value in the series and Rank 2 to the next smallest and so on. In the processing of ranking, the algebraic sign is considered, that is, the lowest rank are assigned to the largest negative numbers, if any. The tied scores are given average rank. Table 27.12 provides the achievement scores of boys and girls. Do boys and girls differ on the measures of achievement?

Table 27.12: Achievement scores of 10 boys and 11 girls

Boys score ($n_1 = 10$)	Girls' scores ($n_2 = 11$)	Ranks$_1$	Ranks$_2$
10	22	1	8
15	32	4	14
11	40	2	17
16	46	5	18.5
20	76	7	21
25	12	9	3
28	18	11	6
30	27	13	10
35	29	15	12
48	40	20	16
	46		18.5
		$SR_1 = 87$	$SR_2 = 144$

$$\text{Check} = \Sigma R_1 + \Sigma R_2 = \frac{N(N+1)}{2} = 87 + 144 = \frac{21(21+1)}{2} = 231 = 231$$

$\Sigma R_1 + \Sigma R_2$ should be equal to $\frac{N(N+1)}{2}$. Thus, it *checks* the calculation so far done.

By formula 27.12:

$$U = (10)(11) + \frac{10(10+1)}{2} - 87 = 110 + 55 - 87 = 78$$

By formula 25.13 :

$$U = (10)(11) + \frac{11(11+1)}{2} - 144 = 110 + 66 - 144 = 32$$

The smaller of these two values, that is, 32 is U and the larger value, that is, 78 is U'.

As a computation check, the obtained U can be verified using the transformation method as under.

$$U = n_1 n_2 - U'$$
$$U = (10)(11) - 78 = 110 - 78 = 32$$

Finally, we test the significance of $U = 32$ and not $U' = 78$ with $n_1 = 10$ and $n_2 = 11$. Entering Table[1] 1 of U test, we find that for two-tailed test at .05 level of significance the critical value of U for $n_1 = 10$ and $n_2 = 11$ is 26. For being significant, our obtained U value should be either 26 or less than it. Since the obtained U is 32, it is not significant. Therefore, we accept null hypothesis (H_0) meaning thereby that boys and girls have been taken from the same population and hence, don't differ.

When n is > 20, the significance of obtained U can be tested through z test by using the following formula.

$$z = \frac{U - \frac{n_1 n_2}{12}}{\sqrt{\frac{n_1 n_2 (n_1 + n_2)}{12}}} \qquad (27.15)$$

Suppose the value U is 180, $n_1 = 15$ $n_2 = 22$, then by formula 27.15, z would be

$$z = \frac{180 - \frac{15 \times 22}{2}}{\sqrt{\frac{15 \times 22(15+22+1)}{12}}} = \frac{15}{\sqrt{\frac{15 \times 22 \times 38}{12}}} = \frac{15}{32.3} = 0.46$$

Since $z = 0.46$, it is not significant at .05 level because it is less than 1.96. Thus, here, the null hypothesis (H_0) will be accepted. The rationale behind using z test when $n_2 \geq 20$ is that when n_2 increases in size, the sampling distribution of U rapidly approaches the normal distribution.

Median Test

The median test is also a non-parametric test and is considered an alternative to the student's t test. The median test is applied to determine whether two or more than two independent samples

---

1. Table for U test may be seen at p. 323 of the book entitled 'Basic statistical Methods' by N.M. Downie & R.W. Heath (Fourth edition, 1970, Harper & Row)

of equal or unequal differ in terms of median. Thus median test compares the medians of two or more independent samples of either equal or unequal sizes. The null hypothesis (H_0) here is that two or more independent samples are drawn from the populations having the same median. Therefore, H_0 proposes that there are equal number of scores in each sample above and below the common median and the observed differences from this distribution is due to the choice factors. The alternative hypothesis (H_1) is that the median of one population differs from that of the other (two-tailed test) or that the median of one population is higher or lower than that of other (one-tailed test).

Major Assumptions of Median test :

Some important assumptions of median test are as under :
- The two or more groups/samples must be independent or uncorrelated.
- Each score should occur in the sample at random or independent of all other scores.

Computation of Median test :

Suppose the Picture completion scale of *WAIS* was administered to a group of normals ($N = 14$) and a group of Psychotics ($N = 13$) Now the researcher is interested in testing the hypothesis of no difference using Median test. The scores are presented in Table 27.13.

Table 27.13: Scores of a group of normals (N) and a group of Psychotics (P) on Picture completion test

N	P
6 (−)	2 (−)
6 (−)	8 (−)
14 (+)	12 (+)
13 (+)	13 (+)
10 (−)	8 (−)
8 (−)	6 (−)
10 (−)	5 (−)
10 (−)	2 (−)
14 (−)	4 (−)
10 (−)	10 (−)
15 (+)	10 (−)
16 (+)	12 (+)
12 (+)	14 (+)
14 (+)	

The first step in computing Median test is to compute a common median of the entire set of 27 scores. Thus

$N = n_1 + n_2 = 14 + 13 = 27$

$\text{Median} = \dfrac{N+1}{2}\text{th score} = \dfrac{27+1}{2} = 14\text{th score}$

For finding the 14th score in the distribution, we arrange all 27 scores in the ascending order as under

$$\text{Median} \downarrow$$

2,2,4,5,6,6,6,8,8,8,10,10,10   10 ,10,10,12,12,12,13,13,14,14,14,14, 15, 16
(14th score)

Therefore, the common median of data of Table 27.13 is 10. In each sample positive signs (+) are assigned to the scores above the common median and negative signs (–) are assigned to the scores at or below the median. Subsequently, a 2×2 contingency table is set up as under:

	N	P	
MdN & Above MdN	7 A	4 B	17 A+ B
Below MdN	7 C	9 D	16 C + D
	14 A + C	13 B + D	27

The chi-square test (X^2) for independence is performed. Here X^2 test from 2 × 2 contingency table incorporating Yates' correction for continuity, that is, formula 27.4 will be used.

$$X_c^2 = \dfrac{N(|AD - BC| - N/2)^2}{(A+B)(C+D)(A+C)(B+D)}$$

$$= \dfrac{27(|9 \times 7 - 4 \times 7| - 27/2)^2}{11 \times 16 \times 14 \times 13} = \dfrac{27(63 - 28 - 13.5)^2}{32032}$$

$$= \dfrac{12480.75}{32032} = .39$$

Thus $X_c^2 = .39$ that fails to reach the critical value of X^2 at .05 level of significant [$df = (r-1)(c-1) = (2 \times 1)(2-1) = 1)$]. The critical value of X^2 at df 1 for .05 level of significance is 3.841. As a consequence, the null hypothesis (H_0) is accepted. Therefore, the Normals and Psychotics don't differ on the the measures of Picture completion test in terms of their common median. Therefore, they come from population having the same median.

Median Test when there are more than two groups :

Median test is also applied when there are more than two groups or samples. Here the procedure is the same. First common median of all the scores taken together is obtained. Then a contingency table is set up showing the number of scores at or above the median and the number of scores below the median for each group. The chi-square (formula 27.1) is applied to the data and appropriate conclusion is drawn.

Let us illustrate the calculation of median test from the following four independent groups of boys (Table 27.14). The problem is to determine by applying median test whether the four samples came from the population with the same median.

Table 27.14: Scores of four independent groups of boys on an intelligence test

Group 1: 10, 16, 12, 18, 15, 20 (n = 6)
Group 2: 15, 19, 30, 20, 22, 24, 36 (n = 7)
Group 3: 8, 10, 14, 16, 23, 28, 30, 34 (2 = 8)
Group 4: 10, 16, 12, 18, 22, 24, 32, 35, 38, 39 (n = 10)

The first step would be to find out common median from this set of ungrouped data. We arrange all the 31 scores in ascending order as under :

8, 10, 10, 10, 12, 12, 14, 15, 15, 16, 16, 16, 18, 18, 19, $\underset{(20)}{\overset{Mdn}{\downarrow}}$, 20, 22, 22, 23, 24, 24, 28, 30, 30, 32, 34, 35, 36, 38, 39

$$\text{Median} = \frac{N+1}{2} = \frac{31+1}{2} = 16\text{th score}$$

The 16th score in the the distribution is 20. Now the 4 × 2 Table is set up as under:

Groups	+ sign Above Median	−sign Median & Below Median	Total
I	1	5	6
II	4	3	7
III	4	4	8
IV	6	4	10
	15	16	31

Calculation of *fe* :

$$\frac{15 \times 6}{31} = 2.90 \quad \frac{16 \times 6}{31} = 3.09$$

$$\frac{15 \times 7}{31} = 3.39 \quad \frac{16 \times 6}{31} = 3.61$$

$$\frac{15 \times 8}{31} = 3.87 \quad \frac{16 \times 8}{31} = 4.13$$

$$\frac{15 \times 10}{31} = 4.84 \quad \frac{16 \times 10}{31} = 5.16$$

Calculation of $fo - fe$

$$\begin{array}{ll}
1 - 2.90 = -1.90 & 5 - 3.09 = 1.91 \\
4 - 3.39 = .61 & 3 - 3.61 = -.61 \\
4 - 3.87 = .13 & 4 - 4.13 = -.13 \\
6 - 4.87 = 1.13 & 4 - 5.16 = -1.16 \\
\Sigma fo - fe = -.03 & \Sigma fo - fe = -.01
\end{array}$$

Ideally, $\Sigma fo - fe$ should be zero. But the above minor fractional differences have occurred due to rounding errors.

Calculation of X^2 :

$$\frac{(-1.9)^2}{2.90} = 1.24 \quad \frac{(1.91)^2}{3.09} = 1.18$$

$$\frac{(.61)^2}{3.39} = .109 \quad \frac{(-.61)^2}{3.61} = .103$$

$$\frac{(.13)^2}{3.87} = .004 \quad \frac{(-.13)^2}{4.13} = .004$$

$$\frac{(1.13)^2}{4.84} = .26 \quad \frac{(-1.16)^2}{5.16} = .26$$

$$X^2 = 1.24 + .109 + .004 + .26 + 1.18 + .103 + .004 + .26 = 3.16$$
$$df = (r - 1)(c - 1) = (4 - 1)(2 - 1) = 3 \times 1 = 3$$

Entering Table chi-square distribution, we find that the critical value of X^2 at $df\ 3$ for .05 level of significance is 7.815. As the obtained value of X^2 falls for short this critical value, it is taken as not significant and the null hypothesis (H_0) is accepted. Accordingly, it is concluded that all the four groups of students do come the population having the same median. As such, they don't differ.

Kendall's Co-efficient of Concordance (W)

The coefficient of concordance proposed by Kendall is symbolized as W and is a measure of correlation between more than two sets of rankings. When the researcher has more than two sets of ranking, he may determine association among them using Spearman P or Kendall's τ between all possible pairs of rankings and then, compute the average of these coefficients to determine overall association. But this will be a tedious task. The computation of W gives the same results and also is very simpler one. W has been considered as a very useful statistic in studies of interjudge or inter-test reliability. One feature of W is that it can be either zero or will be positive. It cannot be *negative*. When there are more than two sets of rankings, there cannot be full disagreement among them. For example, suppose there are three judges like A, B and C who express their opinion about an object or individual. Further suppose that A and B shows disagreement and C also disagrees with A, then B and C must show agreement with each other In fact, the W is an index of divergence of the actual agreement shown in the data from the maximum possible or perfect agreement.

Computation of W:

The computation of W is very simple and it bears a linear relation to the average Spearman's P taken over all groups. The formula for its computation is as under:

$$W = \frac{S}{\frac{1}{12} K^2 (N^3 - N)} \quad (27.16)$$

where,

S = sum of squares of deviations from mean of Rj (Rj is the sum of rank assigned to each individual separately)
K = Number of judges or sets of rankings
N = Number of individuals or objects which have been ranked.

Suppose three teachers have ranked 8 students or the trait of punctuality, Rank 01 indicates very high on punctuality and so on the last rank indicates a very low punctuality. The ranked data are presented in 27.15.

Table 27.15: Ranked data of 8 students by Three teachers on the trait of Punctuality

Teachers	A	B	C	D	E	F	G	I
X	2	1	3	4	6	8	7	5
Y	1	2	4	6	8	7	5	3
Z	2	1	3	6	5	8	7	4
Rj	5	4	10	16	19	23	19	12

$$\text{Mean of Rj} = \frac{5+4+10+16+19+23+19+12}{8} = \frac{108}{8} = 13.5$$

$$S = (5 - 13.5)^2 + (4 - 13.5)^2 + (10 - 13.5)^2 + (16 - 13.5)^2 + (19 - 13.5)^2$$
$$+ (23 - 13.5)^2 + (19 - 13.5)^2 + (12 - 13.5)^2$$
$$= 72.25 + 90.25 + 12.25 + 12.25 + 30.25 + 90.25 + 30.25 + 2.25 = 340.5$$

$$W = \frac{340.5}{\frac{1}{12}(3)^2 (8^3 - 8)} = \frac{340.5}{\frac{9 \times 504}{12}} = \frac{340.5}{378} = .90$$

When N > 7, the significance of W is tested through X^2. The value of W is converted into X^2 with the help of the following formula

$$X^2 = K(N-1)W \quad (25.17)$$

Thus, here $X^2 = 3(8-1).90 = 18.9$ and df would be $N - 1 = 8 - 1 = 7$. At $df = 7$ the critical value of X^2 for .05 level of significance is 14.067. Since the obtained X^2 exceeds the critical value, the X^2 is significant. Accordingly, W is also taken to be significant one. The null hypothesis (H_0) which states that K rankings are unrelated (or independent) is rejected. Therefore, it is concluded that the judges do exhibit agreement regarding ratings of students on punctuality.

When $N \leq 7$, the significance of W is tested with reference to a Table[1] especially developed for the purpose. It should be noted clearly that the degree of agreement among K judges is reflected by the degree of variation among N sums of ranks. If there is a random agreement among K judges, then the various Rj's would be approximate by equal.

Summary and Review

- In psychological and educational researches various types of non-parametric statistics are used.
- The chi-square test is one of the most popular non-parametric statistics used for testing various hypothesis like equal probability hypothesis, goodness-of-fit hypothesis and independence hypothesis.
- The chi-square is used with Yates' correlation in 2×2 table if expected frequency of any cell is less than 5.
- The chi-square test is also used with a table large than 2×2 with small expected frequency with caution. In such situation Yates' correction are not applied.
- The chi-square test has been misused frequently.
- There are two important nominal measure of association based upon X^2. They are phi coefficient (ϕ) and coefficient of contingency (c).
- The ϕ coefficient is applicable with 2×2 table whereas coefficient of contingency is applied with table having equal number of rows and columns such as 2×2, 3×3, 4×4, 5×5 etc.
- Mann-Whitney U test is a nonparametric substitute of parametric t test.
- Median test is also a non-parametric substitute of parametric t test in terms of median of the distribution.
- The coefficient of concordance (W) is a measure of association between more than two sets of ranking.

---

1. See Table T of the book entitled *Nonparametric statistics for the Behavioural Science* by S. Siegel & NJ. Castellan, Second edition (1988) p. 365.

Review Questions

1. Discuss the nature of X^2. Discuss its advantages and disadvantages.
2. Discuss the major assumption of X^2.
3. What is phi-coefficient? with examples, discuss its relationship with X^2.
4. What are the major characteristics of the coefficient of contingency. Discuss its limitations.
5. What is Mann-Whitney U test? Discuss its main features and assumptions.
6. Define Median test ? Discuss its major assumptions.
7. Define coefficient of concordance. Discuss its major features.

■ ■ ■

References

Abrams, S. (1977). *A polygraph handbook for attorneys.* Lexington, M.A.: Health

Ahuja, G.C. (1976) *Manual for group test of intelligence (G.G.T.I.) : 13 to 17th Years.* Agra: National Psychological Corporation

Ahuja, P. (1974) *Manual for group test of intelligence (9 to 13 years)* Agra. National Psychological Corporation

Airasian, P.W. & Madaus, G.J. (1972) Functional types of student evaluation. *Measurement and Evaluation in Guidance, 4,* 221-233

Albrecht, R.R. & Feltz, D.C. (1987). Generality and Specificity of attention related to competitive anxiety and sport performance. *Journal of Sport Psychology, 9,* 231-248

Allen, M.J & Yen, W.M. (1979) *Introduction to measurement theory,* C.A: Brooks/code

Allport, G.W. (1937) *Personality: A psychological Interpretation.* New York : Holt

Allport, G. & Vernon, P. (1931). *Studies in expressive movement.* Boston: Houghton Mifflin

Allport, G., Vernon, P. & Lindzey (1970). *Study of Values* (Revised third edition) Chicago: Riverside

Allport, G., Vernon, P. & Lindzey, G. (1960). *A Study of Vales.* Boston : Houghton Mifflion

Allport, G., Vernon, P. & Lindzey, G. (1960). *A Study of values* (3rd ed.). Manual: Chicago: Riverside

Allport, G.W. (1935). Attitudes. In C. Murchison (ed.) *Handbook of Social Psychology* (pp. 798-884). Woreaster, MA: Clark University Press.

Ambile, T.M. (1983). *The Social Psychology of Creativity.* New York: Springle-Verlag

American Education Research Association, American Psychological Association & National Council and Measurement in Education (2014) *Standards for Educational and Psychological Testing.* Washington, DC : American Educational Research Association.

American Educational Research Association, American psychological Association & National council on Measurement in Education (1985) *Standards for Educational and psychological Testing.* Washington, DC: American psychological Association

American Educational Research Association, American Psychological Association & National Council on Measurement in Education (1999) *Standards for Educational and Psychological Testing*. Washington, DC: American Psychological Association

American Psychological Association (1996) Affirmative Action : What benefits? Washington, Dc : Author

American Psychological Association (1992) Ethical principles of psychologists and code of conduct. *American psychologist 47(12) 1597-1611*

American Psychological Association (2002) Ethical principles of Psychologists and code of Conduct. *American Psychologist.57,* 1061-1073

Anastasi, A. & Urbina, S. (1997) *Psychological Testing* (7th ed.) Delhi: Pearson Education

Anderson, N & Shackleton, V. (1986) Recruitment and Selection: A review of developments in the 1980s. *Personnel Review, 15,* 19-26.

Angoff, W.F. (1971) Scores, Norms and equivalent scores. In R.L. Thorndike (Ed.) *Educational Measurement* (2nd ed.) Washington, DC : American Council on Education

Archer, R.P. Handel, R.W., Ben-Porash, Y.S. & Tellegen, A. (2016) *Minnesota Multiphasic Personality Inventory. Adolescent Restructured Form (MMPI-A-RF)*. Minneapolis; University of Minnesota Press

Aron, A., Aron, E.N. & Coups, E.J. (2006). *Statistics for Psychology*. Delhi: Pearson Education

Atkinson, J.W. (Ed.) (1958) *Motives in fantasy, action and society*, New York: Van Nostrand

Atkinson, L., Quarrington, B., Alp, I.E. & Cyr, J.J. (1986). Rorschach validity: An empirical approach to the literature. *Journal of clinical Psychology, 42,* 360-362.

Atkinson, M.J. (2003). Review of California Psychological Inventory. (3rd edition) 15th Mental Measurement Yearbook. Retrieved from OVID. Mental Measurement Yearbook database

Banerjee, S. & Kundu, R. (1983). *Computer Model of Scoring Rorschach responses. Psychological Studies, 28,* 51-55

Barlow, D.H. & Durand, M.V. (2009). *Abnormal Psychology : An Integrative approach.* Belmont, CA: Wardsworth Cengage Learning

Baron, R.A. (2002). *Psychology*. Delhi: Pearson Education

Bartol, C.R. & Bartol, A.M. (2014). *Criminal behaviour: A Psychological approach.* New Jersey: Pearson

Batey, M., Furnham, A.F. & Safiullina, (2010). Intelligence, General Knowledge and personality as predictor of creativity. *Learning and Individual differences, 20,* 532-535

Bean, K.L. (1953). *Construction of Psychological and Educational Tests.* New York: McGraw Hill

Becker, K.A. (2003). *History of the Stanford-Binet intelligence scales: Content and Psychometrics.* Itasca, IL: Riverside Publishing

Beeby, C.E. (1979). Teachers, Teacher education and Research. In Gardner, R.L. (Ed.) *Teacher Education in developing countries.* London: Institute of Education, University of London. pp. 135-161.

Bellak, L. & Bellack, S.S. (1973). *Senior Apperception Technique.* New York: CPS

Bellak, L. & Bellak, S. (1952). *Children's Apperception Test.* New York: CPS Co.

Bellak, L. & Hurvich, M.S. (1966). A human modification of children Apperception test (CAT-H) *Journal of Projective techniques and Personality Assessment,* 30, 228-242

Bellak, L. (1975). *The TAT, CAT and SAT in clinical use* (3rd ed.) New York: Grune & Sttratton.

Benett, G.K. Seashore, H.G. & Wesman, A.G. (1984). *Differential Aptitude Test. Technical Supplement.* San Antonio, TX: Psychological Corporation

Bennett, G.K, Seashore, H.G. & Wesman, A.G. (1947). *Differential Aptitude Test.* New York: Psychological Corporation

Ben-Porath, Y.S. & Tellegen, A. (2008). *Minnesota Multiphasic Personality Inventory-2-RF.* Minneapolis: University of Minnesota Press

Bernstein, L. (1956). The examiner as an inhibiting factor in clinical setting. *Journal of Consulting Psychology,* 20, 287-390

Beutler, L.E. & Berren, M.R. (1995). *Integrative assessment of adult personality.* New York: Guilford Press

Bhatia, C.M. (1955). C.M.'s *Bhatia's Performance Tests of Intelligence under Indian conditions.* London, England: Oxford University Press

Bilbao, P.P., Lucido, P.I, Ivingan, J.C. & Javien, R.B. (2008). *Curriculum Development,* Quezon City: Lorimar Publishing Inc.

Binet, A. & Simon, T. (1905). Methods nonvelles Pour le diagnostic der niveau intellectual des anorman x. In R.M. Kaplan and D.P. Saccuzzo. *Psychological Testing* (5th edition): Wordsworth

Bloom, B.S. (1960). Learning for mastery: *Evaluation Comment* 1(2).

Bloom, B.S., Englehar, M.D., Furst, G.J., Hill, W.H. & Krathwohl, D.R. (1954). *Taxonomy of Educational Objectives: Handbook I, Cognitive domain.* New York: David Mckay Company

Blum, G.S. (1950). *The Blacky Pictures and Manual.* New York: Psychological Corporation

Blum, G.S. (1968). Assessment of Psychodynamic Variables by Blacky Pictures. In P. McReynolds (Ed.) *Advances in Psychological Assessment* (VA-I). Palo Alto, California: Science and Behaviour Books.

Boake, C. (2002). From the Binet-Simon to the Wechsler-Bellevue Tracing the history of intelligence testing. *Journal of Clinical and Experimental Neuropsychology, 24,* 383-405

Boden, M. (2004). *The Creative Mind: Myths and Mechanisms.* Routledge

Boring, E.G. (1923). Intelligence as the tests test it. *New Republic.* 34: 35-37

Borman, W.C. (1982). Validity of behavioural assessment for predicting military recruiter performance. *Journal of Applied Psychology. 67,* 3-9

Bowman, M.L. (1989). Testing individual differences in ancient China. *American Psychologist. 44,* 576-578

Brittain, H.L. (1907). A Study in imagination. *Pedagogical Seminary, 14,* 137-207

Brody, N. (1972). *Personality: Research and Theory:* New York: Academic Press

Brookhart, S.M. & Nitko, A.J. (2008). *Assessment and Grading in Classrooms.* Upper Saddle River, NJ: Pearson Education

Brown, L, Sherbenou, R.J. & Johnsen, S.K. (2010). *Test of Non-verbal Intelligence-4* (Fourth edition). Austin: Pro-Ed

Buck, J.M. (1948). The H-T-P technique, a qualitative and quantitative method. *Journal of Clinical Psychology, 4,* 317-396

Buck, J.N. (1950). *Administration and Interpretation of the H-T-P test: Proceedings of the H-T-P Workshop at Veterans Administration Hospital, Richmond, Virginia.* Beverly Hills, CA: Western Psychological Services

Buck, J.N. (1992). House-Tree-Person Projective drawing technique (H-T-P): *Manual and interpretative guide.* Los Angels. CA: Western Psychological Services

Burns, R.C. & Kaufman, S.H. (1970). *Kinetic Family Drawings (K-F-D): An introduction to understanding through Kinetic drawings.* New York Brunner/Mazel

Burns, R.C. & Kaufman, S.H. (1972). *Actions, Styles and symbols in Kinetic Family Drawings (K-F-D).* New York: Bruner/Mazel

Burt, C. (1949). The structure of the Mind; A review of results of factor analysis. *British Journal of Educational Psychology, 19,* 110-111, 176-199

Butcher, J.M., Hooley, J.M., Mineka, S. & Dwivedi, C.B. (2017) *Abnormal Psychology*. Noida: Pearson India Education

Butcher, J.M., Williams, C.L., Graham, J.R., Archer, R.P., Tellegen, A., Ben-Porath, Y.S., Kaemmer, B. (1992) Minnesota Multiphasic Personality Inventory-Adolescent (MMPI-A) *Manual for administration, scoring and interpretation*. Minneapolis: University of Minnesota Press

Butcher, J.N. (ed.) (1995). *Clinical Personality assessment: Practical approaches*. New York Oxford University Press

Butcher, J.N., Dahlstrom, W.G., Graham, J.R., Tellegen, A. & Kaemmer, B. (1989). Minnesota Multiphasic Personality Inventory-2 (MMPI-2). *Manual for administration and Scoring*. Minneapolis: University of Minnesota Press.

Butcher, J.N., Gautam, J.R., Dahlstrom, W.G., Tellegen, A.M., & Kaemmer, B. (1989) MMPI-2 manual for administration and scoring. Minneapolis: University of Minnesota Press

Butcher, J.N., Williams, C.L., Graham, J.R., Archer, R.P., Tellengen, A., Ben-Porath, Y.S. & Kaemmer, B. (1992). Minnesota Multiphasic personality Inventory-Adolescent (MMPI-A). *Manual of Administration, Scoring and Interpretation*. Minneapolis: University of Minnesota Press

Cacioppo, J.T., Cries, S.L., Berntson, G.G., & Coles, M.G. (1993). If attitudes affect how stimuli are processed, should they not affect the event-related brain potential? *Psychological Science, 4*, 108-112.

Cacioppo, J.T., Crites, S.L., Gardner, W.L. & Berntson, G.G. (1994). Bioelectrical echoes from evaluative categorizations: I.A. late positive brain potential that varies as a function of trait negativity and extremity. *Journal of personality and social psychology* 67, 115-125.

Cacioppo, J.T., Petty, R.E., Losch, M.E. & Kim, H.S. (1986) Electromyographic activity over facial muscle regions can differentiate the valence and intensity of affective reactions. *Journal of personality and social psychological.* 50, 260-268.

Camilli, G. & Shepard, L.A. (1985). A computer programme to aid the detection of biased test items. *Educational and Psychological Measurement, 45*, 595-600

Campbell, D.T. & Fiske, D.W. (1959). Convergent and discriminant validation by the multitrait-multimethod matrix. *Psychological Bulletin*, 56, 81-105

Campbell, V.L. (1990) A model for using tests in counseling. In C.F. Watkins & V.L. Campbell (Eds.) *Testing in counseling practice*. Hillsdale, NJ: Lawrence Erlbaum.

Campion, M.A., Parsell, E.D. & Brown, B.K. (1988) Structured interviewing: Raising the psychometric properties of the employment interview. *Personnel Psychology 41*, 25-62.

Carroll, J.B. (1993). *Human Cognitive Abilities: A Survey of factor-analytic studies.* Cambridge, MA: Cambridge University Press

Carroll, J.B. (1993). Human Cognitive Abilities: A Survey of factor-analytic studies. New York: Cambridge University Press

Cassell, W.A. (1965). Perception and Symptom localization. *Psychosomatic medicine. 27,* 171-176

Cassell, W.A. (1980a). *Somatic Inkblot Series II.* Anchorage Alesa: Aurora Publishing Co.

Cassell, W.A. (1980b). *Body symbolism and the Somatic Inkblot Series.* Anchorage Alasa: Aurora Publishing Co.

Cassell, W.A. (1984). Computerized Projective Testing as an aid in Medical diagnosis. *Alaska Medicine, 26,* 1-9

Cattell, J.M. (1887). Experiments on association of ideas. *Mind, 12,* 68-74

Cattell, R. (1949). *Culture Free Intelligence Test, Scale I, Handbook.* Campaign, IL: Institute of Personality and Ability Testing

Cattell, R.B. (1950). *Personality: A systematic, theoretical and factual study.* New York: McGraw-Hill

Cattell, R.B. (1998). Where is intelligence? Some answers from the triarchic theory. In J.J. McArdler et al. (Eds.) *Human Cognitive abilities in theory and practice.* Mahwah, NJ: Erlbaum

Cattell, R.B., Cattell, A.K. & Cattell, H.E. (1993). *Sixteen Personality Factor Questionnaire.* Fifth edition. Champaign, IL: Institute for Personality and Ability Testing

Cattell, R.B., Eber, H.W. & Tatsuoka, M.M. (1970). *Handbook for the Sixteen Personality Factor Questionnaire.* Champaign, IL: Institute of Ability and Personality Testing.

Chaplin, W.F., Hohn, O.P. & Goldberg, L.R. (1988). Conceptions of State and traits: Dimensional attributes with ideals as prototypes. *Journal of Personality and Social Psychology. 54,* 541-557

Chaudhary, U. (1960). Modification of Thematic Apperception Test. *Journal of American Social Psychology.* 51, 245-63

Cicarelli, S.K. & white, J.M. (2017) *Psychology* Noida: Pearson India Education.

Coan, R.W. & Cattell, R.B. (1985). The development of an early School Personality Questionnaire. *Journal of Experimental Education. 28(3),* 143-152

Cohen, R.J. & Swerdlik, M.E. (2005). *Psychological Testing and Assessment.* Boston: McGraw-Hill

Cohen, R.J., Montague, P. Nathanson, L.S. & Swerdlik, M.E. (1988). *Psychology tasting: An Introduction to tests and measurements.* Maintain view, CA: Mayfield.

Colman, A.M. (2015). *A Dictionary of Psychology.* Oxford: Oxford University Press

Cone, J.D. (1999) Introduction to the special section on self-monitoring: A major assessment method in clinical psychology *Psychological assessment,* 11, 411-414

Costa, P.J. & McCrae, R.R. (1992). *NEO-PI-R: Professional Manual.* Odissa, Fla: Psychological Assessment Resources

Costa, P.T., Jr. & McCrae, R.R. (1992). Revised NEO Personality Inventory (MEO-PI-R) and NEO Five factor Inventory (NEO-FFI). *Professional Manual.* Odessa, F.L. Psychological Assessment Resources

Costantino, G., Malgady, R.G. & Rogler, L.H. (1988). *TEMAS (Tell-Me-A-Story) Manual.* Los Angeles, CA: Western Psychological Services

Cox, R.H. (1998) *Sport Psychology.* Boston: McGraw-Hill

Craig, R.J. & Olson, R.E. (1995). McMI-II Profiles and Typologies for the patients seen in marital therapy. *Psychological Reports.* 76(1), 163-170

Cronbach, L.J. & Meehl, P.E. (1955) Construct validity in Psychological tests. *Psychological Bulletin.* 52, 281-302

Cronbach, L.J. (1951). Coefficient alpha and the internal structure of the tests. *Psychometrika,* 16, 297-334

Cronbach, L.J. (1970). Essentials of Psychological Testing. (3rd ed.) New York: Harper & Row

Cronbach, L.J. (1988). Five perspectives on validation argument. In H. Wainer & H. Braun (Eds.) *Test Validity.* Mahwah, N.J: Erlbaum

Cronbach, L.J. (1990). *Essentials of Psychological Testing* (5th edition) New York: Harper & Row

Cronbach, L.J., Glaser, G.C., Nanda, H. & Rajaratnam, N. (1972). *The dependability of behaviour measurement: Theory of generalization for scores and profiles.* New York: Wiley

Csikszentmipalyi, M. (1996) *Creativity: Flow and the Psychology ofengagement with everyday life,* New York. Basic Books.

Cunningham, W.A., Preacher, K.J. & Banaji, M.R. (2001). Implicit attitude measures: consistency, stability and convergent validity. *Psychology Science12,* 163-170.

Cureton, E.E. (1957). The upper and lower twenty-seven percent rule. *Psychometrika, 22,* 293-296

Dahlstorm, W.C. (1969). Invasion of Privcy: How legitimate is the current concern over this issue? In J.N. Butcher (Ed.) MMPI: *Research developments and clinical applications.* New York: McGraw-Hill

Dalal, A.K. (2011). A Journey back to the roots: Psychology in India. In H. Cornelisson, G. Misra & S. Verma (Eds.) *Foundations of Indian Psychology.* (pp. 27.57), New Delhi: Pearson

Dana, R.H. (1993). *Multicultural assessment perspective for professional Psychology.* Boston: Allyn & Bacon

Das, J.P. & Naglieri, J.A. (2001). The Das-Naglieri Cognitive Assessment System in theory and practice. In J.J.C. Andrews, D.H. Saklofske & H.L. Janzen (Eds.) *Handbook of Psychoeducational assessment* (pp. 33-63) San Diego, CA: Academic Press

Das, J.P., Naglieri, J.A. & Kirby, J.R. (1994). *Assessment of cognitive processes. The PASS theory of intelligence.* Boston: Ally & Bacon

Defence Institute of Psychological Research (2008). *Development of a new Psychological test battery for the selection and trade allocation of other ranks in the Indian Army* (DIPR Technical report) Delhi: DIPR

Defense Institute of Psychological Research (2007) *Development of a comprehensive battery of cognitive abilities for selection of candidates for commissioned ranks in the Armed Forces* (DIPR Technical Report), Delhi: DIPR

Delhees, K.H. & Cottell, R.B. (1971). *Manual for Clinical Analysis Questionnaire (CAQ).* Champaign, IL: Institute for Personality and Ability Testing

Derogatis, L.R. (1994). *Symptom Checklist 90-R: Administration, Scoring and procedures manual* (3rd edition) Minneapolis, HN: National Computer systems.

Donnay, D.A.C. (1997). E.K. Strong Legacy and beyond: 70 years of Strong Interest Inventory. *The Career Development Quality. 46(1),* 2-22

Drummond, R.J. (1996). *Appraisal procedures for counselors and helping professionals* (3rd ed.) Englewood Cliffs, NJ: Merrill

Dubey, B.L. (1982). *A pragmatic view of Rorschach Inkblot technique.* Agra: National Psychological Corporation

DuBois, P.H. (1970) *A History of Psychological Testing.* Boston: Allyn & Bacon

Durand, V.M. & Barlow, H.B. (2006). *Essentials of Abnormal Psychology* (4th ed.) Belmont, C.A.: Thompson Wordsworth

Dush, D.M. (1985). Review of Holtzman Inkblot technique. *Ninth Mental Measurement Yearbook, 1,* 602-603

Ebel, R.L. (1956). Obtaining and reporting evidence on content validity. *Educational and Psychological Measurement, 16,* 269-282

Ebel, R.L. (1972). 'Why is a longer test usually a more reliable one?' *Educational and Psychological measurement, 32,* 249-53

Edwards, A.L. & Kilpatrick, F.P.A. (1948) A technique for the construction of attitude scales. *Journal of Applied Psychology. 32,* 414-422

Edwards, A.L. (1954) Statistical Methods for Behavioural Sciences. New York: Rinchart

Edwards, A.L. (1954). *Manual for the Edwards Personal preference Schedule.* New York: Psychological Corporation

Edwards, A.L. (1959). *Edwards Personal Preference Schedule.* New York: Psychological Corporation

Edwards, A.L. (1957). *The Social desirability variable in Personality assessment and research.* New York: Dryden

Eisner, E.W. (1991) Should America have a national curriculum? *Educational Leadership, 49,* 76-81

Eisner, E.W. (1994). *The Educational Imagination on design and Evaluation of School programme* (3rd ed.) New York: McMill an

Elifson, K.W., Runyon, R.P. & Haber, A. (1990) *Fundamental of Social Statistics,* New York: McGraw H.K

Ellis, A. (1946). The validity of the personality questionnaires. *Psychological Bulletin, 43,* 385-440

Exner, J.E. & Weiner, I.B. (1995). The Rorschach: A comprehensive System: Vol. 3. *Assessment of children and adolescents* (2nd ed.), New York: Wiley

Exner, J.E. (1962). A comparison of human figure drawings of Psychoneurotics, character disturbances, normals and subjects experiencing experimentally induced fears. *Journal of Projective techniques, 26,* 292-317

Exner, J.E. (1969). *The Rorschach Systems.* New York: Grune & Stratton

Exner, J.E. (1974). *The Rorschach: A Comprehensive System.* New York: John Wiley and Sons

Exner, J.E. (1993). *The Rorschach: A Comprehensive System*: Vol. 1. Basic foundation (3rd edition), New York: Wiley

Exner, J.E. (1999). *The Rorschach: Measurement concepts and issues of validity*. In S.E. Embretson & S.L. Hensberger. (Eds). *The new rules of measurement*. Mahwah, NJ: Erlbaum

Eysenck, H.J. & Eysenck, M.W. (1985). *Personality and individual differences*. New York: Plenum Press

Eysenck, H.J. & Eysenck, S.B.G. (1975). *Manual of the Eysenck Personality Questionnaire*. Hodder & Stoughton

Eysenck, H.J. (1959). Rorschach Review. In O.K. Buros (Ed.). *The fifth mental measurement year book*. Highland Park, New Jersey: Gryphon Press

Eysenck, H.J. (1965). *Fact and fiction in Psychology*. Baltimore, H.D: Penguin Books

Eysenck, H.J. (1982). Development of a theory. In H.J. Eysenck (Ed.). *Personality Genetics and Behaviour: Selected Papers* (1-48), New York: Praeger

Fabiani, M., Gratton, G., Karis, D., & Donchin, E. (1987). Definition, Identification, and Reliability of measurement of P300 Component of the event-related brain potential. In P.K. Acklas, J.R. Jennings and M.G. Coles (eds.) *Advances in Psychophysiology*. Vol. 2, pp. 1-87, Greenwich, CT: JAI Press

Feist, G.J. (1998) A meta analysis of the impact of personality on scientific and artistic creativity. *Personality and Social Psychological Review*, 2, 290-309

Feldhusen, J.F. (1964) Students' perceptions of frequent Quizzes and Post-Mortem discussion of tests. *Journal of Educational Measurement* 1(1), 51-54

Ferguson, G.A. & Takane, Y. (1989) *Statistical Analysis in Psychology and Education*. New York: McGraw-Hill

Fiske, D.W. & Baughman, E.E. (1953) Relationship between Rorschach scoring categories and the total number of responses. *Journal of Abnormal and Social Psychology*, 48, 25-32

Fiske, D.W. (1967) The subject reacts to Tests. *American Psychologist*, 22, (4), 287-96

Foote, J. & Khan, M.W. (1979) Discriminative effectiveness of the Senior Apperception Test with impaired and nonimpaired elderly persons. *Journal of Personality Assessment*, 43, 360-364

Frank, L.K. (1939) Projective methods for the study of Personality. *Journal of Psychology*, 8, 389-413

Fredericksen, M. (1962). Factors in in-basket performance. *Psychological Monographs*. 76 (22).

Friedman, M. & Rosenman, R.H. (1974) *Type A behaviour and your heart*. New York: Knopf

Gardner, E.F. (1993). *Multiple Intelligence: The theory in Practice.* New York: Basic Books

Gardner, E.F. (1999a). *Intelligence redefined: Multiple Intelligences for the 21st century.* New York: Basic Books

Gardner, E.F. (1999b). Who wons intelligence? *Atlantic Monthly.* 67-76

Gardner, H. (1983). *Frames of Mind: Theory of Multiple Intelligence.* New York: Basic Books

Garrett, H.E. (1981). *Statistics in Psychology and Education.* Bombay: Vakils, Feffer and Simons Ltd.

Gay, L.R. (1980). *Educational Evaluation and Measurement* Columbus: Bell & Howell Company

Golden, C.J., Puriksh, A.D. & Hammeke, T.A. (1985). *Luria-nebraska Neuropsychological Battery. Forms I and Forms II. Manual.* Los Angels: Western Psychological Services.

Goldman, L. (1961). *Using Tests in Counselling.* New York: Appleton-Century-Crofts

Goodenough, F.L. (1926). *Measurement of intelligence by drawing.* Yonkers New York: World Book

Gordon, W.J.J. (1961). *Synectics: The development of creative capacity.* New York: Harper & Row

Gottfredson, L.S. (2003). Dissecting Practical intelligence theory. Its claims and evidence. *Intelligence, 31,* 343-397

Gough, H.G.C (1987). *California Psychological Inventory. Administrator's guide.* Palo Alto, CA: Consulting Psychological Press.

Greenwald, A.G., McGhee, D.E. & Schwartz, J.L.K. (1998). Measuring individual differences in implicit cognition: The Implicit association test. *Journal of Personality and Social Psychology, 74,* 1464-1480

Gregory, R.J. (2004). *Psychological Testing: History, Principles and Applications.* Delhi: Pearson Education

Gronlund, N.E. (2000). *How to write and use instructional objectives.* NJ: Prentice-Hall

Grove, J.R. & Prapavessis, H. (1992). Preliminary evidence of the reliability and validity of an abbreviated profile of mood states. *International Journal of Sport Psychology, 23,* 93-109

Gucciardi, D.F., Gordon, S. & Dimmock, J.A. (2009a). Development and preliminary validation of a mental toughness inventory for Australian football. *Psychology of Sport and Exercise, 10,* 201-209

Gucciardi, D.F., Gordon, S. & Dimmock, J.A. (2009b). Evaluation of Mental toughness training programme for youth-aged Australian footballers: A quantitative analysis. *Journal of Applied Sport Psychology*. 21, 307-323

Guilford J.P. (1959) *Personality*. New York: McGraw-Hill

Guilford, J.P. & Fruchter, B. (1981). *Fundamental Statistics in Psychology and Education*. Auckland: McGraw-Hill International Book Company

Guilford, J.P. & Zimmerman, W.S. (1956). Fourteen dimensions of Temperament. *Psychological Monographs*, 70 (10, Whole No. 417)

Guilford, J.P. (1950). Creativity. *American Psychologist*. 5, 444-454

Guilford, J.P. (1967). *The nature of human intelligence*. New York: McGraw-Hill

Guilford, J.P. (1988). Some changes in the structure of intellect abilities. *Educational and Psychological Measurement*, 48, 1-4

Guion, R.M. & Ironson, G.H. (1983). Latent trait theory for organisational research. *Organizational behaviour and Human Performance*, 31, 54-87

Gupta, S., Khandelwal, S.K., Tandon, P.N., Meheshwani, M.C., Mehta, U.S. and Sundram, K.R. (2000). Development and Standardization of Comprehensive Neuropsychological Battery in Hindi (Adult form) *Journal of Personality and Clinical Studies*, 16, 75-109

Guskey, T.R. & Bailey, J.M. (2010). *Developing Standards-based report cards*. Thousands Oaks, CA: Corwin

Guttman, L. (1950). Relation of Scalogram analysis to other techniques. In S.A. Stouffer et al. (eds.) *Measurement and Prediction*. Princeton, NJ: Princeton University Press

Halpern, F. (1958). Child case study. In E.F. Hammer (Ed.). *The Clinical Application of Projective Drawings*. Springfield, IL: Charles C. Thomas

Halstead, W.C. (1947) *Brain and Intelligence*. Chicago: University of Chicago Press.

Harkness, A.R., McNulty, J.L. In Ben-Porath, Y.S. (1995). The Personality Psychology Five (Psy-5): Constructs and MMPI-2 Scales. *Psychological Assessment*, 7, 104.

Harris, D.B. (1963). *Children's drawing as a measure of intellectual maturity: A review and extension of the Goodenough Draw-a-Man Test*. New York: Harcourt, Brace & World.

Harrow, A.J.A. (1972). *A Taxonomy of the Psychomotor domain*. New York: David Mckey

Hartshorne, H., May, M.A. & Maller, J.B. (1929). *Studies in Service and Self-control*. New York: MacMillian

Hartshrone, H. & May, M.A. (1928). *Studies in deceit*. New York: Macmillian

Hartshrone, H., May, M.A. & Shuttleworth, E.K. (1930). *Studies in organization of character*. New York: MacMillian

Hathaway, S.R. & McKinely, J.C. (1943). *Manual for the Minnesota Multiphasic Personality Inventory*. New York: Psychological Corporation.

Hathaway, S.R. & Mckinley, J.C. (1940). A Multiple Personality Schedule (Minnesota): Construction of the Schedule. *Journal of Psychology, 10*, 249-254

Hattie, J. & Coaksey, R.W. (1984). Procedures for assessing the validities of tests using the 'Known-group method'. *Applied Psychological Measurement, 8*, 295-305

Hearst, E. (1979). One hundred years: Themes and Perspectives. In E.Hearst (Ed.) *The first Century of experimental psychology*. Hillsdale. NJ: Erlbaum

Henry, E.M. & Rotter, J.B. (1956). Situational influences on the Rorchach response. *Journal of Consulting Psychology, 20*, 457-462

Henson, R.K. (2001). Understanding internal consistency reliability estimates: A conceptual primer on coefficient alpha. *Measurement and Evaluation in Counseling and Development, 34*, 177-189

Hess, E.H. (1965) Attitudes and Pupil Size. *Scientific American* 212, 46-54.

Hillon, T. (2008) The logic and methodology of Millon inventories. Cross-cultural personality assessment. In G.J. Boyle, G. Mathews, & D.H. Saklofske (Eds.) *Sage Handbook of personality theory and Assessment. Vol-2, Personality theory and assessment* Los Angeles, CA: Sage

Hinrichs, J.R. (1978) An eight-year follow-up of a management assessment center. *Journal of Applied Psychology, 63*, 596-601

Holland, J.L, Powell, A.B. & Fritzsche, B.A. (1994). *The Self-Directed Search (SDS)Professional User's guide*. Odessa, F.L: Psychological Assessment Resources

Holland, J.L. (1973). *Making vocational choices*. Englewood Cliffs, NJ: Prentice-Hall

Holland, J.L. (1985) *Making vocational choices: A theory of vocational personalities and work environments* (2nd ed.) Englewood cliffs, NJ: Prentice Hall.

Holland, J.L. (1985). *Making vocational choices: A theory of vocational personalities and work environment* (2nd ed.) Englewood Cliffs, NJ: Prentice Hall

Holland, J.L. (1997) *Making vocational choices: A theory of vocational personalities and work environments* (3rd ed.). Odessa, F.L.: Psychological Assessment Reserves.

Holland, J.L. Fritzsche, B.A. & Powell, A.B. (1994). *The self-directed search (SDS) Technical Manual Odessa*, F.L.: Psychological Assessment Resources.

Holland, J.L., Powell, A.B & Fritzsche, B.A. (1997) *Self-directed search (SDS). Professional user's guide.* Odessa, F.L.: Psychological Assessment Resources.

Holt, R.R. (1971). *Assessing Personality.* New York: Harcourt, Brace & Jovanovich

Holtzman, W.H (1986). Holtzman Inkblot Technique (HIT). In A.I. Rabin (Ed.) *Assessment with Projective techniques: A concise introduction.* New York: Springer

Holtzman, W.H. (1961). *Guide to administration and scoring: Holtzman Inkblot Technique.* New York: Psychological Corporation

Holtzman, W.H. (1975). New developments in Holtzman Inkblot technique. In P. McReynolds (Ed.) *Advances in Psychological assessment.* San Francisco: Jossey Bass

Holtzman, W.H. (1988). Beyond the Rorschach. *Journal of Personality Assessment.* 52, 578-609

Holtzman, W.H., Thorpe, J.S., Swartz, J.D., & Herron, E.W. (1961). *Inkblot Perception and Personality-Holtzman Inkblot technique* Austin: University of Texas Press

Homstrom, R.W., Silber, D.E. & Karp, S.A. (1990). Development of Apperceptive Personality Test. *Journal of Personality Assessment*, 54, 252-264

Horn, J.L. & Cattell, R.B. (1966). Refinement and test of the theory of fluid and crystallized intelligence. *Journal of Educational Psychology.* 57, 253-276

Horn, J.L. & Hofer, S.M. (1992) Major abilities and development in the adult period. In R.J. Sternberg & C.A. Berg (Eds.) *Intellectual development.* Boston, M.A.: Cambridge University Press.

Horn, J.L. & Noll, J. (1997). Human cognitive capabilities: Gf-Gc. theory. In D.P. Elanagan, J.L. Genshaft & P.L. Harrison (Eds.) *Contemporary Intellectual Assessment. Theories, Tests & Issue.* New York. Guilford Press

Horn, J.L. (1968). Organisation of abilities and the development of intelligence. *Psychological Review*, 75, 242-249

Horn, J.L. (1991). Measurement of intellectual abilities: A review of theory. In K.S. McGrew et al., (Eds.) *Woodcock-Johnson technical manual* (pp. 197-232). Chicago: Riverside

Horn, J.L. (1994). Theory of Fluid and Crystallized intelligence. In R.J. Sternberg (Ed.) *Encyclopedia of human intelligence* (pp. 443-451)

Huck, J.R. & Bray, D.W. (1976). Management assessment center evaluations and subsequent job performance of white and black females. *Personnel Psychology*, 29, 13-30

Huff, D. & Geis, I (1954) *How to lie with statistics* New York: W.W. Norton & Co.

Humphreys, L.G. (1962). The organization of human abilities. *American Psychologist*, 17, 475-483

Hunsley, J. & Bailey, J.M. (1999). The clinical utility of the Rorschach: Unfulfilled promises and an uncertain future. *Psychological Assessment*, 11, 266-277

Indian Council of Social science Research (ICSSR) (1972) *A Survey of Research in Psychology*. Bombay: Popular Prakashan.

Jackson, D.N. & Messick, S. (Eds.) (1967) *Problems in human assessment*. New York: McGraw-Hill

Jackson, D.N. (1977) *Jackson Vocational Interest Survey Manual* Port Huron, MI: Research Psychological Press

Jackson, D.N. (1977) *Jackson Vocational Interest Survey Manual*. Port Huron, MI: Research Psychological.

Jackson, D.N. (1994) *Multidimensional Aptitude Battery* (MAB) Manual. Port Huron MI: Sigma Assessment systems. (1st ed. 1984)

Jackson, S.A. & Eklund, R.C. (2012) Flow In G. Tenebaum, R.C. Eklund & A. Kamata (Eds.) *Measurement in Sport and Exercise Psychology*. Champaign IL: Human Kinetics

James, L.R., Demaree, R.G. & Wolf, G. (1984) Estimating within-group interrater variability with and without response bias. *Journal of Applied Psychology*, 69, 85-98.

Jenkins, C.D., Zyzanski, S.J. & Rosenman R.H. (1979). *Jenkins Activity Survey: Manual*. San Antonio, TX: Psychological Corporation

Jensen, A.R. (1991) General Mental ability" from Psychometrics to Biology. *Diagnostique*. 16, 134-144.

Jensen, A.R. (1993). Spearman's hypothesis tested with chromometric information-processing tasks. *Intelligence*, 17, 47-77

Jenson, A.R. (1980) *Bias in Mental Testing*. New York: Free Press.

Johansson, C.B. (1991) *Career Assessment Inventory: Vocational version*. Minneapolis, MN: National computer systems

Jolles, J. (1952) *A catalogue for the qualitative interpretation of the H-T-P* Los Angeles: Western Psychological services

Jung, C.G. (1910). The Association Method. *American Journal of Psychology*, 21, 219-269

Jung, C.G. (1923). *Psychological Types*. London: Routledge & Kegan Paul

Kaiser, H.F. & Michael, W.B. (1975) Domain valiability and generalizability. *Educational and Psychological Measurement* 33, 31-35

Kamat, V.V. (1934) *Measuring intelligence of Indian children* Bombay: Oxford University Press

Kamiya, J. (1968) Conscious of brain waves, *Psychology Today* (11), 56-60

Kaplan, R.M. & Saccuzzo, D.R. (2002) *Psychological Testing: Principles, Applications and Issues.* Singapore: Wadsworth Thompson Learning

Kaplan, R.M. & Saccuzzo, D.R. (2010) *Psychological Testing: Principles Applications and Issues* (8th ed.) Belmont, C.A.: Wadsworth Language Learning

Karp, SA Holstrom, R.W. & Silber, DE (1990) *Appreciative personality Test Manual.* Orland Park, IL: International Diagnostic Systems

Kaufman A.S. & Kaufman, N.L. (1983a) *Kaufman Assessment Battery for children (K-ABC) Administration and scoring manual* Circle Pines, MN.: American Guidance Service.

Kaufman A.S. & Kaufman, N.L. (1983b) *Kaufman Assessment Battery for children. (K-ABC) Interpretative Manual.* Circle Pines, MN: American Guidance Science

Kaufman, A.S. & Kaufman, N.L. (1990) *Kaufman Brief Intelligence Test (K-BIT): Manual circle Pines*, MN: American Guidance Science

Kaufman, A.S. & Kaufman, N.L. (1993) *Kaufman Adolescent and Adult Intelligence Test (KAIT): Manual.* Circle Pines, MN: American Guidance Service

Kaufman, J.C. & Baghetta, R.A. (2009) Beyond Big and Little: The four C model of Creativity. *Review of General Psychology. 13 (1),*: 1-12

Keith, T.Z., Powell, A.L. & Powell, L.R. (2001) Review of Wechsler abbreviated scale of Intelligence. In B.S. Pleke & J.C. Impara (Eds.) *The fourteenth mental measurements yearbook* (pp. 1329-1331). Lincoln: University Nebraska Press.

Kelley, T.C. (1928) Crossroads in the mind of man: *A study of differentiable mental abilities.* Stanford, CA: Stanford University Press

Kelley, T.L. (1939) The selection of upper and lower groups for the validation of test items *Journal of Educational Psychology* 30, 17-24

Kelly, A.V. (2009). *The Curriculum: Theory and Practice.* Newbury Park, CA: Sage

Kendall, M.G. (1955). Rank Correlation Methods, 2nd ed. London: Griffin

Kent, G.H. & Rosanoff, A.J. (1910) A study of association in insanity. *American Journal of Insanity 67*, 37-96

Kirkland, M.C. (1971) The effects of tests on students and schools. *Review of Educational Research* 41 (4), 303-50

Kochchar, S. (2006) *Educational and vocational guidance in secondary schools*. New Delhi: Sterling

Kohli, A, Kaur, M. & Malhotra, R. (2006) *Practice Manual of Draw-A-Person*. New Delhi: Prasad Psycho

Kohli, A., Kaur, M. & Malhotra, R. (2006). *Draw-A-Person, Practice Manual*, New Delhi: Prasad Psycho.

Kopelman, R.E. (2003) The Study of Values: Construction of fourth edition. *Journal of vocational behaviour*, 203-220

Kozbelt, A., Beghetto, R., Runco, M.A. (2010). Theories of creativity. In J.C. Kaufman & R.J. Sternberg. *The Cambridge Handbook of Creativity*. Cambridge University Press

Kraepelin, E (1896) Der Psychologischa versuch in der Psychiatric. In R.J. Cohen & M.E. Swerdlik's *Psychological Testing and Assessment* (Sixth Edition) Boston: McGraw Hill

Krathwohl, D.R., Bloom, B.S. & Masia, B.B. (1964) *Taxonomy of Educational objectives. Handbook II: Affective domain*. New York: David McKay.

Kuder, G.F. & Diamond, E.E. (1979) *Kunder occupational Interest survey: General Manual*. Chicago: Science Research Associates

Kuder, G.F. & Richardson, M.W. (1937) The theory of the estimation of reliability. *Psychometrika*, 2, 151-160

Kulashrestha, S.K. (1971) *Stanford-Binet Intelligence scale Hindi adaption* (3rd ed.). Allahabad: Manas Sevasanthan

Kuder, F & Zytowski, D.G. (1991) *Kunder occupational Interest survey Form DD: General Manual* (3rd ed.) Monterey, CA: CTB McMillan/McGraw-Hill

Kunder, G.F. & Zytowski, DG (1991). *Kunder occupational Interest survey Farm DD: General Manual* (3rd ed.) Monterey, CA: CTB McMillan/McGraw-Hill

Kuppuswami, B. (1964) *Advanced Educational Psychology*. New Delhi: Sterling

Kurtz, A.K. & Mayo., S.T. (1980). *Statistical Methods in Education and Psychology*. New Delhi: Narosa Publishing House

Lah, M.I. (1989). New validity, normative and scoring data for the Rotter Incomplete sentence blank. *Journal of Personality Assessment*. 3, 678-687

Lajunena, T. & Scherler, H.R. (1999). In the EPQ Lie Scale bidimensional? Validation study of the structure of EPQ lie Scale among Finnish and Turkish University students. *Personality and Individual differences*, 26(4), 657-664

Landy, F.J. (1986). Stamp Collecting Versus Science. *American Psychologist*, 41, 1183-1192

Lang, P.J., Bradley, H.M. & Cuthbert, B.M. (1990). Emotion, Attention, and the Startle reflex. *Psychological Review*, 97, 377-395

Langer, E.J. & Abelson, R.P (1974). A patient by anyother name: clinical group differences in the labeling bias. *Journal of Consulting and Clinical Psychology*, 42, 4-9

Lanyon, B.P. & Lanyon, R.I. (1980). *Incomplete Sentences Tasks: Manual*. Chicago: Stoelting

Lathan, G.P. & Wexley, K.N. (1977). Behavioural observation scales for performance appraisal purposes. *Personnel Psychology*. 30, 255-268

Lawshe, C.H. (1975). A quantitative approach to content validity. *Personnel Psychology*, 28, 563-575

LeDoux, J.E. (1996). *The Emotional Brain:The mysterious underpinnings of emotional life*. New York: Simon & Schuster

Lewandowski, D.G. & Saccuzzo, D.P. (1976). The decline of Psychological testing: Have traditional procedures been fairly evaluated? *Professional Psychology*, 7, 177-184

Lewis, D. & Burke, C.J. (1949). The use and misuse of the chi-square test. *Psychological Bulletin*, 46, 433-489

Libby, W. (1908). The imagination of adolescents. *American Journal of Psychology*. 19, 249-252

Lichtenstein, D., Dreger, R.M. & Cattell, R.B. (1986). Factor structure and standardization of the Preschool Personality Questionnaire. *Journal of Social behaviour and Personality*. 1(2), 165-182

Likert, R. (1932). A technique for measurement of attitude, *Archives of Psychology*, 22:40

Lilienfeld, S.O. (1999). Projective measures of Personality and Psychopathology: How well they do work? *Skeptical Inquire*, 23(5), 32-39

Lindell, M.K., Brandt, C.J. & Whitney, D.J. (1999). A revised index of interrate agreement for multi-item ratings of a single target. *Applied Psychological Measurement*, 23, 127-135

Linn, R.L. & Miller, M. David (2011). *Measurement and Assessment in Teaching*. Delhi: Pearson Education

Linn, R.L. (1994a) Criterion referenced measurement: A valuable perspective clouded by surplus meaning. Annual Meeting of the American Educational Research Association: Criterion-referencedmeasurement: A 30-year retrospective. *Educational Measurement: Issues and Practice*, 13, 12-14

Linn, R.L. (1994b). Fair test use: Research and Policy. In M.G. Rumsey, C.B. Walker & J.H. Harris (Eds.) *Personnel Selection and Classification*. Hillsade, NJ: Erlbaum

Long, L. & Mehta, P.H. (1966). *The first mental measurement handbook for India*. New India. New Delhi: NCERT

Lorr, M. & Mc Nair, D.M. (1988). *Manual for the profile of mood states–Bipolar from*. San Diego. Educational and Industrial Testing Service

Lumbart, T.L. (1994). Creativity. In R.J. Sternberg (Ed.) Thinking Problem solving. San Diego C.A.: Academic Press

Luria, A.R. (1966a). *Human brain and Psychological processes*. New York: Harper & Row

Luria, A.R. (1966b). *Higher Cortical functions in man*. New York: Basic

Luria, A.R. (1980). *Higher Cortical functions in Man* (2nd ed.), New York: Basic books.

Mabon, H. (1998) Utility aspects of personality and performance. *Human performance*, 11, 289-304

Machover, K. (1949). Personality Projections in the drawings of human figures: A method of personality investigation. Springfield IL: Thomas

Malin, A.J. (1969). *Manual for Malin's intelligence scale for Indian children*. Nagpur: Nagpur child guidance centre

Manuele-Adkins, C. (1989) Review of the self-Directed search: A guide to educational and vocational planning. *Tenth Mental Measurement Yearbook* 738-740

Marsch, M.W. & Cheng, J.H.S. (2012). Physical self-concept. In G. Tenenbaum, R.C. Eklund, & A. Kamata (Eds.) *Measurement in Sport and Exercise Psychology*. Champaign IL: Human Kinetics

Martens, R. (1987). *Choaches guide to sport psychology*. Champaign, IL: Human Kinetics Publishers

Maruish, M.E. (Ed.) (1994) *The use of psychological testing for treatment planning and outcome assessment* Hillsdale, N.J. : Erlbaum

Masling J. (1960). The influence of situational and interpersonal variables in projective testing. *Psychological Bulletin* 57, 65-85

Masling, J. (1965). Differential indoctrination of examiners and Rorschach responses. *Journal of consulting Psychology*. 29, 198-201

Matrarazzo, J.D. (1990). Assessment involving objective and subjective components. *American Psychologist*, 45, 999-1017

MC Arthur, D.S. & Roberts, G.E. (1982) *Roberts Apperception test for children: Manual* Los Angeles : Western *Psychological services*

Mc Call, R.B. (1994). *Fundamental Statistics for behavioural Sciences.* (6th ed.) Fort Worth, TX: Harcourt Brace

McCall, W.A. (1922). *Howto measure in education.* New York: Macmillan

McCall, W.A. (1939). *Measurement.* New York: Macmillan

McFall, R.M. & Lillesand, D.V. (1971). Behaviour rehearsal with modeling and coaching in assertive training. *Journal of Abnormal Psychology 77,* 313-323

McGrew, K.S. & Flanagan. D.P. (1998). *The intelligence test desk reference, Gc-Gf cross-battery assessment* Boston: Allyn & Bacon.

McGrew, K.S. (1997). Analysis of the major intelligence batteries according to a proposed comprehensive Gf-Gc framework. In D.P. Flanagan, J.L. Genshaft & P.L. Harrison (Eds.) *Contemporary Intellectual assessment. Theories Tests and Issues.* (pp. 151-180). New York. Guilford.

McMillan, J.H. (2001) Secondary teachers' Classroom assessment and grading practices. *Educational Measurement* : Issue and Practice, 20-32

McMillan, J.H., Myron, S. & Workman, D. (2002) Elementary teachers' classroom assessment and grading practices. *Journal of Educational Research* 95(4), 203-213

McNair, D.M., Lorr, M. & Droppleman, L.F. (1971). *Profile of Mood States Manual.* San Diego, C.A.: Educational and Industrial Testing Service

McNemar, O.W. & Landis, C. (1935). Childhood disease and emotional maturity in the psychopathic woman. *Journal of Abnormal and Social Psychology*, 30, 314-319

Mednick, S. A. & Mednick, M.T. (1967). *Examiner's Manual Remote Associates Test.* Boston, Mass: Houghton Mifflin

Mehl, M. & Pennebaker, J.V. (2003). The sounds of social life. A Psychometric analysis of students daily social environments and natural conversations. *Journal of Personality and social Psychology, 84,* 857-870

Mehta, P. (1962). *Manual for Group Intelligence Test.* New Delhi: Manasayam

Mellenbergh, G.J. (1994). Generalized linear item response theory. *Psychological Bulletin, 115,* 300-307

Messick, S. (1995) Validity in psychological assessment. *American Psychologist,* 50, 675-680

Michael, J.J. (1968). Structure of intellect theory and the validity of achievement examinations. *Educational and Psychological Measurement, 28*, 1141-1149

Milgram, S. Mann, L. & Harter, S. (1965). The lost-letter technique: A tool of social research: Public opinion Quarterly, 29, 437-438

Miller, L.A., Lovler, R.L. & McIntire, S.A. (2013). *Psychological Testing,* New Delhi: Sage

Millon, T. (2008) The logic and methodology of Million Inventories: Cross Cultural personality assessment. In G.J. Byle, G. Mathews & D.H. Saklofske (Eds.) *Sage Handbook of Personality Theory and Assessment.* Vol. 2. *Personality Theory and Assessment.* Los Angeles CA: Sage

Millon, T. (1981) *Disorders of Personality, DSM-III:* Axis II. New York: Wiley

Millon, T. (1990) *Toward a new personality: An evolutionary model.* New York: Wiley

Millon, T. Grossman, S. & Millon, C. (2015) MCMI-IV: *Millon-Clinical Multiaxial Inventory Manual* (1st ed.) Bloomington, MN: NCS Pereson

Minimum, E. (1970) *Statistical Reasouring in Psychology and Education* NewYork: D Wiley

Mischel, W. (1968) *Personality and Assessment.* New York: Wiley

Mitchell, J. (1999). *Measurement in Psychology: Critical history of a methodological concept.* New York: Cambridge University Press

Mitra, S.K. (1972). Psychological Research in India. In S.K. Mitra (ed.) *A survey of Research in Psychology* (pp. xvii-xxxiii). Bombay: Popular Prakashan

Moreland, K.L., Eyde, L.D., Robertson, G.J., Primoff, E.S. & Most, R.B. (1995). Assessment of test user qualifications: A research-based measurement procedure. *American Psychologist,* 50, 14-23

Morgan W.G. (1995) Orgin and history of Thematic Apperception Test images *Journal of personality assessment,* 65, 237-254

Morgan, C.D. & Murray, H.A. (1935). A method for investigating fantasies: The Thematic apperception Test. *Archives of Neurology and Psychology,* 34, 289-306

Morgan, C.D. (1938). Thematic Apperception test. In H.A. Murray (Ed.) *Explorations in personality: A clinical and experimental study of fifty men of college age* New York: Oxford University Press

Mukherjee, B.N. (1993). Needed research in educational Psycho assessment in India. *Psychological studies,* 38, 85-100

Muller, J. (2011). What is authentic assessment? In LA Miller, R.L. Lovler and S.A. McIntire's *Psychological Testing* New Delhi: Sage

Murphy, K.R. & Balzer, W.K. (1981) *Rater errors and rating accuracy* Presented at Annual conference of American Psychological Association. Los Angeles.

Murphy, K.R. & Constans, J.J. (1987) Behavioural anchors as a source of bias in rating. *Journal of Applied Psychology* 72, 573-577

Murphy, K.R. & Davidshafer, C.O. (1994). *Psychological testing: Principles and Application* (3rd ed.). Englewood Cliffs, NJ: Prentice Hall

Murphy, K.R. & Davidshofer, C.O. (1988) *Psychological Testing Principles and Applications*. N.J. : Prentice Hall.

Murray, H.A. & Morgan, C.D. (1938) *Explorations in personality*. New York: Oxford Book company

Murray, H.A. (1938) *Explorations in personality*. New York: Oxford University Press

Murray, H.A. (1943) *Thematic apperception test Manual* Cambridge. W.A.: Harvard University

Murstein, B.I. (1961) Assumptions, Adaptation level and Projective techniques. *Perceptual and major skills* 12, 107-175

Murthy, H.M. (1966). A scale of Bhatia's Performance tests. *IndianPsychological Review*, 2, 133-34

Mussen, S.H. & Slodel, A (1955). The effects of sexual stimulation under varying conditions on TAT sexual responsiveness. *Journal of Consulting and clinicalPsychology, 19, 90*

Myers, I.B. & Briggs, K.C. (1943). *The Myers-Briggs type Indicator*. Palo Alto, C.A. : consulting Psychologist press

Naglieri, J.A. & Das, J.P. (1997) *Das-Naglieri cognitive assessment system: Interpretive Handbook*. Itasca, IL: Riverside

Naglieri, J.A. (1988) *Draw-A-Person:Aquantitativescoringsystem-Manual*. San Antoms, TX: Psychological corporation

Naglieri, J.A., McNeish, T.J. & Bardos, A.N. (1991) *Draw-A-Person: Screeningprocedure foremotionaldisturbance-examiner'smanualAustin*, TX: PRO-ED

National Council of Education Research and Training (1991) *IndianMentalMeasurement Handbook:Intelligenceandaptitudetests:* New Delhi: Author

National council of Educational Research and Training (1998) *Handbookofpersonality measurementinIndia*. New Delhi: Author

National Council of Educational Research and Training (2001) *Handbookofvalueattitude andinterestmeasurementinIndia*. Delhi: Author

Nickerson, R.S. (1999) Enhancing creativity. In R.J. Sternberg's *HandbookofCreativity*. Cambridge University Press.

Neary, M. (2003) Curriculum and concepts. *In curriculum studies in Post-compulsory and adult education: A teacher's and student teacher's study guide*. Cheltenham: Nelson Thornes Ltd.

Neisser, V. (1967) *Cognitive Psychology*: New York: Appleton-century-Crofts

Nideffer, R.M. (1976a) *The inner athlete: Mind plus muscle for winning*. New York: Thomas Y. Crowell company.

Nideffer, R.M. (1976b) Test of attentional and interpersonal style. *Journal of Personally and social Psychology, 34*, 394-404.

Novick, M.R. & Lewis, C. (1967) Coefficient alpha and the reliability of composite measurements. *Psychometrika. 32*, 1-13

Nunnally, J.C. & Berntein, I.H. (1994) *Psychometric theory* (3rd ed.) New York: McGraw Hill

Nunnally, J.C. (1959) *TestsandMeasurements:AssessmentandPrediction*. New York: McGraw Hill

Oliva, P.F. (1997) *The curriculum: Theoreticaldimensions*. New York: Longman

Oliva, P.F. (2009) Development of the curriculum: (7th ed.) Boston, MA: Allyn & Bacon

Osborn, A. (1957) *Applied Imagination* New York: Charles Scriber's sons.

Osgood, C.E., Suci, G.J. & Tannenbaum, P.H. (1957) *The Measurement of Meaning*. Urbana: University of Illinois Press

Oss Assessment staff (1948)- *Assessment of Men: Selection of personnel for the office of strategic services*. New York: Rinehart

Ostrow, A.C. (Ed.) (1996) *Directory of Psychological tests in sport and exercise sciences*. (2nd ed.). Morgantown, W.V.: Fitness Information Technology

Owens, W.A. & Schoenfeldt, L.F. (1979) Toward a classification of persons. *Journal of Applied Psychology, 65*, 569-607

Owens, W.A. (1976) Background data. In M. Dunnette (Ed.) *Handbook of Industrial and organizational Psychology*. Chicago: Rand McNally

Pareek, U. (1980) *ASurveyofResearchinPsychology* 1971-76, Part 1. Bombay: Popular Prakashan

Pareek, U. (1981). *A Survey of Research in Psychology*. 1971-76 Part 2. Bombay: Popular Prakashan

Parker, K.C.H., Hanson, R.K. & Hunsley, J. (1988) MMPI, Rorschach and WAIS: A meta-analysis comparison of reliability, stability and validity. *Psychological Bulletin*. 103, 367-373

Pathak, P. (1987) Draw-a-Man test for Indian children. Pune: Anand Agencies

Pervin, L.A. & John, O.P. (1997). *Personality: Theory and Research*. New York: John Wiley & Sons.

Phares, E.J. (1984). *Clinical Psychology*, Illinois: The Dorsey Press

Phelps, E.A., O'Connor, K.J., Cunningham, W.A., Punayama, E., Gatenby, J., Gore, J.C. (2000). Performance on indirect measures of race evaluation predicts amygdata activation. *Journal of Cognitive Neuroscience*, 12, 729-738

Piaget, J. (1972) Intellectual evaluation for adolescence to adulthood. *Human Development*, *15*, 1-12

Porter, R.B. & Cattell, R.B. (1985). *Handbook for the Children's Personality Questionnaire* (CPQ), Campaign, IL: IPAT

Quenck, N.L. (2000). *Essentials of Myers-Briggs Type Indicator Assessment*. New York: Wiley

Rajdev, R., Melson, W.M., Hart, K.J. & Fercho, M.C. (1994). Creterion-related validity and stability: Equivalence of MMPI and MMPI-L. *Journal of Clinical Psychology*, 50, 361-367

Ramalingaswamy, P. (1974). *Manual of Indian adaptation of WAIS Performance Scale*. Delhi: Managayan

Rankin, R.E. & Cambell, D.T. (1955). Galvanic skin response to Negro and white experimenters. *Journal of Abnormal and Social Psychology*, 51, 30-33

Rao, S.L., Subbakrshna, D.K., Gopu Kumar (2004) NIMHANS *Neuropsychological Battery: Manual*, Banglore: National Institute of Mental Health and Neurosciences.

Rapaport, D., Gill, M.M. & Schafer, R. (1946). *Diagnostic Psychological Testing* (2 Vols.). Chicago: Year Book

Ravizza, K. (2010). Increasing awareness for sport performance. In J.M. Williams (Ed.) *Applied Sport Psychology: Personal growth to peak performance* (6th ed.) Boston: McGraw-Hill

Reber, A.S., Allen, R. & Reber, E.S. (2009). *Penguin Dictionary of Psychology*. England: Penguin Books Ltd.

Reckas, M.D. (1996). Test Construction in the 1999: Recent approaches every psychologist should know. *Psychological Assessment, 8,* 354-359

Reitan, R.M. & Davison, L.A. (Eds.) (1974). *Clinical neuro-psychology: Current status and applications.* Washington, DC: Winston

Reitan, R.M. & Woflsun, D. (1993). *TheHalstead ReitanneuropsychologicalTestBattery: TheoryandClinicalinterpretation* (2nd ed.) Tucson, AZ: Neuropsychology Press

Reitan, R.M. (1969) *Manualforadministrationofneuropsychologicaltestbatteriesforadults andchildren.* Indianapolis. IN: Author

Raven, J., Raven, J.C. & Court, J.H. (1998). *ManualforRaven'sStandardProgressive Matrices,* Oxford: Oxford University Press

Riggs, D.S., Murphy, C.M. & O'heavy, K.D. (1989) Intentional falsification in reports of interpartner aggression. *Journal of Interpersonal violence,* 4, 220-232

Ritzler, B.A. Sharkey, K.J. & Chudy, J.F. (1980) A comprehensive Projective alternative to the TAT *JournalofPersonalityAssessment,* 44, 358-362

Roid, G.H. (2003). *Stanford-Binet Intelligence Scales, Fifth edition. Technical Manual.* Itasca, IL: Riverside

Roid, G.H. (2003). Stanford-Binet itelligence Scales (5th edition) Itasa, IL: Riverside

Rokeach, M. (1973). *The Nature of Human values,* New York: Free Press

Rosenthal, R., Rosnow, R.L. & Rubin, D.B. (2000). *Contrasts and effects sizes in behavioural research: A correlational Approach.* Cambridge University Press

Rosenzweig, S. (1945). The Picture-association method and its application in study of reactions to frustration. *Journal of Personality, 14,* 3-23

Rosenzweig, S. (1960). The Rosenzweig Picture-Frustration Study, Children's Form. In A.I. Rabin & M. Haworth (Eds.) *Projectivetechniquewithchildren.* Orlando, FL: Grune & Stratton

Rosenzweig, S. (1970). Sex differences in reaction to frustration among adolescents. In J. Zubin & A.M. Freedom (Eds.) *Psychopathology of Adolescence.* Orlando, F.L Grune & Stratton

Rosenzweig, S. (1977). *Manual for the Children's Form of the Rosenzweig Picture-Frustration Study,* St. Lowix, MO: Rana House

Rosenzweig, S. (1978a) *Adult Form Supplement to the basic manual of the Rosenzweig Picture-Frustration (P.F.) Study.* St. Lowis, MO: Rana House

Rosenzweig, S. (1978b). *Aggressive behaviour and the Rosenzweig Picture-Frustration.* New York: Praeger

Rosenzweig, S. (1978C) The Rosenzweig Picture Frustration (P.F.) *Study: Basic Manual*. St. Louis, MO: Rana House

Rosenzweig, S. (1981). *Adolescent Form Supplement to the basic manual of Rosenzweig Picture-Frustration (P.F) Study*. St. Lowis, MO: Rana House

Rosenzweig, S. (1988) *Revised norms for the children's Form of the Rosenzweig Picture-Frustration (P.F) Study*. St. Lowis, MO: Rana House

Rotter, J.B. & Refferty, J.E. (1950). *Manual:TheRottenIncompleteSentencesBlank*. San Antonio, TX: Psychological Corporation

Rotter, J.B. & Wickens, D.D. (1948) The Consistency and generality of ratings of social aggressiveness made from observations of role playing situations. *Journal of Consulting Psychology. 12*, 234-239

Rotter, J.B., Lah, M.I. & Rafferty, J.E. (1992). *RotterIncompleteSentencesBlankManual*. San Antonio, TX: Psychological Corporation

Rotter, J.E. & Rafferty, J.E. (1950). *TheManual for the Rotter Incomplete Sentences Blank*.

Rulon, P.J. (1939). A simplified procedure for determining the reliability of a test by split halves. *Harvard Educational Review, 9*, 99-103

Sacchi, C., Richard De Minzi, M.C. (1989). The Holtzman Inkblot technique in preadolescent personality. *British Journal of Projective Psychology. 34*, 2-11

Saettler, G.P. (1980). *TheEvolutionofAmericanEducationalTechnology* Englewood, Colorado: Libraries Unlimited.

Sax, G. (1974). *Principles of Educational Measurement and Evaluation*. California: Wordsworth Publishing Company

Saylor, J.G., Alexander, W.M. & Lewis, A.J. (1981). Curriculum planning for better teaching and learning. (4th ed.) New York: Holt, Rinehart & Winston

Schacham, S. (1983). A shortened version of the Profile of Mood States. *Journal of Personality Assessment, 47*, 305-306

Schuerger, J.M. (2001). *16PFAdolescentPersonalityQuestionnaire:* Champaign, IL: IPAT

Schwartz, L.A. (1932). Social situation pictures in the psychiatric interview. *American Journal of Orthopsychiatry, 2*, 124-132

Scriven, M (1967). The methodology of evaluation. In R.W. Tyler, R.M. Gagne, & M. Scriven. *Perspectives of curriculum evaluation*. Stokie, ill: Rand McNally & Company

Scriven, M. (1967). The Methodology of Evaluation. In R.W. Tyler, R.M. Gagne & M. Scriven (Eds.) *Perspective of Curriculum Evaluation*. 39-83. Chicago. IL: Rand McNally

Seashore, H.G., Wesman, A.G. & Doppelt, J.E. (1950). The standardization of Wechsler Intelligence scale for children. *Journal of Consulting Psychology*, 14, 99-110

Secord, P.F. & Backman, C.W. (1974). *Social Psychology.* 2nd ed. Tokyo: McGraw-Hill

Sharma, T.R. (1972). Measuring intelligence through bicycle drawings. *Indian Educational Review, 7(1)*, 1-30

Shavelon, R.L., Webb, N.M. & Rowley, G.L. (1989). *Generalizability Theory.American Psychologist, 44*, 922-932

Sheffe, H. (1957). *The Analysis of variance*, New York: J. Wiley

Sheldon, W.H. & Stevens, S.S. (1942). *ThevarietiesofTemperament*, New York: Harper and Row

Shepard, L.A. (1993). Evaluating test validity. In L. Darling-Hammard (Ed.) *ReviewofResearchinEducation. 19*, 405-450. Washington: American Educational Research Association

Shiffman, S., Hufford, M. & Hickcox, M. (1997). Remember that? A comparison of real life versus retrospective recall of smoking lapses. *JournalofConsultingandClinicalPsychology*, 65, 292-300

Shneidman, E.S. (1958). Some relationships between thematic and drawing materials. In E.F. Hammer (Ed.). *The clinical Application of Projective Drawings*. Springfield, IL: Charles C. Thomas

Shrivastava, A.K., Tripathi, A.M. & Misra, G. (1996). The status of intelligence testing in India: A preliminary analysis. *IndianEducationalReview*, 31, 1-11

Siegel, S. & Castellan, N.J. (1988). *Non-parametricstatisticsfortheBehaviouralSciences*. New York: McGraw-Hill

Simonton, D.K. (1994). Individual differences, developmental changes and social context. *Behavioural and Brain Sciences. 17*, 552-563

Simpson, E.J. (1972). The classification of Educational objectives, Psychomotor domain. *ILLINOIS: Teacher of Home Economics. 10*, 110-164

Smith, M.K. (2000). Curriculum theory and Practice. The encyclopedia of Informal Education. www.infed.org/biblio/b-curric.him

Spearman, C. (1904). 'General intelligence' objectively determined and measured. *American Journal of Psychology, 15*, 201-293

Spearman, C. (1927). *Theabilitiesofman*. New York: MacMillian

Spearman, C.E. (1923). *The nature of intelligence and the principles of cognition*. London: MacMillan

Sperry, R.W. (1968). Hemisphere disconnection and Unity of conscious experience. *American Psychologist, 29*, 723-733

Spielberger, C.D. & Sydeman, S.J. (1994). State-Trait Anxiety Inventory and State-Trait Anger Expression Inventory. In M. Marwish (Ed.) *The use of psychological testing for treatment planning and outcome assessment*. Hillsdale, NJ: Erlbaum

Spielberger, C.D., Johnson, E.H., Russell, S.F., Crane, R.J., Jacobs, G.A. & Worden, T.J. (1985). The experience and expression of anger: construction and validation of an anger expression scale. In M.A. Chesney & R.H. Rosenman (eds.) *Anger and Hostility in Cardiovascular and behavioural disorders*. New York: McGraw-Hill

Spranger, E. (1928). *Types of Morn*. Halle: Niemeyer

Sternberg, R. (1985). *BeyondIQ: A Triarchic theory of Intelligence*. New York: Cambridge University Press

Sternberg, R.J. & Lubart, J.I. (1999). The concept of creativity: Prospects and Paradigms. In R.J. Sternberg's *Handbook of creativity?*: Cambridge University Press.

Sternberg, R.J. (1984). The Kaufman Assessment Battery for Children: An information processing analysis and critique. *Journal of Special Education*. 180, 269-279

Sternberg, R.J. (1986). *Intelligence Applied: Understanding and increasing your intellectual skills*. San Diego (CA): Harcourt Brace Jovanovich

Sternberg, R.J. (2011). *Cognitive Psychology*: Cengage Learning

Stevens, S.S. (1946). *On the theory of Scales of Measurement Science* 103 (2684). 677-80

Stewards, R.J., Gimeniz, M.M. & Jockson, J.D. (1999). A Study of Personal preferences of successful university students as related to race/ethnicity and sex. Implications and recommendations for training, practice, and future research. *Journal of college student development*. 36, 123-131

Straus, M.A. (1979). Measuring intrafamily conflict and violence, The Conflict Tactics Scale. *Journal of Marriage and Family, 41*, 75-85

Streiner, D.L. (2003). Starting at the beginning: An introduction to coefficient alpha and internal consistency. *Journal of Personality Assessment, 80*, 99-103

Swensen, C.H. (1968). Empirical evaluations of human figure drawings: 1957-1966 *Psychological Bulletin, 70*, 20-44

Taba, H. (1962). *Curriculum development: Theory and Practice* New York: Harcourt, Brace & world

Tanner, D. & Tanner, L.M. (1975) *Curriculum development: Theory and Practice*, New York: Macmillan

Tate, M.W. (1948). *Statistics in Education*: New York: McGraw-Hill

Taylor, M.K., Gould, D. & Polo, C (2008) performance strategies of US Olympians in practice and competition. *High ability studies 19*, 19-36

Terman, L.M. (1916) *The measurement of intelligence*. Boston: Houghton Mifflin

Terry, P.C., Keohane, L. & Lane, H. (1996). Development and validation of a shattered version of the profile of mood states suitable for use with young athletes. *Journal of sport sciences, 14*, 49

Thomas, P.R., Murphy, S.M. & Hardy, L. (1999). Test of performance strategies: Development and preliminary velidation of a comprehensive measure of athletes Psychological skills. *Journal of sport sciences, 17*, 697-711

Thorndike, R.L., Hagen, E.P. & Sattler, J.M. (1986) The Stanford-Binet Intelligence Scale: Fourth Edition *Technical Manual*. Chicago: River ride

Thorndike, R.M. & Thorndike-Christ, T.M. (2015) *Measurement and Evaluation in Psychology and Education* Noida, Delhi: Person India Education.

Thorntone, G.C. & Byham, W.C. (1982). *Assessment centers and managerial performance*, New York: Academic Press

Thurstone, L.C. & Thurstone, T.G. (1941). Factorial studies of Intelligence, *PsychometricMonographs*. No. 2

Thurstone, L.C. (1938). *PrimaryMentalAbilities*. Chicago: University of Chicago Press

Thurstone, L.L (1931). The measurement of social attitude *Journal of Abnormal and social Psychology*, 26, 249-69

Thurstone, L.L. (1938) Primary Mental Abilities. *Psychometric Monographs*. No. L

Thurstone, L.L. (1947) *Multiple factor analysis*. Chicago: University of Chicago Press

Tinsley, H.E.A. & Weiss, D.J. (1975) Interrater reliability and agreement of a subjective judgments. 6, 212-217

Torrance, E.P. (1974). Torrance tests of creative thinking. Lexington, MA: Personnel Press

Trautscholdt, M. (1883) Experimentalle unterschungen uber die association der vorstellunger. In R-J Colen & M.E. Swerdlik's *Psychologicaltestingandassessment* (sixth edition) Boston McGraw-Hill

Tryon, R.C. (1957) Reliability and behaviour domain validity : Reformulation and historical Critique, *Psychological Bulletin, 54*, 229-249

Tutko, T.A. & Richards, J.W. (1971). *Psychology of coaching.* Boston: Allyn & Bacon

Tutko, T.A. & Richards, J.W. (1972) *Coaches' Practical guide to athletic motivation.* Boston: Allyn & Bacon

Twentyman, C.T. & McFall, R.M. (1975). Behavioural training of social skills in shy males. *Journal of consulting and clinical Psychology*, 43, 384-395

Tyler, L.E. (1963). *Tests and Measurements.* Englewood Cliffs, New Jersey: Prentice-Hall

Tyler, R.W. (1949). *Basic Principles of curriculum and instruction* Chicago: University of Chicago Press

Vernon, P.A. (1993) *Biologicalapproachestothestudyofhumanintelligence.* Norwood. NJ: Ablex

Vernon, P.E. (1950) *The structure of human abilities.* New York: Wiley.

Waddell, D.D. (1980). The Stanford-Binet: An evaluation of the technical data available since the 1972 restandardization. *Journal of School Psychology*, 18, 203-209

Wagner, E.E. (1983). *TheHandTest.* Los Angeles: Western Psychological Services

Walker, D.F. (1971) A naturalistic model for curriculum development. *School Review*, 80(1), 51-67

Wallace, S. (2015) (Ed.) *A Dictionary of Education.* UK: Oxford University Press

Watson, C.G., Felling, J. & Maceacherr, D.G. (1967) Objective Draw- A. Person Scales: An attempted cross validation. *Journal of Clinical Psychology*, 23, 382-386

Wechsler, D. (1939) The measurement of adult intelligence Baltimore: Williams and Wilkins

Wechsler, D. (1967). *Manual for the Wechsler Preschool and Primary Scale of intelligence.* New York: The Psychological Corporation

Wechster, D. (1944) *The measurement of Adult intelligence* (3rd ed.) Baltimore: Williams & Wilkins

Weinberg, R. & Forlenza, S. (2012) Psychological skills. In G. Tenenbaum, R.C. Elkland, & A Kanata (Eds.) *Measurement in sport and exercise psychology.* Chamaign, IL: Human Kinetics

Weinert, F.E. & Harry, E.A. (2003) The stability of individual differences in intellectual development In R.J. Sternberg, J. Lautrey & T.L. Lubart (Eds.) *Models of intelligence: International perspectives.* Washington DC: American Psychological Association

Weinstein, G. & Fantini, M.D. (1970) *Towards Humanistic education*, New York: Praeger

Welsh, J.R.J., Wassan, T.W. & Ree, M.J. (1990) *Armed services vocational aptitude Battery (ASVAB): predicting military criteria from general and specific abilities*. Brooks AFB, TX: US Air force Human resources Laboratory

Whipple, C.M. (1910). *Manual of Mental and Physical tests*. Baltimore: Warwick & York

White, R.W., Sanford, R.N., Murray, H.A. & Bellack, L. (1941). *Morgan-Murray Thematic Apperception Test: Manual of directions* (Mimeograph) Cambridge, H.A: Harvard Psychological Clinic

Wickes, T.A. (1956). Examine influences in a testing situation. *Journal of Consulting Psychology*. 20, 23-26

Wiggins, G.P. (1993) *Assessing students performance: Exploring the purpose and limits of testing*, San Francisco, C.A.: Jossey-Bass

Wiggins, J.S. (1973) *Personality and Prediction: Principle of personality assessment*. Reading, M.A. Addison-Wesley

Wilson, G.G. & Vitousek, K.M. (1999) Self, monitoring in the assessment of eating disaders. *Psychological Assessment*, 11, 480-489

Winer, B.J. (1962) *Statistical Principle in Experimental Design* New York: McGraw-Hill

Wirt, P.D. & Lachar, D. (1981). The personality inventory for children: Develpment and clinical applications. In P. McReynolds (Ed.) *AdvancesinPsychologicalassessment*. 5, San Francisco: Jossay-Bass

Wirt, P.D., Lachar, D., Klinedinst, J.K. & Seat, P.D. (1991). *MultidimensionaldescriptionofChildPersonality:AManualforthePersonalityInventoryforchildren* (1990 edition) Los Angeles: Western Psychological Services.

Wolk, R.L. & Wolk, R.B. (1971) *Manual: Gerontological Apperception Test*. New York: Behavioural Publications

Wood, E.S. & Wood, E.G. (1996) *The world of Psychology* Boston : Allyn & Bacon

Wood, J.M., Nezworski, M.T. & Stejskai, W.J. (1996a). The Comprehensive system for the Rorschach: A Critical examination. *Psychological Science*, 7, 3-10

Wood, J.M., Nezworski, M.T. & Stejskai, W.J. (1996b). Thinking critically about the comprehensive system for the Rorschach: A reply to Exner. *Psychological Science*, 7, 14-17

Woodcock, R.W., McGrw, K.S. & Mather, N. (2001). *Woodcock-JohnsonPsycho-EducationalBatteryIII*. Itasca Il: Riverside

Woodrock, C., Duda, J.L., Cumming, J. Sharp, L.A., Holland M.J.G. (2012) Assessing mental skill and technique use in applied interventions: Recognizing and minimizing threats to the psychometric properties of the TOPS. *The Sport Psychologist, 26*, 1-15

Yamamoto, K. & Dizney, H.F. (1965) Effects of three sets of test instructions on scores on an intelligence test. *Educational and Psychological Measurement. 25(1)*, 87-94

Zarrella, K.L. & Schuerger, J.M. (1990). Temporal Stability of Interest inventories. *Psychological Reports. 66*, 1067-1074

Zhang, L.M., Yu, L.s. Wang, K.N. (1997). The physiological psycho. assessment method for pilot's professional reliability. *Aviation, Space & Environmental Medicine.* 68, 368-372

Zubin, J, Eron, L.D. & Schumer, F. (1965) *An Experimental approach to projective techniques.* New York: John Wiley

Major Events in the field of Psychological and Educational Assessment

2200 B.C.E. : Proficiency testing in China where emperor evaluates public officials

1115 B.C.E. : Open and competitive Civil service examination in China during Chan Dynasty and such examination test proficiency in various areas such as writing, arithmetic, geography, agriculture, etc.

400 B.C.E. : Plato suggests that individuals should work consistent with, their abilities and endowments–an idea supported by psychologists through ages.

387 B.C.E. : Plato suggests that the brain is the mechanism of mental processes.

335 B.C.E. : Aristotle suggests that heart is the mechanism of mental processes.

175 B.C.E. : Claudius Galenus (popularly known as Galen) designed experiments to demonstrate that it is the brain and not the heart that is the seat of intellect.

200 A.D. : The Dark Ages started and science takes a backseat to superstition and faith.

1265 : Thomas Aquinas points out that the idea of a human capacity to think and reason should be replaced by the idea of immortal soul.

1550 : The Renaissance witnesses a rebirth in Philosophy and Johaun Weyer, a German physician, writes that those accused of being witches were, in reality, suffer from mental and physical disorders.

1700 : With the writing of French philosopher Rene Descartes, German philosopher Gottifried Leibruiz and a group of English philosophers, namely, John Locke, George Berkeley, David Hume and David Hartley, the cause of philosophy and science advances.

1734 : Christian Von Wolff wrote two books *Psychologia Empirica* (Empirical Psychology, 1732) and *Psychologia Rationalis* (Rational Psychology, 1734) that anticipate psychology as science. Wolff who was student of Gottfied Leibniz also elaborates the Leibniz's idea that there exists perceptions below the threshold of awareness, which anticipated Freud's idea of unconscious mind.

1774 : Frauz Mesmer outlined his cure for some mental illness, originally called *Mesmerism* and now known as *hypnosis*.

1808 : Franz Gall writes about phrenology (the view that an individual's skull shape and placement of bumps on the head can reveal his important traits of personality)

1823	: *The Journal of Phrenology* is founded to study Franz Gall's idea that personality traits, ability and special talents are located in bumps on the head. But the journal folds by the early twentieth century.
1829	: In his famous book *Analysis of the Phenomena of Human Mind* James Mill, English Philosopher argues that the structure of mental life consists of ideal & sensations thus anticipating an approach to experimental psychology to be called *structuralism*.
1834	: Ernst Heinrich Weber published his theory of *Just Noticeable Difference*, now known as Weber's law.
1859	: Charles Darwin publishes *on the origin of species by Means of Natural Selection* detailing his view evolution and expending on the theory of survival of the fittest and natural selection. These ideas greatly influenced Freud who emphasized the importance of instinctual sexual and aggressive urges.
1860	: Gustav Fechner, a German physiologist, publishes *Elements of Psychophysics* in which he explores the various ways through which people respond to stimuli such as sound and light.
1869	: Sir Francis Galton, half cousin of Charles Darwin, published *Hereditary Genius* and argues that the intellectual abilities are biological in nature. He also pioneered a statistical technique that Karl Pearson would later call correlation. Subsequently, Galton makes numerous contributions to measurement with various inventions as well as innovations.
1879	: Wilhelm Max Wundt founded the first formal laboratory of Psychology at the University of Leipzig, Germany. This event is considered momentus because psychology is now being treated as a science and not just a branch of philosophy. The establishment of this laboratory marks the formal beginning of the study of human behaviours, emotions and cognition.
1883	: The first laboratory of Psychology in America is founded at John Hopkins University
1885	: Herman. Ebbinghaus invented nonsense syllabus as a means to study memory processes
1890	: American Psychologist James McKeen Cattell coins the term *mental test* and in 1921 he starts *Psychological Corporation* a company with the goal of useful application of Psychology.
1890	: Sir Francis Galton developed a statistical technique of correlation to better understand interrelationships in his intelligence studies.

1892	:	The famous psychiatrist Emil Kraeplin who studied with Wundt, publishes his work concerning the use of a test that involves word association.
1892	:	American Psychological Association (APA) was founded and headed by G. Stanley Hall, with an initial membership of 42.
1895	:	Alfred Binet founded the first laboratory of Psychodiagnosis and along with Victor Henri publishes articles calling for the measurement of cognitive abilities such as memory as well as other human abilities. Interestingly, Binet also wonders aloud about the possible use of inkblots to study personality.
1896	:	The first psychological clinic was developed at the University of Pennsylvania by Lightner Witmer making the birth of clinical Psychology. In 1907 he founded a journal called *Psychological Clinic*.
1904	:	Charles Spearman who was student of Wundt at Liepzig University, starts to lay the fondation for the concept of reliability of the test. He also begins building the mathematical framework for factor analysis.
1905	:	Alfred Binet and Theodore Simon publish a 30-item intelligence test intended to help identify mentally retarded of Paris School children.
1906	:	The journal of Abnormal Psychology was found by Morton Prince.
1911	:	Edward Thorndike publishes first article on *animalintelligence* leading to theory of operant conditioning.
1912	:	William Stern develops the original formula for Intelligence Quotient (IQ) after studying the scores on Binet's intelligence scale. The formula is (MA/CA) × 100.
1913	:	Jon Watson, publsies '*PsychologyasaBehaviouristviewsit*' which becomes known as the 'behaviourist manifesto: As the behaviourist view it, behavioural observation becomes a key tool of assessment.
1913	:	World war I proves to be a boon for psychological testing because several recruits were to be quickly screened for intellectual functioning as well as emotional fitness.
1916	:	Lewis M. Terman working at Stanford University, publishes the Stanford Revision of Binet-Simon intelligence Scale.
1916	:	First Indian Department of Psychology is established at Calcutta University.
1917	:	Robert Yerkes (President of APA at that time) develops Army Alpha and Army Beta tests for assessing intelligence in a group format.
1921	:	Psychological Corporation launched the first Psychological test development company that not only commercializes psychological testing but allows testing to take place at offices and clinics also.

1927	:	Carl Spearman publishes a two-factor theory of intelligence.
1927	:	German neurologist Kurt Goldstein begins to develop neurodiagnostic tests on the basis of research with such soldiers who suffered brain injury during World War I.
1927	:	The first edition of Strong Vocational Interest Bank is published.
1931	:	L.L. Thurptone publishes Multiple Factor Analysis, a very important work that will have the effect of focusing attention on the study of cognitive abilities.
1932	:	Jean Piaget publishes The Moral judgement of Children beginning his popularity as the leading theorist in cognitive development
1935	:	Christia D. Morgan and Henry A. Murphy Collaborate to develop a test of personality originally called *Morgan-Murray Thematic Apperception Test*. The final version of the test is published in 1943 with the authorship credited to Henry A. Murray and the staff of the Harvard Psychological clinic.
1938	:	L.L. Thurstone proposed that intelligence consists of about seven group factors known as Primary Mental abilities.
1938	:	Bender-Gestalt test, authored by physician Lauretta Bender, is published. Originally the test consists of nine designs which the testtaker is required to copy. The Bender-Gestalt II is released in 2003.
1938	:	Raven publishes Raven's Progress Matrices intended to measure Spearman's of factor.
1939	:	David Wechsler working at Bellevue Hospital in New York City introduces an intelligence test called *Wechsler-Bellevue Intelligence Scale*. Later on, this scale is revised several times and from it, two other tests, that is, children test as well as a preschooler test are developed.
1940	:	World War II prompts a strong need for screening military recruits. At this time, Starke R. Hathaway, a psychologist, and John Charuley McKinley a psychiatrist, collaborate to develop a personality test called Minnesto Multiphasic Personality Inventory (MMPI).
1941	:	Raymond B. Cattell introduces a theory of intelligence based on two general factors called *fluid intelligence* and *crystallized intelligence*.
1942	:	Jean Piaget publishes Psychology of intelligence discussing his theories of cognitive development.
1942	:	MMPI is developed and soon becomes the most widely researched and widely accepted psychological assessment device.

1945	: David Rapaport, Roy Schafer and Merton Gill develops *Diagnostic Psychological Testing* which proved to be a milestone is clinical assessment.
1951	: Lee Cronbach formulated Coefficient alpha for assessing test reliability. In fact, the formula for coefficient alpha is a modification of Kuder-Richardson's twentieth formula.
1952	: American Psychiatric Association publishes DSM I.
1957	: Donald Super publishes *The Psychology of Careers* in which he proposed a theory of careers which was researched by him over next three decades.
1957	: C.E. Osgood describes the Semantic Differential Scale.
1958	: Lawrence Kohlberg publishes the first version of Moral judgement scale.
1959	: Cambell and Fiske formulated a test validation approached popularly called as Multitrait Multimethod Matrix.
1961	: Holtzman Inkblot Technique (HIT) is published. It is different from Rorschach Inkblot test in the sense that it has two parallel forms.
1962	: The beginnings of the practical application biofeedback experientation for monitoring certain types of brainwaves on command can be traced.
1963	: Stanley Milgram publishes the famous *Behavioural study of obedience* and makes a very important contribution to Psychological assessment. Alfred Bandura introduces the idea of observational learning considered important for the development of personality.
1963	: Robert Cattell proposes the theory of fluid and crystallized intelligence.
1968	: American Psychiatric Association publish DSM-II
1968	: Walter Mischel publishes *Personality and Assessment* and psychologists started questioning whether personality traits are consistent across various situations.
1972	: First ICSSR Survey of Research in India edited by Prof. S.K. Mitra is released.
1974	: John E. Exner publishes a system called the *Comprehensive system* for administering, scoring and interpreting responses on Rorschach Inkblot test.
1974	: Friedman and Rosenman popularize Type-A Coronary behaviour pattern.
1975	: John Holland formulates a classification system, which consisted of six personality types based upon the corresponding interest pattern.
1976	: Michael P. Maloney and Michael P. Ward publish *Psychological Assessment: A conceptual Approach* in which a conceptual model of assessment process is

	developed and this is claimed to be in, sharp contrast to 'test-oriented and test controlled' practices.
1976	: Paul T. Costa, Jr. and Robert R. McCrae started a research programme which emphasizes upon the analysis of 16 PF. Later on, they conceive of *Big Five* and develop their own test called *NEO-PI-R*.
1982	: Psychological Research wing of DRDO, India is elevated to a full-fledged organisation and named as Defence Institute of Psychological Research or DPIR.
1983	: Howard Gardner proposes the theory of Multiple intelligence and argued that intelligence is something to be used to improve the life and not something to assess and quantify human beings.
1985	: The American Psychological Association and other groups jointly publish the influential *Standards for Behaviural and Psychological Testing*.
1987	: American Psychiatric Association publishes DSM III -R
1987	: *National Academy of Psychology* (NAOP) in India is founded.
1989	: MMPI-2 is published
1993	: American Psychological Association (APA) publishes *Guidelines for providers of Psychological services to Ethnic, Linguistic and Culturally diverse populations*.
1994	: American Psychiatric Association publishes DSM-IV
1999	: APA and other groups revised *Standards for Educational and Psychological Testing*.
2000	: DSM-IV (TR) is published
2005	: Fifth edition of Stanford-Binet intelligence test (SB-5) is released
2005	: APA Ethical Principles of Psychologists and code of conduct is revised
2009-11	: Fifth ICSSR Survey of Research in Psychology in India edited by Prof. Girishwar Mishra (Four Volumes) is released
2013	: DSM-5 is released

Glossary

[A]

Achievement test : A test that intends to evaluate degree of accomplishment or learning that has taken place usually as a result of exposure to some training.

Age Norms : A type of norms that shows the level of test performance for each separate age group in the normative sample.

Average Proportional distance (APD) : Relatively a new measure for evaluating internal consistency of test where the researcher focuses not on similarity of scores of items of the test rather on degree of difference that exists between test scores.

Alternate-forms reliability : A form of reliabilities in which alternative forms of the same test are administered to a group of heterogeneous and representative subjects. Subsequently, the scores for these two forms are correlated.

Aptitude test : A test that intends to assess one or more clearly defined and relatively homogeneous ability.

Assessment : Involves appraising the level or magnitude of some attribute of an individual. Testing is one small part of assessment that incorporates intention, observation, rating scales, checklists, etc.

Attitude : It is learned cognitive, affective ad behavioural predisposition to respond positively or negatively towards some persons, objects or situations.

Acquiescence : A response style in testing that characterizes agreement with whatever presented.

Age Scale : A test with such items that organise by the age at which most test takers are capable of responding in the way keyed correct.

Alpha level : It is the significance level, that is probability of making Type I error.

Alternative hypothesis : It is the researcher hypothesis about the likely outcome of the study (H_1).

ANOVA : Analysis of variance is the most common inferential statistical tool for analysis of the results of experiments when DV is measured on interval scale or ratio scale.

Arithmetic Mean : It is a measure of central tendency derived by calculating average of all scores in a distribution

Average deviation : It is a measure of variability derived by summing the absolute value of all the scores in distribution and dividing by the total number of scores.

[B]

Behavioural Assessment : A technique of evaluation based upon the analysis of samples of behaviour, including antecedents and consequences of the behaviour.

Bender-Gestalt test : A test for screening neurophysiological deficit that involves copying designs. The test has been developed by Lauretta Bender.

Behavioural checklists : List of behaviours with some predefined operational definitions that the investigators are trained to use in an observational study.

Biased sample : A sample that is not representative of the population.

Bias in construct validity : A type of bias involved in a test when it is shown to measure different traits for one group than another or to assess the same trait but with differing degree of accuracy.

Bias in content validity : A type of bias involved in a test when its items are relatively more difficult for members of one group than for another group despite the fact that intelligence level of the two groups is the same.

Bias in predictive validity : A type of bias demonstrated when the inference drawn by the test scores contains a constant error as a function of membership in a particular group.

Bimodal distribution : A distribution of scores in which the central tendency consists of two scores occurring an equal number of times and both are the most frequently occurring scores.

[C]

C Scale : A variant on the stanine scale with eleven unit or points.

Carryover effect : A form of sequence effect in which systematic changes in performance occur as a result of completing a sequence of conditions rather than different sequence.

Case study : A descriptive method in which an in-depth analysis is made of either a single individual or single organisation or single event.

Ceiling effect : Such effect occurs when scores on two or more conditions are at or near the maximum possible for the scale being used, tendering the impression that no real differences exist between the conditions concerned.

Ceiling level : A stage in test in which subtest items are ranked from easiest to hardest, the level above which the testtaker would certainly fail for all the remaining items or questions.

Classical theory of measurement : A very dominant theory in psychological testing, which assumes that an observed score consists of a true score and measurement error or error score.

Coefficient alpha : An index of reliability that may be thought as the mean of all possible split-half coefficients corrected by Spearman-Brown prophecy formula. Also referred to as Cronbach's alpha.

Coefficient of correlation : An index of the strength of the linear relationship between two continuous variables and expressed as number, which can range from -1 to $+1$.

Coefficient of determination : A value which indicates how much variance is shared by two variables under study. The value is obtained by squaring the value of obtained correlation multiplying by 100 and expressing the result as percentage. If $r = .40$, coefficient of determination would be $(.40)^2 \times 100 = 16\%$, the amount of variance accounted for by the correlation coefficient.

Coefficient of equivalence : An index indicating parallel-forms reliability or alternative-form reliability.

Coefficient of stability : A index indicating the estimate of test-retest reliability obtained during the defined interval.

Cross-lagged panel correlation : A type of correlational research intended to deal with the directionality problem. If X and Y are measured at two different time intervals and if X precedes Y, then X might cause Y but Y cannot cause X.

Correlation Matrix : A table of intercorrelation between all the variables, which is the beginning point of factor analysis.

Component intelligence : In the theory of Sternberg, the internal mental mechanisms that are responsible for intelligent behaviour.

Concurrent validity : A type of criterion-related validity in which the criterion measures are available at approximately the same time as the test scores.

Construct validity : A type of validity which is found when the measure being used accurately assess some hypothetical construct.

Cohort effect : A Cohort is a group of individual born at the same time. Cohort effect can reduce the internal validity of cross-sectional studies because the differences between groups could result from the effects of growing up in different environment.

Confound : An extraneous variable that covaries with the independent variable and can provide an alternative explanation of results.

Content analysis : A procedure used in descriptive research to categorize systematically the content of the behaviour recorded by the researcher.

Control group : A group in experiment not given a treatment that is being evaluated. It provides a means of comparison.

Convenience sample : A type of non-probability sample in which volunteers from a group of people who meet the general requirements of the study are included.

Continuous variable : A variable for which infinite number of values exists.

Content sampling : Refers to the variety of the subject-matter contained in the items. Also referred to as *item sampling*.

Content validity ratio (CVR) : This is used to assess agreement among raters regarding to what extent the individual test item is worth to be included in test. A formula of CVR has been developed by C.H. Lawshe.

Convergent validity : Such validity indicates that a test measures the same construct as another test intending to assess the same construct.

Cross-validation : A revalidation on a sample testtakers different from those testtakers on whom the test performance was originally found to be valid predictor of the same criterion.

Criterion-related validity : A type of validity in which the newly constructed test is correlated against some external criterion.

Criterion-referenced test : A test where the intention of the researcher is to determine where the testtaker stands with respect to a tightly-defined some criteria or objectives.

Crystallized intelligence : In Cattell and Horn's theory a form of intelligence which indicates what one has learnt through the investment of fluid intelligence in the culture.

Culture-free test : A test that is devoid of the influence of any particular culture and therefore, does not provide any special advantage to person belonging to any particular culture. In fact, such is more an ideal than a reality because all tests reflect some favour to a particular culture.

Culture-fair test : A test that is designed to minimize the influence of a culture on different aspects of the evaluation procedures.

Culture loading : An index that shows the magnitude to which a test incorporates the knowledge and feelings associated with a particular culture.

Cut score : A score that is derived as a result of judgement used to divide a set of data into two or more classification, with the intention to draw some inference from such classification. Also called as *cut-off score*.

[D]

Debriefing : A session arranged after the experiment in which the experimenter explains the purpose of the study, tends to reduce discomfort felt by the participants and also tries to answer any question posed by the participants.

Deception : A strategy of researcher in which the participants are not told about all the details of an experiment in the beginning. It is commonly used for avoiding the demand characteristics.

Dehoaxing : A portion of debriefing in which the real purpose of the study is explained to the participants.

Demand characteristics : Any feature of the experimental procedure that is likely to increase the chances that the participants will detect the true purpose of the study.

Dependent variable : The variable about which prediction is made on the basis of the research.

Descriptive statistics: A group of statistics that provide the summary of the main features of a set of data collected from a sample of participants.

Deviation IQ : A type of standard score for reporting intelligence quotient with a mean of 100 and the standard deviation of 15.

Diagnosis : A conclusion arrived on the basis of evidence and opinion through the process of distinguishing the nature of something.

Diagnostic test : A test used to identify the areas of deficit to be targeted for intervention.

Discriminant validity : A type of validity which is demonstrated when a test does not correlate with tests or variables with which it should not because it differs from them.

Discrete variable : A variable in which each level represents a distinct category which is qualitatively different from other category such as male and female.

Domain sampling : A sample of test items from all possible items that could be used to assess a particular construct or a sample of behaviour from all possible behaviours that could be indicative of a particular construct.

[E]

Expectancy table : A table that displays the established relationship between the test scores and the expected test scores and the expected outcomes on a relevant task.

Experiential intelligence : In theory of intelligence by Sternberg, it refers to the ability to deal effectively with novel tasks.

Extravalidity concerns : It is the side effects and unintended consequences of testing.

Emotional intelligence : The term was first formally introduced by P. Salovey and D. Mayer in 1990 to include some groups of competencies like ability to express emotions accurately, ability to access and evoke emotions, ability to comprehend emotional message and ability to regulate one's own emotions for promoting growth and well-being.

Error variance : The component of variance attributable to random sources irrelevant to the trait being measured by the test. Most common sources of error variance include those related to item sampling, test administration, test scoring and interpretation.

Exploratory factor analysis : A type of factor analysis in which mathematical procedures are employed to estimate factors, extract factors or decide how many factors are to be retained.

External validity : The extent to which the results of a study generalize to other settings.

Ecological validity : It is said to exist when the researcher studies psychological phenomena in day-to-day situation.

Effect size : Amount of variance in dependent variable that can be attributed to the independent variable or amount of influence that one variable has on the other variable.

Evaluation apprehension : A form of anxiety that leads the participants to behave so as to be evaluated positively by the experiment.

Event sampling : A procedure in research in which only certain types of behaviours occurring under precisely defined conditions are sampled.

Experimental realism : The extent to which the participants are deeply involved in the experiment.

Extraneous variable : A variable that is uncontrolled and therefore, is not of interest to the researcher but it could affect the results.

[F]

Face validity : It occurs when a test appears to measure a given trait.

Factor analysis : It is a mutlivariate analysis in which several variables are intercorrelated and variables that correlate highly with each other form a factor.

Factorial Design : Any experimental design which includes more than one independent variable.

Field experiment : An experiment that is conducted outside the laboratory in some actual field like hospital, school, college, industry, etc.

Field Research : A research that is done in any location other than a laboratory. It is wider than field experiment.

Formative evaluation : A type of evaluation that is done while the programme is in progress.

Fluid intelligence : In Cattell's theory of intelligence, it consists of non-verbal abilities which are less dependent on culture and formal instruction than what is called crystallized intelligence.

Flynn effect : A fact that intelligence measured by a standardized test rises each year after the test was normed. Thus it is a sort of intelligence inflation.

Forced-choice format : A type of item where each of the two or more choices have been predetermined to be equal in terms of social desirability and the testtaker is forced to choose any one.

Factor loading : In factor analysis the correlation between an individual test and the single factor.

Factor Matrix : A table of correlations between variables and factors. These correlations are called as factor loading.

[G]

General factor : In Spearman's theory of intelligence, the general factor (or g factor) is general intelligence or ability to solve problems or ability to reason. This g factor is measured by tests of intelligence to a greater or lesser degree.

Grade norms : Norms prepared as a reference in the context of grade of the testtaker who achieved a particular score.

Group test : Paper-end-pencil tests which are suited to the testing of large group of persons at the same time.

Generalizability theory : Also referred to as domain sampling theory of reliability that recognizes several alternatives of generalization of test results.

Good Subject role : A type of bias shown by the participants in which they try to guess the experimenter's hypothesis and then, tend to behave in such a way as to conform it.

[H]

Halo effect : A tendency to rate higher or low on all dimensions because of a global impression.

Heritability Index : An estimate of how much of the total variance in any trait is accounted in terms of genetic factors. This index can vary from 0.0 to 1.0.

Homogeneous Scale : A scale in which all items tend to assess the same thing. This is assessed by item-total correlation.

Hypothesis : An inference about a relationship between variables, which are tested empirically.

[I]

Independent variable : A variable that can be directly manipulated by the experimenter (Type-E independent variable) or the participants can be selected by virtue of possessing certain attribute (Type-S independent variable).

Inferential statistics : Statistics that are used to draw conclusions about the broader population on the basis of a study using a sample of that population. t-test and F-test are good examples.

Interaction : In factorial design interaction is said to occur when the effect of one independent variable is dependent upon the level of another independent variable.

Internal validity : It indicates the extent to which a research or study is free from methodological flaws, especially free from confounding variables.

Incremental validity : An index of the explanatory power of some additional predictors over and above the predictors already present in the study. Often used in relation to the predictive validity.

Individual test : Test administered on one person at a time.

Informed consent : A principle, often used in testing, which states that testtakers are made aware in writing so that they can understand the purpose and likely consequences of testing.

Interscorer reliability : A form of reliability that reflects degree of agreement between scorers. Also known as inter raters reliability.

Interval scale : A measurement scale that provides information about relative strength of ranks and is based upon the assumption of equal-sized units or intervals.

Ipsative test : A test in which the average of subscales is always the same for every testtaker. It means then that for the individual testtaker high scores on subscales are balanced by low scores on other subscales.

Interpersonal intelligence : In Gardener's theory of multiple intelligence, interpersonal intelligence is the ability to understand other people, what motivates them and how to work cooperatively with them etc.

Item analysis : A group of statistical procedures designed to explore how individual test items work as compared to other items in the test as well as in the context of test as a whole.

Item-characteristic curve (ICC) : A graphic representation showing item difficulty and item discrimination.

Item-difficulty Index : Such index indicates how many testtaker responded correctly to an item.

Item discrimination index : Such statistic or index indicates how adequately a test item makes discrimination between inferiors and superiors.

Item-endorsement Index : A statistic usually in personality test or other contexts where test items are not keyed correct or incorrect, indicating how many testtakers responded to an item in a particular direction. In ability test and achievement test, this statistic is called as item-difficulty index.

Item-reliability index : Such index provides an indication about test's internal consistency. Higher the index, higher is the internal consistency of the test.

Item-response theory (IRT) : This theory assumes that a trait being measured by a test is undimensional and the extent to which each item measures the trait. This theory is also known as *latent-trait theory* or *latent-trait model*.

Item-validity Index : An index showing the extent to which a test measures what it intends to measure. Higher this index, greater is the test's criterion-related validity.

Item Sampling : It indicates varieties of the themes or subject-matter covered by the item. It is also called as *content sampling*.

[K]

Kuder-Richardson formulas : A series of formula developed by G.F. Kuder and M.W. Richardson designed to estimate inter-item consistency or reliability of the test. Of these series, formula number 20 and 21 became very popular.

Kurtosis : A measure that indicates the nature of steepness (peak vs. flat) of the center of a distribution of scores.

[L]

Likert Scale : A scale that presents the testtakers with response option ordered on an agree/disagree or approve/disapprove continuum.

Local norms : Norms prepared from a local representative sample rather than from a national representative sample.

Longitudinal design : A design of research in which the same set of participants is tested over times.

Levels of measurement : Properties of the numbers that are used in a test-nominal, ordinal, equal interval or ratios.

Linear regression : A statistical process that is used to predict one set of test scores from one set of criterion scores.

Linear transformation : A method of changing raw scores for interpretation purposes and it does not change the characteristics of the raw data in any way. Example, are z scores and T scores.

Literature Review : A systematic and careful examination of published and unpublished reports concerning a topic of interest.

[M]

McCall's T : A standard score with a mean of 50 and standard duration of 10. It can be obtained from a simple liner transformation of z scores a by formula $T = 10z + 50$.

Mean : The arithmetic average of a set of scores.

Measurement error : Variations or inconsistencies in the measurement produced by a test.

Measurement : Assignment of numbers according to rules.

Measures of central tendency : Those statistical measures that provide information about the middle of a set of scores such as Mean, Median and Mode.

Measures of relationship : Statistical measurements that describe the relationship between two sets of scores such as Pearson r, Spearman rho, etc.

Measures of Variability : Statistical measures that reflect the spread of scores in the distribution such as range, standard deviation, Q, variance, etc.

Median : A score point in the distribution above which lies 50 percent of the cases and below which lies the 50 percent of cases.

Mode : The most frequently occurring score in the distribution.

Multiple regression : A statistical process in which more than one set of test scores is used to predict one set of criterion scores.

Multitrait Multimethod design : A design for test validation that accumulates evidence for reliability, convergent validity and discriminant validity into one study.

Multivariate analysis : Analysis that provide information about relationships between combination of three or more variables or groups.

[N]

Neuropsychological tests : Such tests intend to diagnose sensitivity to the effects of brain damage.

Nominal scale : This is the simplest and lowest level of measurement where categories are arbitrary and don't indicate more or less of anything.

Nonprobability sampling : A type of sampling in which not everyone has an equal chance of being selected from the population.

Norm-referenced test : A test in which the performance of each testtaker is interpreted with reference to the relevant standardization sample.

Normal distribution : A frequency distribution with symmetrical, mathematically defined, bell-shaped frequency distribution.

Normalized standard score : A score obtained by a transformation which makes a skewed distribution into a normal distribution.

Norms (Plural norm) : A summary of the performance of a group of individuals on which a test was standardized.

Normal Probability distribution : A theoretical distribution that exists as a perfect and symmetrical distribution. Also known as *normal curve*.

Nonlinear transformation : A process of changing a score that ensures that the new score does not necessarily have a direct numerical relationship to the original score and magnitude of the differences between the new scores and other scores on the scale may not parallel with the magnitude of the differences between the original score and other scores on the scale from which the original scores might have been derived.

Normative sample : A group of individuals presumed to be representative of the population who may take the test.

[O]

Objective criterion : A measurement that is measurable and observable such as number of errors committed while during a work.

Objective test format : A special test format that has one response as correct such as multiple-choice question.

Objective tests : Tests that are structured and requires the testtakers to respond to structured multiple choice items or rating scales.

Odd-even reliability : A method of split-half reliability of a test obtained by assigning odd-numbered items to one-half of the total items of the test and even-numbered items to the other half and subsequently, correlating the two sets of scores.

Ordinal scale : A level of measurement in which all things/objects measured can be rank-ordered and there is no absolute zero point on the scale.

Outlier : An extremely atypical finding in the research.

Orthogonal axes : An important assumption in factor analysis that exhibits that factors are at right angles to one another, which means that they are independent or uncorrelated.

Open-ended question : A question that usually cannot be answered specifically. Such questions require the respondents to produce something spontaneously.

[P]

Parallel forms reliability : A method of estimating reliability in which equivalent forms of the test are developed by generating two forms using the same rules. The correlation between the two forms is the estimate of parallel forms reliability.

Pearson Product Moment Correlation : An index of correlation propounded by Karl Pearson that is used between two continuous variables.

Percentile : The percentage of persons in the standardization sample who scored at or below a raw score, percentile can vary from 0 to 100.

Percentile rank : The proportion or percentage of scores that fall below a particular score. Thus percentile rank refers to the *percentage* and *percentile* refers to a score.

Percentile band : A range of percentiles that are likely to represent a subject's true score.

Performance scale : A test that requires participants to do something rather than to answer questions.

Point scale : A test or scale in which points such as 0, 1 or 2, etc. are assigned to each item. One feature of point scale is that all items with a particular content can be grouped together.

Predictive validity : A type of criterion-related validity in which criterion measures are obtained in future usually after some months or years after the test scores have been obtained.

Projective hypothesis : An assumption that the personal interpretation of ambiguous stimuli reflect the unconscious needs, motives and conflicts of the examinee.

Projective test : A test in which the testtakers face some vague and ambiguous stimuli and resond with their own constructions.

Prophecy formula : A formula developed by Spearman and Brown that is used to correct for loss of reliability when split-half method of estimating reliability is used.

Psychological testing : Refers to all the possible uses, applications and underlying concepts of psychological tests.

[Q]

Quartiles : Points that divide the frequency distribution into four equal parts.

Quantitative item analysis : A statistical analysis of the responses which testtakers give to the individual test questions.

[R]

Random error : Error that is nonsystematic and unpredictable, resulting from some unknown causes. In fact, such error is the unexplained difference between the testtaker's true score and the obtained score.

Range : A rough measure of variability calculated by subtracting the lowest number in a distribution from the highest number in the distribution.

Ratio scale : The highest level of measurement in which numbers are assigned to points with the assumption that there is a true zero point showing an absolute absence of property and that each point is an equal distance from the numbers adjacent to it.

Raw scores : The basic score that results when the testtaker completes a psychological test.

Regression line : The best-fitting straight line through a set of different points in a scatter diagram.

Reliability : Refers to the consistency of true scores. Theoretically, reliability is the ratio of true score variance to observed score-variance.

Resentative sample : A group (or sample) of people composed of individuals with characteristics similar to those for whom the test is to be finally used.

Residual : The difference between predicted and observed values in a regression equation.

Response style : A tendency to respond to a test item in a certain way irrespective of theme or content.

[S]

Sample : A group of individuals/objects selected to represent the entire population.

Sampling error : A statistic that reflects how much error can be attributed to the lack of representation of the target population due to the characteristics of the sample of the respondents.

Scorer reliability : The extent of agreement between or among the persons scoring a test or rating an individual. Also known as *interrater reliability*.

Self-report tests : Tests that depend upon the testtakers' descriptions of their feelings, beliefs, opinions and mental states.

Scatter diagram : A picture of the relationship between two variables, that is, X and Y.

Shrinkage : Sometimes a regression equation is created for one group and is used to predict the performance of another group of respondents. Thus procedure tends to overestimate the magnitude of the relationship for the second group. The amount of decrease in strength of relationship from original individuals to the individuals with which the equation is used is known as *shrinkage*.

Spearman's rho : A nonparametric method of finding the correlation between two sets to ranks.

Split-half method : A method of estimating internal consistency or reliability of the test in which test is divided into two equal halves and subsequently, they are correlated.

Standard deviation : A very important measure of variability that represents the degree to which scores vary from mean.

Standard error of measurement : An index of measurement error which indicates the extent to which the respondent's score might vary over a number of parallel tests.

Standard score : A transformed score in which original score is stated in terms of distance from mean in standard deviation units.

Standardization fallacy : A fallacious view that a test standardized on one population is ipso-facto unfair when used in other population.

Standardization sample : A large and wider group of subjects representative of the population for whom the test is intended.

Stanine scale : A scale in which all raw scores are converted to a single digit system of scores ranging from 1 to 9.

Simple random sampling : A type of sampling in which every member of a population has an equal chance of being selected as a member of the sample.

Single group validity : When a test is validated for one group but not for another, such as valid for Hindus but not for Muslim.

Slope : The expected change in Y for every one unit change in X on the regression line.

Standardized tests : Tests that have administered to a large group of individuals who are similar to the group for whom the test has been developed so as to develop norms.

Sten scale : A 10-unit scale with five units above and five units below the mean.

Systematic measurement error : A type of measurement error that arises, when not known to the taste constructor, a test consistently measures something other than the trait for which it was intended.

[T]

T-score : A transformed score with a mean of 50 and standard deviation of 10.

True score : Testtaker's hypothetical real score on a test. Such score is never known but it can be probabilistically estimated.

Test : A measurement device that qualifies behaviour.

Test administrator : Person giving or administering the test.

Testtaker : Person responding the items/questions of the test.

Test-retest reliability : A method of estimating reliability in which test is administered at two different occasions on the same sample and the two sets of scores are subsequently, correlated.

Third variable : A variable that tends to account for the observed relationship between two variables.

[U]

Univariate analysis : A statistical analysis that summarizes individual question responses.

User's manual : In psychological testing the manual that gives instructions for administration and also provides guidelines for test interpretation.

[V]

Validity : A test is valid to the extent it measures successfully what it intends to measure.

Validity coefficient : The correlation between test and the criterion.

Validity shrinkage : When a test predicts the relevant criterion less accurately with the new sample of the testtakers than with the original tryout sample, validity shrinkage is said to exist.

Variance : An index that indicates the degree of dispersion in the distribution of scores.

[W]

Within-group norming : A practice of administering the same test to each and every testtaker but scoring the test in a different way according to the race or tribe of the testtaker.

Work sample : A small-scale assessment in which the testtakers complete the task such as building a doghouse, etc.

[Z]

Z-score : A standard score that has a mean of zero (0) and standard deviation of 1.

Chapter-1 : Basic Principles of Measurement

1. Assigning numerals to the attribute according to some rules is called :
   (a) Evaluation　　(b) Measurement　　(c) Assessment　　(d) Appraisal
2. Measuring extraversion of a student is an example :
   (a) Physical measurement　　(b) Psychological measurement
   (c) Normative measurement　　(d) None of these
3. The major properties of measurement are :
   1. Magnitude　　2. Equal-intervals　　3. Absolute zero
   A ratio scale has the properties of :
   (a) 1 and 2　　(b) 2 and 3　　(c) only 3　　(d) 1, 2 and 3
4. Measurement of temperature in terms centigrade has :
   (a) Absolute zero　　(b) Arbitrary zero　　(c) Technical zero　　(d) None of these
5. Which of the following is *not* appropriate for interval measurement?
   (a) t ratio　　(b) F ratio
   (c) Pearson r　　(d) Coefficient of variation
6. Which one is considered as the highest level of measurement?
   (a) Ratio scale　　(b) Internal scale　　(c) Nominal scale　　(d) Interval scale
7. The equivalence relationship between two classes of objects or persons is characterized by being :
   (a) Reflexive　　(b) Symmetrical　　(c) Transitive　　(d) All of these
8. If a > b and b > c, then a > c. Such relationship is called as :
   (a) Reflexive　　(b) Symmetrical　　(c) Transitive　　(d) None of these
9. Percentiles and Median are most suited to :
   (a) Interval measurement　(b) Ratio Measurement
   (c) Ordinal Measurement　　(d) Nominal Measurement
10. Ordinal measurement are characterized by :
    1. Equal-interval measurement　　2. Absolute zero point
    3. Magnitude
    codes :
    (a) 1 and 3　　(b) 2 and 3　　(c) only 2　　(d) only 3
11. If 5 students arranged themselves from shortest to tallest and the researcher assigned the shortest a score of 1 and the tallest a score of 5, what level of measurement the researcher is using?
    (a) Nominal measurement　　(b) Interval measurement
    (c) Ordinal measurement　(d) Ratio measurement

12. Which levels of measurements are produced by most of the psychological and educational tests?
    (a) Nominal and Ordinal measurement
    (b) Ratio and Nominal measurement
    (c) Ordinal and Interval measurement
    (d) Ratio and Ordinal measurement

Answers

1. (b) 2. (b) 3. (d) 4. (b) 5. (d) 6. (a) 7. (d) 8 (c) 9. (c) 10. (d) 11. (c) 12. (c)

Chapter 2 : Fundamental of Evaluation and Assessment

1. Which is not a correct statement?
    (a) Evaluation includes value judgement
    (b) Evaluation includes both qualitative and quantitative description of the performance of individual
    (c) Evaluation includes only a qualitative descriptive of the performance of individuals.
    (d) Evaluation is the systematic process of collecting and analyzing data for taking a decision.
2. Annual examination of students in the school is an example of:
    (a) Diagnostic evaluation
    (b) Formative evaluation
    (c) Summative evaluation
    (d) None of these
3. If the teacher is interested in knowing the knowledge, skills etc. of a student before admitting him in an intended course, it becomes an example of:
    (a) Placement evaluation
    (b) Formative evaluation
    (c) Summative Evaluation
    (d) Diagnostic evaluation
4. Pre-Ph.D. Entrance test in the university illustrates:
    (a) Diagnostic Evaluation
    (b) Placement Evaluation
    (c) Internal Evaluation
    (d) External Evaluation
5. Which of the following is a misfit according to the criterion of 'what is being evaluated'?
    (a) School evaluation
    (b) Curriculum evaluation
    (c) Evaluation of personnel
    (d) Diagnostic evaluation
6. Which is not a phase of evaluation?
    (a) Planning phase   (b) Execution phase   (c) Process phase   (d) Product phase
7. Which one cannot be considered as a function of evaluation?
    (a) To recommend for modification in the on-going programme
    (b) To identify the influence of undesirable impact upon the programme or policy
    (c) To formulate goals of a programme
    (b) To accommodate external members as one of the evaluators
8. Which type of assessment indicates that the individuals can perform well when they put forth their best efforts?
    (a) Criterion-referenced assessment
    (b) Norm-referenced assessment
    (c) Typical performance assessment
    (d) Maximum performance assessment

9. According to Airasian and Madaus which is the correct sequence of assessment procedures to be used in the classroom?
   (a) Placement Assessment, Formative assessment, Diagnostic assessment and Summative assessment
   (b) Formative assessment, Diagnostic assessment, Placement assessment and Summative assessment
   (c) Diagnostic assessment, Placement assessment, Summative assessment and Formative assessment
   (d) Placement assessment, Diagnostic assessment, Formative assessment and Summative assessment
10. When a teacher says that Mohan's performance in the classroom is better than 90 percent of his class members, he is obviously referring to :
    (a) Norm-referenced assessment   (b) Criterion-referenced assessment
    (c) Diagnostic assessment   (d) None of these

Answers
1. (c) 2. (c) 3. (a) 4. (b) 5. (d) 6. (b) 7. (d) 8. (d) 9. (a) 10. (a)

Chapter 3 : Principles of Psychological Testing

1. Which one is the origin of the rudimentary form of psychological testing in history?
   (a) 2200 B.C.E.   (b) 1800 B.C.E.   (c) 400 B.C.E.   (d) 200 B.C.E.
2. Who coined the term mental test in psychological testing?
   (a) R.B. Cattell   (b) J.M. Cattell   (c) F. Galton   (d) None of these
3. Who spurred scientific interest in the study of individual differences?
   (a) F. Galton   (b) A. Anastasi   (c) C. Darwin   (d) J.M. Cattell
4. In India which of the following institution is not concerned with developing the psychological tests?
   (a) NCERT   (b) ISI (Indian Statistical Institute)
   (c) NLEPT   (d) UGC
5. Read the following statements and answer on the basis of code provided :
   1. Psychological test is a standardized procedure
   2. Psychological test predicts non-test behaviour
   3. Psychological test measures a sample of behaviour
   Codes:
   (a) Only 1 and 3 are correct
   (b) Only 2 and 3 are correct
   (c) Only 1 is correct
   (d) 1, 2 and 3 are correct

6. Match List I with List II and answer on the basis of code given:

   List I (Types of Tests)
   A. Power Test
   B. Non-Verbal Test
   C. Objective Test
   D. Standardized Test

   List II (Description)
   1. Test having uniform procedure of administration
   2. Tests measuring knowledge
   3. Tests have pictorial form of items
   4. Tests having stems and options for responding

   Codes:
   (a) A B C D
       2 3 4 1
   (b) A B C D
       2 3 1 4
   (c) A B C D
       3 1 2 4
   (d) A B C D
       2 3 4 1

7. Which is not a characteristic of a good Psychological and educational test?
   (a) Reliability (b) Validity (c) Practicality (d) Smoothness

8. Read the statements carefully and answer or the basis of codes provided.
   1. Tests are used to provide motivation
   2. Tests provides feedback or knowledge of results
   3. Tests are used for developing a theory
   Codes:
   (a) Only 2 is correct
   (b) 1 and 3 are correct
   (c) 2 and 3 are correct
   (d) 1, 2 and 3 are correct

9. When the test scores are compared against some specified reference group, it is known as:
   (a) Criterion-referenced test
   (b) Norms-referenced test
   (c) Group dependent test
   (d) None of these

10. Ravens Progressive matrices is one example of
    (a) Non-language test
    (b) Performance test
    (c) Non-verbal test
    (d) None of these

11. Which one was the first widely used personality inventory?
    (a) Woodworth Personal Data Sheet
    (b) Minnesota Multiphasic personality Inventory
    (c) Woodworth Psychoneurotic Inventory
    (d) Cattell Sixteen Personality Factor Questionnaire

12. A test that required the test takers to respond to the test items about their feelings and beliefs can be categorized as
    (a) Behaviour modification test
    (b) Self-report test
    (c) Test of maximal performance
    (d) None of these
13. Who published the Stanford-Binet test?
    (a) Alfred Binet
    (b) Alfred Simon
    (c) Robert Woodworth
    (d) Lewis Terman
14. In which of the following test the role of testtaker is least cleared?
    (a) Self-report test
    (b) Projective test
    (c) Non-standardized test
    (d) Objective test
15. In which type of test, a teacher generally prefers to admission in the classroom?
    (a) aptitude test
    (b) Intelligence
    (c) Achievement test
    (d) Interest Inventory
16. Which test measures the testtakers' ability to perform in the field in which they have not received proper training?
    (a) Aptitude test
    (b) Achievement test
    (c) Test of Maximal performance
    (d) None of these
17. What type of a test a career counsellor is most likely to administer?
    (a) Intelligence test
    (b) Aptitude test
    (c) Interest Inventory
    (d) Personality test
18. What do all psychological tests require you to do?
    (a) Give answers to a Question
    (b) Perform behaviour
    (c) Fill out a form
    (d) All these three
19. Which type of test requires the testtakers to respond about their own feeling and beliefs in general?
    (a) Self-report test
    (b) Projective test
    (c) Observation Test
    (d) Attitude test
20. Which one is best suited to a large group of individuals who are similar to the group for which test was originally designed?
    (a) Objective test
    (b) Projective test
    (c) Standardized test
    (d) Non-standardized test
21. CAT stands for
    (a) Computerized Adaptive Test
    (b) Computerized Adaptive Testing
    (c) Computerized Administrative Testing
    (d) None of these

Answers

1. (a) 2. (b) 3. (c) 4. (d) 5. (d) 6. (a) 7. (d) 8. (d) 9. (b) 10. (c) 11. (c) 12. (b)
13. (d) 14. (b) 15. (c) 16. (a) 17. (c) 18. (b) 19. (a) 20. (c) 21. (b)

Chapter 4 : Item writing and Item Analysis

1. Which of the following is popularly known as the *lowest common denominator* of the test?
   (a) Distractor (b) Option (c) Stem (d) Item
2. When the situation requires the testtakers to recall rather than recognize information and to express his ideas clearly and concisely, the most appropriate decision would be to construct:
   (a) Objective Item (b) Essay Item
   (c) Double-barrelled item (d) None of these
3. A test is called objective test if it is
   (a) Carefully planned (b) Impartially administered
   (c) Scored objectively (d) Culturally unbiased
4. If the testtaker criticizes a test item, the test construction should :
   (a) change it
   (b) dismiss the testtakers' objections
   (c) praise the testtakers for pointing out the limitation
   (d) listen as objectively as possible
5. Option is to distractor as :
   (a) answer key is to raw score (b) possibility is to mistake
   (c) objective is to subjective (d) answer key is to obtained score
6. For which of the following item stems, would options probably be the most difficult to write?
   (a) The author of the Discovery of India is
   (b) Of the following, the largest Indian state in area is
   (c) Which of the following statements best supports the argument that premarital sex is no longer harmful for mental health?
   (d) Who is the first Muslim President of India?
7. The most critical reviewer of a test item is likely to be :
   (a) Item writer himself (b) A subject expert (c) A testtaker (d) A teacher
8. Completion item is an example of :
   (a) Selected - response item (b) Constructed-response item
   (c) Essay item (d) None of these
9. A best distracter in multiple-choice item is one that is :
   (a) responded by all testtakers (b) responded more by high scorers
   (c) responded more by low scores
   (d) responded equally by both high scorers and low scorers
10. In personality test the item difficulty index is best called by the name of :
    (a) item endorsement index (b) item validity index
    (c) item reliability index (d) None of these

11. If a group of testtakers attempted to guess the correct responses on all the items of a true-false test, an estimate of the average percent of correct responses would be :
    (a) 20 (b) 50 (c) 75 (d) 60
12. Which of the following indices of item difficulty is considered most desirable for inclusion of the item in test?
    (a) 0.50 (b) 0.60 (c) 0.75 (d) 0.25
13. For a four-option multiple choice item, the optimal difficulty value will be :
    (a) 0.625 (b) 0.265 (c) 0.725 (d) 0.823
14. If an item is correctly responded by greater number of testtakers from the group of low scorers than from the group of high scorers, the item is said to be making :
    (a) a positive discrimination (b) a negative discrimination
    (c) no discrimination (d) a good discrimination
15. If an item is correctly responded by 20 testtakers from low scorers (N = 50) and by 40 testtakers from high scorers (N = 50), then its item-discrimination index would be :
    (a) 0.60 (b) 0.40 (c) 0.70 (d) 0.30
16. Item-validity index provides an indication to :
    (a) Content validity of the test (b) Criterion-related validity of the test
    (c) Construct validity of the test (d) None of these
17. Which one of the following can be described as objective test format?
    (a) Essay items (b) Multiple choice items
    (c) Interview (d) Projective items
18. Which one of the following format has the easiest scoring?
    (a) Essay items (b) Interview
    (c) Multiple choice items (d) Projective items
19. A rule of thumb is to write items .... more than the researcher wants to retain finally for the test according to test plan.
    (a) two times (b) three times
    (c) two and half times (c) three and half times
20. In which part of test construction, the list of characteristics of testtakers lie?
    (a) Developing the test plan (b) Item writing
    (c) Defining the purpose (d) Defining the target audience

Answers

1. (d) 2. (b) 3. (c) 4. (d) 5. (b) 6. (c) 7. (c) 8. (b) 9. (c) 10. (a) 11. (b) 12. (a) 13. (a) 14. (b) 15. (b) 16. (b) 17. (b) 18. (c) 19. (a) 20. (d)

Chapter 5 : Reliability and validity of the test

1. Reliability means :
   (a) Consistency of test (b) Consistency of test scores
   (c) Consistency of items (d) Consistency in scoring
2. Which of the following is *not* related to the technical or logical meaning of reliability?
   (a) Error variance (b) Total variance
   (c) True variance (d) Residual variance
3. In which method of estimating reliability of the test, heterogeneity of test is a source of error variance?
   (a) Split-half method (b) Test-retest method (c) Coefficient alpha (d) K-R formula
4. Which method of reliability gives an average of all possible split-half reliability coefficient?
   (a) Flanagan formula (b) Scorer reliability (c) Coefficient alpha (d) Rulon formula
5. When a test constructor has developed an attitude scale, each item of which is having five response options for providing the most appropriate answer, he will be safe in calculating reliability if he chooses method of :
   (a) Split-half reliability (B) K-R 20 (c) Coefficient alpha (d) Test-retest
6. Time sampling is an important source of error variance in :
   (a) Split half reliability (b) test-retest reliability
   (c) Interscorer (d) K.R. formulas
7. When a test has the reliability coefficience of .06, its index of reliability would be :
   (a) .87 (b) .77 (c) .85 (d) .70
8. Which of the following is not a validity in technical sense of the term?
   (a) content validity (b) predictive validity
   (c) face validity (d) concurrent validity
9. Which is a correct method of estimating a relationship between predictor scores and criterion scores?
   (a) W coefficient (b) Multiple correlation
   (c) Expectancy table (d) Multitrait-Multimethod matrix
10. Which of the following does not provide evidence for construct validity?
    (a) homogeneity of test (b) pretest posttest differences
    (c) appropriate developmental changes (d) statistical inference
11. Correlational analysis based on multitrait-multimethod matrix reflects evidences for
    (a) Convergent validation (b) Distriminant validation
    (c) both for 'a' and 'b' (d) For neither 'a' nor 'b'
12. The higher values of correlations in multitrait multimethod method reflect :
    (a) Reliability of the test (b) Validity of the test
    (c) Practicality of the test (d) None of these

13. Who proposed multitrait multimethod matrix?
    (a) Campbell & Fiske (b) Anastasi & Urbnia
    (c) Guilford & Lindiquist (d) Spearman and Cattell
14. Monotrait heteromethod correlations provide evidence for:
    (a) Construct validity of the test (b) Reliability of test
    (c) Practicality of test (d) None of these
15. For 6 factors or variables, how any pairs of factors would be arranged?
    (a) 20 (b) 15 (c) 21 (d) 17
16. If reliability of a test is .75, what will be its maximum validity?
    (a) .87 (b) .92 (c) .79 (d) .81
17. Which of the following statement is false?
    (a) Validity requires a very low standard error of measurement
    (b) Validity requires reliability
    (c) Reliability requires validity
    (d) Reliability requires a low standard error of measurement
18. Which of the following aphorism best applies to the relationship between content validity and adequacy of sampling?
    (a) A penny saved is a penny earned (b) Time and tide wait for no one
    (c) A rolling stone gathers no moss
    (d) A rose by any other name would smell as sweet
19. A reliability coefficient is computed on the basis of test scores for a group of 400 testtakers. Subsequently, the reliability is recomputed on the basis of scores of the middle 40% of the group. In the second case, the reliability is expected to:
    (a) remain the same (b) increase
    (c) decrease (d) be indeterminable
20. If any test item displays low inter correlations, the test will tend to have a:
    (a) Low standard error of measurement (b) High reliability coefficient
    (c) Low reliability coefficient (d) Low standard error of estimate
21. Reliability is to validity as:
    (a) accuracy is to error (b) Precision is to flexibility
    (c) Consistency is to truth (d) Consistency is to content
22. $K-R_{21}$ is in part a measure of:
    (a) Concurrent validity (b) Predictive validity
    (c) Homogeneity of test items (d) Standard error of estimate
23. Reliability coefficient are most useful in assessing error resulting from:
    (a) Biased sampling (b) A limited number of items
    (c) Cheating (d) Guessing

24. Which of the following is not considered as a traditional type of validity?
    (a) Criterion-related validity
    (b) Content validity
    (c) Construct validity
    (d) Parallel forms validity
25. A test was administered to each of the following groups and reliability coefficient were computed. Which of the groups would yield the highest reliability coefficient?
    (a) The upper half of the class as indicated by test scores
    (b) The lower half of the class as indicated by the test scores
    (c) All students of class X
    (d) The middle half of the class as indicated by the test scores
26. If the testtaker takes an arithmetic test that required him to perform different types of arithmetical calculations, the test is said to have evidence for
    (a) Validity based on its relationship with a construct
    (b) Face validity
    (c) Validity based on its content
    (d) Validity based on its relationship with a criterion
27. If the testtaker perceives a test as attractive, he is referencing evidence of its :
    (a) Reliability
    (b) Content
    (c) Face validity
    (d) Predictive validity
28. When the reliability of the test is high, the standard error of measurement will be..... As the reliability of the test decreases, the standard error of measurement tends to ...
    (a) low; decreases
    (b) high; increases
    (c) low; increases
    (d) high decreases
29. Which one of the following measures tends to estimate the performance of a test over time and provides an estimate of the stability of test?
    (a) Alternate forms of reliability
    (b) K-R 20
    (c) Coefficient alpha
    (d) test-retest method
30. Who is likely to apply generalizability theory?
    (a) Test developer
    (b) Test user
    (c) Testtaker
    (d) Test administrator
31. Both concurrent validity and predictive validity tend to establish the evidence of :
    (a) Validity based on test content
    (b) Validity based upon the perception of test users
    (c) Both reliability and validity
    (d) Validity based upon test-criteria relationship
32. Multitrait-Multimethod design does not provide :
    (a) Discriminant evidence of validity
    (b) Predictive evidence of validity
    (c) Convergent evidence of validity
    (d) Content validity

33. Sohan proposed a set of some factors underlying intelligence. Subsequently, he administered a test of intelligence and used a statistical technique based on correlation to use these data. Which one of the following he used for the purpose?
    (a) Linear regression (b) Confirmatory factor analysis
    (c) Exploratory factor analysis (d) None of these
34. Sita conducted a factor analysis in which he searched for underlying some theoretical structures in her construct. Which one of the following design she is using :
    (a) Multitrait-Multimethod design (b) True experimental design
    (c) Exploratory factor analysis (d) Confirmatory factor analysis
35. The validity coefficients resulting from the cross validation are usually to be ... the validity coefficient resulting from original validity coefficients?
    (a) Lower than (b) Higher than (c) The same (d) Of no relation to

Answers
1. (b) 2. (d) 3. (c) 4. (c) 5. (c) 6. (b) 7. (b) 8. (c) 9. (c) 10. (d) 11. (c) 12. (b)
13. (a) 14. (a) 15. (b) 16. (a) 17. (c) 18. (d) 19. (c) 20. (c) 21. (c) 22. (c) 23. (b) 24. (d)
25. (c) 26. (c) 27. (c) 28. (c) 29. (d) 30. (b) 31. (d) 32. (b) 33. (b) 34. (c) 35. (a)

Chapter 6 : Norms and Transformation of Scores

1. Average performance of the normative group of a particular class in school is called as:
   (a) Age norms (b) Grade norms
   (c) Standard score (d) Percentile norms
2. Shyam has obtained a score of 50 on a test and this score entitles him to 90th rank in his group. Here the score of 50 is a :
   (a) Percentile rank (b) Percentile (c) Centile (d) Decile
3. The concept of T-score was proposed by :
   (a) Thorndike (b) Cattell (c) McCall (d) Anastasi
4. If Mohan has a score of 20 and mean of distribution of scores is 30 and standard deviation of 8, then, his z-score would be :
   (a) +1.25 (b) +1.66 (c) –1.69 (d) –1.25
5. z-score reflects :
   (a) Linear transformation (b) Non-linear transformation
   (c) Partly linear transformation and partly non-linear transformation
   (d) None of these
6. If z score is $+2.00\sigma$, then T score would be :
   (a) 70 (b) 80 (c) 30 (d) 40

7. Which is a limitation of z score?
   (a) z-score is rough standard score
   (b) z-score has fraction points
   (c) z-score is linearly transformed score
   (d) z-score has no connection with normal distribution
8. Which is *true* about percentile norms?
   (a) Percentile norms are linearly transformed standard score
   (b) Percentile norms have inequality in their units
   (c) Percentile norms are suitable only for intelligence test
   (d) Percentile norms don't differ much from standard score norms
9. Which of the following is the variant of Stanine scale?
   (a) C scale   (b) T scale   (c) Z scale   (d) Interval scale
10. In stanine scale the standard deviation has been set at :
    (a) 1.24   (b) 2.00   (c) 2.24   (d) 2.87
11. In between $\pm 2.00\sigma$, what percentage of cases lie in normal distribution of scores?
    (a) 95.44%   (b) 62.26%   (c) 99.74%   (d) 65.25%
12. What are values of percentile at $Q_1 Q_2$ and Q_3 in the distribution of scores?
    (a) 25th, 50th and 75th respectively   (b) 50th, 25th and 75th respectively
    (c) 75th, 50th and 25th respectively   (d) 75th, 25th and 50th respectively
13. Which one of the following standard scores tend to bring a change in the unit of measurement?
    (a) Stanine score (b) Percentile   (c) z score   (d) T score
14. Which is not a correct statement about the use of norms?
    (a) When normative group is small, it is not considered a representative
    (b) Only one right population is regarded as normative group
    (c) Norms are used for providing a meaningful comparison and interpretation
    (d) Test publishers often develop and publish the results of various normative groups
15. Which of the following standard score always has a mean of zero and standard deviation of one?
    (a) T score   (b) Percentile   (c) z score   (d) None of these
16. Most psychological tests produce which levels of measurement?
    (a) Ordinal and ratio   (b) Interval and ordinal
    (c) Nominal and ordinal   (d) Interval and nominal

Answers

1. (b) 2. (b) 3. (c) 4. (d) 5. (a) 6. (a) 7. (b) 8. (b) 9. (a) 10. (b) 11. (a)
12. (a) 13. (b) 14. (b) 15. (c) 16. (b)

Chapter 7 : Taxonomy of Educational Objectives

1. The educational objectives in cognitive domain were proposed by :
   (a) Simpson and colleagues
   (b) Bloom and colleagues
   (c) Linn and Colleagues
   (d) Harrow and colleagues
2. In Blood's taxonomy of educational objections which of the following lie at the top of cognitive hierarchy of complexity?
   (a) Knowledge
   (b) Evaluation
   (c) Analysis
   (d) Application
3. In Bloom's taxonomy of educational objectives the six categories have been amended in hierarchy of which of the following dimension?
   (a) Simplicity–complexity
   (b) Internalisation–externalisation
   (c) Abstractness–concreteness
   (d) Responding–valuing
4. In Bloom's taxonomy which one lies at the lowest end of hierarchy of simplicity complexity?
   (a) Evaluation
   (b) Synthesis
   (c) Knowledge
   (d) Comprehension
5. In taxonomy of psychomotor objectives proposed by Simpson what is the meaning of origination?
   (a) Well-developed skills that can be modified
   (b) Waiting new movement patterns to fit a particular situation
   (c) Readiness to take a particular type of action
   (d) Performance act where learnt response becomes habitual
6. In taxonomy of educational objectives proposed by Krathwohl what is basis of hierarchical order of the five categories?
   (a) Internalisation
   (b) Externalisation
   (c) complexity
   (d) None of these
7. In taxonomy of educational objectives proposed by Krathwohl at which stage the level of interalisation is at minimum?
   (a) Responding
   (b) Organisation
   (c) Receiving
   (d) Characterisation
8. In taxonomy of educational objectives proposed by Krathwohl responding refers to:
   (a) Students willingness to attend to a particular activities
   (b) Students' active participation in a activity
   (c) Students bid to develop internally consistent value system
   (d) Students effort to attach importance to a particular object
9. Those instructional objectives that are commonly classified under interest are placed under which category of affective domain:
   (a) Receiving
   (b) Responding
   (c) Valuing
   (d) Organisation

10. In taxonomy of affective domain proposed by Krathwohl, which one of the following lies at 3.3 level?
   (a) Commitment
   (b) Conceptualisation of value
   (c) Preference for a value
   (d) Satisfaction in response

Answers
1. (b) 2. (b) 3. (a) 4. (c) 5. (b) 6. (a) 7. (a) 8. (b) 9. (b) 10. (a)

Chapter 8 : Classification Instructional goal and objectives: Educational Decision-making

1. Specific objectives are concerned with
   (a) General intended outcome
   (b) Discrete intended outcome
   (c) Both of the above
   (d) None of the above
2. Some instructional objectives are considered to be required as minimum essentials which must be acquired by all students inspective of their background or ability. These are called as
   (a) Performance objectives
   (b) Behavioural objectives
   (c) Learning objectives
   (d) Mastery objectives
3. When instructional objectives are viewed in terms of learning outcomes, it is called as
   (a) Process of learning
   (b) Product of learning
   (c) Method of learning
   (d) Law of learning
4. Which is not a good criteria for selecting appropriate instructional objectives?
   (a) The instructional objectives should be in harmony with content and general goals of the school
   (b) The instructional objectives should be in tune with the sound principles of learning
   (c) Instructional objectives should not much emphasize upon all important learning .... of the course
   (d) Instructional objectives should honour the abilities of the students.
5. Assessing instructional objectives on the basis of writing a good dramma or poem is an example of
   (a) Performance test
   (b) Product evaluation
   (c) Process evaluation
   (d) Affective measures

Answers
1. (b) 2. (d) 3. (b) 4. (c) 5. (b)

Chapter 9 : Understanding the various domains of curriculum

1. Who said, "curriculum is a tool in hands of artist (teacher) to mould his material (the pupil) according to his ideas (objectives) in his studio (school)."
   (a) Cunnigham (b) Oliva
   (c) Tyler (d) Taba

2. The basic principle that what the school don't teach may be as important as they do teach, is clearly reflected by
   (a) Core Curriculum (b) Activity-Centered Curriculum
   (c) Null Curriculum (d) Learner-Centered Curriculum

3. A curriculum in which its topics are revisited each time in a more detail or depth as the learner acquired more knowledge and skills is known as
   (a) Core Curriculum (b) Concomitant Curriculum
   (c) Subject-centered Curriculum (d) Spiral Curriculum

4. William James is the supporter of
   (a) Naturalism (b) Pragmatism
   (c) Idealism (d) Existentialism

5. Taba's model of curriculum development is an example of
   (a) Deductive model (b) Inductive model
   (c) Partly deductive and partly inductive (d) None of these

6. Tyler's rationale includes
   (a) Four basic questions (b) Six basic questions
   (c) Seven basic questions (d) Seven basic questions

7. In Tyler's model the curriculum planners proceed
   (a) From general to specific (b) From specific to general
   (c) In circular way (d) In some unknown way

8. Which model qualifies for being called as curriculum of affect:
   (a) Tyler's model (b) Taba's model
   (c) Eisner's model (d) Weinstein & Fantini's model

9. Who considered the experiences of the learners, their attitudes and feelings and social context in which they live as the important vehicles of curriculum development?
   (a) Eisner (b) Weinstein & Fantini
   (c) Oliva (d) Smith

10. Which of the following components has not been included in Eisner's model of curriculum development?
    (a) Structural (b) Intentional
    (c) Instrumental (d) Pedagogical

Answers
1. (a)   2. (c)   3. (d)   4. (b)   5. (b)   6. (a)   7. (a)   8. (d)   9. (b)   10. (a)

Chapter 10 : Planning Classroom Tests and Assessment

1. The purpose of testing and assessment during instruction is to make
    (a) Summative assessment
    (b) Formative assessment
    (c) Value assessment
    (d) Baseline assessment
2. The purpose of end-of-instruction testing and assessment is to make
    (a) Baseline assessment
    (b) Formative assessment
    (c) Summative assessment
    (d) Diagnostic assessment
3. Table of specifications generally involves
    (a) Developing a list of instructional objectives
    (b) Preparing the instructional content
    (c) Preparing a two-way chart relating the instructional objectives to instructional content
    (d) All of the above
4. In constructing test items and performance assessment, difficult vocabulary, unclear instructions, etc constitute the examples of
    (a) Construct-relevant factors
    (b) Construct-irrelevant factors
    (c) Construct-validation factors
    (d) None of these
5. Writing an essay is an example of
    (a) Performance assessment
    (b) Performance learning
    (c) Formative test
    (d) Learning test
6. If a teacher uses psychological tests for determining whether students have the knowledge necessary to learn new materials, he is using the test as a
    (a) Summative assessment
    (b) Formative assessment
    (c) Placement assessment
    (d) Authentic assessment
7. When a teacher is to answer the question like which students have mastered the learning tasks to the extent that they should be promoted to the next grade, he is basically using:
    (a) Diagnostic assessment
    (b) Placement assessment
    (c) Summative assessment
    (d) Formative assessment
8. When teachers use psychological tests periodically throughout the year, they are using the test as:
    (a) Formative assessment
    (b) Summative assessment
    (c) Placement assessment
    (d) Diagnostic assessment
9. Which is correct about authentic assessment?
    (a) Authentic assessment usually uses more than one measure of assessment
    (b) Authentic assessment is more reliable and valid than other measures of assessment

(c) Authentic assessment emphasizes upon the measurement of application of knowledge and skills in performing real-world tasks
(d) Authentic assessment is a criterion-referenced test

10. Suppose a teacher wants to answer the questions: On which learning tasks do the students exhibit satisfactory performance progressively? In such situation, the teacher is basically using:
    (a) Placement assessment   (b) Diagnostic assessment
    (c) Formative assessment   (d) Summative assessment

Answers
1. (b) 2. (c) 3. (d) 4. (b) 5. (a) 6. (a) 7. (c) 8. (a) 9. (c) 10. (c)

Chapter 11 : Achievement Tests

1. Which of the following is not a feature of achievement test?
   (a) Achievement tests are present and past-oriented
   (b) Achievement tests measure the degree of learning after exposure to the instruction
   (c) Achievement tests may be standardized or have no standardization
   (d) Achievement tests are used only for classroom achievement of students

2. Which is not a function of achievement test?
   (a) Achievement tests motivate students to learn
   (b) Achievement tests serve as a means of evaluating instructional programme
   (c) Achievement tests help teachers in modifying the curriculum
   (d) Achievement tests help in identifying mentally deficient children

3. Iowa Test of Basic Skills (ITBS) is an example of
   (a) Prognostic achievement test   (b) Diagnostic achievement test
   (c) Single survey test   (d) Survey test batteries

4. If the nature of achievement test is such that permits the interpretation of scores with reference to a large standardization sample, it is called as
   (a) Criterion-referenced achievement test   (b) Non-referenced achievement test
   (c) Individual achievement test   (d) Group achievement test

5. In constructing an achievement test if the researcher has an item with difficulty value of 0.60, he should expect then, that the discrimination index of this item will be, in general,
   (a) Good   (b) Moderate
   (c) Poor   (d) Worst

6. Which of the following is the most important to test in an achievement test?
   (a) Skills   (b) Memory
   (c) Inferential reasoning   (d) Impossible to say

7. A primary reason that could be given for not using standardized achievement test in the lower elementary grades is that:
   (a) Tests for these grade are not available
   (b) Teachers are not interested in test results
   (c) Scores are not reliable
   (d) Tests for these grades must be individual
8. Achievement test batteries given at the primary school level tend to differ from those given at high school level in that the
   (a) High school batteries are mainly diagnostic in nature
   (b) Elementary school batteries are largely nonverbal
   (c) High school batteries are more general in content
   (d) Elementary school batteries are shorter

Answers
1. (d) 2. (d) 3. (d) 4. (b) 5. (a) 6. (d) 7. (c) 8. (c)

Chapter 12 : Aptitude test

1. Which of the following statement is false about aptitude test?
   (a) Aptitude tests are cognitive measures
   (b) Aptitude tests concentrate upon the present-oriented tasks
   (c) Aptitude tests are designed to reflect potential rather than proficiency
   (d) Aptitude test, assess abstract reasoning abilities
2. DAT is a :
   (a) Specific aptitude test   (b) Global aptitude test
   (b) Multiple aptitude test battery (d) None of these
3. In DAT scholastic aptitude is determined by combining the score of :
   (a) Verbal reasoning and language usage (b) Numerical reasoning and space relations
   (c) Verbal reasoning and numerical reasoning (d) Language usage and space relations
4. In GATB, K stands for :
   (a) Manual dexterity   (b) Motor coordination
   (c) Numerical ability   (d) None of these
5. In GATB when the scores on S, P and Q are combined, it yields :
   (a) Cognitive measure   (b) Psychomotor measure
   (c) Perceptual measure   (d) Sensory measure
6. Which is true about GATB?
   (a) It gives nine different scores through twelve tests
   (b) It has ten tests and yields ten scores
   (c) It yields ten scores through fifteen tests
   (d) It gives twelve score through twelve tests

7. Which of the following is most descriptive of GATB?
   (a) It is basically an individual test
   (b) It is culture-free test
   (c) It uses multiple cut off score
   (d) It measures aptitudes for high-level professionals
8. Of the following, which is the best measure of musical talent?
   (a) Rhythm judgement
   (b) Seashore measures of musical talent
   (c) Success of some musical endeavor
   (d) Tone judgement
9. AFQT does not include
   (a) General Science
   (b) Vocabulary
   (c) Arithmetic
   (d) Mechanical ability
10. Which one of the following is not found in MAB?
    (a) Digit symbol
    (b) Comprehension
    (c) Picture arrangement
    (d) Object relation

Answers
1. (b) 2. (c) 3. (c) 4. (b) 5. (c) 6. (a) 7. (c) 8. (c) 9. (a) 10 (d)

Chapter 13 : Theories and Measurement of Intelligence

1. Who defined intelligence as what the intelligence tests test?
   (a) Jensen (b) Boring (c) Terman (d) Thurstone
2. 'Intelligence refers to mental capacity to automatize information processing' is claimed by :
   (a) Wechsler (B0 Baron (c) Terman (d) Sternberg
3. Who for the first time emphasize upon the concept of age differentiation in developing intelligence scale?
   (a) Jensen (b) Binet (c) Spearman (d) Cattell
4. According to Thurstone which one of the following is not included in primary mental abilities?
   (a) Word fluency
   (b) Associative memory
   (c) Divergent production
   (d) Inductive reasoning
5. In Guilford's model of Structure-of-intellect, how many abilities have been included :
   (a) 120 (b) 180 (c) 160 (d) 140
6. Who proposed the concept of associative intelligence and abstract intelligence together to develop a theory of intelligence?
   (a) Cattell (b) Carroll (c) Jensen (d) Wechsler
7. For the solution of a completely new problem, which one will be the most effective?
   (a) Crystallized intelligence
   (b) Fluid intelligence
   (c) Visualization processing
   (b) All of these
8. In Cattell-Horn theory of intelligence which one stands excluded?
   (a) G_{sm} (b) Cds (c) G_v (d) G_n

9. Army Alpha Test and Army Beta Test were developed under the guidance of :
   (a) R.B. Cattell (b) R.M. Yerkes (c) C. Spearman (d) T. Million
10. Which of the following theories does not accept the existence of g-factor intelligence?
    (a) Cattell theory (b) Carroll theory
    (c) Cattell-Horn-Carroll theory (d) Sternberg theory
11. According to Sternberg, knowledge-acquisition component is a part of :
    (a) Existential intelligence (b) Contexual intelligence
    (c) Componential intelligence (d) Fluid intelligence
12. According to Gardener theory of intelligence, a psychologist usually score high on :
    (a) Linguistic intelligence (b) Interpersonal intelligence
    (c) Intrapersonal intelligence (d) Visospatial intelligence
13. PASS Model of intelligence is based upon :
    (a) Luria Model (b) Atkinson-Shiffrin-buffer model
    (c) Information-processing model (d) None of these
14. SB_5 is based upon :
    (a) Spearman's g factor theory (b) Cattell-Horn-Carroll model
    (c) Cattell's theory (d) Thurstone's primary mental abilities
15. The fluid intelligence of SB_5 is assessed by :
    (a) Memory for sentences test (b) Verbal analogies test
    (c) Picture-abtudies test (d) Form-Board test
16. Which one of the following tests is common among WAIS-IV, WISC-IV and WPPSI-III?
    (a) Matrix reasoning test (b) Object assembly test
    (c) Cancellation test (d) Digit Span test
17. Which of the following tests yield independent and comparable scores for both intelligence and achievement?
    (a) WAIS-IV (b) K-ABC (c) CAS (d) None of these
18. Raven's Progress Matrices is :
    (a) Power test (b) Speed test
    (c) Both power and speed test (d) Neither power nor speed test
19. If you come to know that the child is suffering from dyslexia and you are asked to assess the intelligence of the child, which of the following tests you will prefer :
    (a) TONI-4 (b) WPPSI (c) RPM (d) UNIT
20. Which of the following does *not apply* to group test of intelligence?
    (a) Group tests lack flexibility
    (b) Group tests use analogies, similarities, classification type of items
    (c) Group tests provide poor opportunity for establishing rapport with testtakers
    (d) Group tests don't provide better established norms than the individual tests

21. Cognitive Abilities Test is a :
    (a) Group-administered test (b) Multiple-choice test
    (c) Non-verbal test (d) Both 'a' and 'b'
22. Cattell Fair Intelligence Test is a measure of :
    (a) Crystallized intelligence (b) Fluid intelligence
    (c) Interpersonal intelligence (d) Logical-mathematical intelligence
23. Who adapted Stanford-Binet test in Hindi?
    (a) S.K. Kulashrestha (b) B.N. Mukherjee (c) Sohan Lal (d) V.V. Kamat
24. Who completed Indian adaptation of WAIS Performance Scale?
    (a) V.V. Kamat (b) P. Ramalingaswamy
    (c) P. Ahuja (d) C.M. Bhatia
25. Who preferred to induce the concept of Index of Brightness in place of IQ in his tests of intelligence?
    (a) Sohan Lal (b) C.M. Bhatia (c) S.M. Mohsin (d) M. Banerjee

Answers

1. (b) 2. (d) 3. (b) 4. (c) 5. (b) 6. (c) 7. (b) 8. (d) 9. (b) 10. (c) 11. (c) 12. (b)
13. (a) 14. (b) 15. (a) 16. (a) 17. (b) 18. (a) 19. (a) 20. (d) 21. (d) 22. (b) 23. (a) 24. (b)
25. (c)

Chapter 14 : Personality Testing

1. Who defined trait as a "generalized and formalized neuropsychic system that guides consistent forms of adaptive and expressive behaviour"
    (a) Cattell (b) Allport (c) Eyesenck (d) Guilford
2. Which one is considered as mini-manifestation of trait?
    (a) Personality types (b) Personality states
    (c) Personality psychopathology (d) None of these
3. Which one emphasizes a global approach to the assessment of personality?
    (a) Psychometric approach (b) Projective approach
    (c) Humanistic approach (d) None of these
4. Which is *not* an assumption of projective test?
    (a) Projection is usually high to stimulus materials that are similar to the testtaker's physical appearance, gender, occupation etc.
    (b) There is a parallel between behaviour obtained on a projective test and behaviour displayed in social situation
    (c) There is unconscious mind within the individual
    (d) Projection to the stimuli requires a minimum skill

5. In Rorshach test when the testtaker uses a well-defined part of the inkblot for a response, the correct symbol for this response would be :
   (a) W (b) D (c) d (d) M
6. Which of the following test has 45 cards for its administration?
   (a) Holtzman Inkblot test
   (b) Somatic Inkblot test
   (c) Whippe Inkblot test
   (d) None of these
7. Considerable consistency in traits can be observed if one observes the persons :
   (a) Over a long period of time
   (b) Over a short period of time
   (c) Over a limited set of situations
   (d) In a few important situation
8. Match List-I with List-II and answer on the basis of codes provided :

   List-I
   A. Rorschach Test
   B. CPI
   C. MMPI-2
   D. TAT

   List-II
   1. A test used to diagnose psychopathological traits
   2. A test using drawings of ambiguous human situations
   3. A test used to assess normal personality traits
   4. A test using inkblots for assessing personality

   Codes
   (a) A B C D   (b) A B C D
       4 1 2 3   4 2 1 3
   (c) A B C D   (d) A B C D
       3 2 1 4   4 3 1 2
9. The House-Tree-Person test is a kind of :
   (a) Verbal technique
   (b) Expressive technique
   (c) Pictorial technique
   (d) Association technique
10. Match List-I with List-II and answer on the basis of codes given :

    List-I
    A. House-Tree-Person Test
    B. Hand Test
    C. Draw-A-Man Test
    D. Draw-A-Person Test

    List-II
    1. Goodenough
    2. Machover
    3. Buck
    4. Wagner
    5. Halpern

    Codes
    (a) A B C D   (b) A B C D
        3 4 1 2   3 1 4 5
    (c) A B C D   (d) A B C D
        5 4 3 2   4 2 3 5
11. Draw-a-Man test is the measure of .... whereas Draw-A-Person test is the measure of ...
    (a) Intelligence, personality
    (b) Personality, Intelligence
    (c) Intelligence, Interest
    (d) Interest and Adjustment

12. Woodworth Personal Data Sheet was based upon :
    (a) Criterion-group strategy (b) Factor-analytic strategy
    (c) Theoretical strategy (d) Logical-content strategy
13. Esysenck Personality Questionnaire has been derived out of :
    (a) Logical content strategy (b) Factor-analytic strategy
    (c) Criterion-group strategy (d) Theoretical strategy
14. MMPI-2 consists of :
    (a) 567 items (b) 556 items (c) 576 items (d) 565 items
15. Who is not one of the authors of MMPI-2?
    (a) Butcher (b) Graham (c) Archer (d) Tellegan
16. Which one is not included as one of the scales in California Personality Inventory?
    (a) Communality (b) Sociability (c) Empathy (d) Liveliness
17. There are five global factors of Cattell's 16 PF questionnaire and each global factor has some corresponding primary factors. Which one of the following global factors contains *wrong* entries of primary factors?

    Global factors   **Primary factors**
    (a) Extraversion/Introversion   : A, F, H, N and Q_2
    (b) Independence/Accommodation   : E, H, L and Q_1
    (c) High anxiety/Low anxiety   : C, I, O and Q_4
    (d) Receptivity/Tough mindedness : F, G, M and Q_3

18. In Cattell's Personality Questionnaire (CPQ), D6 stands for :
    (a) Guilt and resentment (b) Hypochondriasis
    (c) Zestfulness and Suicidal disgust (d) Anxious depression
19. In NEO-PI-R Big five factors, the facets like actions, ideas, feelings, fantasy, values and aesthetics together constitute :
    (a) Agreeableness (b) Openness to experience
    (c) Conscentiousness (d) None of the above
20. Match List-I with List-II and answer on the basis of codes provided.

List-I		List-II	
A.	NEO-PI-R	1.	Gough
B.	CPI	2.	Myers
C.	MBTI	3.	Costa
D.	MMPI-A	4.	Ben-Porath
		5.	Dahlstrom

    Codes

	A	B	C	D		A	B	C	D
(a)	3	2	5	4	(b)	3	1	2	3
(c)	3	1	2	4	(d)	2	5	4	1

21. Which one is considered as the projective substitute of EPPS?
    (a) CAT   (b) TAT   (c) H-T-P   (d) APT
22. Which test assumes that psychopathology is a homogeneous conditions that is additive?
    (a) MMMI   (b) MMPI-A   (c) MMPI-2   (d) MMPI-2-RF
23. Which test has been developed by psychologists placed under the category of mother-daughter duo?
    (a) MBTI   (b) EPQ   (c) MCHI   (d) None of these
24. Which does not qualify as a technique of behavioural assessment method of personality testing?
    (a) Rating scale   (b) Self-monitoring
    (c) Biofeedback   (d) Cumulative recording
25. MBTI was based upon :
    (a) Spranger's Types of Men   (b) Carl Jung's different psychological types
    (c) Costa & McCrae's CANOE   (d) Holland's R-I-A-S-E-C

Answers
1. (b) 2. (b) 3. (b) 4. (d) 5. (b) 6. (a) 7. (a) 8. (d) 9. (b) 10. (a) 11. (a)
12. (d) 13. (b) 14. (a) 15. (c) 16. (d) 17. (d) 18. (a) 19. (b) 20. (c) 21. (b) 22. (d)
23. (a) 24. (d) 25. (b)

Chapter 15 : Measurement of Interest, Values and Attitudes

1. Which one of the following areas is not measured by the latest version of Strong Interest Inventory?
   (a) Activities   (b) Leisure activities   (c) Subject areas   (d) Artistic
2. Jackson Vocational Interest Survey covers :
   (a) 26 work roles and 8 work styles   (b) 8 work roles and 26 work styles
   (c) 10 work roles and 24 work styles   (d) 24 work roles and 10 work styles
3. Kuder Occupational Interest Inventory does not provide score in :
   (a) Mechanical area   (b) Literary area   (c) Clerical area   (d) Subject area
4. Self-Directed Search was developed by :
   (a) Holland   (b) Jackson   (c) Kuder   (d) Strong
5. As a researcher, which one of the following you would select for guiding a high school student who does not plan to attend college?
   (a) Minnesota Vocational Interest Inventory   (b) Self-Directed search
   (c) Kuder Occupational Interest Inventory   (d) None of these
6. Who constructed SOV?
   (a) Allport   (b) Lindzey   (c) Vernon   (d) All the three

7. Imaginativeness is an example of :
   (a) Instrumental value (b) Terminal value (c) Social value (d) Political value
8. The lost-letter technique is a measure of :
   (a) Attitude (b) Interest (c) Value (d) None of these
9. The base of measurement of attitude by implicit association test is :
   (a) Physiological responses (b) Response latency
   (c) Favourablenss/Unfavourableness (d) None of these
10. Match List-I with List-II and answer on the basis of code provided.

    List-I
    A. Implicit Association Test
    B. Lost-Letter technique
    C. Method of Summated ratings
    D. Semantic Differential Scale

    List-II
    1. Milgram
    2. Greenworld et al.
    3. Likert
    4. Osgood et al.
    5. Cunningham et al.

    Codes
    (a) A B C D   (b) A B C D
         2 1 3 4   2 5 3 4
    (c) A B C D   (d) A B C D
         2 1 4 3   4 3 2 1

11. The Semantic Differential Scale is a measure of attitude through assessing :
    (a) Connotative meaning (b) Denotative meaning
    (c) Cognitive meaning (d) None of these

Answers

1. (d) 2. (a) 3. (d) 4. (a) 5. (a) 6. (d) 7. (a) 8. (a) 9. (b) 10. (a) 11. (a)

Chapter 16 : Measurement of Creativity

1. Which is not a component of creativity?
   (a) Flexibility (b) Fluency (c) Elaboration (d) Familiarity
2. Which involves transformational learning in creativity?
   (a) Big-c (b) Pro-c (c) Mini-c (d) Little-c
3. Who introduced Four'c model?
   (a) Torrance & Osborn (b) Guilford & Gordon
   (c) Kaufman & Beghetto (d) Torrance and Kaufman
4. Abstractness of Titles' is a part of :
   (a) Guilford Divergent ability Test (b) Torrance test of creativity
   (c) Remote Associates Test (d) None of these

5. Which trait of personality is consistently related to the measurement of creativity?
   (a) Extraversion (b) Responsibility
   (c) Self-confidence (d) Openness to experience
6. Who has favoured brainstorming as a technique of enhancing creativity?
   (a) Gordon (b) Guilford (c) Torrance (d) Osborn
7. 'Fantasy analogy' is a technique of :
   (a) Measuring creativity (b) Lowering the level of creativity
   (c) Enhancing the level of creativity (d) None of these
8. Which is *not* assessed by Torrance test of creativity (Third edition)?
   (a) Originality (b) Flexibility (c) Fluency (d) Elaboration
9. If you ask your friend to respond within one minute as many words as possible that begin with letter 'b', you are interested in assessing the dimension of :
   (a) Originality (b) Fluency (c) Flexibility (d) Elaboration
10. Which statement is not true?
    (a) Torrance Test of creativity has only figural part
    (b) Torrance Test of creativity has both verbal part and figural part
    (c) Remote Associates Test has been proposed by Mednick & Mednick
    (d) Following Guilford, originality, fluency, flexibility and elaboration are the aspects of creativity

Answers

1. (d) 2. (c) 3. (c) 4. (b) 5. (d) 6. (d) 7. (c) 8. (b) 9. (b) 10. (a)

Chapter 17 : Neuropsychological Assessment

1. Which is a true statement?
   (a) Neuropsychological assessment provides information about the functioning of frontal lobe
   (b) Neuropsychological assessment is a performance-based technique which evaluates cognitive deficit
   (c) Neuropsychological assessment evaluates the underlying neurological factors of physical disorder
   (d) Neuropsychological assessment does not entertain the use of psychological tests
2. Aphasia refers to :
   (a) Deviation in language performance (b) Deviation in textual performance
   (c) Deviation in normal hearing functioning (d) Deviation in normal visual functioning
3. Which is *not* a characteristic of useful neuropsychological test batteries?
   (a) Adaptability (b) Thoroughness (c) Sociability (d) Reliability

4. The value of GNDS in the range of 41 to 67 on Halstead-Reitan test battery indicates:
   (a) Normal functioning (b) Mild impairment
   (c) Moderate functioning (d) None of above
5. The Halstead-Impairment Index (HII) in the range of 0.8 to 1.0 indicates:
   (a) Mild impairment (b) Severe impairment
   (c) Profound impairment (d) None of these
6. Which of the following test batteries is based upon the work of famous Russian Neuropsychologist Alexander Luria?
   (a) BNT (b) CANTAB (c) HRTB (d) None of these
7. On which of the following scales performance indicates degree of compensation gained by the patient?
   (a) C_8 (b) C_9 (c) S_1 (d) S_3
8. LNMB-C is more suited for children in the age range of:
   (a) 04 to 08 years (b) 08 to 12 years (c) 12 to 16 years (d) None of these
9. Expressive speech is assessed by:
   (a) C_8 (b) C_6 (c) C_{10} (d) C_{11}
10. if you are interested in evaluating the person's ability to identify drawings and unfocussed objects, you will probably concentrate upon the performance of the person on:
    (a) C_1 (b) C_2 (c) C_3 (d) C_4

Answers

1. (b) 2. (a) 3. (c) 4. (c) 5. (b) 6. (d) 7. (c) 8. (b) 9. (b) 10. (d)

Chapter 18 : Grading and Reporting

1. When the teacher makes the formative evaluator of the performance of students, it is called as:
   (a) Reporting (b) Grading (c) Assessment (d) None of these
2. Which is *not* a use or function of grading and reporting?
   (a) Administrative use (b) Instructional uses (c) Guidance uses (d) Selection uses
3. Which is *not* a type of grading and reporting system?
   (a) Letters to parents/Guardians (b) Checklist of objectives
   (c) Pass-fail system (d) Teacher's observation
4. A good letter grade reflects:
   (a) Achievement (b) Effort (c) Study habit (d) Work completed
5. If the grading is based upon the ranking of students in order to overall achievement of student, it is known as:
   (a) Absolute grading (b) Relative grading
   (c) Short-cut grading (d) Long-term grading

Answers
1. (a)　2. (d)　3. (d)　4. (a)　5. (b)

Chapter 19 : Application of Psychological Testing in different fields

1. Which test is used by clinical psychologists during the process of clinical diagnosis?
   (a) MMPI
   (b) Halstead-Reitan Neuropsychological battery
   (c) Luria-Nebraska Neuropsychological Battery (d) None of the above
2. Which is *not* an example of electrophysiological technique?
   (a) EEG　　　　(b) PET　　　　(c) K-ABC　　　　(d) MRI
3. Which one is a self-scoring tool commonly used by a counsellor?
   (a) OVISO II　　(b) SDS　　　　(c) GATB　　　　(d) ASVAB
4. In which model of interest test interpretation, the obtained data are interpreted to answer 'how' and 'why' of client's behaviour?
   (a) Evaluative　　(b) Predictive model　(c) Genetic model　(d) In none of these
5. The famous Purdue Pegboard test is a measure of
   (a) Cognitive abilities　(b) Psycho-motor abilities
   (c) Interest　　　　　　(d) None of these
6. If there are 10 pairs of workers who are to be compared with respect of adoptability to the work environment using the method of paired comparison, how many sets of comparison would be carried out?
   (a) 54　　　　　(b) 45　　　　　(c) 10　　　　　(d) 35
7. Which method assesses the performance of the workers on the basis of frequency of the various critical incidents?
   (c) BOS　　　　(b) BARS　　　　(c) MSS　　　　(d) None of these
8. If the examiner is interested in assessing the mechanical principles involved in the various diagrams of machinery presented, he would most likely to select.... as instrument.
   (a) Bennett Mechanical Comprehension test
   (b) Revised Minnesota Paper form Board test
   (c) General Aptitude test battery
   (d) Differential Aptitude test
9. Which is not assessed by the Profile of Mood States (POMS)?
   (a) Depression　　(b) Vigor
   (c) Confusion　　　(d) Emotional Control
10. In India the most famous Defense Institute of Psychological Research (DIPR) came into existence in the year
    (a) 1985　　　　(b) 1987　　　　(c) 1982　　　　(d) 1998

Answers
1. (d) 2. (c) 3. (b) 4. (c) 5. (b) 6. (b) 7. (a) 8. (a) 9. (d) 10. (c)

Chapter 20 : Basic Ideas in Statistics

1. Statistics that summarizes the basic features of the sample data, is called as :
   (a) Inferential statistics (b) Descriptive statistics
   (c) Inductive statistics (d) None of these
2. According to Ferguson and Takane, Statistics is the branch of :
   (a) Experimental method (b) Scientific methodology
   (c) Scientific theory (d) None of these
3. A certain portion of observations selected from the complete set of observations is called as:
   (a) Sample (b) Subpopulation (c) Population (d) Variate
4. Students enrolled in a particular university constitute the example of :
   (a) Finite population (b) Infinite population (c) Parameter (d) Statistics
5. A measurement that describes a sample data is known as :
   (a) Statistics (b) Parameter (c) Variate (d) Value
6. In a study of the impact of temperature upon work efficiency, 100 female students were selected for participation. In this example, sex is an example of :
   (a) Variable (b) Constant (c) Variate (d) Value
7. In a study of weight of adult females, the weight of Anamika would be an example of :
   (a) Value (b) Variable (c) Variate (d) None of these
8. Number of boys in a school constitutes an example of :
   (a) Dependent variable (b) Independent variable
   (c) Continuous variable (d) Discrete variable
9. A classification of persons into males and females is an example of :
   (a) Nominal scale of measurement (b) Ordinal scale of measurement
   (c) Interval scale of measurement (d) Ratio scale of measurement
10. Regarding Ordinal scale of measurement, which statement is not true :
    (a) Relationship is reflexible (b) Relationship is asymmetrical
    (c) Relationship is transitive (d) Relationship contains equivalence
11. Measurement of height in terms of feet and inches constitute the example of :
    (a) Interval scale (b) Ratio scale (c) Nominal scale (d) Ordinal scale
12. Which is not correct about interval scale of measurement?
    (a) Equivalence relation (b) 'Greater than' relation
    (c) Specification of ratio between differences (d) A true zero point
13. The difference between parametric statistics and runparametric statistics is called as:
    (a) Sampling error (b) Standard error (c) Error variable (d) None of these

14. Median is the example of:
    (a) Parametric statistics (b) Nonparametric statistics
    (c) Pure statistics (d) None of these
15. For analyzing categorical data which one of the following would be most appropriate?
    (a) Parametric test (b) Non-parametric test
    (c) Both 'a' and 'b' (d) None of these
16. t-test is an example of:
    (a) Inferential statistics (b) Descriptive statistics
    (c) Non-parametric statistics (d) None of these
17. If the researcher accepts a null hypothesis, which is really false, he is apt to commit:
    (a) Type I error (b) Type II error (c) Standard error (d) Sampling error
18. Which is not a correct statement?
    (a) Alternative hypothesis is an operational statement of research hypothesis
    (b) Null hypothesis is no difference hypothesis
    (c) Acceptance of null hypothesis when it is true, constitute Type I error
    (d) Rejection of null hypothesis when in fact, it is false constitutes a correct decision.
19. The power of statistical test is best defined as:
    (a) $1-\beta$ (b) $1-\alpha$ (c) $\beta-1$ (d) $\alpha-1$
20. When the level of significance is made more stringent, that is, from .05 to .01, the probability of committing:
    (a) Standard error increases (b) Type I error increases
    (c) Type I error decreases (d) Type II error decreases

Answers
1. (b) 2. (b) 3. (a) 4. (a) 5. (a) 6. (b) 7. (c) 8. (d) 9. (a) 10. (a) 11. (b)
12. (d) 13. (d) 14. (b) 15. (b) 16. (a) 17. (a) 18. (c) 19. (a) 20. (c)

Chapter 21 : Measures of Central tendency

1. Which statement is not true?
   (a) The measures of central tendency provides a single summary value that describes the level of a set of observations
   (b) Such measure indicates a standard of performance not previously known
   (c) Such measure provides a means to compare the level of performance under two (or more) conditions or two or more existing groups
   (d) Such measure provides an estimate about the cluster of scores.
2. What is the midpoint of class interval 60-69 in a distribution?
   (a) 64.5 (b) 65.5 (c) 63.5 (d) 66.5

3. Which is not true about the mean?
   (a) Mean is most sensitive to the presence or absence of extreme score in the distribution
   (b) Mean is thought as a balance point in the distribution
   (c) Mean is the most appropriate measure in case of open-ended distribution
   (d) Mean is responsive to the exact position of each score in the distribution
4. Which one of the following gives equal weight to all scores in the distribution?
   (a) Mean   (b) Median   (c) Mode   (d) None of these
5. Which is the most stable measure of central tendency?
   (a) Mean   (b) Median   (c) Mode   (d) None of these
6. In case of open-ended distribution, the most appropriate measure of central tendency is
   (a) Mean   (b) Median   (c) Mode   (d) None of these
7. If the researcher is interested in finding out the average in case of some qualitative traits such as beauty, which one is the most appropriate statistics?
   (a) Mean   (b) Median   (c) Midpoint   (d) Mode
8. Which is the correct method of calculated mode?
   (a) 3 Mean-2 Median   (b) 2 Mean-3 Median
   (c) $\dfrac{Mean \times Median}{3}$ (d) $\dfrac{Mean \times Median}{2}$
9. Which is the most unstable measure of central tendency?
   (a) Median   (b) Mode   (c) Mean   (d) None of these
10. Which measure of central tendency is equal to P_{50}?
    (a) Mean   (b) Median   (c) Mode   (d) None of these
11. Which is not a true statement?
    (a) There is no mode in the distribution if all scores in the data have the same frequency
    (b) If a constant number is added to each individual score in the distribution, the mean is increased by the value of that constant
    (c) The value of median can be determined graphically
    (d) In a positively skewed distribution the value of mode is greater the that of mean

Answers

1. (d)   2. (a)   3. (c)   4. (a)   5. (a)   6. (b)   7. (b)   8. (a)   9. (b)   10. (b)   11. (d)

Chapter 22 : Measures of Variability

1. Which is not correct about the measures of variability?
   (a) Measures of variability are otherwise known as measures of dispersion
   (b) The standard deviation is the simplest all the measures of variability
   (c) The most reliable and stable measure of variability is the standard deviation
   (d) Measures of central tendency is the single summary value that tells about the scattereduces of scores or about clustering of scores in the distribution.

2. Which one of the following is directly related to increase or decrease of sample size?
   (a) Quartile deviation
   (b) Standard deviation
   (c) Range
   (d) Mean deviation
3. The quartile deviation is equivalent to
   (a) 25th percentile
   (b) 50th percentile
   (c) 75th percentile
   (d) 100th percentile
4. Which one reflects the inter quartile range?
   (a) Distance between Q_1 and Q_2
   (b) Distance between Q_2 and Q_3
   (c) Distance between Q_1 and Q_3
   (d) None of them
5. In case of open-ended distribution, which is the appropriate measure of variability?
   (a) Average deviation
   (b) Quartile deviation
   (c) Standard deviation
   (d) None of these
6. If all scores in distribution have identical values in the distribution, the value of standard deviation will be
   (a) Above 1
   (b) Below 1
   (c) Zero
   (d) None of these
7. When a constant number is added to the distribution, the standard deviation is
   (a) Raised by that constant number
   (b) Lowered by that constant number
   (c) Not affected at all
   (d) Raised by double the value of that constant number
8. When the distribution is asymmetrical, the correct measure of dispersion is :
   (a) Range (b) Standard deviation (c) Quartile deviation (d) Variance
9. In a normal distribution, the most accurate measure of dispersion is :
   (a) Mean deviation (b) Quartile deviation (c) Standard deviation (d) Range
10. Variance is defined as :
    (a) AD^2
    (b) SD^2
    (c) Q^2
    (d) None of these
11. When the measure of central tendency is mean, the most appropriate measure of variability is :
    (a) Standard deviation (b) Mean deviation (c) Quartile deviation (d) None of these
12. When the measure of central tendency is median, the correct and appropriate measure of variability is :
    (a) Quartile deviation (b) Mean deviation (c) Standard deviation (d) None of these
13. In a normal distribution Quartile deviation and Standard deviation have a constant relationship. Which one is correct?
    (a) $Q = 0.6745\sigma$
    (b) $Q = 0.7746\sigma$
    (c) $Q = 0.8745\sigma$
    (d) $Q = 0.7848\sigma$
14. In a normal distribution Q is known as :
    (a) Standard error
    (b) Probable error
    (c) Error variance
    (d) Error variable

Answers

1. (b) 2. (c) 3. (b) 4. (c) 5. (b) 6. (c) 7. (c) 8. (c) 9. (c) 10. (b) 11. (a) 12. (a) 13. (a) 14. (b)

Chapter 23 : Normal curve

1. Which is the other name of Normal curve?
   (a) DeMoivre's curve (b) Curve of symmetry
   (c) Curve of dispersion (d) Curve of asymmetry
2. In a normal curve how many areas fall within $\pm 2\sigma$?
   (a) 96.44% (b) 97.54% (c) 95.44% (d) 95.89%
3. In a normal curve, the degree of skewness is :
   (a) >1 (b) <1 (c) 0 (d) None of these
4. In a normal curve the maximum height of the ordinal at centre is :
   (a) .3998 (b) .3989 (c) .9398 (d) .8398
5. If the distribution is perfectly normal, the difference between mean, median and mode is :
   (a) >2 (b) <2 (c) 0 (d) <1
6. When the scores in the distribution is messed at the low end of the scale, it is said to be :
   (a) Positively skewed (b) Negative skewed (c) Leptokurtic (d) Platykurtic
7. When the value of Kurtosis is zero, the distribution is said to be :
   (a) Leptokurtic (b) Platykurtic (c) Mesokurtic (d) None of these
8. When Kurtosis is negative, the curve is said to be :
   (a) Platykurtic (b) Leptokurtic (c) Mesokurtic (d) None of these
9. When Skewness is zero, the distribution is said to be :
   (a) Platykurtic (b) Leptokurtic (c) Negatively skewed (d) Symmetrical
10. When Kurtosis is positive, the distribution is said to be :
    (a) Leptokurtic (b) Platykurtic (c) positively skewed (d) Mesokurtic
11. In a positively skewed distribution :
    (a) Mode>Median>Mean (b) Median>Mean>Mode
    (c) Mean>Median>Mode (d) None of these above
12. In negatively skewed distribution :
    (a) Mode>Median>Mean (b) Mean>mode>Median
    (c) Mode>Mean>Median (d) None of the above
13. If the value of Kurtosis is .2632, it is considered as :
    (a) Mesokurtic (b) Platykurtic (c) Leptokurtic (d) None of these
14. If the value of kurtosis is greater than .2632, the distribution is considered as :
    (a) Platykurtic (b) Leptokurtic (c) Mesokurtic (d) None of these
15. If the value of kurtosis is less than .2362, the distribution is said to be :
    (a) Leptokurtic (b) Platokurtic (c) Mesokurtic (d) None of these
16. In a normal distribution, the quartile coefficient of skewness amounts to :
    (a) Greater than zero (b) Zero (c) Less than one (d) None of these

17. The moment coefficient of skewness for a distribution of test scores having $m_2 = 200$ and $m_3 = 2225$ would be :
    (a) .786   (b) .825   (c) .792   (d) .925
18. The moment coefficient of kurtosis for a distribution of test scores having $m_4 = 144.94$ and $m_2 = 8$ would be :
    (a) –.740   (b) –.789   (c) –.689   (d) –.882
19. If g_2 is greater than zero it indicates that the distribution is :
    (a) Leptokurtic   (b) Platykurtic   (c) Mesokurtic   (d) None of these
20. If g_2 is less than zero, it indicates that the distribution is :
    (a) Platykurtic   (b) Leptokurtic   (c) Mesokurtic   (d) None of these

Answers

1. (a) 2. (c) 3. (c) 4. (a) 5. (c) 6. (a) 7. (c) 8. (a) 9. (d) 10. (a) 11. (c)
12. (a) 13. (a) 14. (b) 15. (a) 16. (b) 17. (a) 18. (a) 19. (a) 20. (a)

Chapter 24 : Percentile and Percentile Rank

1. The 50th percentile is also known as :
   (a) D_2 (b) Median   (c) D_8 (d) None of these
2. A PR of 80 indicates that 80% of the total cases lie :
   (a) above it   (b) below it
   (c) either below or above it   (d) None of these
3. Deciles divide the total data into :
   (a) Five equal parts   (b) Seven equal part   (c) Ten equal parts   (d) None of these
4. P_{20} is equal to :
   (a) D_1 (b) D_3 (c) D_4 (d) D_2
5. Which is the incorrect statement?
   (a) Percentiles are equal unit of measurement
   (b) Percentiles can be treated arithmetically
   (c) Percentiles can be a help in determining the shape of distribution
   (d) Norms can be reported in terms of percentiles

Answers

1. (b) 2. (b) 3. (c) 4. (d) 5. (a)

Chapter 25 : Correlation and Regression

1. When the scores in one variable show the opposite trend of the scores in the other variable, it is expected that there exists :
   (a) Positive correlation   (b) Negative correlation
   (c) Zero correlation   (d) None of these

2. The range of Pearson r is :
   (a) ±1 to ±1.5   (b) ±1 to ±2   (c) 0 to ±1   (d) ±1 to ±2.5
3. Which of the following measurement is most suited for Pearson Product Moment correlation?
   (a) Ratio and ordinal measurement
   (b) Ordinal and Nominal measurement
   (c) Ratio and Interval measurement
   (d) Ordinal and Interval measurement
4. In Pearson r, the r is a symbol which is short for :
   (a) Regression
   (b) Representativeness
   (c) Replication
   (d) Repeated measures
5. Which is a false statement?
   (a) Pearson r is a product moment r
   (b) Pearson r can be either positive or negative
   (c) Pearson r may be regarded as arithmetic mean
   (d) Pearson r is a mutlivariate statistical method
6. When the coefficient of alienation (or k) is zero, Pearson r will be equal to :
   (a) 0.65   (b) 1.00   (c) 0.85   (d) .75
7. What would be the value of coefficient of determination if r = .90?
   (a) 0.75   (b) 0.85   (c) 0.81   (d) 0.19
8. If $\Sigma D^2 = 25$ and N = 6, the value of Spearman's Rank order correlation (P) would be :
   (a) 0.29   (b) 0.39   (c) 0.44   (d) None of these
9. If $P = .39$ and N = 30, its z would be equal to :
   (a) 2.10   (b) 2.21   (c) 2.39   (d) 2.31
10. When the impact of a third variable is to be eliminated, the most effective method of correlation between remaining two variables would be :
    (a) Multiple correlation   (b) Partial correlation   (c) Pearson r   (d) Spearman P
11. When, in any study two variables are to be controlled and the remaining two variables are to be correlated, it is called as :
    (a) First-order partial correlation
    (b) Second order partial correlation
    (c) Third order partial correlation
    (d) None of these
12. In Y is to be predicted on the basis of X variable, then X is here the example of :
    (a) Predictor variable   (b) Criterion variable   (c) Irrelevant variable   (d) None of these
13. In $Y' = a + bX$, the slope of the line is :
    (a) b   (b) 9   (c) Y' (d) X
14. Which is not a correct statement?
    (a) Correlation is limited to the study of linear relationship between variables whereas regression implies both linear and nonlinear relationship
    (b) Correlation coefficient need not imply cause-and-effect relationship whereas regression implies such effect :
    (c) Both correlation and regression can be simple as well as multiple
    (d) If the value of the b coefficient becomes significant at given level of significance, the value of r is not significant at that level of significance

Answers
1. (b) 2. (c) 3. (c) 4. (a) 5. (d) 6. (b) 7. (c) 8. (a) 9. (a) 10. (b) 11. (b)
12. (a) 13. (a) 14. (d)

Chapter 26 : z test, t test and F test : Significant of Mean differences

1. When N < 30 and the researcher is interested in testing the significance of difference between two sample means, the most appropriate statistic would be :
   (a) z test (b) t test (c) F test (d) None of these
2. Who propounded t test?
   (a) Spearman (b) Fisher (c) Gossett (d) None of these
3. In ANOVA, the total variance is commonly analyzed into :
   (a) between variance and within variance
   (b) between variance and estimated variance
   (c) Within variance and estimated variance (d) estimated variance and real variance
4. Which of the following does not account for between variance in ANOVA?
   (a) Experimental error (b) Individual difference
   (c) Treatment effect (d) Rosenthal effect
5. In testing the significance of mean difference between two sample means, if t = 7.78, then F would be equal to :
   (a) 60.25 (b) 40.75 (c) 49.00 (d) 50.75
6. Tukey's Honestly significant difference (HSD) test is applied after :
   (a) The value of F is not significant
   (b) The value of F test becomes significant
   (c) The value of t test becomes significant
   (d) The value of z test becomes significant
7. The t test is usually known as :
   (a) Teacher's t test (b) Student t test (c) People's t test (d) None of these
8. In one way ANOVA if K = 4 and N = 40, then the total df would be :
   (a) 39 (b) 42 (c) 45 (d) 43
9. In simple or one-way ANOVA, if the between variance is 4.58, and the within variance is 6.23, the resulting F would be
   (a) Significant at .01 level (b) Significant at .05 level
   (c) Significant at .001 level (d) Not significant
10. One assumption underlying ANOVA is that the effects of various factors on the total variance are :
    (a) Multiplicative (b) Additive (c) Divisible (d) None of these

Answers
1. (a) 2. (c) 3. (a) 4. (d) 5. (a) 6. (b) 7. (b) 8. (a) 9. (d) 10. (b)

Chapter 27 : Some Nonparametric tests

1. Which is the correct formula of X^2?
   (a) $\Sigma \dfrac{(fo-fe)^2}{fe}$
   (b) $\Sigma \left(\dfrac{fo-fe}{fe}\right)^2$
   (c) $\Sigma \dfrac{(fe-to)^2}{fo}$
   (d) $S \dfrac{(fo-fe)}{fe}$

2. Which is true about X^2?
   (a) The value of chi square is always positive
   (b) The value of chi-square is always negative
   (c) The value of chi-square is always less than 1
   (d) The value of chi-square is always less than 10.

3. When 'fo' and 'fe' are equal, the value of chi-square will be :
   (a) less than 1   (b) zero
   (c) less than 2   (d) Some value in minus

4. The decision of Yates' correlation for continuity is taken on the basis of :
   (a) Observed frequency   (b) Expected frequency
   (c) Difference between observed and expected frequency   (d) None of these

5. What is the best solution to the computation of X^2 when the expected frequency is less than 5 in 3 × 2 contingency table?
   (a) To apply Yeates' correlation   (b) To increase the number of participants
   (c) To remodel the contingency table   (d) None of these

6. In 2 × 2 table when both variables have genuine dichotomes, the most appropriate method of correlation would be :
   (a) Spearman's P   (b) Pearson r   (c) Phi coefficient   (d) None of these

7. When the Contingency table is available in terms of 3×3 table, the maximum value of c would be
   (a) 0.761   (b) 0.816   (c) 0.824   (c) 0.616

8. When the researcher wants to the differentiate between two groups whose data are available on ordinal measurement, the most appropriate statistic would be
   (a) Median test   (b) Sign Test
   (c) Mann-Whitney U test   (d) None of these

9. In a distribution of ranks of two independent samples, n_2 is 24, the value of Mann-Whitney U test is 48 and $z = 3.24$. Tell whether the obtained value is
   (a) Significant at .05 only   (b) Significant beyond .01 level
   (c) Significant at .01 level only   (d) None of these

10. When the researcher is interested in exploring degree of agreement among four sets of rankings of 6 participants, the most appropriate statistic would be :
    (a) F test   (b) Coefficient of contingency
    (c) Mann-Whitney U test   (d) Coefficient of concordance

Answers
1. (a) 2. (a) 3. (b) 4. (b) 5. (b) 6. (c) 7. (b) 8. (c) 9. (b) 10. (d)